HENRI POINCARÉ

Poincaré at 57. Source: Poincaré, *Oeuvres* (11.2, 295).

HENRI POINCARÉ
A Scientific Biography

JEREMY GRAY

PRINCETON UNIVERSITY PRESS

PRINCETON AND OXFORD

Copyright © 2013 by Princeton University Press

Published by Princeton University Press, 41 William Street,
Princeton, New Jersey 08540

In the United Kingdom: Princeton University Press,
99 Banbury Road, Oxford OX2 6JX

First paperback printing, 2022
Paperback ISBN 978-0-691-24203-3

The Library of Congress has cataloged the cloth edition as follows:
Gray, Jeremy, 1947–
Henri Poincaré : a scientific biography / Jeremy Gray.
pages   cm
Includes bibliographical references and index.
ISBN 978-0-691-15271-4 (hardback)
1. Poincaré, Henri, 1854–1912.   2. Scientists–France–Biography.   I. Title.
Q143.P7G73 2012
509.2–dc23
[B]
2012019641

British Library Cataloging-in-Publication Data is available

This book has been composed in ITC New Baskerville using LaTeX

Cover: Jules Henri Poincaré (1854-1912). Photograph from
the frontispiece of the 1913 edition of his *Last Thoughts*.
Image courtesy the University of St. Andrews, Scotland.
Cover design by C. Alvarez Gaffin

press.princeton.edu

Typeset by S R Nova Pvt Ltd, Bangalore, India

Contents

List of Figures

Preface

I WAS ABLE TO work on this project because of a grant from the Lever-hulme Trust. Without their generosity it would not have been possible for me to write this book, and I thank them heartily for it.

It will be evident in many places how much I have relied on the work of the people at the Archive Henri Poincaré in Nancy, so much of which has been freely available on the web for some time, and in particular, for the more purely biographical passages, how much I have used the writings of Laurent Rollet. I would like to thank him, Philippe Nabonnand for discussions on geometry, Scott Walter for his guidance with the physics and the TEX, Gerhard Heinzmann for pushing me toward Wittgenstein, and all of them for their help and support over many years. The Archive Henri Poincaré is a living resource for the history of mathematics, science, and philosophy, worthy of this distinguished son of Nancy.

I thank my wife, Sue Laurence, for all her support. Particular thanks are also due to June Barrow-Green, my colleague and guide with Poincaré for many years. Thanks too to Benjamin Wardhaugh, who read the whole book and made many useful comments, and to two anonymous referees. I also thank people who were kind enough to read chapters: Tom Archibald (who introduced me to Poincaré's work on Saturn's rings and has a paper forthcoming on the topic), Jed Buchwald, Alain Chenciner, José Ferreirós, Phil Holmes, Al Martinez, Jean Mawhin, John McCleary, Arthur Miller, Saul Schleimer, and John Stillwell, and for helpful and informative discussions, David Aubin, Jeremy Avigad, Philip Ehrlich, Moritz Epple, Jesper Lützen, Colin McLarty, David Rowe, and David Stump.

I also thank Richard Temple of Senate House Library, London and Catherine Harpham at Imperial College London for their help with sources.

Finally, it is a pleasure to thank Alison Durham, Vickie Kearn, and Sara Lerner at Princeton University Press for all their help with this book.

Some Comments on the Text

Poincaré's actual calculations are at once technical and impressionistic, a combination that often caused his readers some distress. To leave them out risks giving the impression that Poincaré's mathematical ability was mysterious at every level when a little detail can make everything more plausible, but to include them can require fussy authorial intervention. I have tried to deal with this by reserving the word "show" for those many occasions when his argument used techniques that were standard for the time and remain unproblematic. When he was more original but less certainly correct I have used the word "argue." But blatantly insufficient rigor also marks some of his work in pure mathematics, notably but not only in his work creating algebraic topology where, in the end, the whole story had to be rewritten: one should not read Poincaré's account as one would a mathematician's today but as an outline.

Poincaré's papers are scattered over several sources. Most of his mathematical and scientific papers were collected in the eleven volumes of his *Oeuvres*, but some were omitted. There are several books and numerous volumes of lecture notes. The situation with his volumes of popular essays is worse. Most, but again not all, were anthologized in one of four volumes, but were often revised before republication; *L'avenir* is an extreme example, the book version is a misleading version of the original. There seems to be no standard edition of these essays, and existing editions vary in pagination. This book being in English, it seems desirable also to refer to the English translations, but they are of imperfect quality and again in numerous editions both in print and on the web. I list the editions I have used below.

I have generally referred to Poincaré's publications in the text with a reference of the form (1898d, v–vi), which in this case identifies two pages in the book (1898d) listed in the first bibliography. Items in the *Oeuvres* are referred to similarly, but (1891d, 485) refers to page 485 in the *Oeuvres*, not the original publication (for the volume, consult the bibliography). Likewise, a reference to an essay generally appears as (1895d, 84, 53), where 84 is the page in the relevant French volume of essays and 53 is the corresponding page in English in Poincaré (2001) (for the volumes, again consult the bibliography), but one of the form (1895d, 638) identifies a page in the original publication, as I hope the context also makes clear, and one of the form (*S&H* 128, 86) refers to those pages in the French and English editions of that work, respectively. Some items seemed important enough to refer to by name; for example, *L'avenir* refers to Poincaré's *L'avenir des mathématiques* (1908c). Most books and articles not by Poincaré are referred to by a reference of the form Smith (2001, 27) or, when the author's name is understood (2001, 27), which refers to page 27 of whatever it is in the bibliography that Smith wrote in 2001.

Almost all Poincaré's correspondence is available online,[1] some of it has been published by the team in Nancy and more will be published by them in the next year or two. This makes it hard to refer to them, but easy for the interested reader to find them, so I have thought it sufficient to give the date of letters I have quoted.

Two topics (Maxwell's electromagnetic theory and the basics of the theory of elliptic functions and Abelian functions) have been given short appendixes, and there is a glossary of those terms for which an explanation in the text would annoy about as many readers as its omission would frustrate.

I have used these abbreviations, and these editions when there is any ambiguity:

- *E&O* for *Électricité et optique*, Poincaré (1890b) and (1901b);

- *S&H* for *La science et l'hypothèse*, 1st edition 1902, Poincaré (1968);

- *VS* for *La valeur de la science*, Poincaré (1905a);

- *S&M* for *Science et méthode*, Poincaré (1908e);

- *DP* for *Dernières pensées*, 1st edition 1913, Poincaré (1963).

I have used the English translations of *S&H* (1905), *VS* (1913), *S&M* (1914), and *DP* (1913), but I have checked all the translations and used mine where necessary. The first three were reprinted in Poincaré (2001) on pages 3–178, 179–353, and 355–572, respectively, which is a good clue to the pages in the separate editions of the books.

A complete bibliography of Poincaré's work, and Rollet's instructive comparison of the original versions of the essays with the versions in the books of essays will be found in Poincaré (2002).

[1] Go to http://www.univ-nancy2.fr/poincare/chp/.

HENRI POINCARÉ

Introduction

ALTHOUGH THERE HAVE been several excellent studies of aspects of Poincaré's work this book is the first full-length study covering all the main areas of his contributions to mathematics, physics, and philosophy. It presents an introduction to his work, an overview of its many fields and interconnections, and an indication of how Poincaré was able to tackle so many different problems with such success. What emerges is a picture of Poincaré as a man with a coherent view about the nature of knowledge, one that he expressed in many of his popular philosophical essays and applied in the conduct of his own research. What he emphasized above all was the act of human understanding. His preferred means of attaining the understanding of a problem was to find the right generalization of its core concepts, often in the form of an analogy, but one measured by its ability to generate new results. He endorsed Ernst Mach's idea of the economy of thought, and he spoke of looking for the "soul of the fact"—the right relationship between the facts that constitutes a productive principle.

He did not disdain rigor, he regarded it as essential, but he observed that rigorous proofs could be too long to be comprehensible, and in mathematical physics they could also fail to capture the way nature seemed to work and be incapable of producing answers useful to the physicist. Formal arguments, as in geometry, could distort the subject by emptying it. His use of the idea of a group underpinned his epistemology and frequently inspired his search for fruitful analogies. He trusted his intuition to make productive contact with calculations in every domain from differential equations to algebraic topology. But he did not put his trust in pure insight—he admitted his own intuitions were frequently wrong—or in finding the "true" meaning of things. On the contrary, he was very aware that what some might call the progress of science, more critical observers would call the accumulating ruins of failed theories. He knew very well that the best theories in physics were inconsistent, he wrote about these problems in several essays and addressed them in his papers and books on electromagnetism and optics, and what he felt was a typical mistake was to infer the existence of objects responsible for the phenomena. For him, theory change was usually a matter of abandoning the objects while refining the relationship between the facts (mathematical and experimental) that had been carefully built up.

The picture of Poincaré that emerges would perhaps have pleased the later Wittgenstein. We have no certainty beyond what shared use and discourse can guarantee, no unmediated access to reality. We do what we can according to our best understanding of the rules of the game (axioms, principles, the best experimental data). Mathematics and physics together offer us a rule-governed way of living in the world, although we may, from time to time, have to change the rules, and our ability to frame these rules is, in some ways, built into how our minds work.

In this book I argue that understanding, thus conceived, was Poincaré's aim in everything he worked on. I trace how he came to the core ideas that animated his theories in many different domains. In addition to the numerous links he found between them, much of his work exhibits standard features that display his sense of the understanding one can have. In the tension between rigor and understanding, acute in both mathematics and physics, there is no doubt that Poincaré frequently failed to provide rigor. But he usually aimed at sharing understanding, expressed in ways in which new knowledge can most efficaciously be acquired. Assessing the extent to which he succeeded is one of the aims of this book.

This is a scientific biography of Henri Poincaré. It is confined entirely to his public life: his contributions to mathematics, to many branches of physics and technology, to philosophy, and to public life. It presents him as a public figure in his intellectual and social world; it leaves the private man alone apart from a deliberately brief account of his childhood and education. A full biography is underway with a team of scholars at the Archive Henri Poincaré at the University of Lorraine, and their book, due out in 2015, will be very valuable.

The book (2006) edited by Charpentier, Ghys, and Lesne can be recommended as an introduction to the implications of Poincaré's work for mathematics today, and I am sure that readers of this book will want to consult Verhulst's *Henri Poincaré: Impatient Genius*, which I have not yet seen.

VIEWS OF POINCARÉ

Pictures are like mirrors: what we see in them reflects aspects of ourselves. The frontispiece shows the famous picture of Poincaré at the age of 57, standing alone on the seashore. He has his back to us, he is slightly stooped, we cannot know what he is thinking, but most likely the picture suggests to the viewers that Poincaré is lost in his own, remarkable thoughts. Some may connect it with Newton's famous remark about picking up a few pebbles on the seashore:

> I do not know what I may appear to the world, but to myself I seem to have been only like a boy playing on the sea-shore, and diverting myself in now and then finding a smoother pebble or a prettier shell than ordinary, whilst the great ocean of truth lay all undiscovered before me. (Brewster 1855, vol. 2, 407)

It may invite us to enter our own speculations faced with the immensity of the oceans. Those who know of Poincaré's profound interest in physics may recognize the connection between the man and his research into the shape of the earth. Others, knowing how visionary and innovative was his mathematics, may see isolation and loneliness. Or just a middle-aged man enjoying a quiet holiday.

How Poincaré was regarded in his lifetime is inseparable from who he was. It is in the community of professional mathematicians that he first emerged, where he made his most profound and lasting contributions, and where his reputation remains most secure. He amazed his contemporaries with his unending stream of intellectual achievements. These began in 1881, when he brought together the mathematical subjects of complex function theory and complex differential equations with an entirely unexpected use of non-Euclidean geometry, to create the theory of automorphic functions. It continued in the middle years of the 1880s with his work on real differential equations and his radically new way of handling celestial mechanics. Despite his ever-deepening involvement with physics, his work in mathematics continued all this time. His most lasting achievement is his creation of the subject of algebraic topology, but he was one of the few to advance the subject of complex function theory in several variables in the 1900s, and he made important contributions to algebraic geometry and Sophus Lie's theory of transformation groups, and even to number theory.

Nonetheless, by the 1890s he had become a professor of astronomy and celestial mechanics, occupied with the fast-moving field of electricity, magnetism, and optics after Maxwell. This brought him into contact with Hertz, Lorentz, and, if only obliquely, Einstein. He was now professionally a physicist and a mathematical astronomer, a man nominated, albeit unsuccessfully, more than anyone else until recently for the Nobel Prize in physics. The physics community greatly appreciated his lectures on a dozen different topics in contemporary physics, and saw him as the leading French authority on electricity, magnetism, and optics, and the author of one of the founding papers in the theory of special relativity—although Einstein's theory was not widely accepted until after the First World War.

The public at large heard of him first in 1890, when he won a prestigious prize for a study of the motion of the planets and the stability of the solar system, and he went on to write two major works on celestial mechanics. He was a loyal French citizen, active in the Bureau des longitudes for many years. He was and remains almost unique among mathematicians and scientists in presenting his ideas to a general audience as well as his peers; he wrote widely on many topics and engaged forcefully in important issues about the nature of mathematics and contemporary physics, and the relationship between science and ethics. As this book shows, he was also highly regarded in his lifetime for a number of interventions in technological discussions.

In many different ways Poincaré exerted a deep and lasting influence, and yet he had few if any pupils working on his ideas, while his contemporaries often kept themselves at a safe distance from what he did and preferred other topics. He was a naturally elegant writer, but his style was often to describe what he had seen in a problem after prolonged contemplation, and he left many details for his readers to fill in. A professor of his time, he was not expected to create a school and his habit of having little to add to his sometimes impressionistic writings did not help. Barely two years after his death Europe plunged into a war in which many young French mathematicians and scientists were killed. When it was over there were few who were eager to develop Poincaré's legacy, and its personal character, with its deep commitment to mathematical physics, was uncongenial to the next generation in France—the young Bourbaki with their orientation toward the abstract pure mathematics that was coming out of Germany. Only after another war, and the full development of the Bourbaki style, did mathematicians begin to rediscover the riches of what Poincaré had set out so many years before.

The modern science, mathematics, and philosophy communities differ in interesting ways, and one is their attitude to the past. A physicist today finds it hard to look back beyond the two great changes in 20th century physics: quantum mechanics and the general theory of relativity. Poincaré's contribution to the special theory of relativity is secure, if inextricably entangled with that of Einstein, but much else in what he did is shrouded in the obscurity of a lost way of thought. Many mathematicians, however, see him today not only as the exemplary creator of new branches of mathematics, but as the source of ongoing topics of research in several areas. As the recent excitement over the solution of the so-called Poincaré conjecture showed, Poincaré's work in topology lives among mathematicians to this day as, to an extent, he does among philosophers of science, who still discuss his philosophy of conventionalism. It fed into the ideas of the Vienna circle, and it has kept his popular essays in print for a century.

Which partial view of Poincaré to present first? The mathematician may always have been the most significant, and Poincaré was an extraordinary mathematician. Most mathematicians would be pleased to produce the work in any one of the first ten volumes of his *Oeuvres*; each one has some papers of remarkable depth and originality: the first two volumes carry the theory of automorphic functions, volume 4 his work on Abelian functions and complex functions of several variables, volume 6 the invention of algebraic topology, volume 7 his work on celestial mechanics and the discovery of chaotic dynamical systems, and volume 10 his work on the partial differential equations of mathematical physics. But by being largely topically organized they give a misleading impression of the man, who switched from one topic to another with great rapidity: to give just one example, in 1905 he published on number theory, geodesics on convex surfaces, the dynamics of the electron, a report on the French geodetic survey in Peru, and a popular philosophical paper on mathematics and logic. However, the last century has not made his mathematics any easier to explain to a broad audience.

Few mathematicians in any period would aspire to preeminence in a branch of physics as well, but Poincaré was one of the dominant figures in the theory of electrodynamics after Lorentz, eminent in celestial mechanics, and capable of publishing in every branch of the subject.

His contributions to physics and astronomy were markedly theoretical, although when space exploration began in earnest NASA brought out an English translation of Poincaré's major work on celestial mechanics, *New Methods of Celestial Mechanics* (1967), but it too belongs chiefly to the experts for whom it was intended. So Poincaré the physicist is more accessible than the mathematician, but his claim on a modern audience has been attenuated by the great changes in physics since his death.

Some mathematicians and physicists might hope to reflect philosophically on their subject, or to assist in its popularization, but starting in 1891 Poincaré did both, often in the same article and certainly in each of his four books of essays. As the public philosopher is the easiest to appreciate, I shall start with him, asking the reader to take on trust that Poincaré's words carried particular weight because he was a highly respected, internationally recognized expert in many fields. In December 1885 he had won the Prix Poncelet of the Institut de France for his mathematical work. In 1886 he became the professor of mathematical physics and probability in the Faculté des sciences in Paris, and was elected president of the Société Mathématique de France (SMF). In January 1887 he was elected a member of the geometry section of the Académie des sciences. This then will be the introduction to this remarkable figure. After some general reflections on Poincaré's ways of working, we move to look at the public intellectual, and then turn to look at his work in mathematics and mathematical physics.

POINCARÉ'S WAY OF THINKING

It also makes sense to start with the philosopher for the good reason that to a remarkable degree, Poincaré was guided in his mathematical and scientific work by his philosophical reflections. There are times in mathematics and physics when the practical mathematician or physicist becomes a philosopher of their subject, whether they, or the later professional community admits it explicitly. As Hilbert said of Dedekind's work on one occasion, "The mathematician was thus compelled to become a philosopher, for otherwise he ceased to be a mathematician."[1]

---

[1] Quoted in Corry (2004, 379).

Whenever the nature of the mathematics changes profoundly, as it did in Poincaré's time with the arrival of characteristically modern mathematics, or physics, as it was to do with the general theory of relativity and with quantum mechanics, some mathematicians and physicists are decisive in the conceptual—indeed, philosophical—reformulation of their subject. The most prominent proponent of the new, highly abstract, modern mathematics was David Hilbert in Göttingen, and it is easy to argue that there was a sea change, even before he articulated it, that many German mathematicians went with and many French opposed: what Cantor advocated, with Hilbert's subsequent blessing, Charles Hermite, the leading French mathematician of his day and an important influence on the young Poincaré, found pointless and disturbing. But Poincaré, for all he spoke against the new set-theoretical foundations of mathematics that were coming out of Germany was not a spokesman for French values, rather, only for his own. He could appreciate the axiomatic approach often favored by Hilbert, he favored the general over the particular in his own work (unlike Hermite). However, what holds his life's work together to a remarkable degree is the tight hold his epistemology had on his ideas of ontology, on what constitutes an answer to a mathematical or physical problem (and what does not), and on what the practice of mathematics and physics consists of.

Poincaré argued that we had every reason to believe the universe was intelligible, but could never know what it was "truly" like. We could only hope to live in it in an effective manner. To do so as a mathematician and a physicist meant, he argued, recognizing the two subjects as parts of the same subject because mathematics was the only language the scientist can speak, and because physics had given mathematics the concept of the continuum, without which little mathematics could be done. As concerns language, this is a theme that runs right through Poincaré's work. In his view, we share the usage of key terms, and we strive for objectivity through discourse, not for truth, a word he seldom used. As for the continuum, this, he said, was chosen by us as a matter of convenience without which science as he knew it could not proceed, but it was not forced upon us—he knew that other continua were mathematically consistent.

Poincaré entertained three levels of working assumptions in our intellectual life. The first was that of hypotheses: these could turn out to be true or false, and scientists make them all the time. At a more fundamental level, certain hypotheses, such as the laws of mechanics, had, over time, been elevated to the level of principles. They could not be verified, but they were extremely plausible and led to a very elegant theory with great range and predictive power. They were conventions, adopted for their efficacy not for their truth, and might one day have to be abandoned in favor of better ones, but until that day they were beyond discussion and controversial evidence should first be assessed on the assumptions that these basic principles hold.

At the third level, underpinning all these conventions, were two epistemological ones. Geometrical conventionalism was his explanation of how knowledge is possible at all. It is a theory of how the individual constructs his or her notion of space and, like cognitive science today, it is a mixture of evolutionary ideas and ideas about early mental development.[2] But it, too, leaves open the idea that other intelligent beings might make a different construction of space, and find it to be, say, non-Euclidean where we find it to be Euclidean. There would be no fact of the matter, only a choice based on convenience, albeit one long since built into the workings of our (and their) minds. In particular, our concept of distance between objects rests on our idea of how to measure, and this is typical of Poincaré: no concept is admitted without a way of evaluating it and deciding upon its correctness. Confronted with an apparent truth, he would always ask: How do you know?

Poincaré's second epistemological convention was his explicitly argued belief that we have a built-in understanding of reasoning by recurrence, from which flows our knowledge of what the natural numbers are, and indeed all the mathematics that does not depend on the continuum. This is not the view of most mathematicians today, but we know now that all rich mathematical theories rest on some assumptions. His loneliest position was his rejection of Zermelo's set theory, on the grounds that the transfinite sets it required were too big to be understood, and if they could not be understood, they could not usefully be talked about. In this sense, Poincaré's preference for our knowledge of the natural numbers over axiomatic set theory is not so strange.

He believed strongly that a knowledge claim had to come with an account of how we can know it. In a long controversy with Bertrand

---

[2] The ultimately flexible, purely utilitarian character of the conventionalism distinguishes it from Kantian intuition, which is, at least in Kant's presentation, an unanalyzed and perhaps even unanalyzable mental grasping.

Russell over what it is to know what distance is, he emphasized that it was imposed on us by our understanding of measurement. Inspired by Maxwell he had confidence in experimental results and rigorous mathematics but no compelling need or logical force in filling the gap between experimental results (measurements, laws) and mathematical theorems with stuff (such as electric fluids, electrons, and the like, that people—Lorentz included—populated their papers with).[3] In part this was a somewhat Kantian recognition that the ultimate nature of things is hidden from us, but it was also a belief that such talk often turned out to be wrong, whereas the theoretical relations between the objects usually prospered. It was the relations we could work with, not the things. He sometimes said he dealt with the form but not the matter of a subject, often capturing the form as a particular mathematical object, the group. But he seldom studied any individual group in detail; for him the crucial fact was the existence of a group. It allowed him to explain how we arrive at a particular convention for understanding space; equally it allowed him to explain why certain functions in number theory exist, and why certain topological spaces are different.

A striking late expression of this idealist strain in his thinking illuminates his ideas about space. In the preface he wrote to the first three-volume collection of his essays he emphasized that he was interested in the language we use to talk about space and how we might have confidence in it without ever having access to some phantom of a fundamental reality. "Space," he remarked, "is only a word that we have believed a thing" (1913, 5). A decade earlier, in a controversy with Le Roy over the nature of scientific knowledge, Poincaré discussed how the brute facts (simple observation statements such as "it is dark") are translated into the language of science, and where the fundamental principles of physics enter the analysis. He was firmly of the opinion that these principles are human creations, and as such capable of revision—which is a good view to hold when theories are likely to change.

Talk, discourse, was crucial to Poincaré. He recognized that we cannot compare our individual sensations, but we nonetheless agree on what is red. Likewise, he argued that in accepting as conventions the

---

[3] Perrin's experiments of 1908 convinced him of the existence of atoms because his analysis showed that it was possible to make definite statements about them that could be consistent with other theories in physics.

fundamentals of Newtonian mechanics, and so moving them beyond the reach of experiment, we are agreeing on what we shall say is true. He made no distinction between saying "the earth rotates" and "it is convenient to say that the earth rotates" because the only way to make the first remark is to subscribe in advance to the second: to unpack its meaning was, for him, to discover that it is a statement in Newtonian dynamics. Truths by convention were part of a shared discourse.

For Poincaré, understanding was central, and it was captured in the ways people were enabled to say what they could not say before.[4] So Poincaré in his own way preferred to speak of use rather than of meaning, even in mathematics. The axiomatic approach promoted by Hilbert does indeed tell us how to use the elements of an abstract structure but never what they are (although we may recognize specific exemplars). But when Poincaré speculated on how to do mathematics he came back to the importance of analogy in guiding one's thoughts, formulating new ideas, and generally doing something new. The rigorous side of mathematics was valuable, for without it there was nothing, but what mattered more was invention, the productive use of old and new ideas. And this was expressed in the steady expansion of our mathematical language. And when Poincaré talked about mathematics he valued new facts only if they united seemingly disparate elements in an unexpected way that, as he put it, "enables us to see at a glance each of these elements in the place it occupies in the whole" (*L'avenir* 375). Elsewhere, in words that not every lesser mathematician might feel comfortable with, he disparaged the mere production of new combinations of mathematical entities as work that can be done by anyone and that is highly likely to be absolutely devoid of interest; the task is to discern the few useful new combinations, which will usually be of ideas drawn from widely separated domains. He said that mathematicians set great store by the elegance of a solution or a proof because of the economy of thought that Mach had identified as a measure of intellectual economy.

Whenever he found it possible, he was rigorous. As he put it in *L'avenir*: "In mathematics rigour is not everything, but without it there is nothing." His education at the École polytechnique raised him to fluency in the theory of differential equations as it stood in the early 1870s, in complex function theory, in the use of Fourier series, and in coordinate geometry.

[4] See the insightful paper, Epple (1996).

This gave him a battery of standard methods to apply, of checks to run before certain theorems can be used, and general tricks of the trade that he rapidly fitted into his way of thinking mathematically. That said, when he himself felt an argument of his was imprecise, as was often the case in his work in applied mathematics, where there is a need to make approximations to get results and rigorous means are not available, he recognized that the proper test is the experimenter's, whose results can guide a theory when rigorous mathematics cannot. And on occasion he offered a vision of how things should be that, in the absence of a proper terminology and body of new results, could only be imperfect, as he did when creating algebraic topology. This is true quite broadly: Poincaré gave his readers good reasons to believe what he said, and where he could be rigorous he usually was, but he was happy to publish when he had convinced himself, and on those occasions his contemporaries had to be content with little more than a shrug.

Poincaré was explicit that understanding was not chiefly a psychological state. When he spoke to the Psychological Society in 1907 he noted that the feeling of having made an insight could be delusive, and the error would only be revealed "when we attempt to establish the demonstration" (1908d, 395). But he did appreciate the power of analogy, and indeed he often sought out analogies and demonstrated their use in various aspects of his work. These were never speculative, groundless analogies, but the result of detailed examinations of stories drawn from different domains that could be used to make productive comparisons. Indeed, his way of working further dismantles the idea that Poincaré pursued a path of pure intuition. He worked regularly from 10 till 12 in the morning and from 5 till 7 in the late afternoon. He found that working longer seldom achieved anything, but that it was not always possible to switch off, which was why he never worked in the evenings. He took a complete rest when on holiday. When reflecting on a topic he liked to walk about, but when preparing he took few notes and very often began to tackle a problem without any clear idea about the solution. He typically felt drawn on by a topic almost like an automaton and did not have the sensation of making an effort of will, but "with Poincaré the feeling of certainty must be rather weak, for every truth appeared to him to be debatable in some respect" (Toulouse 1909, 192). Organizing his ideas generally came easily to him, but if the effort became painful he would give up and abandon the work, which usually happened when he lost interest. On the other hand, he sometimes found that things went well by dropping a subject and coming back to it at intervals.[5]

He set out his views on what it is to understand mathematics in several places, most programmatically in his (1904c) on definitions in mathematics. This was an address at the Musée pédagogique, and chiefly concerned itself with the teaching of mathematics in schools, but Poincaré was concerned with the marked difference between formal and intuitive mathematics. Just as a precise definition might not be understood in a classroom, understanding a proof is not the same thing as checking its logical validity. Today, he said, we know that our predecessors worked with many imprecise definitions that we have made precise, that of continuity, for example, but in the last fifty years logic has sometimes produced monsters: functions that are not continuous, or, if continuous, have no derivatives. Indeed, strictly speaking, these are the typical functions and the ones one finds without going looking for them are a small corner of the whole. But if logic was our only guide, the student would have to begin in a teratological museum, and this would be as futile as studying an elephant only with a microscope. Everything would be correct, but it would not show the true reality. So it is necessary to proceed in the reverse direction, and build up the student's intuition before gradually introducing rigor. "It is not enough to doubt everything," he said in paragraph 7, "one must know why one doubts." Even the future professional mathematician (*géomètre*) needs intuition, "For if it is by logic that one proves, it is by intuition that one invents" (para. 9). The real skill, said Poincaré, is in choosing between the different correct things one can write down so as to see the goal from a long way away, and without intuition the mathematician is like a writer who knows grammar but has no ideas.

The question of *why* something is true was much more important for Poincaré than the question of *what* is true. Even a strict proof would not always answer the question of why a result holds: there had to be what he called the "soul of the fact" (*L'avenir* 376), the central idea that explained why it was true and also how it could be proved. This did not necessarily make the proof entirely short or obvious, but it told him how to tackle the problem and which standard techniques would resolve it.

[5] See Toulouse (1909, 145–146).

This accounts for the longevity of his popular essays. It is quite remarkable that so many of them are worth reading well over a hundred years after they were written, and they remain fresh because they address issues that will always come round in the education of mathematicians, scientists, and the general public. Perhaps non-Euclidean geometry has become a surrogate for some other belief about space, but the issue of how a rigorous mathematical theory can be interpreted in physical terms remains pertinent. Certainly Poincaré is not the best person to read about special relativity, but his reflections on how the best science of his day seemed to be crumbling in the face of almost inexplicable laboratory results speaks to anyone concerned about how theories change (as they do). His hostility to logicism and his deep distrust of the rising axiomatic set theory leave Poincaré's strictures about how to define mathematical objects with limited appeal—except that to this day a number of eminent voices raise concerns about the cavalier acceptance of the infinite. Very likely his predictions for the future of mathematics look no better than most gamblers' best estimates, and no better than Hilbert's selection— and Hilbert was in a much better position to make his predictions come true—but read in their entirety they say valuable things about how mathematics does, or might, develop. And no one has written better about how to be an inventive mathematician. Other volumes in the series Poincaré wrote for tried for topicality, and now they pay the price of being yesterday's news. Poincaré tried to speak about human understanding, and that does not date. He might even be presented, with his theory of knowledge, as a precursor of cognitive science.

Poincaré's Achievements

Poincaré first emerged on the mathematical scene in 1879 and 1880 with a number of small papers on number theory after the manner of Hermite, who was pleased with them, and a couple on differential equations. He turned 26 in 1880, so he was no prodigy. But in 1881 he began to publish the work that had led him to write the essay and its supplements which placed him second in a prize competition of the Académie des sciences, and a stream of new ideas began that transformed the study of three areas of mathematics: complex function theory, differential equations in the complex domain, and non-Euclidean geometry. By 1884, when Poincaré's interests began to embrace yet

Figure 0.1. Charles Hermite (1822–1901). Source: Arild Stubhaug, *The Mathematician Sophus Lie* (Springer, 2002).

more fields he had rewritten the theory of Riemann surfaces, created new classes of functions that solve a large class of hitherto intractable differential equations, and placed at the center of it all the topic of non-Euclidean geometry that had previously been merely exotic. The new functions he defined, variously called Fuchsian or Kleinian functions after other investigators or more generally, automorphic functions, were a generalization of elliptic functions, a subject of considerable importance in its own right although one that did not detain Poincaré.

It is possible to trace Poincaré's progress quite closely in these years (as chap. 3 below describes) and we do not see a sudden flash that illuminates the whole. Rather, we see the gradual emergence of a governing family of ideas, built around the deepening appreciation of the group idea. Once Poincaré saw how non-Euclidean geometry entered the story he had a program that he could pursue that raised questions that, mostly, he could solve and that, in what was both a cooperation and a competition with the German mathematician Felix Klein, led eventually to a brilliant insight that had to remain an unproved conjecture for twenty-five years.

Thereafter, his work displayed no particular pattern. Unlike most of his contemporaries he did not stay in one field and deepen his understanding of it. Nor, like some more restless souls, did he simply switch fields from time to time. He took up new interests, but seldom dropped any. His earliest interests remained his last—his very last paper, as fate was to determine it, was on Fuchsian functions and number theory. But a significant shift came when he began to develop theories that applied to planetary astronomy: the shape of planets (described in chap. 5 below), their orbits, and the long-term stability of the solar system.

These were traditional questions going back at least as far as Newton, and they were central to an establishment that revered Laplace, but Poincaré invigorated them. Once again he soon reached a governing idea, in this case that for such problems the long-term behavior of the solution curves was what had to be understood. This marked a complete contrast with the astronomers' incremental tradition in which prodigious amounts of calculation were deployed to calculate the ephemerides for only a few years ahead. Poincaré succeeded to a remarkable degree with a preliminary study of differential equations and their solution curves on surfaces—which was a further way for him to appreciate their topology—and then embarked on what became a lifelong involvement with planetary motion. What remains one of his most celebrated discoveries is his demonstration that there is a deep reason for the failure of traditional methods to resolve even the simplest nontrivial problem, the three body problem: even three bodies moving under their mutual gravitational attraction can display chaotic motion and have orbits extremely sensitive to the initial conditions, thus making long-term predictions almost impossible.

When Poincaré became a professor of mathematical physics and probability in 1886 his interest in physics deepened, and no topic was more important and exciting than the theory of electricity, magnetism, and optics. To the British this meant the theory presented by James Clerk Maxwell, who had died in 1879, but this theory was distasteful to French scientists who found it lacking the elegant mathematical sophistication they were used to in their own tradition. Poincaré even found it inconsistent, but he also admired it for its depth, its mathematics, and its appreciation of the fact that there will not be a unique explanation of nature if there is any explanation at all. In the 1890s Poincaré became the French expert on the theory, the man who could indeed provide an elegant exposition of the ideas of Maxwell, Helmholtz, and Hertz, point out their strengths and weaknesses, and in due course do the same for Lorentz's contributions. He also became the adjudicator of a number of disputes in the subject, contributed to the technological exploitation of the new ideas, and, in 1905, the author of one of the lasting ideas in what is now the subject of special relativity: what he modestly called the Lorentz group. Finally, in 1911 his grasp of Max Planck's new theory of quanta was influential in the acceptance of the new ideas with a speed that Planck had feared impossible.

From 1890 until his death Poincaré retained an interest in the theory of real and complex functions in one and several variables and worked successfully on a number of outstanding problems. He made lasting contributions to Sophus Lie's theory of transformation groups and to algebraic geometry. But his major contribution to mathematics in those years was undoubtedly that of topology. It was one of his abiding beliefs that a qualitative analysis of a problem ought to precede a quantitative one, and the pioneer of qualitative methods in mathematical analysis was Bernard Riemann, who had died in 1866 leaving behind such a profound reorganization of the subject that it was take at least a generation to assimilate. Poincaré's involvement with Riemann surfaces early in his career educated him in the power of Riemann's ideas—in many ways Riemann and Poincaré were kindred spirits—and his formulation of the three body problem had led Poincaré to contemplate problems in extending Riemann's ideas to three dimensions. What he accomplished here essentially created a new branch of modern mathematics: algebraic topology. It may have done so in part because his methods were so visionary that they had more or less to be done again and differently in order to be rigorous, but also set out an attractive topic and ways of approaching its problems. He outlined several ways of defining three-dimensional manifolds, sketched what later would be called a Morse-theoretic decomposition of them, and described the two natural algebraic objects that are associated to a manifold, their first homotopy and homology groups, with enough precision to establish a profound problem, one that grew in successive interpretations to become the Poincaré conjecture.

This great range invites the question: Was there one Poincaré or many? If, trivially, there were many—the Poincaré of (sometimes) rigorous pure mathematics, the applied mathematician and lecturer happy with heuristic arguments, the scientist immersed in the details of geodesy or celestial mechanics—there was also only one. Not just because he took a firm view of what his task was, which was to develop the understanding of everything he looked at, but because of the many analogies and links he found between the subjects he worked on. Among the links he mentioned explicitly were ones between celestial mechanics and problems in the theory of complex functions of several variables; between physicists' intuitions and the often contrived methods for solving the partial differential equations of mathematical physics; and between geometry, physics, and philosophy. The right approach for a book such as this one is surely to follow his own, and to seek to explain what he discovered by conveying why it has to be described in particular ways, to aim to reilluminate the radiating centers of his own systems of ideas.

His way of working explains why Poincaré had rather distant relations with his contemporaries and no real students. As his nephew Pierre Boutroux explained to Mittag-Leffler, Poincaré was willing to be very patient with students, but when it came to expressing an opinion his standards were very high: either they had really grasped the idea, or they had not. Add to that the fact that the French system was much more closely tied to the old model of young independent inventors making their way in the world than the German graduate school approach, and the fact that most mathematicians in the 19th century worked on their own anyway, and his isolation is less surprising. But it did not spring from any reluctance to express himself, or from an "ivory tower" mentality: he served energetically on numerous committees and editorial boards.

Another measure of the man is afforded by the work of others that excited and impressed him. The first of these seems to have been Georg Cantor's work on point-set topology, which he applied to his own work in the 1880s. He was impressed by Lie's theory of transformation groups when he met Lie in Paris, but he did not work on the subject until 1900, after Lie was dead. Hill's new approach to the study of the motion of the moon he regarded as an insight into dynamical systems that was likely to be very useful in numerous ways. Among the physicists, the ideas first of Hertz and then Lorentz impressed him and drew him to the frontier of electromagnetic theory. Hilbert's *Foundations of Geometry* he recognized as presenting a profound and radical challenge to his own ideas, and this seems to have impressed him more than Hilbert's work on integral equations, where Poincaré always gave the palm to Ivar Fredholm's contributions. He appreciated Hermann Minkowski's *Geometry of Numbers* as a breakthrough in number theory, a topic Poincaré regarded as particularly difficult, and he seems to have appreciated the work of Italian geometers on the theory of algebraic surfaces sufficiently to produce his own, complex analytic, version of one of their most incisive results. His last enthusiasm was for Planck's insight into the quantum nature of radiation. On the other hand he never learned much from Einstein's theory of special relativity and seems not to have fully grasped it, despite coming up with the Lorentz group at the same time. He did little with the work of his contemporaries, whether he got on with them personally (as he did with Paul Appell) or not (Émile Picard), and he seems to have not cared about the younger generations of French analysts—Émile Borel, Maurice Fréchet, Henri Lebesgue, and Paul Montel—and not to have taken up even the idea of measure theory, despite his interest in probability theory. He may even have thought their interest in the strange behavior of functions somewhat misplaced. Even those who strayed into his territory, like Jacques Hadamard and Paul Painlevé, do not seem to have become mathematical confidantes. He disliked what he saw of the attempt to reduce mathematics to logic, and while he remained polite he was doubtful that any attempt to reduce mathematics to axiomatic set theory would succeed.

In Context

It would be impossible, and therefore absurd, to summarize the state of France in the second half of the 19th century, but a few perspectives are most relevant here. The defeat by the Prussians in the Franco-Prussian War and the loss of parts of Alsace and Lorraine in 1870–71 charged national feelings with potent emotions of shame, rivalry, and renewed patriotism. These burst out in the Dreyfus affair, which began in 1894, but they permeated scientific life in the form of repeated comparisons with whatever was happening in Germany.

Education prospered in the aftermath of the defeat. As Gispert (1991) has described, the higher education budget more than doubled between 1877 and 1883, new university professorships were created, especially in the provinces, research became professionalized, student numbers increased. The French Mathematical Society, founded in 1872, was initially dominated by graduates of the École polytechnique, but as the status of teaching rose so too did that of the École normale supérieure at the expense of the École polytechnique until by 1900 Poincaré was a marked exception in having attended the École polytechnique: Appell, Borel, Cartan, Darboux, Goursat, Hadamard, Painlevé, and Picard were all normaliens. New journals were founded—as they were in all academic subjects and in all types of journalism. Even so, Germany with its twenty-two universities and its productive emphasis on research, was the object of many a nervous glance. Poincaré, as we shall see, made numerous contributions to several of these journals, and his popular reputation was lucratively sustained as a result.

The Belle Epoque, as the period in French history from the 1890s to 1914 has become known, was a period when foreign travel took off among the middle classes, and champagne, high fashion, and operettas were the height of fashion. As a child Poincaré had traveled widely with his parents, and he continued this habit in later life. He also liked music, Wagner especially, but did not play an instrument although he had been taught piano briefly as a child.[6] But it was also the period of Zola's novels, when the gap between the rich and the poor widened considerably, and fashionable pleasures were contrasted with the miseries of the growing slums. On occasion Poincaré spoke out against what he saw as the mindless accumulation of wealth, although his involvement in explicitly political activity was sporadic, and he was a moderate on the side of Dreyfus (see chap. 2, sec. "The Dreyfus Affair" below).

French scientists were among those who hoped to improve the quality of life, Pasteur most famously, but the half century also saw the arrival of the telegraph, wireless, street lighting, and a number of other technological breakthroughs, and Poincaré was to contribute to the discussion of many of these issues. The ideology that accompanied these contributions to the creation of modern life was usually positivism in the form that Auguste Comte had given it in the 1850s but without his oddly religious overtones. This was a belief that the evidence of the senses, perhaps refined through simple scientific theories, was the only reliable form of knowledge, superior to any form of metaphysics or organized religion.

---

[6] When Poincaré died, *Le monde artiste* (20 July 1912, p. 458), recorded that Poincaré liked to go alone to matinees at the Opéra-Comique to listen to Pélleas et Mélisande and with his young daughters to La trompette, where he was the most attentive and assiduous listener.

Positivism was the default position of most scientists, and sometimes spilled over into a view that scientists were therefore particulary suited to advise government, perhaps best from a position of studied neutrality that had the unintended side effect of keeping the state funding of science at low levels, with detrimental effects on the growth of physics as a discipline in France. It often went with a naive progressivism, a belief that science dispassionately and sensibly applied would lead to steady, and ultimately remarkable improvements in the quality of life. Its greatest weakness was a failure to understand politics, where, of course, it was opposed by the powerful Catholic Right, and played into the lasting divisions created in the traumatic conflict of the French Revolution. Poincaré was far from being a positivist, but when he was drawn into defending his views of science against the arguments of Le Roy who sought to present science as little more than the inventions of scientists, he was arguing against opinions intended to bolster the theology of the Church, as is discussed below (see chap. 1, sec. "Science, Hypothesis, Value").

Émile Boutroux, the most prominent philosopher of his generation, and who was only nine years older than him, was his brother-in-law and it is often speculated that some of the Kantian aspects of Poincaré's thought derive from Boutroux. One of his central concerns in Boutroux's major book (1874/1895) was to bring a respect for contemporary science into philosophy without, as a result, introducing a determinism that would shut out ethics and free will. To do this he invoked a layered nature of science that put logic at the lowest level, with mathematics upon it, then mechanics, and so on until physiology and finally sociology was reached, and argued that no layer could be reduced to the one below because an element of contingency always intervenes. Mathematics could not then have an a priori claim on our knowledge of the world, whatever its rigorous character, and moreover, as he put it "the law of causality is but the most general expression of the relations arising from the observable nature of things."[7] The word "relations" here alerts us to the later ideas of Poincaré, and in fact Boutroux went on to distinguish two kinds of laws in science. The first kind are mathematical, abstract, and almost necessary although capable of revision, but they are remote from reality. The second kind are intuitive, observational, and completely empirical, but are not deterministic at all. The former operate when science explains,

---

[7] See É. Boutroux (1874, 25–27), quoted in Heidelberger (2009, 123).

the latter when it describes. But the mathematical laws are rooted neither in reality nor in the fundamental nature of the intellect: "mathematics is necessary only with respect to postulates whose necessity cannot be demonstrated, and so is hypothetical after all."[8] As Heidelberger points out, following Pierre Boutroux at this point, the element of contingency gives a pragmatic edge to Émile Boutroux's analysis that brings it close to Poincaré's conventionalism. On Boutroux's formulation, as on Poincaré's, the mathematical laws of nature cannot be made to apply except by treating them as free choices of the mind, taken with as much pragmatism as is worthwhile.

But it is also the case that Poincaré thought through anything that interested him for himself. He was never in the grip of a philosophical orthodoxy that drove him to explain everything in somebody else's language as if it had greater epistemic depth. He rather resembled Helmholtz in his struggles with the Kantians: sometimes sympathetic, but never a party member.[9] This was because his expertise was in mathematics and physics, topics the philosophers (Boutroux excepted) had largely ignored, but which he knew raised genuine problems in epistemology and ontology and on which he knew he had something original and important to say. The reflections of mathematicians and physicists on their subjects in the years 1880–1914 in fact created philosophies of their subjects well in advance of anything the professional philosophers deigned to contribute.

Habits and Customs

We have some evidence of how Poincaré actually worked on a daily basis. Like all really good mathematicians, Poincaré, kept a structured account or story of mathematics in his mind, one that placed the key concepts, methods, and theorems in a coherent way. He read in the fashion of some of the best mathematicians, as Pierre Boutroux observed,

> He did not force himself to follow long chains of deductions, the closely-woven net of definitions and theorems that one usually finds in mathematical memoirs. But going straight away to the result that

---

[8] See É. Boutroux (1895, 215), quoted in Heidelberger (2009, 135).
[9] I do not find Poincaré's conventionalist epistemology as close to Kant's synthetic a priori as does Folina (1992).

lay at the centre of the memoir, he interpreted it and reconstructed it in his own way; he took control of it in his own way and then, taking the book up in his hands once again he looked rapidly through the propositions, lemmas, and corollaries, that furnished the memoir....Instead of following a linear route his mind radiated from the centre of the question he was studying to the periphery. As a result, in his teaching and even in ordinary conversation he was often difficult to follow and could even seem obscure. When he expounded a scientific theory, or even told a story, he almost never began at the beginning but, ex abrupto, he set forth at once the salient fact, the characteristic event or the central person, someone he had absolutely not taken to time to introduce and whose name his interlocutor did not even know. (P. Boutroux 1914/1921)

He added, "All his discoveries my uncle made in his head, most often without the need to check his calculations in writing or setting his proofs down on paper. He waited for the truth to strike him like thunder, and counted on his excellent memory to remember it."

We may note a number of approaches to topics that Poincaré frequently tried, among them the following.

1. He looked whenever possible for transformations of a problem, not just to simplify it but because groups of transformations were at the heart of every place where they arose. As we shall see (for example in chap. 3) whenever he studied geometry, the corresponding transformation group was the key to all the important issues.

2. In problems on mathematical physics he looked for the basic principles: conservation of energy and of momentum, least action, and so forth.

3. When problems depended on some parameters, as for example many problems in dynamics do, he looked at the effect of varying the parameters on the quantities he was trying to find. If some important function vanished for a certain value of the parameters, would it vanish for a range of parameter values?

4. He frequently made a distinction between the qualitative and the quantitative aspects of a problem, arguing that the broad

features had to be understood first before the details could be fitted in, and that understanding the broad features aided the more detailed quantitative explorations.

5. He appreciated the close connection between real harmonic functions and complex analytic functions that Riemann had pioneered in the 1850s, and extended the intuitive properties of harmonic functions to the more complicated setting of functions of several variables.

6. He had a liking for naive geometrical analyses involving curves and surfaces, and in that setting for finding a way of reducing the dimension of the problem, for example looking at a flow on a surface by seeing how the flow repeatedly crossed a curve, and by reducing questions about three-dimensional manifolds to their two-dimensional boundaries. But he was not a visualizer who relied on what he could draw or see with his mind's eye.

7. He was comfortable with the fundamental features of finite-dimensional vector spaces, such as spaces of solutions of certain equations, or objects specified by parameters, and willing to consider infinite-dimensional analogues.

8. He analyzed problems in mathematical physics by looking phenomenologically at very small regions, where the physical process could be expected to be linear, and from this deducing an infinite system of linear equations that might then produce a linear partial differential equation. But he abandoned this final step with vigor when he saw that the new quantum theory was necessarily discontinuous.

Of course, the precise problem area would then determine how much detail Poincaré would think fit to give. The discovery of a new, and solvable, class of differential equations in pure mathematics, or of the way to generalize an important function in number theory, did not call for much detail, but when he was occupied with problems in new technology, as he was with radio waves, he was much more willing to show how the fundamental theory delivered useful information.

In 1909 Poincaré had allowed himself to be subjected to a battery of medical and psychological examinations conducted under the direction of Dr. Étienne Toulouse, the director of the Laboratoire for experimental psychology at the École des hautes études in Paris. Toulouse was interested in the psychology of exceptional people, and had already published a similar investigation of Zola in 1896 and some notes on Berthelot in 1901. Among other matters, Poincaré also answered questions about his personal views. He had believed in religion when he took his first communion, but found that by the age of 18 he had ceased to believe. He believed in freedom of thought, the right to research and to tell the truth, and for that reason he opposed clerical intolerance. He was a republican in politics, and thought that the state should not intervene very much, except in certain matters of health. He favored political equality for all, and had no theoretical objection to judicial or political rights for women, although he feared the influence of the church upon them. He was indeed a man of his time in his progressive beliefs in the merits of science for the public good, the role of the scientific intellectual, his republican dislike of the anti-intellectual positions of the Catholic Church and its insistent attempts to shape opinions and extend their influence. This was the milieu he grew up in, and most likely because of the damage caused by the Franco-Prussian War he was quite patriotic and very willing to serve his country in the best ways he could. He was never bellicose, but he was not averse to military rhetoric (he had, after all, studied at the École polytechnique) and there is a hint from time to time that he did not cite German authors when perhaps he should have done (he read and spoke German fluently, but he seldom cited widely). He had principled reasons for staying away from the world of politics, and was only drawn into the Dreyfus affair when he saw he had a contribution to make, but had strong views on education and on ethics that he expressed quite frequently as his popular esteem grew.

We also learn from Toulouse's account, which was published in 1909, and is in part a record of facts that anyone who knew Poincaré in person would probably have known. He was 1.65 m tall (5′ 5″) and weighed about 70 kg (154 pounds), which was regarded as moderately corpulent.[10] He had begun to put on weight when he married. His face was colored and his nose large and red. His hair was chestnut, his mustache fair. He had, it seemed, a share of minor ailments: digestion took some two

---

[10] It gives him a body mass index of 25.7.

to three hours during which time he could not think productively and indigestion frequently interfered with his sleep. For that reason he kept regular habits, with breakfast at eight, lunch at noon, and dinner at seven (he ate meat quite a lot), and never had coffee after dinner. He went to bed at ten and rose at seven. He did not smoke and never had, and disapproved of the habit. He had never exercised systematically, but he liked walking and would willingly go fifteen kilometers. He seldom had headaches. He made no mention to Toulouse of the illnesses that had affected him in public on several occasions, and only mentioned a disabling bout of rheumatism when he was 32.

Toulouse reported that Poincaré either made his mind up quickly or found it increasingly difficult to do so; and he noted, as many did who knew Poincaré, that he seemed almost permanently distracted. He relayed the story (p. 27) of Poincaré discovering to his surprise one day on a walk that he had a birdcage in his hand and having to retrace his steps to find the place where he had inadvertently picked it up. He also reported that Poincaré would answer questions even when it seemed that he had not been listening. This compares well with Boutroux's memory that Poincaré

> thought in the street, when he went to the Sorbonne, when he went to take part in a scientific meeting, or when he went on one of the long walks he was accustomed to take after his dinner. He thought in his room at home or in the lecture theatre at the Institut, when he wandered about, pulling a face and playing with his key ring. He thought at the table, at family reunions, even in the salons, often interrupting brusquely in the middle of a conversation to force his interlocutor to follow a chain of thought he had come up with. (P. Boutroux 1914/1921, 197–200)

Toulouse also noted that when animated by a topic Poincaré liked to walk about with his hands behind his back, his brow furrowed and his eyes blinking. He was good with languages, spoke clearly and correctly but rather timidly, so outside of the lecture theater he spoke on public occasions only after careful preparation, and often by reading from notes. But he had never thought to learn a speech by heart and give it from memory. Writing quickly and clearly came easily to him, although he did not have particularly fine handwriting and never had. But although right handed he could write quickly with his left hand, and recalled being ambidextrous up to the age of eight; he also had difficulty distinguishing his left from his right.

The most interesting finding Toulouse made concerned Poincaré's visual abilities. Poincaré had taken to wearing spectacles in his early thirties, but had good vision in both eyes, and he did reasonably well on the tests of short term visual memory. He was not good at recognizing faces, and was helped by hearing people's voices. Strikingly, he did not think visually, and claimed to have no long-term visual memory and to rely on his motor memory; when asked to copy simple figures from memory, which he did quite well, he did so by recalling the motion of his eyes. Toulouse put it this way: "Poincaré analyses the objects that he sees and looks at; it is by analysis that he reproduces them. The lucidity of his observations is remarkably clear." But he noted that Poincaré did draw occasionally. Altogether a curious set of abilities for someone who is rightly regarded as one of the great geometers.

1

The Essayist

POINCARÉ AND THE THREE BODY PROBLEM

On 20 January 1889 the ambitious Swedish mathematician Gösta Mittag-Leffler went to the Court of King Oscar II of Sweden to announce the judges' decision concerning the prize competition the King had announced in 1885. Mittag-Leffler had administered it, and would have been feeling very pleased, for the competition had been a success: the result would surely play well with the King and would add, as intended, to the celebrations for the King's 60th birthday. Moreover, it would be popular with professional mathematicians; none of this could do other than advance Mittag-Leffler's own career.

King Oscar II was an enlightened monarch who proposed prize competitions from time to time. He had studied mathematics at Uppsala University and retained an interest in it all his life. He had given financial support to Mittag-Leffler's new journal, *Acta Mathematica*, occasionally sponsored individual mathematicians, and in 1884 had asked Mittag-Leffler to run the prize competition whose result he was shortly to discover (it is not known if he had had the original idea or if it was a suggestion of Mittag-Leffler's). Together they took the risky decision of calling for essays on specific topics, rather than merely awarding a fine recent piece of work. The four topics they chose, and which were published in *Acta Mathematica*, some German and French journals, and in English translation in *Nature* (30 July 1885) were all substantial, but the winning one addressed in suitably technical language the most publicly accessible topic:

A system being given of a number whatever of particles attracting one another mutually according to Newton's law, it is proposed, on the assumption that there never takes place an impact of

Figure 1.1. Gösta Mittag-Leffler (1846–1927). Source: Arild Stubhaug, *The Mathematician Sophus Lie* (Springer, 2002). Image courtesy the Picture Collection, The National Library of Norway.

> two particles to expand the coordinates of each particle in a series proceeding according to some known functions of time and converging uniformly for any space of time.

As every mathematician and astronomer could see, a successful resolution of this problem would enable one to describe the position of every planet in the solar system at every past and future moment of time. For this reason the problem is often referred to as asking for a proof of the stability of the solar system, although an alarming range of life-destroying orbits for the earth are compatible with the stated requirements.

Mittag-Leffler had ensured that the entries, all of which had to be received by 1 June 1888, would be judged by a very distinguished panel: Weierstrass, Hermite, and, in a lesser capacity, Mittag-Leffler himself. Karl Weierstrass in Berlin and Charles Hermite in Paris were the leading mathematicians in each country in their generation, the most powerful influences behind the scenes, and, not coincidentally, people whom

Mittag-Leffler had already used to help him in his career. All three worked chiefly or exclusively in an area of mathematics called complex function theory, which was one of the dominant fields of mathematics at the time and taught at least to the better students in all leading universities. Entries were to be submitted anonymously, tied to their authors by a system of sealed envelopes, although it is hard to believe that the panel would not recognize the handwriting of most of the people capable of entering the competition. And there was a prize: a gold medal bearing his Majesty's image and having a value of one thousand francs, and a sum of 2,500 crowns (about 2,600 francs, equivalent to a professor's salary for half a year).

To minimize the risk that no one would enter, or that no entries of sufficient quality would be received, the four questions were carefully tailored to current interests, specifically the work of Lazarus Fuchs, a German mathematician, the French mathematicians Charles Briot and Jean-Claude Bouquet, both professors at the Sorbonne and known for their work on complex function theory and differential equations, and Henri Poincaré. In fact, as the judges knew very well, Poincaré's name could have been mentioned in connection with the work of Fuchs, for he had astounded the mathematical community with his extension of it in the years from 1880 to 1884 and made the reputation of *Acta Mathematica* in so doing. It could also have been mentioned in connection with the problem of the stability of the solar system, because Poincaré had also written extensively and with remarkable success on a simpler system of equations but always with an eye on the problem of planetary motion.

It was, of course, Poincaré who had won the competition. The King could be assured that his generosity had brought forth new work from the most talked about new member of the mathematical community. Mittag-Leffler could feel pleased that the man whose career he had encouraged by opening the pages of his journal to him would now repay him with a new work of great importance, one that was also going to be published in *Acta Mathematica*. All in all, a most satisfactory outcome. And indeed, the news played very well. Hermite was able to persuade the editors of the scientific pages of *Le Temps* and *Le journal des débats* to publicize the news, and they were followed by the *Journal officiel* and *Le correspondant*.[1] Since

---

[1] For an account of Poincaré drawn from extensive coverage in the journals of his day, see Ginoux and Gérini (2012).

a second prize had gone to another French mathematician, Paul Appell, the whole thing was turned into something of a French celebration. Both men were made Knights of the Légion d'honneur for their work[2] and, as one of Poincaré's obituarists was to note (Darboux 1916, xxxi–xxxii), once the prize was announced:

> The name of Henri Poincaré became known to the public, who then became accustomed to regarding our colleague no longer as a mathematician of particular promise, but as a great scholar of whom France has the right to be proud.

It helped, of course, that astronomy was a popular subject and celestial mechanics had a significant place in French scientific life, marked, for example, by Le Verrier's discovery of Neptune in 1846, an achievement commemorated by his being one of the 72 mathematicians, scientists and engineers to have their names inscribed on the Eiffel Tower in 1889.

Poincaré's account of what he had done in tackling the problem of the stability of the solar system formed the content of his first essay for the general public, his (1891d) in the *Revue générale des sciences pures et appliquées*.[3] The *Revue générale* had been founded by Louis Olivier. It was published twice a month, and its first issue had come out on 15 January 1890. From the start, Olivier aimed to make it preeminent among the many journals which proliferated in France at this time, and he filled its pages with articles on every kind of science from agriculture to mathematics. He intended to speak to working scientists—his own training was in microbiology and he had studied with Pasteur—as well as educated general readers, and to convey a progressive message about the benefits of science and rational thought in solving problems of national importance while educating its citizens in the virtues of honesty and integrity. He edited his journal vigorously, and worked hard to bring in the best authors, and he seems to have succeeded: when he died in 1910, at the age of only 56, Nobel laureates were among those who wrote obituaries of him.[4]

---

[2] As was Mittag-Leffler, but at the lowest rank, which he felt was an insult and tried unsuccessfully to decline (Stubhaug 2010, 383).

[3] He was to return to this theme in his (1898e), by which time he was established as a leading astronomer.

[4] On Olivier, see Della Dora (2010) which concentrates on Olivier's successful pioneering work on creating the scientific cruise.

Poincaré was sympathetic to the cause of the small journals, and wrote a preface (1911b) to an otherwise almost valueless account of the *Revue bleue* and the *Revue scientifique* by Jacques Lux in 1911/12. Lux observed that the two revues had been founded in 1863 to revive French intellectual life, and had dedicated themselves anew to this task after the defeat in the Franco-Prussian War. The *Revue scientifique* in particular aimed not so much to popularize science as to make the methods of science better known—this was a call that Poincaré was to answer with more skill than most. In his preface, Poincaré noted the role these journals had had since their founding in breaking down the gulf between "those charmed by Letters and those passionate to research into scientific truths." He ascribed this gulf to the university system, and to the greatly varied nature of scientific work, and he dismissed the idea that some preestablished harmony could be relied upon to bring many muddled accounts into something "solidly established, free of error, and rich in beauty." Instead, he said, it was possible to know the whole world only by train, and "the revues are our trains. No doubt," he went on, "the trips are rather quick, and one can only see the essential things—but, as with good guides, one does see the essentials."

In his essay (1891d), Poincaré carefully explained that Newton's laws permit the completely accurate solution of two bodies orbiting around their common center of gravity: as Kepler had been the first to suggest, each body moves in an ellipse. But no one had been able to solve the problem of three or more bodies exactly. The best that could be done was to approximate the motion by expressing the coordinates of the separate bodies in polynomials of arbitrary degree. But this was not without its problems, he said. Suppose, he went on, that the motion is actually described by a periodic function with a very long period, and that the masses of the various bodies enter this function as parameters and are very small. Even so, increasing powers of these masses appear in the approximation, and these powers will grow indefinitely although the function being approximated does not (being periodic). For example, to approximate the function $\sin mt$ one can use the expression $mt - \frac{m^3 t^3}{6}$. This is a very good approximation for small values of t but it increases indefinitely with t while $\sin mt$ remains between -1 and $+1$.

That said, he went on, this method of approximations is reliable, and it is already used to calculate tables of ephemerides accurate for navigational purposes or for tracking asteroids several years ahead. However, the final aim of mechanics is altogether more grand: it is to see if Newton's laws describe all astronomical phenomena. One can imagine, said Poincaré, any amount of improvement in these approximative methods, and indeed many people had invented new ones, but no one had been able to show that the infinite series that expressed these solutions converged and so could be used to establish the behavior of the solar system at all future times. This is what the Prize problem had called for when it asked that the solutions be expressed as "known functions of time...converging uniformly for any space of time."

But, said Poincaré, could not complete rigor be achieved in some new way, say for the problem of three bodies moving under their mutual gravitational attraction? It had long been known that there were some extremely special cases that could be treated this way. There were results about three bodies constrained to lie in a plane. The American astronomer George William Hill had found a periodic solution in which the three orbits are all nearly circular (Hill 1877) and had used it to give a new and improved theory of the motion of the moon.[5] Other special cases allowed for the bodies to move in different planes. There were solutions in which the planets slowly spiraled outwards to a closed orbit, and others in which, far back in the past, they had departed from such a closed orbit and were now spiraling into another; Poincaré called these the singly and doubly asymptotic solutions. But the general case seemed very far from being amenable to any kind of analysis.

In his work, he said, he had exploited another of Hill's insights: the actual motion, although not periodic, can sometimes be very well described by an approximation in terms of periodic functions. Laplace had shown this a century before in analyzing the motion of the moons of Jupiter. However, the existence of the asymptotic solutions already mentioned raised the worrying possibility that the power series solutions of the equations of motion would in fact diverge for large enough values of the time variable t (and maybe for all values of t). This was not, after

[5] On Poincaré's use of Hill's work, see Barrow-Green (1997, 22–28). For what will surely prove to be the definitive account of Hill's work, see Wilson (2010).

all, a fatal objection, but it suggested a need to look for other ways of tackling the problem. And in fact he had been able to show that as far as the stability of the solar system was concerned, it was possible to give new results. He had looked at a special case of the three body problem, in which a small body of negligible mass orbits two larger bodies (such as the sun and Jupiter) which are going round in circles about their center of gravity in a common plane (it is further assumed that the small body moves in this common plane). In this case, he said, he had been able to show that under certain conditions the small body will return infinitely often indefinitely close to any position it has ever occupied, and in this sense its orbit is stable. (This form of stability was called Poisson stability by Poincaré in the prize-winning memoir (1890c, para. 8).) The asymptotic solutions mentioned before are not stable in this sense, so there are both unstable and stable solutions, and, said Poincaré, he had been able to show that the initial conditions that lead to an asymptotic orbit are exceptional and so stability holds in general.

This short account concealed as much as it revealed; there was a much more complicated and mathematically deeper story going on behind the scenes (see chap. 4 below). It is nonetheless informative. The problem at stake is to establish the stability of the solar system, not so much because anyone doubts it as because it would help confirm the insight that Newton's laws apply to the entire physical universe. Poincaré gave a reasonable hint as to the methods used, and a good indication of their imperfections. Even when the answer is very simple, say when the orbit is periodic, the approximations may lack that simplicity. The solutions will generally be given as infinite power series, but there is no proof that these make numerical sense because convergence may not be established, or even true. Nor were these problems a reflection of how poor had been the attempts to solve the problem: they were intrinsic to the problem as the variety of orbits, especially the asymptotic ones, showed. Then, rather modestly, Poincaré indicated that where everyone had been blocked he had managed to find a way forward, and show that there were good reasons to believe, at least in the first mathematically nontrivial case, that most orbits were stable. It was both an explanation of why the problem is so difficult and an indication that Newton's laws do indeed lead to plausible conclusions.

POINCARÉ'S POPULAR ESSAYS

Controversies over a New Geometry

In 1891 Poincaré also published another substantial essay in the second volume of the *Revue générale des sciences pures et appliquées*. This was his "Les géométries non euclidiennes" (1891b). Poincaré's success with the three body problem is what brought him into the public eye, but the essay on non-Euclidean geometry proved to be the start of a controversy that was to keep him there. It drew on what was not only one of his earliest but also one of his lifelong interests, it spoke to a contemporary concern about the nature of space and geometry, and Poincaré had something important and unexpected to say.

Geometry in 19th-century France was dominated by the legacy of two mathematicians, Gaspard Monge and Adrien-Marie Legendre, and was perpetuated through the influential École polytechnique. Monge, who had played a major role in establishing this École and its sister establishment the École normale supérieure during the French Revolution, had created the discipline of descriptive geometry. This was an ingenious way of coupling the depiction of objects in three-dimensional space on two planes (plan and elevation) with some simple algebra. This made it possible to calculate lengths and angles for masonry and it became an essential tool for military and civil engineers. Since in principle the job of the École polytechnique was to provide two years basic training for candidates to the Écoles d'application (such as Ponts et Chaussées, Mines, Artillerie, Arts et Métiers) this gave descriptive geometry a permanent place in the syllabus there, which indeed it retained until after the First World War.

Adrien-Marie Legendre was a more classical mathematician. He saw himself in the tradition of Euler and worked to extend his work, writing on number theory and the creation of new functions in mathematics. But unlike Monge, who prospered during the Revolution and was briefly a minister for the Navy and went on to enjoy the support of Napoleon, Legendre lost his inherited wealth in the Revolution, and turned to textbook writing to make money. In this he was very successful. He decided to write a geometry book that candidates for the École polytechnique would have to read—each year there were between 400 and 700 applicants for about 200 places (Belhoste 2003, 46)—and it ran to several editions, continuing after his death in 1833. He also took the opportunity to abandon the previous generation's enthusiasm for intuitive geometry—loosely inspired by a Cartesian-derived enthusiasm for clear and distinct ideas—and restore the rigors of a mostly Euclidean treatment. His failure to succeed completely in this is part of the story of non-Euclidean geometry.[6]

Euclid's *Elements* had survived to the modern era in the form of a handful of editions together with Arabic versions and commentaries. These were edited in various ways and by the 16th century had become accepted as the foundations of geometry. Famously, the material is presented in the form of some definitions, some postulates or axioms, and then an impressively rigorous derivation of a sequence of results. With few exceptions the theorems are not difficult to believe; what impressed people was the quality of the deductions. And because the deductions were so convincing it came to be believed that the results—the theorems—were all true. In this way, starting from simple and undeniable premises it was said that genuine knowledge of the world was obtained by nothing but pure thought. Euclid's *Elements* became a paradigm of logical reasoning and the great, perhaps the only exemplar, of certain knowledge.

Of course, as many who reflected on the *Elements* said, the conclusions were only as good as the premises, and among those one stood out as more difficult to accept than the others. This was the parallel postulate, which, in the way it is stated in book I of the *Elements* says "That, if a straight line falling on two straight lines makes the interior angles on the same side less than two right angles, the two straight lines, if produced indefinitely, meet on that side on which are the angles less than the two right angles."

It was worrying that the parallel postulate made assertions about the behavior of lines arbitrarily far away; there was no possibility of checking it empirically. But, how could it not be true? Assume all the postulates of the *Elements* except the parallel postulate. Now consider a plane, a line ℓ in the plane, a point P in the plane but not on the line, and the perpendicular from P to a point Q on the line ℓ. On the face of it, three

---

[6] See, for example, Gray (2010).

things can happen to lines in the plane and passing through the point P:

- The Euclidean case: all the lines through P meet the line ℓ on one side or the other of PQ except for one, which never meets ℓ and is the unique parallel to it through P;

- All the lines through P meet the line ℓ on one side or the other of PQ except those that meet it on both sides; there are no lines that never meet ℓ and is there is no parallel to it through P;

- All the lines through P meet the line ℓ on one side or the other of PQ except those that never meet it on either side; there are infinitely many lines that never meet ℓ and they are all parallel to it, so the parallel to ℓ through P exists but is far from unique.

Of these possibilities, the second one has the distressing consequence that if m is a line through P that meets ℓ on both sides of the perpendicular PQ, say at points A and B, then there is no unique direction from A to B in the plane—both ℓ and m will do—and the lines ℓ and m enclose an area. This possibility had never been explicitly ruled out by Euclid, but it is inimical to the spirit of the *Elements*. Happily, it was shown in the 18th century to be incompatible with the other postulates of Euclid on other grounds, and so it can be discarded.

The third possibility was much harder to deal with. As every history of this topic shows, from Greek times, through several centuries of investigation in the Islamic world, to the modern West, many a mathematician tried and failed to show that this possibility also contradicts the other postulates of Euclid. They all failed because there is no contradiction: a geometry can be built on precisely the postulates in Euclid's *Elements* with the parallel postulate itself removed and replaced by the postulate that there are many lines through P that do not meet ℓ. In particular, all of Legendre's several attempts were flawed, and he gradually replaced each one by another in different editions.

The new geometry is the one known as non-Euclidean, or, less provocatively but more obscurely, as hyperbolic geometry. These failures, some more public than others, gave the parallel postulate an air of mystery: attempting to refute the alternative and so establish the truth of Euclidean geometry was quite a popular pastime, and by the 1820s some adventurous spirits were beginning to contemplate the shocking thought that the alternative might be true.

The non-Euclidean geometry of a space of two or three dimensions was known to the Hungarian mathematician Janos Bolyai and the Russian mathematician Nicolai Ivanovich Lobachevskii independently by the mid-1820s. Confident that there were two logically possible but mutually inconsistent descriptions of geometry, both men suggested how astronomical observations might resolve the matter, although observations were too imprecise to do so. Their published accounts of it met with nothing but resistance, and they died (Bolyai in 1860 and Lobachevskii in 1856) without ever knowing that their conclusions would come to be accepted. The only recognition either received came from Carl Friedrich Gauss, who was widely recognized as one of the greatest mathematicians ever, but who confined his enthusiasm for their work to his correspondence with friends and the Göttingen Scientific Society.[7]

When Gauss died in 1855 mathematicians discovered many unpublished mathematical ideas in his papers, among them his investigations of non-Euclidean geometry. The fact that the great Gauss had accepted it had a powerful effect on them, and the climate shifted toward accepting the new geometry. What proved decisive, however, was the work of another Göttingen mathematician, Bernhard Riemann. Riemann, who knew Gauss personally and conversed with him on many subjects (but not, it would seem, the foundations of geometry), presented a paper to the Philosophy Faculty at the University of Göttingen in 1854 as part of the requirements for the Habilitation degree, which was a necessary and sufficient qualification to teach in a German university at the time. Its initial premises are very different from any of Euclid's, very intuitive, and very profound. They were ultimately to do much more than dethrone Euclidean geometry.

Riemann proposed that all one needed to be able to do in order to do geometry was to be able to talk about lengths and angles. He sketched a way of doing that which required nothing more than a set of points and a way of talking about infinitesimally small distances. The rest could be done using the traditional calculus. Remarkably, such geometries can be in any number of dimensions—Riemann was even willing to contemplate spaces of infinitely many dimensions—and the geometries that result need not be Euclidean. Their existence depends on a vast generalization of a deep theorem of Gauss's from 1827, which said that it is possible

---

[7] On this episode in Gauss's life, see Gray (2010).

to consider the geometry on a curved surface without regard to any ambient space. Riemann extended this point of view to his much more general setting. By defining geometry intrinsically, no appeal was made to Euclidean geometry, and its paramount role in defining geometric concepts came to an end.

News of Riemann's work, and then of the earlier achievements of Bolyai, Lobachevskii and Gauss, spread steadily among professional mathematicians. Riemann died in 1866 at the age of 39, and his paper "On the hypotheses that lie at the foundations of geometry" (Riemann 1867) was published posthumously the next year. In 1868 news of it reached a young Italian mathematician, Eugenio Beltrami, who had come to some of the same ideas but was held up by his professor's worry that Beltrami's use of the calculus in his work might implicitly rely on Euclidean geometry. Once he saw that he was not basing his work on a contradictory set of ideas he published his own account of the new non-Euclidean geometry (Beltrami 1868), and it proved very convincing.

Non-Euclidean Geometry Comes to France

The years 1870–71 were, of course, momentous in France. Defeat in the Franco-Prussian War, and the surrender of territories in Alsace and Lorraine, were a national trauma (one that marked the young Poincaré, growing up in Nancy in the Lorraine), and they brought to a head a growing feeling that the sciences that had sustained the French Army at the start of the century had been allowed to wither. Reforms were energetically set in place to improve the place of mathematics and science in France. The Société Mathématique de France (SMF) was established in 1872 with a new journal, and two young Frenchmen, Gaston Darboux and Jules Hoüel set up a journal of their own, the *Bulletin des sciences mathématiques* in 1870 with the deliberate policy of bringing mathematical discoveries from abroad to the attention of French mathematicians. Germany was the main place to look, because mathematics and the sciences were visibly prospering there, but Hoüel saw to it that Beltrami's work was speedily translated into French, as was the most accessible of Lobachevskii's attempts; his booklet of 1840 (French edition 1867), and Bolyai's "Appendix" was also translated by Hoüel and published in 1867.

In this way non-Euclidean geometry came to France. Generally speaking it is safe to say that mathematicians who bothered to look into the matter were convinced, but it is not clear that it was anything more

Figure 1.2. Jean-Gaston Darboux (1842–1917). Source: Arild Stubhaug, *The Mathematician Sophus Lie* (Springer, 2002).

than a curiosity, given that no one had given a convincing reason to prefer non-Euclidean geometry to traditional Euclidean geometry, which is much simpler to use. All this was to change when, in 1880, Poincaré (see chap. 3, sec. "The Competition" below) showed that the new non-Euclidean geometry was the natural geometry for treating a range of problems in mathematics, and that in that setting Euclidean geometry was a small, singular case. By 1884, when the torrent of Poincaré's papers on the subject ceased, no mathematician could doubt that there were two important geometries in mathematics.

Popular knowledge of non-Euclidean geometry spread in various ways. One can imagine it diffusing among students at the École polytechnique and École normale supérieure as they heard about it from their professors one way or another. It also came from Germany, where the eminent German scientist Hermann von Helmholtz had been writing about it since 1868. In Helmholtz's first publication on the subject (1868, translated into French by Hoüel) he noted that the geometry on a sphere was also a logically consistent geometry but omitted non-Euclidean geometry; Beltrami wrote to tell him about it and thereafter Helmholtz discussed it too. His essays were translated into French and published in such journals as *Le moniteur scientifique* in 1870 and *Revue scientifique* for 1877.

Professional mathematicians could accept non-Euclidean geometry because they could accept an essentially Riemannian view of the subject, but many mathematically literate people (scientists, engineers, teachers) who could not share that standpoint found it much more difficult. The new geometry destroyed the absolute character of mathematical truth, it contradicted the supposed transparency of the postulates of Euclid. If indeed there were two geometries with different theorems then only one could be true, and its truth would presumably be a matter of empirical verification. So the first question that needed to be answered was: Is there a geometry other than Euclid's? Then, of course, if the answer is yes, the next question is: Which is true, the old Euclidean or the new non-Euclidean geometry? It is this debate that Poincaré joined publicly in 1891, and he did so by trying to educate his readers in the Riemannian way of thinking.

In his article "Les géométries non euclidiennes" (1891b) he explained that no deductive science is any better than its premises, and among those of Euclidean geometry are some he regarded as a priori analytic judgments (a noticeably Kantian turn of phrase) such as "two quantities each equal to a third are equal to each other." He said he would not be concerned with these. But there are premises of a different kind, which are special to geometry, such as "there is a unique line through any two points," and the parallel postulate. It had first been shown by Bolyai and Lobachevskii he said, that there was geometry based on the same assumptions as Euclid's except for the parallel postulate. The matter had been entirely reformulated by Riemann, whose work inspired Beltrami and Helmholtz. Poincaré then ran through some of the most important points in non-Euclidean geometry: the angle sum of a triangle is always less than two right angles and the difference between the two right angles and the sum is proportional to the area of the triangle; it is impossible to make a similar copy of a figure that is different in size.

He then asked his readers to imagine two-dimensional creatures living on the surface of a sphere. They could describe their world in geometric terms, it would be finite in size, and it would have no boundary. Riemann, said Poincaré, had proposed that our physical space is exactly like it but, of course, three-dimensional. However, the two-dimensional space of the imaginary creatures had a property not shared with Euclidean geometry: there are infinitely many distinct lines through any two (antipodal) points.

The Riemannian (spherical) geometry was in some ways the opposite of Lobachevskian geometry: in spherical geometry there are no parallels; in Lobachevskii's geometry there are infinitely many. In spherical geometry the angle sums of triangles exceed two right angles, in Lobachevskii's geometry it is always less. But it was still possible to object: Despite the appearance of consistency in each geometry, could there not lurk a hidden contradiction? Evidently not in the case of spherical geometry, which was a branch of ordinary geometry. But what about non-Euclidean geometry?

Poincaré asked his readers to imagine a surface and figures drawn on that surface in flexible inextensible cloth in such a way that when the cloth is moved over the surface the figure can deform without any of its lengths changing. For most surfaces this will prove to be impossible, but for some it can be done. These are the surfaces of constant curvature: spheres (which have constant positive curvature), the plane and the cylinder (which have zero curvature), and the surfaces of constant negative curvature, upon which the geometry of Lobachevskii holds. To explain what this is, Poincaré asked his readers to use a dictionary, such as those that convert between two languages (for example, French and German). In this dictionary, he said, we find these words and their meanings:

- space: the region above a fixed fundamental plane

- plane: a hemisphere meeting the fundamental plane at right angles

- sphere: sphere

- circle: circle

- angle: angle

- and so on.

With this dictionary one could translate every statement in Lobachevskian geometry into a statement in Euclidean geometry. Were two mutually contradictory statements to arise in Lobachevskian

geometry, their translations would be two mutually contradictory statements in Euclidean geometry, and no one doubts that Euclidean geometry is free from contradiction. What that certainty is derived from, he
added, is another question, very interesting and not insoluble, but not
one he was going to treat here. The dictionary also shows that non-
Euclidean geometry is not a merely logical exercise but one that can have
applications, and indeed, he said, he and the German mathematician
Felix Klein had already given applications in the study of differential
equations.

Could it still be maintained that the axioms of Euclidean geometry
were the sole foundations? Clearly not. The English philosopher John
Stuart Mill had claimed that every definition in mathematics concealed
an axiom (the axiom that asserted the existence of the object being
defined). While Poincaré thought this went too far, and the existence
of some objects in mathematics can be deduced from the existence of
others, there was nonetheless some truth in Mill's remark. For example,
what is one to make of the claim that two figures are equal? Two figures
are said to be equal when one can be moved to fit exactly on top of
the other, but this presupposes that the moving figure remains equal to
itself—plainly, a vicious circle beckons. The truth of the claim can only
be verified by assuming some property of the figures, such as that they
are rigid, a property we expect of solid bodies in our world. It is not an
analytic a priori judgment, but only a postulate like many other postulates
of Euclidean geometry.

Motion in geometry was (as we shall see shortly below) one of Poincaré's
most fundamental concerns. Here he confined himself to describing a
theorem recently proved by the Norwegian mathematician Sophus Lie.
This theorem says that given a space of n dimensions, in which rigid
body motion is possible, and in which the position of a body is described
by a finite number, p, conditions then there are only finitely many
possible geometries that describe this space. In particular, there are only
finitely many geometries that would describe a space of three dimensions
under these conditions. Most of the geometries introduced by Riemann
therefore cannot admit rigid body motions, and will therefore never be
described by axiom systems akin to Euclid's.

What then, he asked, are axioms in geometry? They cannot be
synthetic a priori judgments, as Kant had suggested, for then they
would impose themselves upon us with such force that non-Euclidean

geometry would be literally unthinkable, and that is not the case. Nor can they be experimental facts, because the lines and planes of elementary geometry are not physical objects and therefore cannot be experimented upon. Even if one found a way around that objection there is a more fundamental objection: geometry is exact but no science founded upon experiments can be exact—the sciences are subject to eternal revision, and geometry is not. Therefore, said Poincaré, he regarded the axioms of geometry as conventions. We are guided by experience, but have nonetheless some freedom to choose. Our choices, not being matters of logic, can only be conventions, and the corresponding "axioms" of geometry (but not those of arithmetic, he noted) are really disguised definitions.

Finally, which geometry is *true*? The question, Poincaré announced, was meaningless. The only question to ask of matters of convention is: Which is the most convenient? The answer, he said, will be Euclidean geometry. It is the simplest, and it agrees with our experience of rigid bodies. As for the supposed astronomical tests, Poincaré observed that what one calls a straight line in astronomy is the path of a ray of light. Faced with an experiment which seemed to show, for example, that the angle sums of triangles were less than two right angles, there were two conclusions possible. The first was that space was non-Euclidean, the other was that space was Euclidean but that rays of light were not, after all, straight and new laws of optics are required. Since people would surely always choose the second option, he said, Euclidean geometry had nothing to fear from experiment.

If this early foray into popular writing was widely understood it would be impressive testimony to the level of debate in France, but it is more likely that it was too densely written and raised too many issues. Riemann's name is invoked in two different contexts. In the first use of the term "Riemannian geometry" it is merely as a synonym for the usual geometry on a sphere. The more general Riemannian geometries are all the geometries of constant or variable curvature, in which the free motion of rigid bodies is either constrained or indeed impossible. It is Riemannian geometry in this second sense that allowed mathematicians to construct rigorous models of non-Euclidean geometry, and to exhibit it, as Beltrami did, as a geometry on a space of constant negative curvature. The novel idea of the dictionary notwithstanding, Poincaré took it for granted that non-Euclidean geometry was coherent. Nowhere in the essay did he explain what it is for people who have no knowledge of it. It is left unstated that there is a notion of rigid body motion in non-Euclidean geometry which would be every bit as natural for creatures brought up believing in non-Euclidean geometry as the notion is for us with our deeply Euclidean education. These leaps in the exposition Poincaré would return to in later essays and explain more fully.

The unexpected reply that there is no way of telling whether it is Euclidean or non-Euclidean geometry which is true, because both are matters of convention, is another matter Poincaré was to return to. It is an original point, and one, as we shall see on several occasions, that is deeply entwined with his thoughts about contemporary physics. Throughout his working life he was going to take the side of multiple interpretations and remain reluctant to say that any particular statement was *true*. It is a position with some philosophical advantages, but it may also mark a gap between him and members of the physics community of his day.

Poincaré's essay drew a reply from G. Mouret, ingénieur en chef at Ponts et Chaussées, in volume 3 of the journal (pp. 39–40). Mouret was the author of "L'égalité mathématique," a long article in the *Revue philosophique* in 1891 and reviewed in the *Revue générale* (vol. 2, p. 826), the theme of which is that mathematics is always a matter of relations between objects in the world. It was precisely in this spirit that he replied to Poincaré by dismissing non-Euclidean geometry as purely speculative, factitious, and akin to doing thermodynamics with negative temperatures. For him, Euclid's was the only geometry. Poincaré replied very politely in the same volume (pp. 74–75) that he did not wish to have the last word on such an old and difficult question, but he had something new to say. In seeking to show the role of experience in shaping mathematics he had also to say that this role was limited, and that the fictions of Riemann and Beltrami were also valuable because they helped the imagination to break the bounds of habits created by daily experience. For example, he said, consider creatures perfectly adapted to living inside a sphere of radius R in which the temperature and index of refraction varied with distance ρ from the center according to the rule that the temperature was $R^2 - \rho^2$ and the index of refraction was $\frac{1}{R^2-\rho^2}$. Such creatures would believe that their world was infinite, and what they called straight lines would be what we call arcs of circles perpendicular to the boundary sphere. They would do so because such curves are the paths of rays of light in the sphere, because they are geodesics in this space, and because solid bodies rotate with such curves as axes, and as a result they would adopt the non-Euclidean geometry of Lobachevskii. This is the first appearance of Poincaré's cooled sphere model of non-Euclidean geometry, which he expanded upon in his essay "L'espace et la géométrie" (1895d), published in the *Revue de métaphysique et de morale* (vol. 3), shortly to be discussed below.[8]

But, he admitted, this remark took him beyond Mouret's point, and so he asked again if experience could be the sole source of mathematics. This he said he doubted because the mathematical and physical continua were so different. He cited Fechner's recent experiments to argue that the physical continuum is one in which one experiences quantities such as weights obey the paradoxical rules $A = B$, $B = C$, $A < C$, whereas in the mathematical continuum one has $A < B < C$. This is another idea he was to expand upon in later essays (for example, 1903a and 1912h). But he felt that Mouret had gone too far in his essay when he had sought to derive the primordial notion of equality from experience. Mouret, he said, had written much that he agreed with: space is not a simple idea, all mathematical ideas reduce to the categories of relation, resemblance, difference, and individuality. However, he could not help but remark that the most characteristic of these ideas are already to be found in Helmholtz's essay "Messen und Zahlen"—"only the conclusions are different." But he could not agree that the proposition "two quantities equal to a third are equal to each other" was a fact of experience that more precise experiments might refute. Rather, it seemed to him that here Helmholtz had been right to conclude that when we call objects in the world equal we do so in conformity with our preconceived idea of mathematical equality.

From this point onwards Poincaré took his responsibilities as an author very seriously, writing two or three articles each year for a general audience, or, perhaps one should say, a variety of audiences. Four broad themes can be discerned in these essays, in which Poincaré steadily set out a sophisticated epistemological position, one that was at times more modern and at times more traditional than his opponents. The first is the nature of geometry, which broadened into an account of the nature of mathematical knowledge, how it relates to logic and to arithmetic as

---

[8] This journal, founded by Xavier Léon and Élie Halévy with the assistance of Léon Brunschvicg in 1891, took up a sophisticated idealist position against popular positivism and excessive religiosity. Émile Boutroux was influential in getting Poincaré to write for it (Rollet 1999, 72).

well as its connections to geometry and the study of space. The same question of how the human mind can know things, and what it can know, is apparent in his writings on physics: several of his essays on this second theme are concerned with the role of facts, experiments, hypotheses, and mathematics in the buildup of science. The third theme is topics that were currently at the forefront of research, such as the problematic connections between electricity, magnetism, and optics that also animated Lorentz and in due course Einstein. Finally, he was not afraid to be drawn into the vexed topics of science and morality.

In 1895 Poincaré wrote about geometry, what it is and how one can know about it, in the essay (1895d). He began with what he called a small paradox. If we met creatures who through the evidence of their senses were as convinced that space is non-Euclidean as we are that it is Euclidean, neither side would have any problem describing their experience in terms of their own geometry. Space, then, is different from the representations we make of it. Space—geometric space— he said everyone agreed was continuous, infinite, three-dimensional, homogeneous (all points of it are identical) and isotropic (it looks the same in every direction). As for what he called our representative spaces, first among these is pure visual space, formed by the images on our retina. This space is only two-dimensional, and not only is the retina far from being homogeneous, the impression that it is continuous is very likely, said Poincaré, to be shown to be an illusion by further research. Our visual sense that space is three-dimensional comes from the accommodation and convergence of our eyes, and this muscular sensation gives us what Poincaré called the complete visual space, which is not isotropic. There are yet other spaces, the space revealed by our sense of touch— tactile space—and motor space, learned through our movements. The spatial senses combine to produce a representative space that is neither homogeneous, isotropic, or even three-dimensional. It is a very distorted image of geometric space, therefore "we do not *represent* external bodies in geometric space, but we *reason* about these bodies as if they were in geometric space" (1895d, 82, 50). When we say we have located an object in space we mean that we have represented to ourselves the movements that are needed to get us there. As Poincaré pointed out explicitly, this formulation does not commit us at all to the preexistence of a notion of space, and indeed we shall see that he was hostile to this notion all his life. For him, the concept of motion in various formulations was fundamental.

But we do not simply receive impressions of space through our senses, we move parts of our bodies and we move around altogether. Some impressions of the motion of external objects can be mimicked entirely by our own motion, others cannot. Those than can be mimicked are the motion of solid, rigid bodies, and therefore *"if there were no solid bodies in nature there would be no geometry."* (1895d, 86, 53; italics in original)

Poincaré then investigated how the behavior of solid bodies generated our sense of geometric space. To show that nothing yet precluded non-Euclidean geometry, he asked his readers to imagine sentient creatures living in a homogeneous sphere of radius R. Everything in the sphere, including the creatures, is made of the same material, and the temperature at a point a distance r from the center of the sphere is proportional to $R^2 - r^2$. Suppose that rigid bodies as they move adapt instantly to the ambient temperature. Suppose too that light travels as if the index of refraction a distance r from the center of the sphere is proportional to $R^2 - r^2$. Then, he said, these creatures would deduce from the behavior of rigid bodies in their space that the geometry of space was non-Euclidean.

Poincaré did not explain why this is the case. He had, in fact, described the model of non-Euclidean geometry that he had used in his own work in the 1880s.[9] In it, geodesics—curves of shortest length—between two points inside the sphere look to us like arcs of circles perpendicular to the boundary sphere, and it is easy to see that the parallel postulate does not hold. While the detailed confirmation of these claims requires some mathematics, we can at least see that geodesics will look curved. Consider, for example, what looks like a straight line joining two points, and another path joining them that is slightly bowed toward the center. As the creatures measure the length along these paths, they will find that the second path is shorter because the rulers they use expand along the second path, and so fewer of them are needed to measure the length.

In the end, Poincaré concluded,

Experience guides us in this choice, which it does not impose on us. It tells us not what is the truest, but what is the most convenient geometry. It will be noticed that my description of these fantastic worlds has *required no language other than that of ordinary geometry.*

---

[9] The formula that appears was taken from Beltrami's presentation of 1868, but in the form Poincaré had given it in his own mathematical work in 1880–81.

Then, were we transported to those worlds, there would be no need to change that language. Beings educated there would no doubt find it more convenient to create a geometry different from ours, and better adapted to their impressions; but as for us, in the presence of the same impressions, it is certain that we should not find it more convenient to make a change. (1895d, 94, 59; italics in original)

His anthropomorphic picture of creatures living on a solid ball that gets cooler as they move away from the center proved to be a popular one. It can be found, for example, in Feynman's *Lectures in Physics* (vol. 2, chap. 42). Above all, it precisely illustrates his point from the earlier essay of 1891: the creatures would naturally say that light rays were curves of shortest length between any two of their points, and that the geometry of space was non-Euclidean. We would naturally say that on the contrary, space was Euclidean and light rays were bent by this strange behavior of the background temperature. Upon reflection, both sides would have to agree that there was no way of resolving the matter logically, but each found their own decision the most convenient and simple to use. Such, one might say but Poincaré did not, is the hidden power of education.

The essay "On the foundations of geometry" (1898c) in the American journal the *Monist* is unusual in being published in English.[10] The *Monist* is itself interesting. It was founded in 1888 by Paul Carus, who was born in Germany in 1852 and emigrated to the United States in 1882, settling in La Salle, Illinois. There he fell in with Edward Hegeler, another German immigrant, who had made a fortune out of zinc, and in due course Carus married Hegeler's daughter. Hegeler gave money to enable Carus to set up his journals, the *Open Court* and the *Monist*; the journals were billed as being devoted to the "scientific study of religion" and the "religious study of science" respectively. Carus's own views are evident in the title of the *Monist*, which derives from the then-fashionable philosophy of monism, a philosophy that promoted the idea that "all is one" and that there is a unity behind appearances. But Carus was not mindlessly tolerant; he sought to sweep out mystical fog by adopting a scientific

---

[10] It was translated from Poincaré's manuscript by T. J. McCormack. It seems that the original French version is lost, and the essay was translated back into French by Louis Rougier for publication in 1922; see Rollet's account in Poincaré (2002, xxii).

approach to religion. He had a lifelong admiration for the German scientist Ernst Mach, who developed a philosophy called neutral monism that was to influence James, Carnap, and Russell. Mach's intention was to find a common ground for objects and sensations that will permit an explanation of how knowledge of the external world is possible. Carus frequently published articles by Mach in the *Monist*, and Mach often found himself in agreement with Carus's own views.[11] Mach's interests, and many of his conclusions, were close to Poincaré's, which is surely how the invitation to write for the *Monist* came about.

Poincaré took the opportunity to present his most systematic view of the nature of geometry. The essay can be read as an early example of cognitive science. It is an answer to the question of how can we find our way round in the world, and it is a clear straightforward argument that joins two difficult positions. The starting point is the idea that we construct our sense of space by making sense of our sense impressions. We do this as infants, and we have done it as a species. The end point is a summary of the results in mathematics that determine the possible spaces we could construct. In between comes an argument about the innate abilities of the mind that can be read as a sort of Kantian transcendental argument (these capacities are necessary if knowledge of the external world is to be had) or as evolutionary (the human mind has this capacity built-in, and it is honed by experience).

The notion of space, Poincaré began, cannot be derived from our sensations alone. Rather, "it is built up by the mind from elements which pre-exist in it, and external experience is simply the occasion for its exercising this power, or at most a means of determining the best mode of exercising it."[12] To establish this, he asked his readers to imagine the world that would be detected by a single, immovable eye. Such an observer would not be able to ascribe a sense of distance or size to the world, and indeed could not say that two points were contiguous; all these observations require either a mobile eye or two eyes. The single, fixed eye would not see a homogeneous, isotropic, three-dimensional space. More controversially, as he admitted, Poincaré denied that any sensation gave a sense of direction, and he concluded that "sensible space has nothing in common with geometric space." In fine Kantian tones he declared,

---

[11] For more information see Gray, *Plato's Ghost* and the references there.
[12] The most accessible edition is the reprint in Ewald (1996, vol. 2).

Geometric space, therefore, cannot serve as a category for our representations. It is not a form of our sensibility. It can serve us only in our reasonings. It is a form of understanding. (para. 3)

Poincaré attributed our ability to create geometry to our appreciation that our sensations vary, and they do so in two distinct ways. In one, a chemical reaction changes what we are looking at (a blue liquid in a vase turns red—as in a litmus test). In the other a sphere turns red and then blue as it rotates, because one hemisphere has been painted red and the other blue. The first type of change is an alteration, the other a displacement, and it is displacements that help us create geometry. They are further divided into external changes, which are independent of our will and our muscular sensations, and internal changes, which are voluntary and are accompanied by muscular sensations. Finally, some external changes can be corrected by internal ones—we may instead walk round the sphere—and the external changes that we can correct are the displacements. An immovable being could never, therefore, create a geometry. Moreover, the correcting process is only crude and never exact; it is therefore an action of the mind that recognizes a common character to some displacements although they are not exactly identical. As a result, the mind is able to recognize a repeated displacement, and so, and here it seems Poincaré departed from the sharp Kantian distinction between space and time (para. 5): "It is this circumstance that introduces number, and that permits measurement where formerly pure quantity held sway." The reference to pure quantity is surely to the idea of a purely topological substratum to our construction of space.

Next (para. 6) Poincaré indicated informally that these displacements form what is mathematically known as a group: one displacement followed by a second yields a displacement, and every displacement can be undone. He had already raised this idea in his (1895d), but his new argument was more thorough. He said it was trivially true that one external change followed by another corresponded to a third external change: in symbols, if A and B are changes then A followed by B, which Poincaré wrote as $A + B$, is a change. But if the changes are displacements, then there must be internal changes correcting A and B, say A' and B', respectively, and one requires that the external change $A + B$ must be corrected by the internal change $B' + A'$, which was by no means obviously correct. It might be, for example, that the imprecision involved in saying two slightly different changes were identical would break down after a sequence of changes.

Poincaré argued that the conclusion that displacements form a group could not be established by a priori reasoning. Was it then the result of experience? In that case it could be modified by experience, and if so, "What, then, will become of geometry?" He answered his rhetorical question by explaining that geometry was indeed safe from all experience, because it was supported by an "artificial convention": borderline cases would be handled by regarding the change as a displacement accompanied by a small alteration (such as small amounts of bending or expansion due to heating). Subject to the operation of these small ad hoc modifications, we can say that the displacements also form a group.

The consequences of this argument (para. 7) were crucial for our knowledge of space. It is the operation of the group that constructs a homogeneous, isotropic space, because we suppose there are displacements that take us anywhere with any orientation, and which leave us otherwise exactly as capable as we were before. So, said Poincaré, we must now study the properties of this group with which we are equipped, taking care to recognize the same group when it is described in different ways (such groups are said to be isomorphic, and are the same for all practical purposes).

He started this analysis with the idea of indefinitely repeating a single displacement, which generates a certain group of displacements. Next, he said, we recognize that any displacement may be regarded as the repetition of a certain number of smaller displacements. Here Poincaré gave a contrived argument (para. 9) to support the belief that there are displacements of every size given by a real number. This corresponds to the mathematical description of the real line as formulated by Cantor, Dedekind, and Meray in the 1870s, and Poincaré indicated that he would explain later why it was the only one compatible with other properties of the group known from experience. In fact, only this hypothesis allows the use of the mathematical theory upon which Poincaré was relying, and he admitted that a proper justification for this part of his theory remained to be supplied.

In paragraph 10 Poincaré introduced the notion of subgroups of the group of all displacements. He carefully explained the difference between a subgroup of a group and a subset: a subgroup is a subset with the property that the composition of any two elements of the subgroup is also in the subgroup, as is the inverse of every element of the subgroup. Of particular interest to Poincaré were the subgroups that correspond to rotations about a fixed point, but of course at this stage in his argument Poincaré had only produced the group which is to construct our sense of space. So he drew attention to groups of displacements that preserve a fixed sensation, such as rotations about a common axis, or, more precisely, he considered the corresponding internal changes corresponding to these displacements. Then, by considering different axes he isolated the subgroup of rotations about a point. Again, starting from a knowledge of space rather than the group, we would say that there are many such subgroups, one for each point of space, and that all these groups were isomorphic (after all, that is what is meant by space being isotropic). We would also say that if R is a rotation about a point P and D is a displacement taking P to P' with inverse D' then the displacement $D' + R + D$ is a rotation about the point P' and that every rotation about P' arises in this way, so the subgroup of rotations about one point is isomorphic to the subgroup of rotations about any other point. Poincaré now indicated how this argument can be reversed, and made to express the concept of the displacement from one point to another in the language of groups, and in this way obtain the concept of a straight line joining two points.

Poincaré was now close to explaining how our innate grasp of groups of displacement not only yields our concept of space, but also disposes us to construct space as Euclidean space. Here (para. 18) he appealed to the purely algebraic results obtained by Sophus Lie. Lie, whose work Poincaré much admired, had established that there were not many groups of displacements that corresponded to a space of a given dimension, although he started from the concept of a space and derived his groups from it.[13] In particular, Lie showed that the groups that gave rise to a homogeneous, isotropic space of three dimensions were limited in number. Poincaré once again ran the argument backwards and showed how to obtain the spaces from the groups. Here he confronted an amusing problem. We can tell points from lines, and feel strongly that

---

[13] Lie returned the compliment, writing in the highly personal and sometimes embittered pages of the preface to Lie (1893, xxiii) that Poincaré "had repeatedly made interesting applications of my group theory. In particular I am grateful to him that he, as Picard was to do later, took my side in my struggles over the foundations of geometry, while my opponents sought to leave my works on that subject ignored."

the primitive element of space is the point, not the line. Put it another way: we think of a straight line as made up of points, but we almost never think of a point as being defined by all the lines through it. But Poincaré could only observe that if you emphasized the sort of subgroup that corresponded to rotations about a point you would construct a three-dimensional space, whereas if you emphasized the sort of subgroup that corresponded to lines joining two points you would construct a four-dimensional space (a line in space can be specified uniquely by four coordinates, but not three).

Poincaré now recapitulated his whole argument to make it clear how specific sensations, specific changes in sensation, and specific internal changes are involved in the statements we make about points in space being close, or on a line, so that we can form a concept of space by imagining the corresponding changes. In particular, we can construct the idea of a rigid body and discuss its movements in space by creating in our mind the motions that are the corresponding internal changes.

At this stage, various arguments current in Poincaré's time and since start to fall into place. If you take space as the primary concept and assume that rigid bodies can be moved around in it arbitrarily, you can prove mathematically, as Lie had done, that there are not many three-dimensional candidates, of which Euclidean, non-Euclidean, and spherical stand out (there are a few others, which can be eliminated on other grounds). Poincaré started with the groups, and noted that they each permit the user to impose a sense of distance on the space being constructed. Indeed, Poincaré argued that this is what Euclid had done in covert form with his definition of congruence: two figures are congruent if one can be *moved* onto the other so as to fit exactly. For there to be a concept of length it has to be the case that a pair of points cannot be mapped onto any other pair of points.[14] Poincaré might have added that Euclid appreciated this point: his *Elements* speak of congruence of segments but never of lengths. As Poincaré did show, one can define angle in the same way.

Now Poincaré had to remove an obstacle that a well-educated reader might make, to the effect that professional mathematicians currently thought of projective geometry as the fundamental geometry, precisely

[14] One then says that an arbitrary pair constitutes the unit length and moves them around in the usual way to measure lengths, construct subunits, and so forth.

because it does not have a concept of length (this view was to be the nub of Russell's disagreement with Poincaré over the nature of geometry; see sec. "Russell" below). It has instead a concept of the straight line, and a complicated invariant that concerns four points on a straight line, from which a concept of length can be derived by letting two of the points recede to infinity in opposite directions, and Poincaré brushed it aside (para. 17) on the grounds that we had plainly not derived our concept of distance in this way. This is surely true, but there is something unsatisfactory about this dismissal; can it be argued that evolution went the way it did without there being spatial considerations involved all along?

One reason for the brusqueness with which Poincaré dismissed projective geometry was that his real interest was in the metrical geometries (Euclidean, non-Euclidean, and spherical). In his view, the axioms of any of these geometries are conventions, and we could be sure they were free of contradiction because they could be studied entirely analytically. (Strictly speaking, this only pushes the freedom from contradiction back to the nature of arithmetic, but Poincaré was only to consider that argument much later.) Spherical geometry can be excluded if we believe space to be infinite, and that leaves Euclidean geometry and its modern rival. We must choose between them, because experience cannot be decisive, and

we choose this geometry rather than that geometry, not because it is more true, but because it is the more convenient. . . . Unquestionably reason has its preferences, but these preferences have not this imperative character. It has its preferences for the simplest because, all other things being equal, the simplest is the most convenient. Thus our experiences would be equally compatible with the geometry of Euclid and with a geometry of Lobachevskii which supposed the curvature of space to be very small. We choose the geometry of Euclid because it is the simplest. If our experiences should be considerably different, the geometry of Euclid would no longer suffice to represent them conveniently, and we should choose a different geometry.

Let it not be said that the reason why we deem the group of Euclid the simplest is because it conforms best to some pre-existing ideal which has already a geometric character; it is simpler because certain of its displacements are interchangeable with one another, which is not true of the corresponding displacements of the group of Lobachevskii. Translated into analytical language, this means that there are fewer terms in the equations, and it is clear that an algebraist who did not know what space or a straight line was would nevertheless look upon this as a condition of simplicity. In fine, it is our mind that furnishes a category for nature. But this category is not a bed of Procrustes into which we violently force nature, mutilating her as our needs require. We offer to nature a choice of beds among which we choose the couch best suited to her stature. (para. 22)

Poincaré's remark about the interchangeable nature of certain Euclidean displacements and the noninterchangeable nature of the corresponding non-Euclidean displacements presumed a well-educated public. It can be elucidated as follows. Displacing a plane first north by a certain amount and then east by another amount has the same effect as making the easterly displacement first and then the northerly one; either way the result is a displacement along a diagonal heading in some measure north-east. To use Poincaré's symbolism, $N + E = E + N$. But the same is not true in non-Euclidean geometry: the two combined displacements $N + E$ and $E + N$ result in displacements in different directions. In the Euclidean case the order in which the displacements are performed does not matter and they are interchangeable; in the non-Euclidean case the order matters and the displacements are not interchangeable.

It is a strange peg on which to hang a criterion of simplicity, but Poincaré had little choice. Once he was satisfied that space is an empty concept, that we are not forced to recognize its essential nature (because it does not have one, there being at least three geometries that were three-dimensional, homogeneous, and isotropic) and geometric space is rather a construct of our minds, it would follow that it is to the nature of our reasoning about geometry that he had to look. So the fundamental distinctions could not be left at the level of the different spaces described by these geometries, but had to be pushed back to the groups. No matter that the distinction is unfamiliar to most people who hear it, the point, for Poincaré was that it was anchored in an explanation of how we can have knowledge of the external world at all.

For the same reason, Poincaré insisted on an analysis of the concept of distance that was the opposite, as he admitted, of the one held by Helmholtz, Lie, and almost everyone else. He regarded it resting on a distinction between matter and form (para. 21). These mathematicians said that the matter of the group existed before its form, the matter being the three-dimensional manifold of space. Whereas for himself, said Poincaré, the form existed before the matter, by which he meant that the isomorphism class of the group we use to construct space comes first (isomorphic groups do not differ in their group-theoretic properties), and they can be defined and used without reference to any kind of matter.

The general belief of those who put matter first, is that objects have sizes (allowing for certain predictable alterations, such as slow growth, heating and cooling, and so forth). When we return home in the evening we expect to find everything in our house the same size that it was when we left in the morning. So when we imagine creating a mathematical description of space, we automatically endow it with a concept of distance. Then, when we think about moving around in this space, which we formalize as a group of transformations mapping the space to itself, we automatically think of the group of isometries (the distance-preserving maps) of the space. If we know of other groups of transformations we can then ask if they contain only isometries. For example, many familiar properties of Euclidean geometry apply to a figure and its scale copies and are invariant under changes of scale (called similarities). We do not discuss the size of a right-angled triangle when applying Pythagoras's theorem to it, and we do not expect to apply a similarity transformation (other than an isometry) when returning to our house in the evening.

Poincaré rejected this argument, because he believed that what came first in our untutored exploration of the world was motion—our motion, the motion of our eyes, our hands, our bodies. When we elicit that class of motions for which we can compensate (the rotating sphere, the rolling train) we activate the mental abilities to imagine carrying around a rigid body. This is an external object with the property that all of the varying sensations we have of it can be compensated for by internal changes. Now we can imagine carrying it around and measuring distances and assigning sizes to objects, but without the existence of such a group in our heads we would not be able to talk of distance.

Here, not for the last time, we encounter a characteristic tacitly held belief of Poincaré's. He did not like the idea that objects could have a property but we could not, even in principle, determine it. Confronted with a claim to a piece of knowledge, he routinely asked, "How do we know this?" and if there seemed to be no way of knowing he would dispute the claim, or at best regard it as a convention or a heuristic device. This will be seen in his ideas about physics and in the study of set theory. It is in this spirit that he denied any sense to the concept of distance independent of the ability to determine it. For Poincaré, the mind behind an immovable single eye would not have a concept of distance but confess to an inability to get it right; such a mind would inhabit a space without the concept altogether. This is a shrewd principle for a scientist to have, because one does not want completely mysterious, elusive concepts actively at work in a laboratory, but it leads to problems with the concept of truth. The straightforward view is that if a statement is true it is timelessly true. To say that dinosaurs roamed the earth in the Cretaceous period is to make a true statement whenever it is made. Poincaré would rather say it was meaningless if made by Plato, because Plato could have had no means of verifying it.[15]

This philosophical position was clearly on display in another essay Poincaré published in 1898, "La mesure du temps" (1898a) in the *Revue de métaphysique et de morale*. The question of the measurement of time was much debated among physicists at the time, because it had become urgent in the study of electromagnetism, but it had been known for some years to raise philosophical problems (see Martinez 2009). For example, Auguste Calinon in his (1886) had dismissed absolute motion as a metaphysical concept that was mathematically meaningless. But when he turned to the study of relative motion he found that it involved measuring durations, which in turn required understanding uniform motion, and so there was a circularity that could only be broken by an arbitrary stipulation. There was another problem in saying that two separated events took place at the same time, but the simultaneity of distant events was necessary to the measurement of velocity, so Calinon said our minds gave us the ability to perceive simultaneity. After Calinon and Poincaré met in August 1886 Calinon sent him a copy of his (1886), which Poincaré promptly read and sent comments on back to

---

[15] Folina (1992, chap. 8), argues that Poincaré's fundamental position was a theory of meaning that came down to "verifiability in principle." A philosophical analysis leads her to find this position coherent and, together with what she sees as Poincaré's "background theory of the synthetic a priori," capable of fending off strict finitism.

its author. That letter is lost, but we have Calinon's reply (15 August 1886), where he admitted to some disagreements on minor points but believed they were in agreement on his major ideas, which were to move all the notions hitherto associated with matter (the measurement of time, speed, force, acceleration, and mass) into geometry, and to distinguish between those theories in which the time variable t plays a role other than simplifying the formulas and those where it does not (i.e., so to speak, when t rather than any function of t enters the theory). He said he had found that even some distinguished professors obstinately opposed his ideas, and he hoped that Poincaré would help him advance them.

In his essay, Poincaré started from our psychological sense of time, and discussed how we construct what he called scientific and physical time. The first, and in his opinion, less original opinion consisted in making clear what is involved in the claim that the hour from noon until one has the same length as the hour from two to three. Any such claim in the end depends on a convention about the equipment used, even if that is the earth itself as it rotates (sidereal instruments being the most accurate measurement of time available). He then turned to a question that he thought had been neglected: What does it mean to say that a physical phenomenon, which happens outside every consciousness, is before or after a psychological phenomenon? In his example, Tycho Brahe's supernova is said to have occurred well over 200 years before he saw it in 1572, so what does it mean to say that it occurred before Columbus's visual experiences on the Hispaniola? What would it mean to say that distant events happened at the same time? To answer these questions, Poincaré invoked the example of writing a letter and sending it to a friend. The writing is the cause of the reading, and we agree that it came first. A peal of thunder is attributed to an electric discharge, and again we say the discharge came first because we believe it is the cause of the thunder. Indeed, whenever two events occur repeatedly in the same order we say that the first caused the second. But in the first two examples the cause (which we assume we know) allows us to decide which event came first, whereas in the general claim about two events the temporal order allows us to infer the cause. Rightly, Poincaré asked how are we to escape this vicious circle?

He answered his question by returning to astronomy. Astronomers, he said, date distant events by assuming that the velocity of light is a constant and the same in all directions. They justify this by pointing to Roemer's measurements on the four main satellites of Jupiter, which harmonize so well with Newton's laws if this assumption is made. Likewise, geodesists measure longitude in a variety of ways: the transport of accurate clocks, astronomical measurements and, more recently, the telegraph. All of these are approximate to some degree, and we reconcile them, and ourselves, to this complexity, by a convention that we use the simplest combination.

PARIS CELEBRATES THE NEW CENTURY

In 1900 Poincaré participated in no less than three of the displays of centennial enthusiasm that the French hosted in Paris. The first of these was the International Congress of Philosophers, which took place from 1 to 5 August. Poincaré read extracts from his memoir *Les principes de la mécanique* (1901e) in which he addressed the question of how can mechanics be an experimental science when its principles are only empirical and approximate. Some of the concerns he raised here were to grow in importance as the opening years of the century went by and the crisis in the foundations of the theories of electricity and light deepened, until by the time he went to the St. Louis Congress in 1904 they were dominating his thoughts. Here we see them as it were in domesticated form. The principle of inertia, he said, was not an a priori truth, nor was it even an experimental law, because it cannot be verified. The principle of reaction doubled as an axiom (uniform rectilinear motion when no forces act) and a definition (of mass). The principle of relative motion seemed well established, but like the principle of conservation of energy it could be neither proved nor disproved, verified nor falsified. All these principles were suggested by experience, they were *conventions* and not absolutely arbitrary, but convenient, which was why experience could suggest them but never reverse them. This view was not original to Poincaré, hints of it can be found in earlier mathematicians reflecting on the applicability of mathematics. Thus Jacobi in 1847 remarked, "From the point of view of pure mathematics, these laws cannot be demonstrated; [they are] mere conventions, yet they are assumed to correspond to nature.... There are, properly speaking, no demonstrations of these propositions, they can only be made plausible."[16] After discussing a particular example, Poincaré commented,

> We see that we began with a particular, and rather crude, experiment, and ended with a very general law, perfectly precise, the certainty of which we regard as absolute. It is us who have conferred this certainty, one may say quite freely, by regarding it as a convention.[17] (S&H 128, 86)

That the laws of mechanics are here regarded as true by convention will be discussed below (see chap. 11, sec. "Geometric Conventionalism").

In the subsequent discussion (1900a) the mathematician Paul Painlevé insisted on the arbitrary character of the principles in Poincaré's presentation. They could always be adapted to any new facts by introducing new hypotheses. He found this an excessive skepticism, and argued that the principles imposed themselves on us, they were the quintessence of innumerable experiences/experiments, and any hypotheses used to rescue them had always turned out to be scientific (i.e., subject to the law of causality). In a word, they were true, and all attempts to replace them had failed because they introduced innumerable complications. In his view science proceeded by successive approximations from empirical origins and under the guidance of these principles. Convergence was not guaranteed a priori, but justified by the success with which theories agreed more and more naturally with reality. To this Poincaré replied that he did not really disagree, science had indeed always progressed that way. But he wished to draw attention to the fact that by a series of more or less conscious artifices the first approximation had been transformed not into a provisional truth capable of revision but into a definitive and rigorous truth.

The mathematician Jacques Hadamard then spoke. In his opinion, which he said agreed with Kirchhoff's, mechanics did not aim to explain but to describe simply and exactly the principles of that science. Apparently contradictory facts could be validly explained away by hypothesizing, for example, a new force if this led to a simple theory, whereas changing the principles raised the problem of avoiding a contradiction

---

[16] From Jacobi (1999, 3, 5) quoted in Pulte (2000, 61n).

[17] The English translation has the journey from experiment to law running backwards, which makes no sense.

with other established facts. He therefore was inclined to agree with Duhem that it is not by a single hypothesis but by a set of them that mechanics can be tested experimentally. Poincaré said he agreed with this. One could say that each experiment gave us a new equation between very many unknowns; we would always have fewer equations than unknowns, but every new experiment reduces the underdetermination of the problem. However, the situation was different with the "unknowns introduced by geometry (the curvature of space) or rational mechanics." These can never be determined by experiment because, one might say, they enter experiments as extra variables.

To other speakers, Poincaré replied that he found questions about existence to be as lacking in sense as those about the truth or objectivity of mechanics, and that the question of the reality of the outside world (presumed by his questioner to obey objective and eternal laws) belonged elsewhere, in the section on metaphysics.

Next Poincaré gave an address to the International Congress of Physics, which ran from 6 to 12 August, on the relationship between experimental and mathematical physics. The essay (1900f) on hypotheses in physics began with the observations that rampant empiricism—a theory that was no theory at all and nothing but a list of empirically verified facts—might seem to be the best kind of science, but it would in fact be of no use whatsoever. It would fail to allow us to predict, and to generalize. Generalization is needed because no experiment can be repeated exactly; it allows the scientist to correct the data because the curves given by the appropriate laws of physics will not pass exactly through the data points. It is the generalization that allows the scientist to predict, and, said Poincaré, "It is far better to predict without certainty than never to have predicted at all." The aim of a good experiment is to maximize the number of predictions having the highest degree of probability.

These generalizations are the domain of mathematical physics, but on what do they rest? It is often said, Poincaré remarked, that they rest on the simplicity of nature, but this is not certain. This puts the scientist in a quandary, because the only way round or through the triviality that every fact has infinitely many generalizations is to make a choice guided by simplicity. To make matters even more difficult, the history of science offers examples of simple processes hiding underneath apparent complications and examples of simple appearances concealing highly complex realities. For example, Boyle's (Mariotte's) law relating the pressure, temperature, and volume of a gas could hardly be simpler, but the motion of the constituent molecules of the gas are vastly complicated: the simplicity is only apparent.

Generalization in science takes the form of hypotheses, and to be sure a hypothesis must be abandoned if its consequences cannot be verified. But a failed hypothesis is nonetheless of positive value. It was entertained because it seemed to work, and its failure means some new, little understood process is at work and a discovery can be made. "It might even be said that it has rendered more service than a true hypothesis." In fact, there are several criteria that govern the use of hypotheses. Not only should they be few in number and consistent, they come in types. Some, said Poincaré, are those one might call natural and necessary, such as the assumption that the motion of distant bodies is irrelevant. Other hypotheses are indifferent: it might not matter whether the experimenter assumes matter is continuous or discrete. But some hypotheses are real generalizations, they can therefore be confirmed or refuted by experiment.

Mathematical physics proceeds, said Poincaré, in three characteristic ways. It assumes that at any moment of time a physical process develops in a way that refers only to the immediate past; for this reason it is typically expressed mathematically by a differential equation. Second, the explanation focuses on tiny regions of space—the spread of heat from one point in a body to its neighbors, the bending of a rod as a deformation of the very small elements of the rod. Third, these localized explanations generally have very simple expressions. Therefore, insofar as this way of analyzing a problem produces localized laws that can all be treated by the same mathematics—the simultaneous solution of the differential equations that arise—the success of mathematical physics can be explained.

However, it seemed to Poincaré that the preconditions for this approach (homogeneity, relative independence of distant parts, simplicity at the local level) were running out, and the student of natural science will be compelled to find other modes of generalization.

The second part of the essay began by confronting the melancholy ruins of failed theories, which lead the man of the world to predict that all theories will fail and that science is bankrupt. To refute this view Poincaré argued that the equations in which the old theories were expressed are still true, the relations they capture preserve their reality. But as for what is related:

> These are merely names of the images we substituted for the real
> objects which Nature will hide for ever from our eyes. The true
> relations between these real objects are the only reality we can
> attain....When theories seem to contradict each other, it is likely
> that it is the images we have supplied which stand in contradic-
> tion. The mechanical world view regards atoms as if they were
> very small billiard balls: that is a metaphor. It is useful, it is an
> indifferent hypothesis. And as such may even be revived. (*S&H* 174,
> 123)

Poincaré gave the example of the fluids of Coulomb, once old-fashioned,
"yet here they are re-appearing under the name of electrons."

Another example was the concept of energy. The principle of conservation of energy signifies that something remains constant, said Poincaré,
but since we cannot give a general definition of energy the principle
cannot be verified. The principle remains meaningful because we can
assert a true relationship between the things we call energy, and the
application of the principle can be extended until it ceases to be useful,
until it no longer correctly predicts new phenomena.

Poincaré laid stress on the fact that most mathematical theories
permit many interpretations, a point of view he illustrated with Maxwell's
analogy about bell ringers. In his review of Thomson and Tait's famous
work, *A Treatise on Natural Philosophy* (1879), Maxwell had argued that it
might never be possible to give an explanatory model of electrodynamics
because of:[18]

> The scientific principle that in the study of any complex object, we
> must fix our attention on those elements of it which we are able to
> observe and to cause to vary, and ignore those which we can neither
> observe nor cause to vary.
>
> In an ordinary belfry, each bell has one rope which comes down
> through a hole in the floor to the bellringers' room. But suppose
> that each rope, instead of acting on one bell, contributes to the
> motion of many pieces of machinery, and that the motion of each
> piece is determined not by the motion of one rope alone, but by that

---

[18] In *Nature* (20); see Maxwell (1890, 776–785).

of several, and suppose, further, that all this machinery is silent and utterly unknown to the men at the ropes, who can only see as far as the holes in the floor above them.

Supposing all this, what is the scientific duty of the men below? They have full command of the ropes, but of nothing else. They can give each rope any position and any velocity, and they can estimate its momentum by stopping all the ropes at once, and feeling what sort of tug each rope gives. If they take the trouble to ascertain how much work they have to do in order to drag the ropes down to a given set of positions, and to express this in terms of these positions, they have found the potential energy of the system in terms of the known co-ordinates. If they then find the tug on any one rope arising from a velocity equal to unity communicated to itself or to any other rope, they can express the kinetic energy in terms of the co-ordinates and velocities.

These data are sufficient to determine the motion of every one of the ropes when it and all the others are acted on by any given forces. This is all that the men at the ropes can ever know. If the machinery has more degrees of freedom than there are ropes, the co-ordinates which express these degrees of freedom must be ignored. There is no help for it.

For Poincaré this point of view was reinforced by his experiences with contemporary theories of electricity and optics. He noted that physicists, even of the calibre of Lorentz, had no difficulty explaining away apparent contradictions between theory and experiment, but this was not enough for Poincaré: "Explanations are what we lack the least."

The International Congress of Mathematicians, where Poincaré was president, also came to Paris from 6 to 12 August 1900, the same week as the physicists. He used the occasion to publish a paper on logic and intuition, his (1902a). The paper partly revived an earlier essay of his, on the nature of mathematical reasoning (1894d), where he had discussed the question of whether mathematics is nothing but a gigantic tautology or if mathematical knowledge grows. He had argued that it is not analytic, as Leibniz had claimed, and therefore tautological, but rather it grows, which it can do because the quintessential mathematical argument is reasoning by recurrence. This time Poincaré began with some conventional remarks for the period about types of mathematical temperament, such as the logical or formal (exemplified by Hermite and Weierstrass) who disdain pictures and strive for precision, and the geometric who like pictures and are animated talkers even when doing analysis (Bertrand, Lie). But, he then noted, intuition fails. It thinks all curves have tangents, it accepts the Dirichlet principle, it talks about infinitesimals—all known to be wrong or incoherent by 1900, when much of analysis and geometry had been given rigorous foundations in arithmetic in a largely German program known as the arithmetization of analysis.

Poincaré's use of the word "intuition" was avowedly loose, and in paragraph 3 he illustrated it with four examples.

1. Two quantities equal to a third are equal to each other.

2. If a theorem is true for the number 1 and it is proved that is true for $n + 1$ when it is true for n then it is true for all positive integers.

3. If on a line the point C is between A and B and the point D is between A and C then D will be between A and B.

4. Through a point one can draw only one parallel to a given line.

All four, he said, can be attributed to intuition, but the first, he said, states one of the rules of formal logic; the second is a true synthetic a priori judgment, being the foundation of rigorous mathematical induction; the third is an appeal to the imagination; and the fourth is a disguised definition. Appeals to the senses, or to the imagination cannot yield certainty—but who doubts the third? Who doubts arithmetic? Poincaré may have regarded those questions as entirely rhetorical, but at about the same time his German rival David Hilbert was beginning to take them seriously and investigate whether foundations could be given to arithmetic itself.

Poincaré was content to regard the rigorization of analysis as both valuable and more or less complete, but this was far from being the end of the story. It is a familiar problem in the philosophy of knowledge that logic cannot increase what we know, because all it does is bring out information already present in the premises. How then can knowledge increase?—as plainly it does. These were themes that Poincaré was to address at length elsewhere, but briefly his view was to be that logic cannot define the numbers, only some Kantian recognition that counting is an ability of the mind can do that. As he put it in paragraph 4,

> In becoming rigorous, mathematical science takes a character so artificial as to strike everyone; it forgets its historical origins; we see how the questions can be answered, we no longer see how and why they are put. This shows us that logic is not enough; that the science of demonstration is not all science and that intuition must retain its role as complement, I was about to say, as counter-poise or as antidote of logic.

He also turned to the philosophers' objection that rigor was attained at the price of objectivity, and that in becoming absolutely precise mathematics ceased to be about anything, and could never be shown to be about the objects it purported to describe. He replied that the problem of how mathematics can be applied had been divided into two, and confidence in the rigor of the mathematics was now assured. This was a great advance, but it did not answer the philosophers directly. However, there never could be a mathematical demonstration that a real and concrete object corresponds to an abstract definition, so this part of the question was perhaps unreasonable.

In science, too, intuition had a major role to play. "We seek reality," said Poincaré in paragraph 5, "but what is reality?" It is not just atoms, say the chemists, but the arrangement of the atoms. So, too, in mathematics it is not the individual lines of a proof and the rule of deduction that joins each line to the next. Rather, as with chess—a game Poincaré admitted he played badly—it is far from enough to know the rules, one must also know the inward reason for a sequence of moves. "Logic," he concluded, "which alone can give certainty, is the instrument of demonstration; intuition is the instrument of invention." He now recalled his article on mathematical reasoning from 1894, and the role of mathematical induction in generating new knowledge, which showed that mathematics was not a matter of logic alone. But nor should he claim that mathematicians always proceeded rigorously: they had visions of the goals they hoped to reach, they were guided by analogy. And then there were people like Hermite (the honorary president of the congress) whose mode of intuition was that of the analyst, but who could not be called a logician. At the end Poincaré retreated—there are many kinds of mathematician. . . .Poincaré could be too polite, at times, but the ever-loyal Mittag-Leffler apparently found the speech inspirational.[19]

SCIENCE, HYPOTHESIS, VALUE

It could be argued that up to this point in his career Poincaré's popular essays could have been collected as a book with the title Russell was to use in 1914, "Our knowledge of the external world." Some of the essays would have described new scientific knowledge of the world, rather more would have reflected on the nature of that knowledge and how we can have it all. There would have been a clear insistence on the possibility of several different interpretations in the science of electricity, magnetism, and optics, and the absence of any logical grounds for separating them—the philosophy of conventionalism. This would have extended to geometry, or perhaps one should say the interpretation of geometry, as Poincaré again gave only conventional grounds for deciding between Euclidean and non-Euclidean geometry. But when he came to pure mathematics, and more precisely to arithmetic, Poincaré's position was different. The plurality fell away, and we are said to have an extralogical ability supported on almost Kantian, transcendental, grounds, that of carrying out mathematical induction or reasoning by indefinite repetition. It was insistence on the extralogical character of this ability that was to bring Poincaré his next share of controversy.

In fact, the first edition of the first volume of his popular essays was published by Flammarion in 1902, his *La science et l'hypothèse* (1902b). This book and its three successors were remarkably successful. The first printing of *S&H* was 1,650 copies and sold out within weeks, and it was rapidly reissued. A second, corrected edition was published in 1906; 20,900 copies of the book had sold by 1914. *La valeur de la science* came out in 1905, and it too sold 21,000 copies by 1914. *Science et méthode* came out in 1908 and sales reached 12,100 copies, and *Dernières pensées* (1913) sold 7,700 copies. The books were also translated; by 1910 *S&H*, for example, was available in English, German, Spanish, Hungarian, and Japanese.[20] The royalties that Poincaré accrued were considerable,

[19] Stubhaug (2010, 452).
[20] See Rollet's essay in Poincaré (2002, ix, no. 2), with figures from Marpeau (2000).

the money that Le Bon earned from the series made his fortune. Sales continued well after the First World War, and the books remain in print in French and English.[21]

S&H was the first book in the series *Bibliothèque de philosophie scientifique* that the dubious figure of Gustave Le Bon edited for Flammarion, where Ernest Flammarion was keen to revive the family business by moving into the field of popular science. Le Bon was a self-made intellectual who openly espoused racist views and wrote an influential work on the psychology of crowds that was to influence Fascism and National Socialism, but he also wrote popular medical works, a book on insanity, and books on archaeology and ancient civilizations, and he took every opportunity to expand his social network. When his idea of a series of scientific popularizations was turned down by Felix Alcan, he took it to Flammarion, who accepted it at once. The aim was to produce a series of largely philosophical books that would also appeal to a scientific and an educated general audience. Le Bon commissioned the titles, and intervened actively to persuade authors to reflect his own views in their works. In the case of Poincaré the decision was taken to reprint a selection of articles and prefaces to some of his more technical works, and Poincaré modified some of the texts. As noted, in the event the publication was a runaway success, and the same business model was employed for the later books.

S&H opens with an introduction Poincaré wrote for the occasion that gives the book a degree of coherence. He highlighted the role of hypothesis in science, which he suggested required an analysis of the nature of mathematical reasoning, and resulted in an appreciation of the conventional nature of much of our knowledge of the world. There followed two essays on mathematics: one on arithmetic and induction (based on 1894d) and one (based on 1893b) on the continuum (more precisely, the mathematical continuum and the physical continuum) to which Poincaré added material on continua of higher dimensions. Then come the three essays on geometry and the nature of space. The first two of these, "Non-Euclidean geometries" and "Space and geometry" are extensively reworked from the articles (1891b) and (1895d), and the third,

---

[21] Lippmann apparently remarked that "Poincaré's philosophy has become popular, which shows how difficult it is to understand"; quoted in Bellivier (1956, 21). Bellivier's book drew on the unpublished memoir of Henri's sister Aline Boutroux (A. Boutroux 2012).

"Experiment and geometry" draws on those articles, his exchanges with Russell, and his article (1898c) in the *Monist*.[22]

The second half of the book is on topics in science. Two essays discuss conventionalism in mechanics (and are modified versions of what had originally been presented in one piece for the ICP), and finally six focus on topics in physics, often drawn from the prefaces of his more technical books and lecture courses, such as the one on thermodynamics and the one on optics and electricity.[23] Most of the modifications here were directed to removing mathematical formulas. For the second edition Poincaré added the essay "La fin de la matière" (1906c) originally published in the *Athenaeum*; as a result it does not appear in the English translation of *S&H*. The essay deals with the implications of the electromagnetic theory of mass, and overlaps very much with the essay "Mechanics and radium" in *S&M*.

The introduction did more than give *S&H* its shape, because Poincaré decided to direct it at the views of Édouard Le Roy, which he considered to be too skeptical and anti-intellectual. So he noted the apparently unassailable position of science and mathematics, and then observed that once the role of hypotheses is taken into account the whole of science and mathematics could seem to be insubstantial and a matter of whim. He attributed this view explicitly to Le Roy (1901a, b), and to it he opposed the view that science visibly does achieve things, whereas a system based on whims would be powerless. Therefore science must somehow provide knowledge of reality, and the only knowledge it could provide, said Poincaré, was that of relationships between things. The book was therefore arranged to show how this can be done. Mathematical knowledge is fruitful not because it is deductive, as is commonly thought, but inductive. But when mathematics is imposed on nature a steadily increasing element of choice gradually creeps in: in the imposition of a structure on the physical continuum, in the choice of a geometry, in the choice of fundamental principles for mechanics. Nature requires none of these choices, he said, but our science does.

---

[22] Indicative of the changes is that in (1893b) some French editions of *S&M*, and the English translation, wrongly credit Kronecker with Dedekind's definition of irrational numbers; other French editions do not make this mistake.

[23] The essay on electrodynamics, from the start to most of section VI (Lorentz), comes from Poincaré's (1901a).

Édouard Le Roy had edited Poincaré's lecture notes on the theory of the Newtonian potential (1894–95) and written a doctoral thesis extending some of Poincaré's ideas in his (1890d) in 1898 before going on to teach mathematics in a Parisian lycée. In an essay of 1899 called "Science et philosophie" published in the *Revue de métaphysique et de morale* and in the paper he presented to the International Congress of Philosophers in Paris in 1900, he pushed his radical conventionalism to its limits. His vitalism, and his radical Catholic beliefs, led him to claim that knowledge is only true if it is confined to one's authentic, immediate relationship with one's surroundings. This may be compatible with faith, but, as he insisted, it regards all theoretical knowledge as a matter of invention, and the intellect as a force deforming all it touches through the medium of discourse. This is not far from Boutroux's neo-Kantianism, as he admitted, but Le Roy went further and argued that there are no facts in science, only entirely arbitrary inventions, which might be necessary on pragmatic grounds. Le Roy even gave several provocative examples of scientific "facts" that were in reality only inventions: the atom, the phenomenon of eclipses, and the rotation of the earth. And this was no merely academic matter: Le Roy argued that because this rotation was only an invention with no *truth* in it all the Catholic Church had done nothing wrong in condemning Galileo; rather, Protestant and anticlerical criticisms of the Church that accused it of bigotry and hostility to science were themselves invalid.[24] It was also an attack on Poincaré who in his address to the ICP had said that the "affirmation: 'the earth turns round,' has no meaning, since it cannot be verified by experiment, . . . or, in other words, these two propositions 'the earth turns round,' and 'it is more convenient to suppose that the earth turns round,' have one and the same meaning."[25]

Le Roy was explicitly the target of Poincaré's criticisms in his next volume of essays, *VS*, which he published in 1905, but his route toward it was far from direct. In 1902 Poincaré had published the essay that was to

---

[24] Later, in his book *Dogme et critique* (1907), Le Roy turned his friend Bergson's vitalism into a modernist philosophy of Catholicism, according to which dogma could be a source of moral values that did not have to contradict rational knowledge, and for this he was attacked by Pope Pius X in his encyclical of 1907, when the Pope turned on the Catholic Modernist movement. Le Roy eventually succeeded Bergson as professor of modern philosophy at the Collège de France in 1921; see Howard (1998).

[25] Reprinted in *S&H* (133, 90).

lend its name to the title of his second book: "Sur la valeur objective de la science" (1902h). In it, he accused Le Roy of anti-intellectualism, of being a nominalist in doctrine but a realist at heart who only escaped absolute nominalism by a desperate act of faith. Le Roy, he said, reduced science to rules of action; scientific knowledge was no knowledge at all. But whereas the rules of a game can be altered, to produce a new game, the rules of science cannot, because they serve to make predictions. Indeed, science can only provide rules of action because, in the main, it can correctly predict.

To refute Le Roy, Poincaré attacked his sharp distinction between brute facts and all the scientific facts that Le Roy labeled inventions. Where Le Roy had distinguished between the darkening of the sky during an eclipse and anything to do with predicting the time, appealing to Newton's laws to construct a prediction, and affirming the rotation of the earth as the cause, Poincaré (para. 3) saw a succession of gradations, each with their own claim to compelling assent, and so numerous as to dissolve Le Roy's hard and fast distinction. The role of convention was restricted to the choice of units of length and time in physics, of definitions and postulates in mathematics. Once those were established, much else followed inevitably, so much so that scientific facts were merely the translation of brute facts into the language of science. Newton's laws follow from some simple assumptions, or may indeed be taken as assumptions and their consequences tested—that, for Poincaré, was all a matter of facts. But he stopped short of saying that the rotation of the earth should be spoken of as a simple fact (p. 231), because it not comparable to the statement that at such and such a time the sky will darken and an eclipse occur.

Poincaré now turned to consider what remained of Le Roy's case. Do scientists create scientific facts? No, he replied (para. 3). Certainly not out of nothing, because they start with the brute facts. Thereafter, all they create is the language in which they express these facts; true to his conventionalism, Poincaré set great store by the ability to converse effectively.[26] There is a greater creative role, he allowed, when it comes to scientific laws (or, rather, those laws that have been raised to the status of principles). But when the nature of scientific laws is considered

---

[26] Naturally, Le Roy's ardent supporter Duhem was not convinced; see Duhem (1954, 149–152).

any substantial disagreement will be settled by appeal to convention: principles are neither true nor false but conventional and convenient (para. 4). The failure of nominalism, he argued, lay in failing to see that science always proceeds in this way. There is an observed relation between two brute facts A and B. To explain this, the scientist introduces an abstraction C (such as the law of gravity) and asserts the principle that there is a rigorous relation between A and C, and another between B and C which is a law capable of revision. For example, if astronomers ever observed a satellite of a planet whose behavior did not follow Newton's laws, they would surely suppose that some other force than gravity also acted on the satellite. In this example A and B are the planet and its satellite, the observed relation is the orbit of B around A, C is the force of gravity and the unalterable principle is that gravitation obeys Newton's laws. The anomalous relation between C and B is explained away by hypothesizing a second force acting on B. The experiential element that operates between B and C is what saves science from being nothing but principles and therefore from being nothing but nominalism.

For Poincaré science was objective because it rested on communication between people. In paragraph 4 he returned to his hypothetical confrontation between Euclideans and non-Euclideans, and argued that if they have analogous senses and accept the same logic—did he briefly imagine there might be another?—then it would be possible to translate their language into ours. And in every case where translation is possible, there is an invariant (which is what is being said in each language) and these invariants are laws which in turn are relations between crude facts, expressed differently in each language. In paragraph 6 he defended the idea of discourse that Le Roy so distrusted, and argued for the conclusion: "No discourse, no objectivity."

He explained that while each person's sensations of red may differ, these differences between people wash out in the process of agreeing that this poppy and this cherry have the same color. So what is objective in sensations can be isolated—Poincaré briefly compared to the idea of "money of exchange" capable of being passed from one mind to another. Science, he went on, speaks only of relations between sensations, and once the role of conventions is understood it is objective precisely because it is a system of relations. Moreover, something in these relations, survives the passage from one theory to another one. He gave the example of the transition then underway from Fresnel's wave theory of light to the electromagnetic theory, and claimed that the objectivity of the surviving relations was as good as that which underpins our belief in the existence of external objects, namely a profound coherence in the sensations by which we experience them (which is so unlike our dreams). Similarly, he said, the ether exists because of the natural kinship of all optical phenomena. But, he insisted, science was not about objects in themselves. Indeed, to say that science cannot be objective because it can speak only of relations and never of things "in themselves" or "as they really are" is absurd. Nothing can reveal the true nature of things, and, in words surely chosen to hint at Le Roy's theology, Poincaré added (VS 267, 347–348) that if some god did know the true state of things, "he could not find the words to express it. Not only would we not be able to guess the answer, but if one gave it to us we would not be able to understand it."

It must have seemed to Poincaré that his reply to Le Roy in 1902 had not done enough to refute him, because he added a section "La rotation de la terre" to the essay when he reprinted it in VS in which he quoted the awkward words and protested at their strange interpretations: the rehabilitation of Ptolemy's system of astronomy and the justification for the condemnation of Galileo. He had not claimed, he said, that "the earth rotates" was a statement of the same kind as the parallel postulate. Rather, he had said that "the earth rotates" and "it is convenient to say that the earth rotates" were a pair of statements comparable to the statements "the external world exists" and "it is convenient to say that the external world exists." Challenged to say why the earth rotates, one would have to admit that deciding between the claims that it does and it does not cannot be made on kinematic grounds, because there is no absolute space. But if the convention is adopted that a dynamical theory covering the apparent motion of the stars, Foucault's pendulum, and much else that would be disparate phenomena on a Ptolemaic theory, is adopted then it is true that the earth rotates. The same sort of argument applied to the motion of the earth around the sun. So Poincaré could conclude,

> The truth, for which Galileo suffered, remains the truth, although it has not quite the same sense as it popularly does and its true sense is much more subtle, profound, and rich. (para. 7)

Poincaré was well aware of the political context. When things had calmed down he observed (1909c, 171) that at the time "every reactionary journal in France had asked me to prove that the Earth goes round the Sun." Indeed, appalled by the controversy his remarks had caused, Poincaré published a reply, "La terre tourne-t-elle?" (1904b), in the form of a letter to Camille Flammarion. In it Poincaré explained that he had yoked the two contentious phrases together in the spirit of modern metaphysics, that doing so in no way implied that Galileo had been in error, and that among the many proofs that the earth did rotate it was enough to note that if it did not, the stars would traverse every twenty-four hours a journey it would take light several centuries to complete.

Another, almost complementary, paper in *VS* (1897d), was one Poincaré was to have given to the first ICM, which had been held from 9 to 11 August 1897 in Zurich to prevent arguments about whether France or Germany should have priority, which he also published in *Acta Mathematica* and the *Revue générale*.[27] It was entitled "On the relationships between pure analysis and mathematical physics," but it opened with a political remark about the dangers of following those who merely want to make money. This was a period of great affluence and a widening gap between the rich and poor in France, and the time of Zola's political critiques, and to those who seem to want only to make money, Poincaré replied that they should be asked what was the good of accumulating so much wealth, when only art and science gave us the possibility of enjoying life. And science could not be done solely for utilitarian reasons, because it required insight in order to proceed. In fact, said Poincaré, mathematics has these aims: to study nature, assist in the fundamental tasks of philosophy, and, as with the arts, to be done for its own sake. And physics, said Poincaré (para. 2), cannot be done without mathematics; mathematics is the only language a physicist can speak, and "the mathematical spirit disdains matter to cling only to pure form." Having thus got round to his stated theme, Poincaré looked beneath the comfortable surface.

From experimental work, the physicist distills laws. But, said Poincaré, there are problems with this. Experiments are particular—laws are general; experiments are approximate—laws are precise: How can this gap be bridged? He answered his own question: by means of analogy, a feeling for symmetry, and said that this capacity for analogy was the contribution of mathematics to physics. Then he turned to physics, and

[27] Poincaré was unable to attend the ICM because of the death of his mother on 15 July 1897. His paper was read by J. Franel; see the ICM Proceedings, p. 30.

argued that "the aim of mathematical physics" was "to reveal the hidden harmony of things," by which he meant a harmony of relations between facts. These relations are to survive changes of theory, although their (physical) interpretation may change, a reference to his well-advertised antirealist position in physics. In 1905, in his controversy with Le Roy, he returned to the theme of harmony and dismissed it as lying beyond any mind that conceives it (*VS* 9, 193). Rather "what we call objective reality is, in the last analysis, what is common to many thinking beings;... this common part,... can only be the harmony expressed by mathematical laws. It is this harmony then which is the sole objective reality." So for Poincaré the "universal harmony [which] is the source of all beauty" is not so much an aesthetic sensation as the condition for the intelligibility of the world.

He then turned in paragraph 3 to the influence of physics upon mathematics. Physics, he said, gave mathematics its best problems, because it was cast in differential equations and relied upon the recently elucidated mathematical continuum. As he put it,

> The only natural object of mathematical thought is the integer... It is the external world that has imposed the continuum upon us, which we would have invented without doubt, but we have been forced to invent.
>
> Without it there would be no infinitesimal analysis all of mathematical science would reduce to arithmetic or to the theory of groups.

Without it, that is, mathematicians would be restricted to number theory (in the manner of the German mathematician Kronecker, who had just died) and the French mathematician and group theorist Camille Jordan. Instead, thanks to physics, we have Hermite's number theory, Fourier and Fourier series, and (here he gave a recent example) Kovalevskaya's work on partial differential equations.[28] But Poincaré was well aware that there were other mathematical models of continua, such as the one proposed by du Bois-Reymond which contains infinitesimals and that he had noted approvingly in his (1893b). Poincaré then added the idea that physics suggests not only problems, but solutions to the problems, and he

---

[28] The steady if subtle French patriotism at work here was also reflected in Poincaré's attitude to sources in his professional work.

mentioned the Dirichlet problem, where he had recently grappled with the failure of mathematicians to make rigorous the methods that physics suggested, and Klein's recent work on function theory on Riemann surfaces, which Klein claimed followed directly from electrostatics.

The paper is not as flimsy as it might at first seem. Its presence in *VS* not only puts it into a context where his conventionalism is again distinguished from that of Le Roy, it puts it explicitly on the progressive side and against the narrow-minded businessman. But it also speaks to the divide in French mathematical and scientific life, at a time when the best French physicists were experimentalists and the emerging new generation of mathematicians were largely pure, by arguing that the distinction between mathematics and physics makes no sense intellectually. The two subjects owe so much to each other, in Poincaré's view, that they simply cannot profitably be pulled apart. Interestingly, it does so by once again describing mathematics as a language.

POINCARÉ AND PROJECTIVE GEOMETRY

In 1899 Poincaré began a long disagreement with Bertrand Russell with his (1899c), in the *Revue de métaphysique et de morale*. The occasion for the article was Louis Couturat's review of Russell's *An Essay on the Foundations of Geometry* (1897), which Poincaré found too uncritical. Russell replied, and Poincaré continued the argument next year in his (1900e), and it is this part of their debate that is the most interesting.[29]

Russell

Russell had an undergraduate education in mathematics at Cambridge that had not satisfied him at all, and he eventually switched to the Moral Sciences Tripos, graduating with first class honors. He entertained several possibilities for his future career, before settling on writing a dissertation in philosophy that could qualify him for a Fellowship at Trinity College, Cambridge. One of his teachers, James Ward, suggested the topic of the philosophy of non-Euclidean geometry, and Russell wrote it up, successfully, in 1894–95. Conversations with Ward and the mathematician

---

[29] For analyses of this debate, see Griffin (1991) and Nabonnand (2000).

Alfred North Whitehead, who were his thesis examiners, left him so disappointed that it is likely Russell destroyed the only copy of the thesis, but a year later he was persuaded to rewrite it for publication, and it became the book he published in 1897.

It bears on its face several signs of Russell's rapid philosophical development. It is quite Kantian in spirit, reflecting Ward's influence, but it is dedicated to McTaggart, the Cambridge Hegelian idealist "to whose discourse and friendship is owing the existence of this book," whose philosophy Russell by then briefly embraced before emerging as a logicist (with Whitehead) and then as an Anglo-Saxon empiricist. Unsurprisingly, Russell dismissed the book in his *My Philosophical Development* as "somewhat foolish," but largely because it had been swept away by Einstein's theory of general relativity. This is fair, and a fine example of Russell's commitment to honesty in philosophy, but it is not a criticism anyone could have made before 1916, and as noted the book drew high praise from Couturat, which drew him and Russell into a friendship that lasted until Couturat's death at the start of the First World War.

Couturat's review had praised Russell for presenting a philosophy of geometry shaped by the most recent discoveries that was also in the great critical spirit. "One can define the spirit and the scope of this important work," he said, "by saying that it was the transcendental aesthetic of Kant revised, completed, and corrected in the light of Metageometry [non-Euclidean geometry]." The burden of Russell's book is that there are known to be several plausible metrical geometries, so distinguishing between them requires an empirical investigation. But all these geometries are special cases of one purely qualitative, nonmetrical, projective geometry, whose primitive terms are the most elementary terms in geometry and in any description of the external world. So it makes sense to offer a Kantian, transcendental argument to show how these concepts can be known; briefly, that without accepting these synthetic a priori forms no knowledge of the external world is possible.

Russell gave three axioms for projective geometry that, he argued, not only capture the concepts a mathematician needs in order to do the subject, but are enough to establish the philosophical conclusion that one is dealing with pure externality. This, he explained in good Kantian fashion, means that one has a conception of a form of externality, which can be dealt with by pure mathematics and as such could be purely hypothetical, but, since some form of externality is necessary for any sense of experience it follows that something is actually asserted in talk about externality, and that means the form must deliver all the familiar features of real projective geometry of some dimension.

Poincaré was willing to grant that the form of externality was appropriate to a homogeneous space. As for the rest, he said that he understood not a word of Russell's reasoning even when he agreed with the conclusions. In particular, Russell had said that any two points determine a unique figure, called a straight line, any three in general determine a unique figure, the plane. Poincaré accepted the mathematics but not the philosophy. He denied that the relation between two distinct points captured the experience of a line joining those points, given that there were so many things to be said about a line segment joining two points that were not true of many relations, and likewise denied that a relation between three points was all there was to a plane, so "the axioms of projective geometry are not indispensable for all experience" (p. 255).

Poincaré then turned to Russell's axioms for metrical geometry. He pronounced himself very happy with Russell's discussion of distance and the ability to move figures around without distortion, but less happy with Russell's idea that distance was a relation between two points. It is one thing, he said, to be able to say of two points that they have such and such a distance between them. It is another to say of two pairs of points, AB and CD, that the distance between A and B is less than, equal to, or greater than the distance between C and D. To see this, Poincaré argued rather abstractly. It may help to pick up Russell's remark (1899, 698) from later in the exchange that we know before we begin that the distance from London to Paris is greater than a millimeter. Let us be more precise: we specify exact locations in London and Paris, and by a millimeter we mean two extremities of a piece of metal somewhere (let us say in neutral Berlin). Then a conventionalist might reply that if we map the whole of Northern Europe onto a flat page in a suitably cunning way and say that the distance between points in Europe is precisely the distance they are apart on the page we may well discover that the distance from London to Paris is less than the length of the Berlin millimeter. This feels wrong, because we tacitly assume that we measure distances by moving the Berlin millimeter wherever we wish and counting how often it fits into a given segment (say, London–Paris). But this ability to move it around without it altering is exactly the property Poincaré wishes us to see contains a conventional element. Now suppose that we drive the millimeter around

Europe. On our definition of distance we would have to agree that it (the car, and ourselves) were changing size alarmingly, being much smaller in Paris than we are, allegedly, in London. We would presumably reject this idea of distance on those grounds.

Poincaré also complained that Russell had not explained carefully enough that any two configurations that are equivalent in metrical geometry are also equivalent in projective geometry. Still, he conceded, this could be done; indeed it could be done neatly in mathematical language by saying that the group of metrical geometry is a subgroup of the group of projective geometry. Indeed, Russell, he went on, had sought to ground metrical geometry in a notion of equality that derived from the allowed motions of figures, but the transformations of projective geometry could be thought of as motions, and the only way Poincaré could see for preferring one group of motions to another was the behavior of rigid bodies. He had also tried to suggest that projective geometry was a purely qualitative geometry, but there was another topic with a better claim to that title: "analysis situs," which made no distinction between the circle and any other simple closed curve.

Poincaré brought this long (29-page) review to an end by saying that he did not know if Russell would want to reply, or even to read all of the review, but if he did said he would most like to hear from Russell of a way in which experience could demonstrate the parallel postulate, and to be given a definition of distance and the straight line independent of that postulate and that was unambiguous and did not contain a vicious circle.

Russell did indeed reply. He admitted the justice of many of Poincaré's criticisms, and said he would not now try to justify views which he no longer held, but on the subject of Euclid's postulate he would continue to defend his former views and on the axioms of projective geometry he would like to try again. Russell summarized Poincaré's conventionalist position this way: Either Euclidean geometry is purely conventional, or we can ask if it is either true or false. And if the latter, can one decide if it is true or false? So, he said, the testability of Euclidean geometry is a purely philosophical question, independent of any experience, and depends on what we take the concept of distance to entail. He then focused on what he took to be a confusion in Poincaré's thought between a property and the measurability of that property. He argued that just as America existed before Christopher Columbus, so do objects have lengths before we can measure them, which is why we can agree that some operations we carry out with objects are indeed measuring lengths at all, or doing it well. Poincaré, he said, by defining measurement before defining length, had matters backwards. And were he to say, "No, we measure objects, not geometric abstractions," he would have to admit that he had some primitive concepts in mind that allowed him to speak of rigid bodies that do not change their shapes as they move.

Russell then put the point another way. To say that the distance between points A and B is a meter is indeed to compare two points with two others, the extremities of a meter stick. But to speak of the distance AB is merely to speak of a relation between two points. If we could not define the relation between two points we could not speak of a meter in the first place. It follows, Russell suggested, that some measurements can decide between Euclidean and non-Euclidean geometry and so nominalism fails.

Russell, without apparently being aware of it, was straying into territory for which he had already been criticized by the American mathematician and booster of non-Euclidean geometry, George Bruce Halsted. In his (1897, 491) Halsted had written thus:

> In projective geometry any two points uniquely determine a line, the straight. But any two points and their straight are, in pure projective geometry, utterly indistinguishable from any other point-pair and their straight. It is of the essence of metric geometry that two points shall completely determine a spatial quantity, the sect. If our author had used for this fundamental spatial magnitude this name, introduced in 1881, his exposition would have gained wonderfully in clearness. . . . It is the misfortune of our author to use the already overworked and often misused word "distance" as a confounding and confusing designation for a sect itself and also the measures of that sect.

Poincaré was to agree. When he replied in his turn he first praised Russell for his "perfect scientific loyalty," which encouraged him to believe that even if this exchange did not end in agreement it was not being conducted in vain. But of course he was not convinced by Russell's views about distance. For a start, he said, a convinced non-Euclidean could agree with all of them, so they offered no way of deciding between Euclidean and non-Euclidean geometry. More importantly, it was true that any definition decomposed into two parts, one which was a theorem or axiom implying the existence of what was being defined, and one that was a mere matter of words. So Russell's argument really was, he said (correctly) that one does not have the right to define distance in any but the Euclidean way, because it is a primitive concept. But while a strict Kantian might proclaim that their intuition was prior to all experience, Russell's intuition has to be completed with some experience, and since that varies from person to person, said Poincaré, it can only be refined by convention. Agreeing on a concept of distance, he suggested, was like agreeing on a concept of yellow: there was a spread of interpretations and no one had the right to single out just one (as it might be, the Euclidean one). A sect, he observed, noting Halsted's above remarks, can be represented, but it is not a quantity because it cannot be measured. So for Russell to claim that a distance AB is nothing more than a relation between the points A and B was meaningless to Poincaré without a way of measuring that distance.

As Poincaré went on to observe, what was at issue here was conflicting views about what it is to define something. In mathematics, he said, he knew of no way of defining terms except in relation to other terms; if something was abstracted from everything else, nothing remained of it. Whereas, for Russell, it seemed that he wanted the fundamental objects of mathematics defined in terms of their simplest components. Russell sought to define distance, or perhaps to elucidate our familiar concept of distance by saying what it *is*, while Poincaré could only do so by saying how the concept of distance is *used*, that is, measured.

Russell had gone on to argue (p. 687) that two objects were bigger or smaller than each other before we could measure them, but that for Poincaré, such a claim made no sense until we could measure them, so "equality and inequality cannot exist with measurement," and therefore without measurement equality and inequality were words devoid of sense, which, he suggested, was a consequence that Poincaré had not seemed to notice. But no, said Poincaré on the contrary that was exactly his point of departure: the ability to compare two different distances was precisely what his talk of convention in measurement was designed to explain. And to show that it did not require a prior notion of distance, but only some experience of solid bodies, he indicated how some experience with bodies (bringing or failing to bring three points of one into coincidence with three points of the other) would be consistent with precisely one of the groups of notions associated with the Euclidean or the non-Euclidean group. But this, because the experience was of bodies, not of space, would not establish whether space was Euclidean or non-Euclidean.

Griffin, in his assessment of the exchanges between Russell and Poincaré, finds that

> as not infrequently happens in philosophy, Poincaré's position is left attractively ambiguous between a truism and a stronger claim which is difficult to defend. It is difficult to determine, however, whether Russell's objection hits this weak spot in Poincaré's position. (Griffin 1991, 177)

He takes the weak spot to be over whether it makes sense to speak of the actual size of spatial figures, in advance, that is, of any convention. As Russell was to put it elsewhere (Russell 1899, 396, in Griffin 1991, 179), if "distances are there to be measured, they must be there before measurement." As Griffin says of a later point in Russell's argument this reply is "commonsensical and rather good." It is what mathematicians do when they define a metric space: they assign a distance to each pair of points in a way that obeys certain rules, but there may or may not be a group of transformations of the space that moves a measuring stick around. Poincaré, in his writings about space, insisted that we have a group first and derive our sense of distance in that spirit, because after all we have a strong belief that we live and move in a world full of moving rigid bodies, but his philosophy here, as elsewhere, is perhaps too willing to collapse considerations of what there is in the world into considerations of how we can know it.

The related issue of whether there is something called space that has properties (for example, curvature) or whether there is nothing to say about space except talk about bodies was not directly addressed by Russell, but it lies at the basis of the disagreement between Poincaré and one of his later, more perceptive, and certainly better-educated mathematical critics, the Italian mathematician Federigo Enriques.

Enriques

Federigo Enriques was only 22 when, in January 1894, he began to teach projective and descriptive geometry at the University of Bologna. His approach, as presented in his *Lezioni di geometria proiettiva*, combined Italian formalist attitudes with a sensitivity to the historical origins and psychological dimension of how geometry can be understood:

> We have tried to show how projective geometry refers to intuitive concepts, psychologically well defined....On the other hand, however, we have warned that all deductions are based only on those propositions immediately inferred from intuition, which are stated as postulates. (Enriques 1898, 347–348)

The theme of intelligibility was one Enriques was to develop over the next few years. He shared it with Poincaré and the other great popularizers of mathematics of the period, and indeed with Felix Klein, who had always argued for the intuitive and geometrical in mathematics. But Enriques disputed Poincaré's geometrical conventionalism head-on in his *Problemi della scienza* of 1906.

In many ways it is a one-sided dialogue with Poincaré. He saw himself as a positivist, and when he disagreed with Poincaré he called Poincaré's approach transcendental. By this he meant that it relied on completed infinite processes, such as abstracting away all irrelevant features (and not just more and more of them). Enriques agreed with Poincaré that points, lines, and planes do not exist in reality; we do not learn what the words mean by encountering examples of them. Instead, they are abstractions expressing relations which are stated "by means of the propositions of geometry" (1914, 177). But Enriques did not conclude that the propositions of geometry do not apply to any real facts except by convention, because, he said, these symbols find an "approximate correspondence" in the physical world "in certain objects for which they stand." The concept of a straight line is derived from such sources as solid bodies rotating about an axis, and the path of light in a homogeneous medium. The first type of experience leads to metrical geometry, the second to a projective geometry, and the agreement between the various experiences (each picks out the same object as a straight line) allows us to create a single geometric representation. Further analysis shows that this agreement is a symmetry in the phenomena that implies the homogeneity of space.

In Enriques' view, the cooled ball metaphor was rigged to head off the reply that the changes in length of rulers in the plate are due to changes in temperature. In the Poincaré model, we cannot say many of the things that characterize what we say about heat (it is localized, different bodies respond differently), so, said Enriques, heat is not playing the role of a physical concept but a geometrical one. Light rays similarly depart from straightness in inhomogeneous mediums. Therefore, Enriques concluded, "In this other world, geometry would be really and not merely apparently different from ours" (1906/1914, 178). To think otherwise is to make the contrast between appearance and reality into something transcendental.

Where Poincaré had argued that all talk of space was meaningless, because there is nothing (no object) to which the word refers, Enriques replied that

> to Kant's thesis denying the existence of a real object corresponding to the word "space," we, together with Herbart, shall oppose the view that "spatial relations" are real. And to the nominalism recently maintained by Poincaré, which declares that these relations do not possess a real significance absolutely independent of bodies, we oppose a more precise estimate of the sense in which geometry is a part of physics. (Enriques 1906/1914, 174)

Poincaré, as Enriques reminded his readers with several quotations from *S&H*, would only talk about the mutual relation of bodies. Enriques, in contrast, believed that there is "an actual physical significance belonging to the *spatial relations or to the positions of bodies*, whose totality may well be denoted by the word 'space'" and so Poincaré had misunderstood what it was to talk about space. In Enriques' opinion, it was possible to conclude that space is homogeneous, and many geometrical problems reduced to claims about measurements. Among these was the true geometry of space; Enriques argued that claims about the angle sums of non-Euclidean triangles could be tested astronomically, and shown to depend, within stated limits of accuracy, upon rather infeasibly large regions of space (1906/1914, 192).

The *Problemi della scienza* is a somewhat anti-Kantian attempt to explain how we can talk about reality, particularly how talk about mathematical and scientific concepts makes sense from a realist perspective. Where Kant had assigned a priori status to geometrical knowledge, Enriques saw geometry emerging "as a part of physics, which has attained a high degree of perfection by virtue of the simplicity, generality, and relative independence of the relations included in it" (1914, 181). The core of Enriques' whole approach was that this process of development, insofar as it is meaningful, is necessarily and forever incomplete, and subject to a continual process of revision, so that what may be "known" at one time may be found to be false later on.[30] This formulation precisely separated Enriques' approach from that of Poincaré, who could not agree to separate what was true from what could be said to be true by a suitably informed mind.

Klein and Hilbert

Poincaré's essay (1902c) in the *Journal des savants* is an indication of how traditional his position was by comparison with some others entering the field. It is his response to the German mathematician David Hilbert's remarkable reformulation of elementary geometry, and an important test of Poincaré's epistemology. But it also marks the entry of the one contemporary he was ever to have of comparable intellectual force and range.

German interest in the foundations of geometry was curiously situated. The influence of Riemann's profound ideas was steadily being promoted by Felix Klein. Klein had emerged in the 1870s as the potential leader of the next generation of German mathematicians, and that, at the time, would mean a dominant position in the world of mathematics. He was made a professor at the small university of Erlangen at the remarkably early age of 23—he found it a sleepy backwater with impoverished facilities, but that should not dent his achievement in the hierarchical world of German academia. He made a success of his appointment and in 1875 moved to the Technical University in Munich and then to the more important University of Leipzig in 1880. He emphasized the role of geometry in every area of mathematics, and most of his research was geometrical, but with a novel twist: he brought out the importance of groups of transformations whenever he could. This was a topic he had gone to Paris to discuss with Camille Jordan, the leading French exponent of the study of finite groups; Klein and his friend Sophus Lie extended Jordan's message to geometry.

When Klein became professor at Erlangen in 1872 he gave an address on mathematics education that is now largely forgotten, and distributed a pamphlet on the nature of geometry that has become only too well remembered. His "A comparative review of recent researches in

---

[30] In *Plato's Ghost* I argued that "Enriques practised what has been called 'meaning finitism': the idea that the meaning of a term of concept is established upon only finitely many instances and is therefore necessarily vague (see his discussion of how one learns the meaning of legal terms, pp. 113, 119). Infinite classes are defined (1906/1914, 129) by considering the conceivable objects falling under a finite number of headings."

Figure 1.3. Felix Klein (1849–1925). Source: Arild Stubhaug, *The Mathematician Sophus Lie* (Springer, 2002). ©Herman Lie.

geometry" (1872), now known as the Erlangen program or as the Kleinian view of geometry, did two things. First, it described how all the principal geometries could be seen as special cases of projective geometry, although the account of non-Euclidean geometry was clearer in two publications from 1871 and 1873, and strictly speaking the case of affine geometry was omitted. This unified the subject, which had been looking as if it was breaking up into many different cases. Second, it showed how several topics of a seemingly more algebraic character could be translated into geometrical problems, thus increasing the means available to treat them and solve outstanding problems. However, the pamphlet was distributed informally, and there is little sign that it had much impact until Klein's reputation had risen so much by the late 1880s that he could find people to translate it into several languages and he was willing to claim how prescient he had been. Indeed, the work of Lie and Poincaré had by then amply demonstrated the importance of group theory in many areas of geometry, but it is unlikely that this owed anything to the Erlangen program.

That said, the view that projective geometry is the fundamental geometry was widely shared and most precisely and forcefully expressed by Klein. But what is projective geometry? This question is surprisingly difficult to answer. Klein's view was that it concerned a space (projective space) made up of points, lines, planes, and so forth, together with a group of transformations of this space to itself that mapped points to points, lines to lines, and preserved a quantity called the cross ratio of any four collinear points. This space was none too precisely defined, but it was understood that projective two-dimensional space for example (the projective plane) was the familiar plane with the addition of a line "at infinity," which is where parallel lines meet. Several responses to such a description are possible. It can be found hopelessly vague, or good enough to get on with. One is meant to think of the simple model of single focused perspective or photography with a pinhole camera, according to which a figure in one plane is projected onto another by a system of rays emanating from a point. If you take the second of these views then you accept a space that is very like the usual Euclidean space, in which you can draw straight lines, and in which any two straight lines meet in a point. You can also transform any geometric figure made up of lines and points into another one, and some features will be preserved: those captured by the cross ratio of any four collinear points.

What makes the geometry interesting is that in this setting all ellipses, parabolas, and hyperbolas are projectively equivalent. This is not too surprising. Projective geometry can be thought of as the study of those properties a figure shares with its shadows. The shadow a straight line in one plane casts on another plane is a straight line, if two lines meet in a point then so do their shadows, and so on. In particular, any ellipse, parabola, or hyperbola can be seen as the shadow of a circle, as a few minutes experimenting with a lamp shade will suggest. This unification of the study of conic sections was the great selling point of projective geometry. It was also attractive because it seemed to be about properties of figures that were as elementary as possible: it made no mention of lengths or angles, it was about the barest properties of straight lines and other geometric figures. It was, as one of its founders had called it, a nonmetrical geometry—a paradoxical name that deftly hinted at its foundational position. Klein's unification of geometry, with projective geometry as the mother geometry, was the clearest, and mathematically richest, demonstration of the importance of the subject.

If you do not find talk of lines at infinity sufficiently clear, and this was a minority opinion among anyone with a taste for geometry at all, what then? There were two competing alternatives. The mathematician von Staudt had published two books, *Geometrie der Lage* (1847) and the three-volume *Beiträge zur Geometrie der Lage* (1856–60), in which he essentially did for projective geometry what Euclid had done for metrical geometry. This gave a self-contained account of the subject with, necessarily, a few primitive terms that were nonetheless intuitively intelligible, but the book was hard going and little read. Klein had a friend at university, Otto Stolz, who talked him through it. Or you could pass to a fairly clear algebraic account, and do algebraic geometry in this new setting. It was this version that gave projective geometry its other appeal: the opportunity to study plane curves of any degree, and surfaces in space that are defined by polynomial equations. In short, it promised the greatest expansion of Descartes's ideas in over a century. The pioneers here were Ludwig Otto Hesse and Julius Plücker. Klein had gone to Bonn University at the age of sixteen and a half to study with Plücker, thinking that he was still working on physics, the topic of his middle years, but Plücker had switched back to geometry, and so Klein learned geometry instead. When Plücker died unexpectedly in 1871 with his major new research only half-published, Klein was the only one who had discussed these ideas with him, and he was therefore brought in to edit Plücker's unpublished work and plunged into high-level mathematics at an early age.

This was the field that Klein had staked out as his own, but his hopes of leading the world of mathematics were to be dashed. In 1880, with his own work doing well but still without a major breakthrough to his name, Klein came across the young Poincaré. It was soon apparent that both men were working in the same area, although from very different points of view. Among the points of overlap was non-Euclidean geometry, although Klein had missed it, perhaps because of his preference for projective over differential geometry. A correspondence ensued, which more than anything else showed Klein just how fast the new man could move. Klein prided himself on being well read; Poincaré, it seemed, was making it up as he went along. Klein dug deep, and the work he did over the next two years was his best, but, as he himself said, it cost him his health. In the summer of 1882 he had a nervous breakdown. It took him a year to recover, and he felt that he was never to recover his earlier force for research. He attracted a stream of very good students who could be entrusted to work out the details of his ideas, but by the 1890s, when he became professor at Göttingen, he had embarked upon another route. He set about becoming the leader of the most powerful, and most diverse empire in mathematics that the world had ever seen, and he succeeded. Among his inspired hirings was David Hilbert.

Hilbert had had a slow, quiet mathematical education just like Poincaré. He had impressed his teachers and earned a reputation as a man of promise without doing anything dramatic at an early age, but in his midtwenties he completely reformulated a major branch of algebra, the study of invariants of polynomial expressions. It is not necessary to know what this is to appreciate what he did. His brilliant idea was to exploit the view that, in mathematics, something exists if denying its existence leads to a contradiction. The important questions in invariant theory were existence questions, and hitherto the attempts to answer them were all explicit. As a result these attempts were full of enormous expressions, laborious to write and almost impossible to work with. The deepest questions asked for the existence of a finite list of invariants (for polynomials in a given number of variables) out of which all the others could be constructed, and only with great effort had this question been answered in the case of two variables. Hilbert's methods answered the question for any number of variables, and did so in very few pages and with very few calculations, and it made his name. There was a price: his methods showed there was a finite list in every case, but did nothing to populate it. Hilbert went on to tackle that problem too with some success, but it was his method of nonconstructive existence proofs that generated the excitement that surrounded him.

Hilbert, however, moved on. All his life he liked to work intensively on one field of mathematics for a few years and then switch to another. By 1902 he had become famous as a number theorist. The extensive report he published on this subject in 1897 is often taken to mark the end of 19th-century number theory with almost the force with which Gauss had begun it in 1801. But meanwhile, and to the surprise of everyone who knew him, Hilbert had come out as someone deeply interested in the foundations of projective geometry. In fact, a number of less well-known German geometers had been picking away at the subject throughout the 1880s and 1890s, bringing out more clearly what did and what did not depend on what. This clarified the subject, and hinted more and more at a vision of projective geometry that did not rely on any trace of its origins as the study of, as it might be, plane geometry with an extra line. When Hilbert had to teach the subject in his first position in Königsberg what caught his attention was exactly this foundational aspect: What were the fundamental terms, how were they related? In this he was not alone. Unknown to him a number of Italian mathematicians were already at work, and indeed publishing, but it was to be Hilbert's vision that succeeded, partly because by then he belonged to the all-powerful Göttingen enterprise, and partly because he saw that this approach would work in many other, and more important, areas of mathematics.

Hilbert shared with Poincaré the great mathematicians' ability to stand back from a problem and locate what is really needed. His insight here was that a mathematician does not need to know what a particular mathematical object is, but only what he or she can do with it. Take the raw ingredients of projective geometry: points, lines, and planes. Euclid, in his *Elements*, wrote of these objects as if we really know what they are: a point is "that which has no size," a line "lies evenly along itself." Hard to argue with, but only intelligible if you somehow know what points and lines are and realize that you agree with these remarks about them. Such statements are useless as definitions. But it is not necessary to say what something *is* in order to work correctly with it. One might distill such insights into some rules and thereafter work with the rules; this was the approach taken to elementary geometry by Hilbert's predecessor Moritz Pasch in his *Vorlesungen über neuere Geometrie* (1882), a work we know Hilbert read. Or one might take the view that knowledge of these essences was unattainable and all we could have was workable rules. In Hilbert's case there is no evidence that he got this quasi-Kantian idea from Poincaré; it is much more likely that he got it from the source: Kant himself. Hilbert came from Kant's home town of Königsberg and was fond all his life of making allusions to his great predecessor, and although there is no evidence that he made a deep study of the *Critique* a certain amount of exposure to philosophy was part of every German mathematician's education.

In Hilbert's reformulated elementary geometry there were some basic undefinable terms (points, lines, and planes) and various things you could do with them. He spelled out these possible activities as axioms. For example, given two distinct points one axiom said that there was a line joining them, and a second axiom that there was only one such line. Theorems had to be established using the undefined terms according to the axioms and in no other way, and Hilbert proved the theorem that two distinct lines in a plane have either one point in common or none (indicating, incidentally, that he was not reworking projective geometry). Then he presented axioms that express how the concept of "betweenness" can be used, and proved theorems using just the axioms he now had. Then came axioms for congruence, axioms about parallel lines, and finally axioms that he called axioms of continuity. These relate to the seemingly harmless fact that in ordinary geometry, given two line segments a and b, it is always possible to make enough copies of either so that laid end to end these copies exceed the other. For murky historical reasons this is known as Archimedes' axiom, and Hilbert realized that it was necessary to spell this out as an axiom in his formulation of geometry. This is most interesting because, as Hilbert showed at some length, it is possible to discuss when coordinates can be introduced into these geometries. The crucial axioms here are the axioms of congruence, which should admit something like moving a line segment around, the parallel axiom, and the Archimedean axiom, which implies that every segment can be duplicated so often it reaches arbitrarily far. So one should expect what Hilbert indeed found, that Archimedean and non-Archimedean geometries can be very different. In any case, by coordinatizing his geometries, Hilbert explicitly reduced any consideration of the consistency of his axiom systems to questions about their coordinates, that is, essentially to questions of arithmetic.

It is often observed in accounts of Hilbert's work on geometry that it is a paradigm of "meaning as use."[31] We know what points, lines, and planes are not because we have been told what the words mean but because we have been told how to use them, and it is possible for any two mathematicians to compare usages. Indeed, every argument in this type of geometry should spell out the rule of deduction or the axiom being quoted at every stage. Several things flow from this approach. It is applicable to other areas of mathematics, so whenever it is unclear what the objects are one can try to capture the rules for their use. It is possible to study interesting objects as if they are points in some projective space (a tactic already used in mathematics by 1900). And it is possible to study strange objects that yet fit the rules for being treated as lines,

---

[31] An approach to meaning philosophers associate with the later Wittgenstein several decades after 1900; see, for example, Coffa (1991).

the purpose here being to establish which theorems really do depend on which axioms. It was this aspect of Hilbert's work that most impressed Adolf Hurwitz, his former professor, who praised Hilbert for having invented a new branch of mathematics, the mathematics of axioms, and indeed he had.

A modest distortion of the historical record provides a vivid illustration of what Hilbert did. The French mathematician Girard Desargues had shown in the early 17th century that two triangles ABC and $A'B'C'$ are in perspective from a point O (which means that the points O, A, A' lie on a line, as do the points O, B, B' and the points O, C, C') and the lines AB and $A'B'$ meet at R, the lines BC and $B'C'$ meet at P, and the lines CA and $C'A'$ meet at Q, then the points P, Q, and R lie on a line. He even gave two proofs. In the simple one it is assumed that the triangles ABC and $A'B'C'$ lie in two different planes; in this case the points P, Q, and R lie on the line common to those planes. If the theorem is taken to be about two triangles in a common plane Desargues gave a more complicated proof. Now, the statement of Desargues' theorem is entirely about plane projective geometry, and it might seem that it should have a simple, purely projective proof—as Desargues showed it does if one is allowed to cast it as a theorem about three-dimensional projective space. But the more mathematicians looked at the strictly two-dimensional theorem in the late 19th century, the more they began to doubt that it was true.

In the event, Hilbert was the first to give an example of a plane geometry in which Desargues' theorem was false, but it was convoluted and a much simpler one was soon produced by the American astronomer Forest Ray Moulton in 1902, which Hilbert included in later editions of his book.[32] Moulton divided the familiar plane into two by a horizontal line which he called the Axis, as in figure 1.4. He then defined lines to be either a conventional line parallel to the Axis, or a conventional line crossing the Axis but sloping to the left, or a vertical line, or a new kind of line obtained from a line crossing the Axis and sloping to the right. If such a line makes an acute angle, α, with the Axis at a point P, it is replaced by a broken line composed of the half of the original line below

[32] Moulton learned Hilbert's work by attending a seminar organized by Eliakim Hastings Moore in Chicago on Hilbert's *Grundlagen*; Moore had picked it up when visiting Göttingen in 1899.

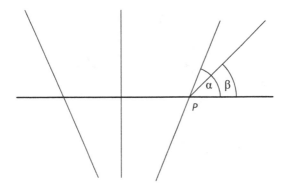

Figure 1.4. Straight lines on Moulton's definition. Source: redrawn by Michele Mayor Angel after David Hilbert, *Foundations of Geometry*, transl. by Leo Unger, 10th ed., rev. and enl. by Paul Bernays, Open Court, La Salle, Illinois (1971, p. 75).

the Axis and a line segment above the Axis making an angle β at P with the Axis, where $\tan \beta = \frac{1}{2} \tan \alpha$.

All that is important from Hilbert's axiomatic point of view is that the lines on this definition obey all the usual axioms of projective geometry, as they do; for example, two lines meet in at most one point—but in this geometry, Desargues' theorem is false, as Moulton's figure, figure 1.5, shows. Because, as we noted above, the incidence axioms for projective geometry in three dimensions are strong enough to imply Desargues' theorem it follows that Moulton's plane cannot be embedded in any three-dimensional space. This startling result is a vindication of Hilbert's axiomatic approach.

This was the radical reformulation of geometry that Hilbert presented, and to which Poincaré had to respond. He certainly did not mistake Hilbert's intent. After noting that many familiar objects in mathematics had evolved quite far from their classical conceptions in recent decades, he observed that Hilbert's proposals did the same for geometry, and the supposed barriers that had marked out spherical, Euclidean, and non-Euclidean geometry as the only conceivable geometries have been broken down at every point. Hilbert, he said (p. 4), had sought "to put the axioms into such a form that they might be applied by a person who would not understand their meaning because he had never seen either point or straight line or plane." It would then be possible to reduce reasoning to purely mechanical rules, and geometry could be created by slavishly following these rules without knowing what the axioms mean.

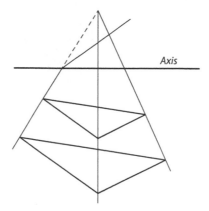

Figure 1.5. Desargues' theorem fails in the Moulton plane. Source: redrawn by Michele Mayor Angel after David Hilbert, *Foundations of Geometry*, transl. by Leo Unger, 10th ed., rev. and enl. by Paul Bernays, Open Court, La Salle, Illinois (1971, p. 75).

The logical connections would be grasped, Poincaré could not bring himself to say "without understanding", "but at any rate without seeing it at all."

Poincaré then surveyed Hilbert's five families of axioms. He noted that Hilbert might have gone further and attempted to ground the betweenness axioms in topological considerations, that the parallel axiom now came ahead of metrical axioms, and that at every stage Hilbert had been careful to shut off any need for an appeal to intuition. Poincaré found Hilbert's treatment of Archimedes' axiom the most original part of the book. "At first blush," he wrote (p. 11), "the mind revolts against conceptions like this. This is because, through an old habit, it is looking for a visual image. It must free itself from this prejudice if it would arrive at comprehension..." Then he went on to praise the treatment of coordinatization. In fact, his verdict was almost entirely positive. Hilbert, he said, was the first to present a geometry which did not, even implicitly, suppose that one was working with the real plane in mind. He noted that the idea of coordinate transformations forming a group was seemingly missing, but that it could be supplied via the study of congruence (an idea to be taken up by Emil Artin a generation later). This deprived Hilbert of the opportunity to compare the groups in his geometries: "Professor Hilbert seems rather to slur over these inter-relations; I do not know why" (p. 22). What Poincaré missed most of all was any consideration of the psychological origins of the axioms: "The axioms are postulated; we do not know where they come from.... His work is then incomplete; but this is not a criticism which I make against him. Incomplete one must indeed resign one's self to be" (p. 23).

These final remarks are not a criticism of Hilbert; rather, they are a self-criticism. Although Poincaré knew perfectly well that there are geometries other than the famous trio, some of which had been described by Lie, whose work Poincaré admired, none had caught his interest, and the whole subject of axiomatic geometry and the comparative study of axiom systems would have come new to him. Not only had Hilbert written in the German top-down approach, disliked by the French generally, undoubtedly the most disturbing fact for Poincaré in all these new geometries was their lack of meaning, their psychological inaccessibility. Poincaré's geometrical conventionalism was rooted in his sense that the behavior of solid bodies taught us which of a limited range of geometries was the simplest to apply. Now Hilbert had disposed of the consideration of rigid bodies, obscured the group idea that corresponds to the allowable motions of these bodies, and produced a family of novel geometries that defied intuition. They were impeccable mathematics, but what did they mean for our ability to know about the external world? And even if the answer was nothing, they brought to mind a series of thoughts about mathematics that Poincaré was not comfortable with. Could or should mathematics be done with such a flagrant disregard for the real world (whatever the real world might turn out to be)? Was mathematics in its heart nothing but a colossal rule-following exercise? Interestingly, Poincaré did not pick up the implication that knowing the meaning of terms might be a matter of correct use and nothing more.

Poincaré had placed mathematics on two pillars. One was the transcendental grasp of the process of mathematical induction, which enabled us to do arithmetic. The other was experience, which, through evolution, guided us to appropriate conventions not only for geometry but also physics. It was the physicists' educated sense of experience that many mathematical concepts had indeed come about in the deeply intertwined and mutually indebted developments in mathematics and physics. Hilbert's book brought back an old view, with a wholly contemporary and even exemplary sense of rigor, that mathematics might be nothing more than an elaborate exercise in following rules, and arbitrary rules at that. It might seem odd that this idea could have come as something of a shock to Poincaré but he had not discussed it before. He knew perfectly well, of course, that at some stage a mathematician has to sit down and write down their arguments according to the agreed rules, but he probably dismissed as utterly trivial and fundamentally wrong the view that there was nothing more to mathematics than that. He was to remark in *L'avenir* that there were always many things one could deduce from what was already known, so the issue was to find out what was really worth saying. That Hilbert's view of mathematics can strike us over a century later as being an important truth about mathematics is a measure of how successful it was to be. It is no wonder Poincaré found it incomplete.

In his (1903a) Poincaré returned to the theme of geometry and space and addressed a topic he had taken for granted some years before: its dimension. This paper can be seen as a continuation of his earlier investigations into how we acquire our concept of space (discussed in chap. 1, sec. "Non-Euclidean Geometry Comes to France" above). There he had noted that pure visual space has only two dimensions, but motor or tactile space has as many dimensions as we have separate muscles, so there is an issue about how we construct a space of three dimensions. It can also be seen as a response to Hilbert's *Grundlagen*, which confined itself to two- and three-dimensional questions but nonetheless left the impression that this was an arbitrary restriction. It was in any case clear that pure geometry could concern itself with spaces of any dimension. As will be seen, it also amplifies issues raised in the debate with Russell. Finally, one could note that the dimension of space had become a popular topic in science fiction and speculative discussion: Could time be a fourth dimension? Were there other dimensions accessible to spiritualists or inhabited by the departed?

Poincaré began with a summary of his ongoing explanation of how our notion of space is constructed from the behavior of rigid bodies, and could lead in principle to Euclidean or non-Euclidean geometry. Therefore space is not a form imposed on our sensibility, and the two geometries rest upon an amorphous continuum. But this raises the question (in para. 2) of whether this amorphous continuum is not itself a form imposed upon us, and we have "enlarged the prison in which our sensibility is confined, but it remains a prison." As he explained to his readers, this continuum, although it lacks any idea of distance, had already been studied by mathematicians such as Riemann and Betti under the name of analysis situs. Theorems in this subject remain true even when the figures to which they apply are subject to the grossest distortions, indeed arbitrary point transformations (Poincaré forgot to insist upon continuity here), so it is the true qualitative geometry. Does this mean that its theorems are the result of deduction? Are they experimental facts? Are they disguised conventions? or imposed upon our sensibility or our understanding?

As he observed, the last two possibilities are mutually exclusive: one cannot simultaneously believe that a fourth dimension cannot be imagined and that experiment tells us that space has three dimensions. If no alternative to three dimensions was possible one would not even contemplate the experiment. On the other hand, empiricists have one disadvantage in their usual position removed: if experiment could show that the dimension of space was certainly greater than 2 and less than 4 we would know that it was exactly, and not merely approximately, 3.

For Poincaré the physical continuum was distinguished by a certain vagueness (see again chap. 1, sec. "Non-Euclidean Geometry Comes to France" above). One can distinguish, to give his example in paragraph 3, 10 grams from 14, but not 10 from 12 nor 12 from 14. So in exploring a continuum one can build up chains $\ldots, E_{j-1}, E_j, E_{j+1}, \ldots$ in which each pair E_{j-1}, E_j and E_j, E_{j+1} is made up of indistinguishable items, but E_{j-1} and E_{j+1} are distinguishable. In this way one can ask if there is a chain of this kind starting at one point A and finishing at another B. If a continuum is such that every pair of points A and B can be joined by such a chain, Poincaré said it was in one piece. Now, he said, imagine that some chains of this kind are removed from the continuum and ask if it remains in one piece. It might be that by removing a finite number of points the continuum falls into two or more pieces; in this case the continuum was one-dimensional. Or, it might be that it fell into pieces only when some one-dimensional continua were removed; in this case the continuum was two-dimensional. If it was necessary to remove two-dimensional continua the continuum was three-dimensional, and so on. This gave Poincaré a working definition of the dimension of a continuum. As he observed, it differed very little from the usual definition of a surface as the boundary of a solid and a curve as the boundary of a surface, but it was applicable to our representation of physical space.

It might seem, he said, that the task of showing that space has three dimensions was done: start with a point and build up from there. But, he said (para. 4), it is necessary to represent to ourselves what a point is, and this is not as easy as people think. We habitually have recourse to the idea of arbitrarily small dots, and that can be made to work, but the harder task is to be able to distinguish one point from another. How can we tell if the point occupied by object A at time α is the same as that occupied by object B at time β when an object untouched and at rest in front of us for a second has traveled though 30 km because of the rotation of the earth? We cannot speak of absolute space, only of relative positions, and must distinguish changes in us from changes of position in the external object. Since he had described how to do this before, he now pressed on to discuss how, when two objects remain at rest with respect to us, we deduce—but cannot observe directly—that they are at rest with respect to each other. Then he looked at (pure) visual space, and concluded that it was only a part of the true, motor space; then he considered space as it is known to us through our sense of touch and the multidimensional space of muscular sensations that reflect how we come to touch something. Then once again he described how all these different "spaces" are put together and we create a description of the physical continuum, drawing on our long personal experience "and the still longer experience of the race" (1903a, 425, 263).

In 1907 he returned to the theme of how we can know geometry and wrote "La relativité de l'espace" (1907a) for, interestingly enough, the journal *Année psychologique*. This journal had been founded by Henry Beaunis and Alfred Binet in 1894 as the first international journal devoted exclusively to scientific psychology, including cognitive psychology[33], which was the theme of Poincaré's contribution. The article was in some ways a repetition of familiar themes, but to a new purpose. Space, he argued, was a meaningless concept: everything we say about it is a statement about objects in it; statements about distance are relative because we could not tell if everything in the universe increased its size a thousandfold one night, and we can even accept the Lorentz–Fitzgerald contraction, which converts spheres to ellipsoids. The value in this last idea was that it established that measurements of lengths made with rods and made with light would always agree. But it is not a simple change of scale, and indeed, Poincaré went on, there is a sense in which any two spaces which are continuous images of each other are geometrically indistinguishable: inhabitants of the one universe will

---

[33] From the website http://www.necplus.eu/action/displayJournal?jid=APY.

regard the inhabitants of the other as laboring, perhaps, under terrible misapprehensions, but neither group would ever be right.

Space, then "is in reality amorphous, and it is only the things in it that give it form" (para. 1)—oddly, one might note, like the earth of *Genesis*, book I, which was without form, and void, and where darkness was upon the face of the (seemingly preexisting) deep, until light, and then the firmament were created in it. But since the intuition of the unintelligible concepts of distance, direction, and straightness was so strong, Poincaré had to account for its source, and (para. 2) he located it in a complicated system of nervous impulses in the depths of the unconscious that allows us to construct our notion of space and position in space. Our ability to move allows us to separate out the rigid from the flexible bodies, to construct a sense of our immediate surroundings, and generalize from there. But why should it produce a three-dimensional space, he then asked? He answered his own question by a further discussion of how we associate our own motions, say of our arm, with points in "space," so that it is our mental construction of space that endows it with three dimensions, so it is derived from our experience with solid bodies. Everything we say could be rewritten in a four-dimensional geometry—here (para. 5) he hinted that the physicist Heinrich Hertz had attempted such a thing—but it would be artificial and feel like a translation into another language for no good reason because our experience of solid bodies leads most naturally to the construction of three-dimensional space.

He returned to the three-dimensional nature of space in his (1912h). Now he used the language of topology which he had developed in his mathematical papers to observe that homeomorphic spaces are indistinguishable, and so it is at this most fundamental level that intuition operates. Now, at this level it is possible to assign a dimension to a space by thinking of cuts in the space. Naively, a space comes in one piece if we cannot, with our senses, distinguish points that are too close together (say the weights of 10 and 11 grams, even though 10 and 12 grams are distinguishable). Such a space Poincaré called a continuum (in para. 2), and familiar examples include the straight line, which falls into two pieces when a point is removed, and simple closed curves, which fall into two pieces when two points are removed. So, said Poincaré, we think of a point as having dimension zero and these curves as having dimension one. Similarly, a surface may be said to be a two-dimensional continuum because it falls into pieces when one or more curve is removed, and a continuum has three dimensions when some surfaces are removed.

Turning to our construction of space, Poincaré went over old ground (para. 3) with our visual, tactile, and muscular spaces of various dimensions, and then examined how we assign different sensory experiences to the same event, for example touching something with different fingers. He argued that we can systematically translate one set of sensations into the other by imagining what happens when, by an actual or imagined motion, we replace one finger with the other. After making this compensation we find that the continuum we construct is, in fact, a three-dimensional one. To the objection that this may be true of our sensory impressions, but could it be that physics itself took place in a space of dimension other than three, Poincaré answered (para. 4) that one could try to carry out the corresponding mathematical analysis in such spaces. The mutual relations of phenomena take a particularly simple form when space is presumed to be three-dimensional—or rather, that there is a six-dimensional space of positions for coordinate axes, which for Poincaré was the same thing better explained. So our senses, and our mathematical physics, are adapted to the behavior of solid external objects which are most simply explained by constructing a space of three dimensions.

Poincaré concluded the essay in paragraph 6 with a reply to Hilbert's idea of betweenness. That, he said, was intended to be solely logical, and to be used by someone with no geometrical sense of what it involved. The reason this seemed unnatural to us was that our intuition of betweenness is rooted in the idea of cuts that contributes to our giving a particular dimension to space. But that intuition, as his own difficult work on topology over the years had taught him, was not restricted to three-dimensional continua. So we did not ascribe three dimensions to space because our minds cannot function otherwise, but because the external world induces us to do so.

POINCARÉ'S POPULAR WRITINGS ON PHYSICS

Poincaré's popular essays in recent discoveries in physics began with two essays in 1896 in the *Revue générale des sciences pures et appliquées* (1896c; 1896d). The first of these is on X-rays, and is illustrated with several of the eerie photographs that have become well known: a wire in a wooded box, a damaged femur, two hands. Poincaré, who plainly shared the contemporary excitement in X-rays, confined his commentary to a few simple statements: the rays travel in straight lines (which is why they can produce such interesting photographs); they are not light rays propagating in the ether (because they cannot be refracted or reflected— this was incorrect); and they are not cathode rays (they are not affected by a magnetic field). The second, an obituary of Félix Tisserand, the distinguished French astronomer who had died suddenly in October 1896, was a heartfelt account of a modest man and a good friend, who had died at the age of only 51. He had been elected to the Académie des sciences at the early age of 33, and become director of the Paris Observatory in 1892.[34] He had also just published the final volume of his *Traité de mécanique céleste* which, in Poincaré's view, stood favorable comparison with the great work of Laplace. Its first volume covered work on perturbation theory, the second the shape and orbit of the planets, the third the motion of the moon, which Poincaré found most interesting, the fourth the motion of satellites (here Tisserand had made some comments on Poincaré's method of periodic orbits).

Poincaré's more reflective essays began with his (1893c), published in the first volume of the *Revue de métaphysique et de morale*. The background story is typical of how Poincaré operated. He had lectured at the Sorbonne in 1892 on thermodynamics (see chap. 10, sec. "Thermodynamics" below). The lectures were fairly elementary, but when they were published they were roundly criticized by the Scottish physicist Peter Guthrie Tait for their supposed neglect of the probabilistic formulation of the subject. The two men had traded a few letters in the pages of *Nature*, and now Poincaré was stating his opinion in a newly founded journal of ideas.

In his four-page essay Poincaré first recalled the mechanical world view, which sees the world as made up of atoms in motion, perhaps also subject to an inverse square law of gravity. He lacked, he said, the authority to raise metaphysical questions about this view, but he could nonetheless discuss an aspect of the view that was perhaps less well known

[34] Tisserand had promoted the use of photography in astronomy, and in his theoretical work found an important criterion for deciding if a comet visible after a long interval is a known comet whose orbit has been perturbed by Jupiter. The criterion is still in use today, and is used to design orbits that use gravity assists to boost the speed of artificial satellites by flying them close to a planet.

but also interesting: the difficulties mechanists have in reconciling their view with facts that might seem to refute it. Chief among these was the problem that according to the mechanical world view all physical processes are time reversible: they make just as much sense run one way as the other. The natural world, however, is full of counterexamples. A hot body loses heat to a colder one, never the other way round.

As he had done in his paper on the subject (1889c), Poincaré first reviewed the existing literature, but only to reject it—the theory of dynamics seemed inadequate to deal with the fact that there is no way heat passes from a colder body to a hotter one. He therefore preferred the English suggestion, which was probabilistic, and merely claimed that the processes we never see do not contradict the mechanical view but are highly improbable. Among these explanations, Maxwell's demon, as it was known, had pride of place. Such a demon, with eyes sharp enough to see individual molecules and hands small enough to grasp them, could arrange to make heat pass from a colder body to a hotter one.

The kinetic theory of gases had been formulated precisely to deal with these difficulties, said Poincaré, but he did not find it completely successful. It was easy to show, he said (referring, without being explicit, to one of his own deeper insights into Newtonian physics, published in his memoir on the stability of the solar system), that mechanism requires that every physical system returns infinitely often arbitrarily close to any state it has occupied. The kinetic theory of gases, however, requires that any system runs down and eventually reaches a state in which every body has the same temperature. To avoid this apparent contradiction, the English reply that the system would indeed return to any earlier state, but with such a small probability that one might have to wait for an enormously long time, millions and millions of centuries. One would not need a demon, on this theory, to make heat pass from a colder body to a hotter one—just a very great deal of patience.

Poincaré ended his article by saying there were yet other problems, which he did not specify, and that Maxwell had struggled most ingeniously against them, but in any case there could be no hope of exact solutions to a problem of such complexity, and so one was driven to use simplifying assumptions. Were these assumptions legitimate, and consistent? He thought not, but in any case the main problem remained for mechanists: how to have a theory with reversibility in its premises and irreversibility in its conclusions.

This drew responses from Georges Lechalas and Couturat. Couturat found that Poincaré had exaggerated the problem, and Lechalas sought to cool the debate down by going back to an earlier generation of ideas, but Poincaré, who replied in the next issue with his (1894c), thought that he had been misunderstood as objecting to all forms of mechanism, and contented himself with observing that Lord Kelvin had recently proposed another solution that was worth considering.

In 1898 Poincaré published a short article on Hertz.[35] Hertz had died in 1894 and his posthumously published book (Hertz 1894) had generated much discussion. In it he had tried to eliminate the concept of force from mechanics, so Poincaré began his article with a lengthy review of why it was in fact impossible to give a satisfactory idea of mass and force, and why the "laws" usually said to govern them were really definitions. He was to repeat this in later accounts; we have already seen that he was to air them in Paris in 1900. He then noted that Hertz had rejected the Newtonian system of mechanics on the grounds that it permitted one to hypothesize motions consistent with Newton's laws but which were never seen in nature, and had then turned to a more restrictive system based on two laws, the conservation of energy and the principle of least action. Poincaré admitted that these made a better starting point but were nonetheless unsatisfactory, and observed that Hertz had rejected them too.[36] So, to formulate a system of mechanics in which the possible motions of a mechanical system and the observable ones coincide, Hertz had supposed each mechanical system was accompanied by a system of hidden masses that accounted for the constraints nature observes. The mathematical elegance of this approach was considerable, but however much Hertz remonstrated that hidden masses were no more mysterious than forces and energy, alike invisible, few were persuaded to adopt them, and certainly Poincaré was not.

Poincaré's essays had always drawn strength from his deep involvement in scientific research, and in 1904 there was one issue that dominated his chosen field to the exclusion of all others: the emergence of microphysics and the electromagnetic world view. The suggestion was that matter was made up of tiny particles that somehow were responsible for the electrical

[35] Poincaré (1897a); see Lützen (2005, 288).
[36] On complicated grounds; see Lützen (2005) for a thorough account of Hertz's mechanics and its reception.

and magnetic fields, and even that the most fundamental constituents of nature might be electrical. The Dutch physicist Hendrik Antoon Lorentz was the leading exponent of this view. He sought to derive Maxwell's laws on the hypothesis that matter was made up of what he called ions, and sometimes regarded as spheres of some small radius and capable of exerting a force through their interactions with the ether, which he regarded as fixed. He did not necessarily equate ions with the particles on display in cathode rays, which were coming to be regarded as electrons, but he was aware that there was disturbing evidence that the mass of these mysterious particles could vary with their velocity, and even that Newton's laws of motion might not apply to them. The ether, which was supposed to sustain all electromagnetic wave phenomena, was of course no better understood then it ever had been, in view of the failure of the Michelson–Morley experiment to detect the motion of the earth through the ether.

1904 was also the year of the International Congress of Arts and Sciences in St. Louis. This was a major event, designed to bring together the arts and sciences at the start of the new century and to celebrate the centennial of the Louisiana purchase of 1803, when the French had sold an amount of land to the new republic under President Thomas Jefferson that doubled the size of the country. It signaled the end of French ambitions in America, and was done in part to limit British influence. In the event the congress opened late, on 30 April 1904, and so it also commemorated the Lewis and Clark expedition to the Pacific Ocean, which had set off from north of St. Louis in May 1804. Twenty million visitors came to the exhibition for the usual range of displays, and some one hundred foreign delegates attended the intellectual side of the congress. Poincaré chose to travel to the United States and attend the congress, rather than go to Heidelberg for the ICM, and there is no doubt that he was given star treatment; the whole of his *La science et l'hypothèse* was added to the congress proceedings.[37] Apart from Poincaré, also Darboux, Picard, Boltzmann, and Ostwald came, and they met a spectrum of young American mathematicians and physicists.[38]

Poincaré's contribution (1904e) is rightly remembered, for he took the occasion to come as close as he ever would to producing a theory of

[37] See *Aesthetics and Mathematics* (vol. II, 629–748)—the volume starts on p. 391.
[38] For accounts of the congress see *Plato's Ghost* (pp. 172–176) and, in greater detail, Zitarelli (2011).

electrodynamics that would have rivaled Einstein's a year later.[39] Here, Poincaré proposed a principle of relativity in the form "the laws of physical phenomena must be the same for a fixed observer as for an observer who has a uniform motion of translation relative to him" (*VS* 176, 294). From this he deduced that "from all these results would arise an entirely new mechanics which would above all be characterized by the rule that no velocity could exceed the velocity of light" (*VS* 197, 314). On the other hand, Poincaré was also willing to entertain the idea that gravity propagated a million times faster than light, and to contemplate, with more reluctance, the end of Newton's law on the equality of action and reaction. His address is an eloquent testimony to the perplexity felt by the best physicists as they contemplated the evidence of fast-moving electrons.

The paper is an important one for understanding Poincaré's views on physics at a time when it was about to undergo one of its most significant changes in over a century. His often-stated opinion was that while the fundamental principles of physics might be better regarded as conventions than truths, they were firmly suggested by experience and could not be refuted by it. Now some of these beliefs were being questioned, and Poincaré described the situation as if an army was under attack. There were, he said, six beliefs that are fundamental to elementary physics. But, for example, the theory of thermodynamics, upon which he had recently lectured, had no rigorous derivation of its fundamental insights. There seemed to be no way to detect the earth's motion through the ether, although Lorentz had some ideas, but they contradicted Newton's third law (action and reaction are equal and opposite) because in Lorentz's theory the ether affected the electron but not the other way round. There were now apparently two types of material objects in the world: solid matter obeying gravity, and electricity obeying Maxwell's laws. If the nature of electricity was becoming understood, thanks to Lorentz's introduction of the electron, the relationship of the two types of stuff was still unclear, and a widespread opinion was that all matter might be electromagnetic in nature, in which case it might indeed not obey Newton's laws. That was Lorentz's view, but Poincaré was not attracted to it. Worse, the study of cathode rays, now understood as the motion

---

[39] Poincaré's St. Louis address was also published in the congress proceedings (1905h) and again in the *Monist* in 1905.

of high speed electrons, suggested that their mass was electromagnetic in nature and depended on their velocity—in which case all of Newton's physics collapsed. Finally, the phenomenon of spontaneous radioactivity, for which the French physicist Becquerel had won the Nobel Prize in 1903 and which the Curies were to get for their study of radium, posed a seemingly impossible challenge to the principle of conservation of energy.

How then should one proceed? Poincaré was cautious. One should loyally defend present principles and not give up everything. Modification in the light of new experiments might yet restore harmony. According to his philosophy of science one should accept well-established experimental findings and rigorous mathematical theory. In between, where objects are introduced with the requisite mathematical properties (electrons, straight lines, ...), these interpretations are not constrained by logic but are arbitrary, and the simplest is chosen by convention. However, he admitted, if even the best established experiments are to be overthrown, it is not clear what was left of his philosophy of science:

> Have you not written, you might say if you wished to seek a quarrel with me—have you not written that the principles, though of experimental origin, are now unassailable by experiment because they have become conventions? And now you have just told us that the most recent conquests of experiment put these principles in danger.
>
> Well, formerly I was right and today I am not wrong. (*VS* 207, 312)

But it might be that a new approach, comparable to that which took physics from a theory of central forces to a physics of principles, would be created, in which recognizable traits of the old view would still be visible. For example, thermodynamics could become based on the laws of chance, and the physical law would no longer be a differential equation but a statistical law. This view, which the 20th, and still more it seems the 21st century only confirm, was prescient. So too was his conclusion:

> Perhaps, too, we shall have to construct an entirely new mechanics that we only succeed in catching a glimpse of, where, inertia increasing with the velocity, the velocity of light would become an impassable limit. The ordinary mechanics, more simple, would

remain a first approximation, since it would be true for velocities not too great, so that the old dynamics would still be found under the new. We should not have to regret having believed in the principles, and even, since velocities too great for the old formulas would always be only exceptional, the surest way in practice would be still to act as if we continued to believe in them. They are so useful, it would be necessary to keep a place for them. To determine to exclude them altogether would be to deprive oneself of a precious weapon. I hasten to say in conclusion that we are not yet there, and as yet nothing proves that the principles will not come forth from out of the fray, victorious and intact. (*VS* 211, 314)

In view of the importance of what we see as his contribution to the solution of these problems and the creation of relativistic mechanics, namely the introduction of the Lorentz group in 1905, it is useful to bring forward his later reviews of these issues. In 1908 he presented in simplified form one of his own, best, recent discoveries: "La dynamique de l'électron," (1908a), in the *Revue générale des sciences pures et appliquées*. Poincaré began this paper by observing that there had been much talk of late of the possible need to abandon the principles of mechanics that had served us since Newton, all because of the properties of the newly discovered element radium. He counseled caution, but reviewed the basic principles of dynamics before turning to the strange behavior of radium. It emitted, he said, three kinds of radiation, called α, β, and γ radiation, and concentrated on the β rays, which were, he said, analogous to cathode rays (they are indeed both electrons). These electrons travel at enormous speeds, an appreciable fraction of the speed of light, and their mass displays unexpected behavior (para. 2): it varies with the velocity of the ray. Moreover, this dependence depends on whether the magnetic field that exerts a force on the ray acts along the line of flight or at right angles to it. Indeed, one experimenter, Kaufmann, had been led to speculate that the velocity of the electron was entirely a function of its velocity, and that it had no mass when at rest. Plainly, such behavior contradicted the laws of Newtonian mechanics.

Next (para. 2), Poincaré considered the problems posed by the motion of light: stellar aberration, and issues to do with the ether. He proposed to show that these problems disappear when Lorentz's theory of local time is used. In this theory, he explained, the ether is at rest. Two stationary observers, A and B, wish to measure the passage of time, and each has a clock. To synchronize their observations, B sets his clock to time zero and immediately sends a signal to A who, on receiving the signal starts his clock and sends a signal back to B. On the assumption that light travels at the same speed in each direction, they can agree that the time of an event is correctly recorded as the average of the time on their separate clocks. Now suppose, however, that A and B are moving (Poincaré implied, but did not say, that they move along the line AB from B to A), and assume, in accordance with experiments, that light always travels through the ether at the same speed. Then the time taken for the light to go from A to B is less than the time taken for the light to travel from B to A, but the two observers have no means of detecting this, and their clocks cannot be synchronized. Instead, they will each display their local times: A's time, and B's time.

Now, after Michelson's experiments, it was possible to measure motion with respect to the ether with such precision that a further phenomenon had been introduced: the Lorentz–Fitzgerald contraction. Poincaré did not give a formula for the amount of contraction as a function of velocity, but an argument about how the spherical wave front emitted by a moving point source of light should be regarded by an observer so that the compensation was always exact. As he admitted,

> This hypothesis of Lorentz and Fitzgerald will appear most extraordinary at first sight. All that can be said in its favour for the moment is that it is merely the immediate interpretation of Michelson's experimental result, if we define distances by the time taken by light to traverse them.
>
> However that be, it is impossible to escape the impression that the principle of relativity is a general law of Nature, and that we shall never succeed, by any imaginable method, in demonstrating any but relative velocities; and by this I mean not merely the velocities of bodies in relation to the ether, but the velocities of bodies in relation to each other. (para. 2)

At this point in the story, however, the classical structure began to fall apart. Radiation emitted from an object exerts a pressure on what it hits, and this is the explanation for the direction of the tails of comets. But it cannot exert an effect on the ether, and so Newton's principle of the equality of action and reaction must be given up. The

Lorentz–Fitzgerald contraction applies to moving electrons, and there was quite some controversy as to how this should be handled, but Poincaré simply reported (para. 4) that the only known, consistent theory that incorporated the principle of relativity was Lorentz's, according to which electrons have no mass other than the part that varies with their velocity, and all forces are electromagnetic in origin. As a consequence, no body can travel faster than light.

Poincaré ended his survey of recent developments with some considerations about gravity. Moving electrons create an electromagnetic field, and if mass is electromagnetic in origin one can ask if there would not be a force on moving bodies from this source, and the answer, he said, is that there will be one. But it will be negligible on all but the fastest moving massive body we know, Mercury, and even there, although it will cause its perihelion to precess it cannot account for the known disparity between the predicted and measured velocities of this precession.

The curiously distant relationship between Poincaré and Einstein will be considered later (see chap. 6, sec. "Poincaré and Einstein") but some things can be said at once. This account, although written in 1908, not only does not mention Einstein by name, it is strikingly free of his way of thinking. In particular, it retains the concept of the ether, and its discussion of time is oddly like an explanation of how all moving observers get the time wrong. It also lapses into sounding as if the Lorentz–Fitzgerald contraction is real. The argument is, indeed, a defense of the principle of relativity, but not in the sense of Einsteinian special relativity.

Poincaré's (1909c) is a refreshed version of the same opinion. One improvement is an exploration of what two observers separating in opposite directions at very nearly the speed of light would make of each other. They cannot allow that the other one is receding from them at a speed faster than that of light, for they know that on the new mechanics this is impossible. They would have to conclude, after sending signals backwards and forwards to indicate their positions, that the only time they could measure would be local time. As a result, two observers like A and B in the earlier paper who are traveling in the same direction at the same speed, cannot tell whether they are moving relative to the ether or not for, as Poincaré indicated, the apparent time the light takes to go between these observers would be the difference in their local times, and this is independent of their motion. But, they would also have to concede that the Lorentz–Fitzgerald contraction takes place along the line of flight.

Poincaré's views on space and time were summarized for the last time in a lecture he gave in London on 4 May 1912 (1912f). He set himself the task of assessing the impact of "the principle of relativity, as conceived by Lorentz" on previous ideas. Not the ideas of Einstein, which had finally begun to attract the support of other physicists, nor his own (which could have been modestly packaged with those of Lorentz), perhaps because this reflected a commonly held opinion, perhaps because Lorentz's formulation struck Poincaré as the deepest insight into the physics. The problem for Poincaré was, of course, that he had previously put our knowledge of the geometry of space, whatever its origins in experience, beyond revision, yet here it was seemingly being revised.

In dealing with this challenge Poincaré saw no reason to retreat from his view that our knowledge of space is constructed from our representation of the sensations that accompany certain movements in space. We are dependent on measuring instruments, starting with our own bodies, in which an element of convention enters when we talk of perfect instruments. The same considerations apply to time. Physical relativity (as Poincaré called it) concerns the different observations two observers in a state of relative uniform motion make of space and time, and Poincaré now distinguished between the actual observations and the laws of motion derived from them by differentiation. Observational values are changed by a change of coordinate axes; the differential equations are not. The principle of relativity applies, said Poincaré, to the equations; their invariance under the appropriate coordinate changes is assured because they are second-order differential equations (and rotating axes can be handled by passing to third-order equations).

Now, said Poincaré (*DP* 105, 21), consider a small piece of the universe distant from the rest and which is visibly rotating with respect to the rest. We know how to take account of this rotation and deduce the laws of motion suitable for that part of the universe, thus vindicating "the convention habitually adopted by the students of mechanics." Relativity in this sense "is no longer a simple convention. It is verifiable, and consequently it might not be verified." It differs from relativity in the broader psychological sense that draws on our sense of time, and cannot, for example decide of two events, one on Earth and one on Sirius, which came first except by a convention. Physical relativity incorporates the idea that widely separated worlds may be treated independently, and is therefore not a necessity of the intellect but an experimental truth holding within limits.

What, then, has changed with the work of Lorentz? Poincaré's answer was that there are now two principles that could serve to define space: the old one involving rigid bodies, and a new one to do with the transformations that do not alter our differential equations. They are not essentially different, because both are statements about the objects around us, but the new one is an experimental truth. To make geometry immune to revision by experimenters, physical relativity must become a convention concerning distant objects. Then, whereas our conventional knowledge of geometry was formerly rooted in the group of Euclidean isometries, it could now be rooted in the Lorentz group. (The group is dead, long live the group!) Now it would be the Lorentz group that preserves our equations, at the price of placing us in a four-dimensional space. And not only that. In an unmistakable reference to the ideas of Minkowski, although he did not mention his name, Poincaré went on,

> Time itself must be profoundly modified. Here are two observers, the first linked to fixed axes, and the second to moving axes, but each believing themselves to be at rest.[40] Not only will any figure, which the first one considers as a sphere, appear to the second as an ellipsoid; but two events which the first will consider as simultaneous will not be so for the second.
>
> Everything happens as if time were a fourth dimension of space, and as if four-dimensional space resulting from the combination of ordinary space and of time could rotate not only around an axis of ordinary space in such a way that time were not altered, but around any axis whatever. For the comparison to be mathematically accurate, it would be necessary to assign purely imaginary values to this fourth coordinate of space. . . .the essential thing is to notice that in the new conception space and time are no longer two entirely distinct entities which can be considered separately, but two parts of the same whole, two parts which are so closely knit that they cannot be easily separated. (*DP* 108, 23)

---

[40] Not, as the Bolduc translation has it, believing the other to be at rest.

So, Poincaré concluded,

> What shall be our position in view of these new conceptions? Shall we be obliged to modify our conclusions? Certainly not; we had adopted a convention because it seemed convenient and we had said that nothing could constrain us to abandon it. Today some physicists want to adopt a new convention. It is not that they are constrained to do so; they consider this new convention more convenient; that is all. And those who are not of this opinion can legitimately retain the old one in order not to disturb their old habits. I believe, just between us, that this is what they shall do for a long time to come. (*DP* 109, 24)

And indeed, Poincaré chose to be a Galilean to the end.[41]

THE FUTURE OF MATHEMATICS

In 1908 he published the text of an address he had intended to give at the ICM in Rome in April 1908 but had been prevented from giving by a severe illness, "L'avenir des mathématiques" (1908c). The text was circulated as a brochure at the meeting and Darboux read the paper on Poincaré's behalf. The full address is significantly different from the much-reduced version published in *Science et méthode*, and can be seen as a response to Hilbert's famous ICM address in Paris in 1900, where he presented twenty-three problems that he hoped would allow mathematicians to glimpse the future of mathematics.[42]

Poincaré began by stating that "To predict the future of mathematics the proper method is to study its history and its present state," leaving it to his audience to decide if in this way he agreed or disagreed with Hilbert's approach eight years before (Hilbert 1901). He then asked, more rhetorically, if this was not what professional mathematicians did anyway, before observing that in the past people had often had gloomy

---

[41] For more on this, see Walter (2009a).
[42] A complete translation of the paper and a further discussion of it appear in Gray (2012, to appear). For an enjoyably speculative evaluation of the whole paper as it was seen a century later, see Davis and Mumford (2008).

visions of the future: all problems had been solved, or were about to be solved, and all that would remain were insoluble problems. But "the meaning of the word solution has been extended," the doomsayers driven back repeatedly, and now, he said, he believed they were all dead. These days the commonplace prediction was rather of progress on all fronts, but that prospect alarmed Poincaré, who feared it "would soon produce a mass just as impenetrable as the unknown truth was to the ignorant." So, he argued, one was forced to make a selection of the facts, especially if one is a mathematician who creates these facts. But just as physicists are not constrained on narrowly utilitarian grounds but built their theories in advance of the electrical technologies which could not have been discovered without them, so too the mathematicians cannot and should not take their instructions from scientists.

As was to be expected, given Poincaré's perspective on understanding mathematics, isolated facts had no appeal for him, but, he said, a class of facts held together by analogy brings us into the presence of a law, and here he endorsed Mach's idea that science proceeded by bringing about economies of thought. "The importance of a fact is measured by the return it gives—that is, by the amount of thought it enables us to economise" (*S&M* 23, 374). The elegance that mathematicians find in a good proof reflects an underlying harmony that in turn introduces order and unity and "enables us to obtain a clear comprehension of the whole as well as its parts. But that is also precisely what causes it to give a large return." Elegance may result from the surprise in finding an unexpected association of ideas, or between the apparent complexity of the problem and the simplicity of its solution, and reflecting on the reasons for this may lead to an unexpected law. So the emotional satisfaction derives from a conformity between the sought-for solution and the necessities of our mind. Similarly, a lengthy calculation that has led to a striking result is not satisfying until we understand why at least the characteristic features of the result could have been predicted. And because it is not order per se, but only unexpected order that has a value, the mechanical pursuit of mathematics would be worthless: "A machine can take hold of the bare facts, but the soul of the fact will always escape it" (*S&M* 22, 373). The aesthetic response to mathematics was regarded by Poincaré as a sign of its efficacy, and this pair of ideas then shaped the rest of his address.

Poincaré now noted that standards of rigor had risen steadily in mathematics, and that this would rightly continue.[43] Indeed, "a demonstration that lacks rigour is nothing." But Poincaré hoped that well-chosen terms, such as "uniform convergence" would encapsulate progress and prevent rigorous proofs from becoming almost incomprehensibly too long, and indeed that such terms would capture a productive unity. He gave the example of "energy" as such a term in physics, and in mathematics he offered the terms "group" and "invariant," which were among his lifelong themes.

Mathematics, said Poincaré, is bordered by philosophy on one side and physics on the other:

> On the one side, mathematical science must reflect upon itself, and this is useful because reflecting upon itself is reflecting upon the human mind which has created it; the more so because, of all its creations, mathematics is the one for which it has borrowed least from outside. This is the reason for the utility of certain mathematical speculations, such as those which have in view the study of postulates, of unusual geometries, of functions with strange behaviour. The more these speculations depart from the most ordinary conceptions, and, consequently, from nature and applications to natural problems, the better will they show us what the human mind can do when it is more and more withdrawn from the tyranny of the exterior world; the better, consequently, will they make us know this mind itself. (S&M 31, 379)

"But it is to the opposite side, to the side of nature," he went on, "that we must direct our main forces." Here, said Poincaré, we find that ninety-nine percent of problems cannot be solved exactly in terms of known functions, and the convergence of power series solutions is often too slow to be of use, so progress must be made qualitatively. Our standards have changed over the decades, and

> there are, therefore, no longer some problems solved and others unsolved, there are only problems *more or less* solved. (S&M 34, 380)

After these generalities, Poincaré turned to specifics, starting with arithmetic (1908c, 934–935). Here he felt that progress was slow because

---

[43] The French text has "rigueur," the English translation renders this only imperfectly as exactness.

there could be no appeal to continuity, and therefore that arithmetic should be guided by the numerous analogies with algebra.[44] There are analogies between the theory of congruences and that of algebraic curves, and he predicted that the solution of problems about congruences in several variables would lead to considerable progress in indeterminate analysis. Then came the analogy between fields and ideals on the one hand and curves on surfaces on the other, which could be explored to mutual advantage. Likewise the theory of quadratic forms was intimately connected with the theory of ideals. What unity there was in arithmetic had come about through the use of transformations, and geometry offered examples of many types of transformation, not all of them linear, that could be applied in arithmetic. The groups that would arise would without doubt be discontinuous, and should be analyzed by looking at their fundamental domains, where the most important contribution was to be found in Minkowski's *Geometrie der Zahlen*. It is amusing to see how late in his speech the name of a mathematician first appears. Next Poincaré returned to one of his own earliest interests, Hermite's theory of forms with real coefficients, before concluding with some rather more perfunctory remarks about the study of prime numbers, where he could find no unity in the few asymptotic laws that were known.

In algebra (1908c, 935–936), he found that the subject had long been dominated by the theory of equations, which was best approached by the theory of groups, but there was also the question of the numerical solution of equations.[45] Then the theory of invariants had taken over for some forty years, but recently it had been abandoned, although the subject was far from exhausted. There was, for example, the problem of determining the invariants of an arbitrary group (this is Hilbert's 14th problem), and despite Hilbert's happy simplification of Gordan's work there were still problems to be solved, and he raised the prospect of a study of the algebra of polynomials modeled on arithmetic. This, although Poincaré did not say so, was already the hope and intention of several German mathematicians (Kronecker, Dedekind, and Hilbert) who looked for analogies between fields of functions and number fields.

---

[44] At this point the text in *S&M* is much reduced from the original, as specific examples were omitted, obscuring the fact that Poincaré was alluding to his own early work.

[45] *S&M* ceases transmission during the algebra section.

As for ordinary differential equations (1908c, 935), if linear equations were by now well understood, nonlinear equations were not. The study of their singular points was a start, but Poincaré held out hope that there would be a fruitful analogy with the use of Cremona transformations that simplify the study of the singular points of algebraic curves.[46] He did not mention that Picard had discussed the topic in the 1890s in his *Traité d'analyse* (vol. 3, chap. IV). Instead, he noted that there were important unsolved problems in the uniformization of the solution functions, and the qualitative study of the solutions, which even in the case of first-order, first-degree equations led to delicate questions about limit cycles, and had analogies with the study of the real roots of algebraic equations, (and with Hilbert's 16th problem, which Poincaré did not mention).

The study of partial differential equations had, he said (1908c, 935–936), recently been greatly advanced by the work of Fredholm. This work had caught Poincaré's attention, and he offered some thoughts on the underlying reason for Fredholm's success, which he connected to the interesting question of why so many of the partial differential equations that were used in physics were linear. He suggested that this arose from the way the modeling arguments worked that were used to construct the equations, and which turned the behavior of infinitely many discrete molecules into a few continuous processes, perhaps via the theory of infinite determinants, which appeared in both Fredholm's work and, earlier, in that of the American lunar theorist Hill (1877). But much remained to be done in this area, and, for example, on the Dirichlet problem on convex domains, and here progress was being made thanks to Hilbert's initiatives. As his audience knew, some of these initiatives had been signaled in Hilbert's ICM address: problems 19, 20, and 23 are on the calculus of variations and boundary value problems, including the Dirichlet problem.

The study of Abelian functions, and especially those that do not arise from algebraic curves, had been blocked for some time, but had recently been almost completely solved, said Poincaré (1908c, 937), by

---

[46] It is not possible to survey here what came of this idea. Poincaré's nephew Pierre Boutroux lectured on the subject and Painlevé contributed to it at the Collège de France in 1908 (see P. Boutroux 1908) and Dulac wrote his paper (Dulac 1908), but then the topic seems to have gone quiet until it was revived in the 1980s with the study of singular points of holomorphic differential equations in the Brazilian school; see Cerveau (2006/2010) and the articles cited there.

the work of Castelnuovo and Enriques. This, together with the work of Wirtinger (no mention of Picard or Painlevé!), resolved many of the outstanding difficulties. Poincaré himself was to publish a major paper in this area in 1910, and one wonders if he already knew where he could make a contribution. The general study of functions of two complex variables, however, as Poincaré knew very well, seemed different at every turn from the theory of functions of a single variable, and among many disparities Poincaré mentioned (1908c, 937–938) the lack of a uniformization theorem in two or more variables.[47] His remark that "conformal representation is usually impossible in a domain of four dimensions" is probably not a reference to an old result of Liouville's so much as an allusion to his own paper on holomorphically inequivalent but topologically equivalent domains in $\mathbb{C}^2$ that implies the failure of the Riemann mapping theorem in two or more complex dimensions.

The theory of groups, interestingly removed from algebra, was so large that Poincaré restricted himself to Lie's continuous groups and Galois's discrete or finite ones. In Lie's theory the way forward had been to generate the transformations of the group infinitesimally, and much had been done by Lie, Killing, and Cartan that now needed to be simplified and coordinated. The study of finite groups lagged behind, for the same reason that arithmetic lagged behind algebra: the lack of continuity, but Poincaré hoped that the analogy between the two kinds of group would help.

Poincaré now turned to geometry (1908c, 938–939).[48] It might appear, he said, that geometry was nothing but algebra and analysis in a different language, but even if that were true it would imply a failure to recognize the importance of a well-formed language. Geometry posed new problems, and for as long as it confined itself to three dimensions it could call on the aid of our senses. In recent years that language had been found to be useful in higher dimensions as well, and the analogy with three dimensions had proved good enough to lead to new, qualitative theorems. Moreover, the geometry of position, which was purely qualitative, was of immense importance, as the work of

---

[47] This follows from the fact that many complex surfaces have vanishing fundamental group, a fact that was only discovered later.

[48] *S&M* resumes transmission here, but only for a while.

Riemann alone showed. It was the geometry of position that unified what otherwise would have appeared, eventually, as isolated problems in analysis.[49] Another unifying feature of modern geometry was the use of transformation groups, the Brill–Noether of algebraic curves merited a mention, as did Darboux's work on infinitesimal geometry.

The penultimate topic Poincaré chose was Cantorism, the mathematical study of infinity, as he put it (1908c, 939).[50] It started in complete generality, which had caused it to provoke horror in Hermite's mind, but most people today had come to terms with it. However, there were now

> certain paradoxes and apparent contradictions, which would have rejoiced the heart of Zeno of Elea and the school of Megara. Then began the business of searching for a remedy, everyone in their own way. For my part I think, and I am not alone in so thinking, that the important thing is only to introduce such entities as can be completely defined in a finite number of words. Whatever be the remedy adopted, we can promise ourselves the joy of the doctor called in to follow a fine pathological case.

This address on the future of mathematics is the supposed source of a famous remark that Poincaré did *not* make.[51] The only-too-familiar quote runs: "Later generations will regard *Mengenlehre* as a disease from which one has recovered." It implies that Poincaré was strongly opposed to the study of the theory of sets. As we can see, Poincaré's position was much more interesting, if less colorful.

Poincaré concluded with some remarks on what he called the search for postulates, and we might call the axiomatic method, noting that "in this direction Mr. Hilbert has obtained the most brilliant results," and offered the hope "that these examples will have been sufficient to show the mechanism by which the mathematical sciences have progressed in the past, and the direction in which they must advance in the future."

This address, unlike the much truncated version in *S&M*, offers an insight into Poincaré's approach to mathematics now he was a statesman of science, as well as an interesting comparison with Hilbert's ICM

---

[49] At this point transmission again breaks off.

[50] *S&M* now continues to the end.

[51] For an account of how E. T. Bell came to make this particular error, see Gray (1991) .

address in Paris in 1900 on the problems of mathematics. Hilbert had not attended the ICM in Rome, but his friend Minkowski reported back that Poincaré's address had been nothing but a "weak remake of your Paris Lecture."[52] This is a predictable criticism, with some justice, but it also reflects a difference of style: where Hilbert had been explicit, Poincaré preferred to be elliptical.

The order in which Poincaré presented the topics is instructive. He began with arithmetic, one of his earliest interests, where he had quite definite things to say about the few unifying features of the subject. By comparison, his few remarks about algebra are rather impersonal. This may be because the theory of ideals has gone to arithmetic and the theory of groups has been put on its own in a later section (which, of course, says something interesting about how those subjects were regarded in 1908). Matters are more vivid when the core French discipline of differential equations is brought into view. He might have mentioned Painlevé by name when talking about singular points of nonlinear equations, but the hope that there might be a productive analogy with algebraic curves and their transformations was particular to Poincaré. He had less to say about the theory of groups, perhaps because his contributions had been slight, but could not resist a discussion of the paradoxes of set theory, where his language was much softer than in his polemical essays against the logicists.

In fact, the address follows quite well the arc of his own career, from his early work under the influence of Hermite to his clashes with Couturat, Russell, and Zermelo. He occasionally acknowledged the work of others: Minkowski, Hilbert of course, Lie and Killing, Castelnuovo and Enriques. Among the French he mentioned Hermite and Darboux by name, but none of his contemporaries and only Cartan among the next generation. Whatever his colleagues might have thought, and they could well have felt slighted, the talk is evidence of who had most forced their way into his attention. And the idea that mathematical understanding is expressed in the productive organization of facts and the recognition of analogies, is indeed an expression of how Poincaré had guided his own work for almost thirty years. As so often with his publications, here too the interest is focused at a certain, high level of generality; unlike most of Hilbert's

---

[52] See his letter to Hilbert of 4 May 1908 in Rüdenberg and Zassenhaus (1973, 144). I thank David Rowe for this information.

problems there are very few specific questions to be answered, although more topics were mentioned than is usually realized. In what was surely a temperamental difference, and probably a difference of priorities, Hilbert had started with foundational questions and Poincaré finished there.

There is another way in which these two addresses differ. Hilbert's is a view of what mathematics *is*: certain theories in which certain problems are presently central. It is openly a wish list of problems whose solutions would mark breakthroughs in the subject. Poincaré's address is a view of what it is to *do* mathematics. It is about the best ways to think so that mathematics can be done productively. In the main, Hilbert selected problems that others found important and mathematics did advance when they were solved. Poincaré offered advice, however elusive, that whenever it can be taken would always enable mathematics to advance. Hilbert expressed his reflections on the grounds of the validity of mathematics in his theory of finite expressions, and Poincaré in terms of the capabilities and limitations of the human mind.

Mathematical Invention

In 1908 Poincaré talked to the Société de psychologie in Paris about the psychology of discovery in mathematics. The published version in *L'enseignement mathématique*, "L'invention mathématique," (1908d), became one of his more famous essays.

Charles-Ange Laisant and Henri Fehr, the editors of the journal, hoped to establish an international journal devoted to mathematical education. The first issue had appeared in 1899, with, among others, Appell, Picard, and Poincaré's names on the editorial board, and it carried an essay (1899d) by Poincaré on logic and intuition in mathematics and teaching. His strongly held view was that the emphasis on rigor in the teaching of mathematics in schools was misplaced. To be sure, the mathematicians of fifty years ago had said many things we found in severe need of proof these days, and those proofs had been supplied, but to deny a role to intuition and to rely solely on rigor was, in his view, a great mistake. A liking for rigor had led to a slew of "bizarre functions that resemble as little as possible honest functions that one can use: not continuous, or continuous but without a derivative. . . . One used to invent new functions for a practical purpose, today to refute the reasoning of one's fathers and one can deduce nothing from that."[53] It would require teaching the range of bizarre functions first, so as to acquaint students with the teratological museum and to ensure that they follow all the steps.

Poincaré said that he for one would not be willing to sacrifice everything to logical rigor. Instead, he thought that a student's understanding broadly followed the historical path of discovery, and that teaching was intended to develop certain habits of mind among which intuition was not the least precious, because it was through intuition that the mathematical world came into contact with the real world. "Practitioners always need it, and for each pure mathematician there should be a hundred practitioners." After all, most people who learn mathematics will become engineers, and only some will become the next generation of leading mathematicians. And they too will need intuition, "for if it is by logic that one proves, it is by intuition that one invents."

At some stage Laisant and Fehr had the idea of sending a questionnaire asking mathematicians about their working methods, and the responses they got were written up in various essays that appeared in volumes 7 to 10 of their journal. However, it appears that not many of the best mathematicians had bothered to reply, and Poincaré is often said to have stepped in to give a speech and thereby help out in some way. This may be true, but quite certainly Poincaré had a lifelong interest in how the mind acquires knowledge, and was quite familiar with contemporary thinking on that question. He belonged to a dining club that Gustave Le Bon had established in 1893 with his friend Théodule Ribot, the *Banquet des XX*, which met on the last Friday of every month (see Poincaré 2002, 152–153). Henri Poincaré attended, as did his cousin Raymond, Prince Roland Bonaparte[54], the publisher Camille Flammarion, and Paul Painlevé.[55] Ribot was a leading figure in the development of psychology in France, a professor of experimental psychology, first at the Sorbonne and then at the Collège de France. His book *La psychologie allemande contemporaine*, which had run to thirteen editions by 1898, traced the

[53] The same words appear in the essay "Les définitions mathématiques et l'enseignement," in *S&M*, which is partly a reworking of this essay.

[54] He was a grandson of Lucien Bonaparte, the Emperor Napoleon I's brother; see Simonetta and Arikha (2011).

[55] Rollet (1997, 16) also found that Poincaré frequented other fashionable salons, including that of Princess Marie Bonaparte, the daughter of Prince Roland Bonaparte, who later acquired the dubious distinction of being at least partly responsible for bringing psychoanalysis to France.

subject of psychology from the contemporary work of Wundt back to Herbart's and Fechner's disagreements with Kant, and thereby raised the status of the new discipline in both France and America, by giving it a philosophical aspect that was to prove congenial to many in other fields. His chapter on the psychology of spatial knowledge was to influence Poincaré. In it, Ribot advanced the theory that people experience a tactile and a visual space, and that motion, our own as well as that of objects, is essential to our construction of "Space," and opposed the idea that a concept of three-dimensional space is built into the retina. He was also among the first to espouse Darwinian evolutionary thinking in France, notably in his *L'Hérédité* of 1873, an approach that Poincaré also shared.

In his essay on mathematical discovery Poincaré explored how the mathematical mind might work in terms that were entirely psychological rather than philosophical. He asked first of all how can it be that there are people who do not understand mathematics, when mathematics is based on simple rules of logic that only a madman would attempt to deny? How can people make mistakes in mathematics, when a healthy mind is incapable of making an error in logic? He answered that he himself was incapable of making the simplest calculation without error, and that almost everyone, forced to recall the results of many long and complicated arguments will find they have faulty memories. What sustained him in mathematics, he suggested, and let him down in chess, was a good sense of the general march of an argument. So much so that he often had the illusion about good mathematical ideas he had been taught, that he had invented them himself. It was this feeling or intuition of mathematical order that he presumed some people had and others did not, and which enabled those who had it to be creative mathematicians.

Anyone, he said, could put bits of mathematics together in new ways, but almost all of these would be of no interest. Discovery consists of finding useful combinations: "to invent," he proclaimed, "is to discriminate and to choose." Or rather, that is the work of the unconscious. This he proceeded to illustrate by drawing on his own experiences: the coffee at night, the famous bus journey, and the walk on the cliffs. His conscious activity in 1880 was being directed to showing that some functions could not exist, but one night his unconscious prompted the opposite thought and "I had only to write up the results, which just took me a few hours" (1908d, 51, 392). Then a period of deliberate work followed, in which the new functions were investigated by means of an analogy with existing functions of a similar kind. Then his unconscious spoke again as he boarded the bus at Coutances, this time to reinterpret a web of triangles in a new, and richer, context.

It seemed to Poincaré that the unconscious played a crucial role in his work. It must first be fed by a period of conscious work. Ideas came to him in bed, he said, when he was in a semihypnagogic state. They were not mechanically analyzed, the unconscious did not work like a machine, it exercised a selective role much as the conscious did; in fact it can divine better than the conscious mind. Here Poincaré referred to Émile Boutroux's discussion of the work of William James in his recent book (É. Boutroux 1908). Perhaps, suggested Poincaré, ideas had hooks, like the atoms of Epicurus, and were stirred up after a period of conscious work and fastened themselves together to form productive combinations. In any case, he observed, the results of unconscious work never gave the results of long calculations ready-made, although it may give a verification. The unconscious provides points of departure for calculations that must be made consciously, but operates by chance. And one must be careful, for the unconscious presents these ideas with a feeling of certainty even when, on rational analysis, they prove to be worthless.

POINCARÉ AMONG THE LOGICIANS

In 1905 Poincaré found his view of the nature of mathematical knowledge challenged by a vigorous school of young philosophers who hoped to derive all of mathematics from logic. This revived Leibniz's hope that all intellectual problems could be reduced to a suitable form of calculation, and two of the leading exponents of this view, Couturat in France and Russell in England, had partly come to it through a study of Leibniz's previously unpublished writings. The view was therefore also anti-Kantian in its understanding of knowledge, which was another reason for Poincaré to oppose it. But the attempt to derive mathematics from logic had other deep contemporary roots. The profusion of new geometries had directly challenged the simple idea that Euclidean geometry was true, and destroyed any simple idea that there was only one geometry that, therefore, had to be true. It seemed even more perilous to entrust the certainty of mathematics to the hands of physicists, who were by no means certain of the foundations of their own subject, and in any case such a move would displace mathematics from its paradigm position in the pyramid of knowledge.

How, then, was the certainty of mathematics to be assured? One view was that it should be derived from logic. The kernel of this belief was the feeling that it did not make sense to deny the everyday workings of logic. Statements cannot be both true and false at the same time; simple deductions from premises hold (as in the established syllogisms of Aristotelian logic). The most forceful proponent of this view was the irascible Gottlob Frege, who from his tenured position in the small German university of Jena pursued all who contested his views with a cutting mixture of sharp insight and personal venom. Frege had realized early on that what constitutes a logical argument must be scrutinized with extreme care to make sure that it is rigorous. Even mathematical arguments need to be phrased with extreme precision, and this leads to the fundamental problem in deriving mathematics from logic. It is not enough to observe that mathematicians follow rules, such as those of everyday logic, and point to theorems to vindicate crucial deductions. They may well be behaving correctly, but this does not mean that they are doing logic. They are doing mathematics and for that to be nothing more than logic everything they use must be either a primitive term or defined logically out of primitive terms.

The simplest case of sound mathematical arguments is in elementary number theory, such as $7 + 5 = 12$. Every symbol here has to be construed logically. The numbers $5, 7$, and 12 must be defined, the operation of addition, "$+$" and the symbol for equality, "$=$". There also has to be an account of how, when this has been done, the logically defined objects can play the roles required of them in ordinary life, and enter propositions about the number of words on a page, pages in a chapter, and chapters in a book. Defining numbers is not as hard as defining time (which St. Augustine famously characterized as a word one knows the meaning of until one is asked to explain it[56]), but it is not easy, and Frege enjoyed taking apart the definitions offered by several prominent mathematicians, Weierstrass among them. In this he was joined by another group of mathematicians who had their own reasons for distinguishing between counting and saying what numbers actually are. These were Dedekind and Cantor in Germany, and Peano in Italy.

---

[56] See Wills (2011, bk. 11, chap. 14).

Frege's own position was closest to Cantor's, which he welcomed although not without some critical remarks. Cantor began with the insight that we can distinguish the size of two collections of objects by pairing them off one by one. If we find that one collection can be paired exactly in this way with the other collection so that no object of either collection is left out or used more than once, then we say that the collections have the same number of objects. This pairing, called a one-to-one correspondence, is the fundamental insight into the nature of number. It does not depend on which object in one collection is paired with which object in the other. We can use it to sort every collection of objects we come across into boxes, where the collections in each box are in one-to-one correspondence with each other and with no collection in any other box.

What then is a number, for example, the number "five"? We could pick a collection we recognize intuitively as having five objects in it, such as the letters in the English word "prime", and say that a collection has five objects in it if and only if it can be put in a one-to-one correspondence with the collection $\{p, r, i, m, e\}$. But this seems rather arbitrary, and Frege and Cantor found it more natural to say that the box itself is the object corresponding to the mathematical term "five" and a collection has five objects in if and only if it can be put in a one-to-one correspondence with any collection in that box, and so is in the box. This seems reasonably satisfactory: the notion of a one-to-one correspondence is clear and it enables us to reduce the number concept to the use of a more basic concept. One can ask if the collection of boxes has the right properties: Is every number found in this way, and does every box correspond to a number? Let us postpone the first question until we reach the Peano axioms (below) and take up the second one.

Here the surprising answer is that indeed there are boxes that correspond to no numbers at all—unless, of course, one is willing to stretch the concept of number. The first such box consists of collections that can be put in a one-to-one correspondence with the collection of natural numbers $\{1, 2, 3, \ldots\}$. This is an infinite set, but nothing seemingly precludes it from defining a box and being a number. Cantor soon discovered that there was no reason to stop here. First he showed that although the collection of rational numbers belonged in this box, the collection of all decimal numbers does not. This collection is strictly larger. Then he showed that given any collection, the collection of its subcollections is strictly larger. This is clear for finite collections. For example, one can form eight subcollections of the collection with three objects $\{a, b, c\}$: they are $\{a, b, c\}$ itself, $\{a, b\}$, $\{a, c\}$, $\{b, c\}$, $\{a\}$, $\{b\}$, $\{c\}$, and the empty subcollection. The same result is true for infinite sets, but it is harder to prove. In principle, this gave Cantor an infinite collection of infinite collections all of different sizes, by passing from an infinite collection to the collection of its subcollections, forming the collection of its subcollections, and so on. The corresponding boxes formed by sorting infinite collections according to their one-to-one correspondences would then be numbers, but a new kind, which Cantor called transfinite. The only alternative would be to find a good reason to reject these infinite collections, but in the heady early days no clear criterion for doing so suggested itself, and instead the idea of a number embraced the finite and the transfinite alike. Mathematicians in the French school around Hermite were not pleased, but their objection was not that Cantor's arguments were not rigorous, rather that they were pointless because they addressed no interesting problem.

But even if you granted that objection and confined your attention to finite numbers, the infinite had already crept in, because each box plainly contains infinitely many collections. Should this matter? Cantor, the mathematician, did not think so, but his friend Dedekind, who was much more concerned with the foundations of mathematics, raised the stakes. He wanted to identify the fundamental objects of all mathematical thought, and believed that these objects could all be thought of as objects in a collection or, in terms we use today, elements of a set. The bare minimum is that these otherwise featureless elements can be objects of thought, distinguished one from another, and can be collected into sets. One can form the set of all elements of a certain kind, for example, such as the set of all positive integers. It is Dedekind, rather than Cantor, who raised the idea that mathematics is about nothing more than sets and their elements.[57]

As we noted, if mathematics is to be reduced to logic, the positive integers have to be defined using, presumably, nothing more than the idea of sets. Dedekind succeeded in doing this, in his *Was sind und was sollen die Zahlen?* (1888) and so, rather more superficially, did Peano in his *Arithmetices Principia* (1889), but since his solution is accordingly easier

---

[57] For a history of the rise of set theory, see Ferreirós (2007).

to describe, we can look at it first. We have to start somewhere, he said, let it be with the number one, 1 (modern presentations usually start with zero, 0). We notice that every positive integer has a successor, and we define $s(n)$ to be the successor of the integer n, and write $2 = s(1)$, $3 = s(s(1)) = s(2)$ and so on, in the hope that will give us all the positive integers and nothing else. To bring this about, Peano stipulated that 1 is not the successor of anything, and that if two integers have the same successor then they are the same. It remains to show that every positive integer arises as a successor, and for this Peano asserted the principle of induction. This asserts that if S is a set containing 1, and for every integer n in the set, $s(n)$ is in the set then the set contains all of the positive integers. These postulates, he showed, are enough to construct a theory of addition and multiplication.

How should we feel about these statements? One reaction is that they neatly capture the fundamental property of the sequence of positive integers, namely that they come obediently one after the other. On this view we are happy with the positive integers but appreciate the succinct way they have been characterized. Another reaction would be that these postulates give us something to aim for, and if we can find sets with these properties we shall have defined the integers in terms of sets. This would be a way of reducing the arithmetical and mathematical concept of a positive integer to a logical, or at least set-theoretical object. That was Dedekind's view and he showed how to do it the year before Peano wrote, in his *Arithmetices Principia* (1888). Or there is an intermediate view, which says that these postulates, together with all the other rules of deduction needed to handle them, do indeed construct, create, define the positive integers, and indeed Dedekind's approach essentially replaced the successor relation with the idea of a set N being mapped one-to-one onto a subset of itself, N', in such a way that N contains exactly one more element than N' (Dedekind 1888, para. VI). On any view, the postulates are easy to accept, except for the principle of induction. If the positive integers are "out there" then there is a question about how we can know that the principle of induction holds. If they are defined or created by these axioms, there is a problem with making the principle of induction an axiom: Why should it be accepted? One cannot make any collection of axioms at will: they must at least be self-consistent.

Whether one sympathizes with the endeavor to reduce mathematics to logic or not, it is at least clear that the task will be difficult. What changed everything was a chance meeting between Russell and Peano in Paris at the International Congress of Philosophers that Russell later described as "a turning point in my intellectual life" (*Autobiography* 1998, 147). Russell gave a solidly old-fashioned paper, and discovered that the Italians, led by Peano, had meanwhile changed the terms of the debate. Peano's axioms were only part of an attempt to devise a unambiguous way of writing all of mathematics that would make it impeccably rigorous. This would not, of itself, reduce mathematics to logic, but it seems to have suggested to Russell that the right formalism was now available for just that task. By this time Russell was engaged on a ferocious attempt to catch up with the last fifty years of pure mathematics, which his Cambridge education had not prepared him for, and this included reading some of the works of Frege. This, and his reading of Leibniz, suggested to Russell that he now had a much better intellectual position from which to provide a philosophy of mathematics. The first significant result of his embarking on this ambitious enterprise was, paradoxically, to make it much harder.

Russell quickly mastered the new ideas, and came to identify a crucial fault with them. On 16 June 1902 he sent Frege a postcard that reads, in part,[58]

> Let w be the predicate: to be a predicate that cannot be predicated of itself. Can w be predicated of itself? From each answer its opposite follows. Therefore we must conclude that w is not a predicate. Likewise there is no class (as a totality) of those classes which, each taken as a totality, do not belong to themselves.

This is a little clearer if it is stated in the equivalent language of naive set theory. Call a set normal if it is a member of itself, and abnormal otherwise. Now consider the set of all abnormal sets. If it is normal it is, by definition, a member of itself, so it is abnormal. If it is abnormal it is not a member of itself, so it is normal. This is a paradox.

It is greatly to Frege's credit, as Russell was the first to remark, that he published the bad news, for it seemed to say that his life's work was in ruins. It was crucial to his arguments that the collection of all objects of such and such a kind was a set: in the loose language used above, that each box was a collection. Russell's paradox made it clear that not every collection was a set, and so one could not casually collect all such and

---

[58] See van Heijenoort (1967, 125), and for Frege's reply, van Heijenoort (1967, 127–128).

suches together and speak of the set of all of them. The naive idea of a set needed to be refined to head this problem off, but it was not clear how to do that. Russell himself also admitted publicly to the problem, in his *The Principles of Mathematics* (1903), and it was by far the largest obstacle on the reduction program that he now so passionately espoused.

The paradox was soon joined by others: Richard's paradox, the paradox of the liar, and paradoxes in Cantor's theory of the transfinite. Some mathematicians took a robust attitude to them. There are many occasions in research mathematics where one is led to a highly implausible conclusion, even a result one cannot believe. Cantor himself, when he wrote to Dedekind to tell him that the points of an interval and the points of a square were in one-to-one correspondence, said that "I see it, but I don't believe it," so strongly did this result seem to contradict the intuition of dimension. But he and Dedekind came round to the view that the conclusion was correct, and a way would have to be found to defend the intuitive concept of dimension, as indeed there was. It was in this spirit that Hilbert and those around him in Göttingen, who had found some of these paradoxes themselves, contemplated them. They saw the difficulties as lying in the theory of sets, which, as a result, would have to become more complicated and less naive, and Hilbert charged Zermelo with sorting this out. It was far harder for those, like Russell and Frege, who sought to reduce mathematics to logic. For them the paradoxes suggested that there might be something wrong with the whole idea. What they could do was to set forth what could be done without any contentious use of the concept of a set, isolate precisely those rules of deduction that could not create an illegitimate "set," and proceed carefully along a path once imagined to be a routine stroll.

Of the paradoxes that came up, a variant of the one attributed to Jules Richard is the most important here.[59] It asks the reader to contemplate (I paraphrase) the smallest positive integer than cannot be named in English in less than fifty syllables. Plainly some names are much longer than that, so there are such positive integers, and in any set of positive integers there must be a least one. So the phrase "the smallest positive integer than cannot be named in less than fifty syllables" defines a unique positive integer—but it does so in only twenty-two syllables! The solution proposed was to exclude definitions of this kind on the grounds that

[59] Richard's paradox is stated more precisely later in this section.

they refer to the whole set of objects that are in the process of being defined, and cannot therefore be a definition of one of the objects. Russell proposed that as each mathematical object is defined it is assigned a level, much as one might think of the copies of a book printed yesterday, today, or tomorrow. This level, called its "type" constrains its use and arguments are invalid if they assume an object is of a lower type than it really is. That avoids the paradoxes, but at a terrible price for a mathematician to pay. For it is common in mathematics to speak of a number that is the greatest lower bound of all the numbers in a set (it may or may not belong to the set itself). This number is of a higher type than the numbers in the set, because its definition rests on theirs, but mathematicians want to treat it in the same way (say that it is the root of an equation or whatever).

This is the reductionist philosophy of mathematics, at once dramatic in its implications, hostile to Poincaré's form of Kantianism, and yet faced with deep problems of its own that Poincaré turned to challenge in 1905 with his (1905c), (1906f) and (1906e).[60] The enemy was once again Couturat, the author of a series of articles entitled "Les principes des mathématiques," later published as a book under the same title. Here, Poincaré said, Couturat had argued that the long struggle between Leibniz and Kant had recently been decided in favor of Leibniz, because of the decisive contributions of Russell and Peano. Synthetic a priori judgments had apparently been shown to be impossible, mathematics was entirely reducible to logic, and intuition played no part in it at all. When Couturat had stated these opinions at the Kant Jubilee in 1904 he did so with such vigor that Poincaré wrote, "I overheard my neighbour whisper 'It's quite evident this is the centenary of Kant's *death.*' "

Poincaré then spelled out to his readers why they should not subscribe to this decisive condemnation. He reminded them that, indeed, Hilbert had exposed the formal character of reasoning in geometry, but, he said, even if the same was done for arithmetic and analysis mathematics could not be reduced to an empty form without mutilating it, because the origin of the axioms would still have to be investigated, however conventional they were taken to be. Nor was logical correctness all there was to mathematics. Judicious selection among possible deductions, guided by instinct, alone "gives a value to the constructed edifice." But perhaps the new logic can demonstrate the truths of mathematics without any appeal

[60] Reprinted in *S&M* with a two-page introduction. There is also Poincaré's (1906a).

to intuition? This was a possibility Poincaré had considered only to reject in an earlier essay (1894d). There he had appealed to the principle of mathematical induction, and so he had to consider the proposed rebuttals of the logicists. They considered this principle as essentially defining what a number is. And, just as Poincaré had considered the parallel postulate as a disguised definition—something presented as an axiom which in fact defines what it is to be a parallel line—so they viewed the principle of mathematical induction as a disguised definition.

Now, said Poincaré, to define something in mathematics, which implies that it exists, it is enough to establish freedom from contradiction, and this could be done in two ways: either objects could be produced that satisfy the definition; or it has to be shown that there are no contradictions among the implications of the putative definition. But, if these implications are infinite in number it is difficult to do this without recourse to the principle of mathematical induction, and Poincaré promised to show that in this respect the logicians had failed. But before he did so he considered another argument. When a term is defined it must be shown that the uses to which it is then put follow from the definition, and this can be very tricky when the definition is of an object—such as number—that we have used and been familiar with for a long time. It has to be shown that no tacit appeal is being made to an existing definition or, if such an appeal is made, then the second definition agrees with the first. This, he said, the logicians had also failed to do.

As an example of the second failing, Poincaré recalled the definition of the number 1 offered by another of the logicists, Burali-Forti:

$$1 = \iota T'\{Ko \cap (u, h) \in (u \in One)\}.$$

This is written in a notation devised by Peano, and indeed in what Poincaré called the "Peanian" language. But, said Poincaré with an affected degree of naivety,

> I do not understand Peanian well enough to venture a criticism, but I am very much afraid that this definition contains a *petitio principii*, seeing that I notice the figure 1 in the first half and the word *One* in the second. (*S&M* 168, 471)

Matters were a little better with Couturat's definition of the number 1, which Poincaré paraphrased as the number of elements of a class in which any two elements are identical. But here Couturat used the word "two", "and I am afraid that if we asked M. Couturat what two is, he would be obliged to use the word *one*."[61] A later generation would put the objection differently: the singletons and doublets already invoked are calling up all the one-element and all the two-element sets, and these collections do not form sets in the current foundations of mathematics, Zermelo–Fraenkel set theory.[62]

Poincaré now turned to recent developments in logic, especially those introduced by Russell. Russell had recast Boole's ideas in the notation and spirit of Peano, and redefined logic as the syntax of the words "if", "and", "or", and "not". This took logic beyond its old, syllogistic form, and Poincaré proposed to see what was the fate of principles that had been beyond the scope of the old logic and accepted although indemonstrable. He noted (see (1905c, 82), a passage dropped from *S&M*) that Russell had also brought in the useful idea of a propositional function: a function $\varphi(x)$ of x that takes the value true or false depending on x. This was used by Russell to handle the theory of relations. It also, Poincaré noted, can be used to make infinitely many statements at once, those of the form "$\varphi(x)$ implies $\psi(x)$" which are valid for infinitely many x; this was beyond the power of the old logic. Now, said Poincaré, of course the new rules for handling these new logical objects had to be accepted as axioms, because they could not be shown to be free from contradiction. In particular, they could not be shown to be free from contradiction by any argument that appealed to the principle of mathematical induction, because that principle had not been admitted by Russell. So they were not disguised definitions but genuine axioms, and there were, on Russell's own count, nine undefined notions and twenty indemonstrable propositions governing their use. Each of these, Poincaré announced with relish, are new independent acts of intuition that might as well be termed a priori synthetic judgments. It does not matter if these judgments allow others to be proved that formerly were accepted a priori; the crucial point, contrary to the claims of the logicists, is that synthetic a priori judgments are still present in the foundations of mathematics.

---

[61] Goldfarb (1988) sees Poincaré's objections as psychologistic and blind to the claims of logic after Frege; my view is that they are rooted in a different epistemology. This is also the view in Tieszen (2010).

[62] Zermelo's presentation is his (1908b).

The original essay now carried three further sections that Poincaré dropped from *S&M*, on cardinal arithmetic. The point at issue is the analogy between logical addition or multiplication and ordinary addition or multiplication, which Poincaré once more would allow, but only as a further act of intuition. But now, he said, came the crucial dividing issue: Could the logicians proceed to derive the rest of mathematics, or could they only offer a small fragment, containing no theory of numbers, no analysis, and no geometry? If so, of course, secondary school teaching would be lightened considerably (1905c, 832).

So Poincaré turned to see whether logicism could generate arithmetic, more precisely, the arithmetic of ordinals. Couturat, said Poincaré, had accepted the Peano axioms as a definition of number. But this will not do. The axioms cannot be shown to be free of contradiction by finding examples of them, and any attempt to show that they were contradiction-free by examining the totality of their implications would require the very principle of mathematical induction Couturat believed they implied. For (in a further passage dropped from *S&M*) either one assumed the principle in order to prove it, which would only prove that if it is true it is not self-contradictory, which says nothing; or one used the principle in another form than the one stated, in which case one must show that the number of steps in one's reasoning was an integer according to the new definition, but this could not be done (1905c, 834).

Poincaré then looked for a proof in the work of Russell and Couturat that these difficulties could be overcome, if only to refute it, but found none. In wondering why this was, he traced the omission back to an earlier account of the principle of mathematical induction by Russell which had glossed over it saying that every number can be obtained by successively adding 1 to the initial 0. However, said Poincaré, the principle of mathematical induction rather says that one can reason by recurrence on numbers (which have been, perhaps, obtained by recurrence). This confusion in Russell's work explained, to Poincaré's satisfaction, why Russell had, without realizing it, put forward a definition that he was incapable of justifying by proving it was exempt from contradiction.

Poincaré then turned to the work of Hilbert, chiefly his ICM address at Heidelberg in 1904 on the foundations of logic and arithmetic (1905a). Poincaré's tone of voice was now much modified. He noted that Hilbert concerned himself only with marks on paper, and with what could be said about such things, unlike Russell, who cheerfully considered all manner of objects comprehended by a proposition. More importantly, in another passage dropped from *S&M* (1906f, 19), Poincaré noted that Hilbert took note of the collapse of Frege's naive notion of a set. This, said Hilbert, showed that "the ideas and methods of ordinary logic lack the precision and rigor required for a theory of sets." But still Poincaré suspected that even Hilbert must have failed, somewhere, to ground the principle of mathematical induction in anything other than an a priori intuition, and he prowled the pages of Hilbert's text looking for the crucial slip. He found it in the proposed proofs that the very simple axiom system Hilbert studied in his paper, which is essentially a stripped down version of the Peano axioms, is contradiction-free. There Hilbert argued that the formulas generated by his axioms were of a certain kind because the axioms were formulas of that kind and the deduction rules ensured that the consequences of the axioms were also of that kind. In this way, all the consequences of the axioms were surveyed— but, said Poincaré —in a way that required the principle of mathematical induction.

Poincaré moved on to consider the consequences Hilbert drew from his claims so far. He was rightly puzzled by the claim that the axiom of complete induction could simply be formalized in the above fashion and added to the earlier axioms without fear of contradiction, thus leading to the existence of the "so-called smallest infinity," the ordinal type of the positive integers. He was willing to suppose that this could be done, but he observed that Hilbert nowhere proved from his axioms that every finite ordinal has a successor that is a finite ordinal, thus leading to the construction of the set of all finite ordinals. Indeed, Poincaré noted with mounting surprise, Hilbert instead went on to claim that the existence of the smallest infinity allowed one to prove that every finite ordinal has a successor. Which, then, came first Poincaré demanded to know: the existence of the smallest infinite set, or the validity of the axiom of complete induction? Hilbert, like Russell before him, had been defeated by the task of proving that defining the natural numbers by means of the axiom of complete induction is contradiction-free.

But even if the principle could be proved, would it follow that it could be used within the axiom system as it had customarily been used before? Hilbert's argument to that effect was a proof by contradiction. He supposed there was an inconsistent statement deduced from the axioms, and traced it back to the original axioms by unraveling its proof. This yields a finite succession of statements that itself can be regarded as a finite ordinal—but what, asked Poincaré, ensured that statements about a certain kind of finite ordinal applied to this finite ordinal? Nothing, he replied, and if on the contrary that could be proved then the principle of mathematical induction would no longer be a principle but it would be a theorem and all this discussion would have been a waste of time. Hilbert's proofs, he added, applied only to certain kinds of marks on paper, not to discussions of proofs.

I omit a summary of Poincaré's analysis of how Hilbert tried to patch up his argument in various places, and pass to his account of the infinite, which was also dropped from *S&M*. Couturat had claimed that the theory of infinite cardinal numbers was purely logical, independent of the theory of ordinal numbers, of considerations of the finite or the infinite, and of the principle of mathematical induction; naturally, Poincaré was not convinced. He cited the central theorem in this regard, Bernstein's theorem, which says that if two sets A and B, say, have the property that a subset of A has the cardinality of B, and a subset of B has the cardinality of A, then A and B have the same cardinality.[63] Poincaré quickly gave the proof of this theorem, not only for its own interest but to show that it invoked the principle of mathematical induction. Then he challenged Couturat, if he knew of a different proof to publish it at once, because it would be an important mathematical result, but if he did not to cease claiming that the theory of cardinal numbers was independent of the principle of mathematical induction. Poincaré was in no doubt of the outcome: "Alas," he wrote, "this is a theory that mathematicians are still discussing without coming close to understanding it."

Next Poincaré turned to geometry, and this discussion does appear in *S&M*. It would seem, he said (1906f, para. 28), from the most thorough, and thoroughly Hilbertian account yet published of elementary geometry, Halsted's *Rational Geometry* (1904), that intuition had been banished entirely from the subject, where only a few years earlier it had held undisputed sway. But no, the most fundamental theorem in the subject was that the axioms of a geometry are free of contradiction, and this Hilbert established by producing objects that satisfied those axioms, and

---

[63] See Poincaré (1906f, 27–29); Poincaré gave a fuller account of this theorem, to head off some criticisms, in his (1906e). For a history of this result, see Ferreirós (2007, 239).

these objects were defined using arithmetic. Therefore the principle of mathematical induction was required after all.

Poincaré now moved to his conclusion: "*The principle of induction cannot be regarded as the disguised definition of the natural number*" (his italics). He reviewed his arguments, and ended the essay by writing,

> To sum up, Mr. Russell and Mr. Hilbert have both made a great effort, and have both of them written a book full of views that are original, profound, and often very true. These two books furnish us with subject for much thought, and there is much that we can learn from them. Not a few of their results are substantial and destined to survive.
>
> But to say that they have definitely settled the controversy between Kant and Leibnitz and destroyed the Kantian theory of mathematics is evidently untrue. I do not know whether they actually imagined they had done it, but if they did they were mistaken. (*S&M* 191, 485)

Such a forceful article naturally drew replies. Hadamard (1906) agreed with Poincaré that Couturat's account of number was inadequate, not least because the definitions of "one" and "two" were circular, but he worried that more might be required of existence than freedom from contradiction, which at the very least should be rigorously established and not simply asserted. He found it self-contradictory of Couturat to oppose the idea, however, and even more worrying that logisticians could not escape the paradoxes while mathematicians, as Poincaré had showed, could identify where the error lay. Hadamard, therefore, put his trust in mathematics and not in a reductionist logical program. Poincaré, in his next (two-part) article, (1905c) and (1906f), considered some of his opponents. In a passage not repeated in *S&M* he referred to essays by Couturat and Pieri, noting that Pieri had done a much better job of saying the axioms were consistent if and only if objects existed that satisfied the axioms, before turning to a new approach offered by Russell, and a letter he had received from Zermelo.

Couturat continued to draw his scorn. Logistics, Couturat had claimed, had lent "stilts and wings" to discovery in the ten years since Peano had published the first edition of his *Formulario*. In (1906f, para. 2) Poincaré dismissed this as a childish illusion: "What! You have stilts and wings for ten years and haven't flown yet!" He had the greatest esteem for Signor Peano, he said, "but, after all, he has not gone any further, or higher, or faster than the majority of wingless mathematicians, and he could have done everything just as well on his feet." And in logistics Poincaré found nothing but shackles. Worse, he observed that logistics had resulted in self-contradiction, and that Russell was proposing to remedy this failure by changing the fundamental notion of class, thus abandoning Peanian logistics.

Couturat had tried to argue that existence in mathematics was not freedom from contradiction but the fact that a certain class is not empty. Insofar as this was not playing with words it meant that noncontradiction had to be established by producing suitable objects. But this, Couturat went on, could be an unreasonable demand, and so to insist on freedom from contradiction was arbitrary and improper; postulates should be assumed consistent unless and until they could be shown not to be. Poincaré barely bothered to refute this specious pleading. He was amused to see that Couturat no longer considered Hilbert to be a logicist (indeed, he was not) but of course this had no bearing on his criticisms of Hilbert's position.

It was more interesting to turn to Russell's attempt to get out of the Cantorian antinomies (today more weakly called paradoxes). Poincaré cited three: Burali-Forti's, the Zermelo–König antinomy (which concerns definability and will not be discussed here; see van Heijenoort 1967, 145), and Richard's. The Burali-Forti antinomy concerns the class of all ordinals: it would be an ordinal, and indeed the largest, but one could add 1 to it and obtain a larger ordinal. Richard's antinomy concerned all the decimal numbers definable in a finite number of words. This is a countable set, but a modified Cantor diagonal argument then produces a decimal number definable in finitely many words but not in the set. Russell had produced a variant of this: the smallest natural number that cannot be defined in less than one hundred English words.

Russell himself had offered no less than three hints at ways round antinomies of this sort, which Poincaré summarized. One called for a restriction to simple definitions, but gave no workable criteria of simplicity. The second called for a limitation on the size of any class worthy of being called a set. This was to prove a fruitful idea, but Poincaré dismissed it, validly at the time, as offering no workable criteria of size. The third called for abandoning the notion of class altogether, but had no clear alternative to put in its place. This left Poincaré with the task of escaping the antinomies in his own way, and he inclined to Richard's original suggestion that an object cannot be defined by a proposition that refers to the object itself. Such putative definitions Poincaré called impredicative, and rejected them on the grounds that they contained a vicious circle: in order to define an object they speak of it before it has been defined.[64] Such definitions, it seemed to Poincaré, included Cantor's definition of Aleph-one, the first uncountable ordinal, and so he said "I am therefore not sure that Aleph-one exists."[65]

Russell's latest theory of cardinal numbers was still incomplete, as its author recognized, so Poincaré turned to the most fully developed attempt by logicists to establish the principle of mathematical induction, Whitehead's (1902). This was a technical article, one Poincaré was not sure he fully understood, and the account in *S&M* was somewhat reduced from the one he originally gave, but the conclusion was the same: Whitehead's argument rested on a vicious circle.

Of course, Poincaré did not convince his opponents. One of Couturat's replies was translated by Jourdain and printed in the *Monist* in 1912, and a glimpse of the enthusiasm with which Jourdain introduced it says enough:

> It is, however, unfortunate that his airy remarks on modern logic... have been taken so seriously by so many....The views of a well-known football player on the science of anatomy would no doubt be widely read, and the views of M. Poincaré on the philosophical questions at the root of mathematics are not, in essentials, of a very different nature....M. Poincaré is one of our greatest mathematicians, and centuries have proved that a man who is a great mathematician need be neither a great philosopher nor a great logician. We do not expect such a combination of qualities, ... nor, as a rule, do we find them.[66] (pp. 481–482)

The most stimulating reply that Poincaré received came from Zermelo, whose letter seems to have been a summary of his deduction of the

---

[64] This is discussed in detail in Folina (1992, chap. 7).

[65] This remark is not in *S&M*.

[66] The very next essay was Halsted's translation of Poincaré's "The latest efforts of the logisticians" (1906f).

principle of mathematical induction from the much less obvious assumption that there are well-ordered sets (Zermelo 1904).[67] Zermelo had also produced an argument based on what he called the axiom of choice, which says that one can always choose at random an element from a set, and even from each set in an infinity of sets. This claim was the source of one of the most famous storms in mathematics. As Poincaré noted, some people, such as Émile Borel, denied it. Among those who did, it was sometimes discovered that they had earlier used it in their work without demur, Borel among them. Russell, who was neutral about the axiom of choice, pointed out that without it the theory of the multiplication of transfinite cardinals collapsed, and hoped that the theory could be restored one way or another. This struck Poincaré as hopeless; in his opinion there could be no self-evident starting point for any proof or refutation of the axiom of choice, because it concerned infinite sets. Zermelo, who was quite robust in his own defense, said of Poincaré's critique that "it has not met with any assent whatever" (Zermelo 1908a; 2010, 159).

Poincaré's conclusions remained firm: logic without intuition was sterile, and logistics was brought down by self-contradictions. More controversially, he argued that the Cantorians, as he called the adherents of transfinite set theory, had forgotten that there are no infinite sets. Hilbert's theory could escape contradictions because it was not based on the idea that infinite sets exist, but Russell's logic was so based, and therefore failed. This position of Poincaré's was the first time that he was to offer an opinion about mathematics that later generations have mostly been unable to share.

Later Thoughts on Mathematics and Logic

In 1905 and 1906 he had tried to say goodbye to his confrontation with logic, but he was trying to halt a rising tide, and the issue was to occupy him for the rest of his life. He was brought back to it in 1909 when he published an essay on the logic of the infinite, (1909b), which is part of an argument now not only with logicians but also with the new breed of set theorists. Much of this essay goes over old ground, notably the

---

[67] A set S is well ordered if it admits a strict order relation $<$; that is, for any two distinct elements a, b of the set either $a < b$ or $b < a$ and $a < b$ and $b < c$ implies $a < c$ and every subset A of the set has a least element, an α in A such that $\alpha < a$ for all a in A.

rejection of impredicative definitions, but now he had Russell (1908) to respond to. In that paper Russell argued that the familiar antinomies may be avoided by sorting the objects under discussion into layers or orders. In the lowest order are the elements of the class of elements satisfying some proposition that can be defined without reference to the proposition itself. To assert that the proposition is true of these elements is to assert a proposition of the first order. Elements of the second order are those whose definition could involve this first order proposition, and so on. Epimenides, the mendacious Cretan, is a liar of the first order if he always lies except when he says he is a liar of the first order, a liar of the second order when he lies even when he says he says he is a liar of the first order but not when he says he is a liar of the second order, and so on. Confronted with him and his claim that he is a liar, the way to proceed is to ask: "Of what order?" Only when he answers this question will his claim have any meaning.

Evidently, a theory of orders or types presupposes a theory of at least the finite ordinals and therefore the integers. That alone would make it seem unsuitable for the foundations, but a more unfortunate feature is that it is necessary to introduce these orders when defining, for example, the integers, and the result is that the sum of two integers of order K is an integer of order $K + 1$.[68] This is not, however, a matter of bookkeeping, but a grave inconvenience across mathematics, and recognizing this to be so Russell proposed an axiom of reducibility that Poincaré found unclear. It said, "We assume that every function is equivalent for all its values to some predicative function of the same argument." Whatever this may mean, it is unquestionably an axiom, as Poincaré remarked. Its purpose is to reassign the types of objects, but at this stage in Russell's development of the theory how it did so, and why it should be accepted, were indeed obscure.

The more lasting way forward had by this time been identified by Zermelo. Prodded by Hilbert, he had axiomatized the naive theory of sets (Zermelo 1908b); his imperfect axiomatization was to be refined by Abraham Fraenkel in 1922 into the Zermelo–Fraenkel axioms for set theory that are still used today. Zermelo proposed seven axioms, and showed how they could dispose of the known antinomies. Poincaré observed that since these axioms were intended to provide a foundation

---

[68] Not $K - 1$ as both the French and English editions of *DP* state.

for all of mathematics it would be impossible to prove them consistent and the only hope was that they could be made to seem self-evident. But since outright contradictions had appeared in the naive theory of sets it would be necessary to make it clear why certain objects were sets and others were not.

To highlight the problem, Poincaré translated every word of Zermelo's axioms for sets into French except his word for a set, which Poincaré left in German: "Menge." The first six axioms seemed perfectly acceptable to Poincaré when Menge was given its intuitive meaning and restricted to finite sets. The seventh asserted the existence of an infinite set, and Poincaré put it aside for the moment. For, as he remarked, the claim that "Every collection of objects of any kind is a Menge," leads to contradictions as soon as infinite collections are admitted. So, said Poincaré, it must be that Zermelo has distinguished between the six legitimate axioms and his seventh axiom on the one hand and the illegitimate remark on the other by attributing to the word "Menge" an intuitive meaning, albeit not exactly the usual one. Poincaré traced this to its source in a deceptively simple definition given by Zermelo (which he left in German in his essay): "A question or assertion *E* is said to be '*definite*' if the fundamental relations of the domain, by means of the axioms and the universally valid laws of logic, determine without arbitrariness whether it holds or not."[69] This was exactly right—the concept of definiteness in Zermelo's paper was to occasion a great deal of discussion.

Poincaré explored the distinction between being predicative and being definite, and came back to the problems he had with infinite sets of any kind. In his view, an infinite set could never be completely surveyed, and so new elements were always arising. Zermelo, he suggested seemed to feel that even an infinite Menge was completely known. This unease led Poincaré to one of his most quoted positions (*DP* 31, 63):

1. Never consider any objects but those capable of being defined in a finite number of words.
2. Never lose sight of the fact that every proposition concerning infinity must be the translation, the precise statement of propositions concerning the finite.
3. Avoid non-predicative classifications and definitions.

[69] This translation is in Zermelo (2010, 193).

Zermelo replied, in his (1909a, b), with a derivation of the principle of mathematical induction from his axioms for set theory, but he did not say whether that made the principle analytic, which he regarded as a purely philosophical question about the axioms. But he did claim at the end of the first article that Poincaré, in his (1909e), which was printed immediately after Zermelo's article in *Acta Mathematica*, now agreed with him about predicative definitions.[70]

On 3 May 1912 Poincaré gave a lecture in London entitled "La logique de l'infini" (1912c) in which he tried to explain why he and his opponents had not been able to reach an agreement about the legitimacy of certain ways of reasoning about the infinite. The problem, as he saw it, was our essentially finite nature, and he asked his listeners to consider two schools, which he called the pragmatists and the Cantorians. The former, to which he belonged, will only speak of objects that can be defined in a finite number of words, "the others, on the other hand, think that objects exist in a sort of large store, independently of any mankind or any divinity that could talk or think about them . . . And from this initial misunderstanding all sorts of divergences in details result":

> Let us take for example Zermelo's theorem according to which space is capable of being transformed into a well-ordered set.[71] The Cantorians will be charmed by the rigor, real or apparent, of the proof. The pragmatists will answer:
>
> "You say that you can transform space into a well-ordered set. Well! Transform it!"
>
> "It would take too long."
>
> "Then at least show us that someone with enough time and patience could execute the transformation."
>
> "No, we cannot, because the number of operations to be performed is infinite; it is even greater than Aleph-zero."
>
> "Can you indicate how the law which would permit space to be well-ordered can be expressed in a finite number of words?"
>
> "No."
>
> And the pragmatists conclude that the theorem is devoid of meaning, or false, or at least not proved. (1912c, 87, 67)

[70] Parsons, in his comment on this paper in Zermelo (2010, 234) finds Zermelo's argument fallacious.

[71] Published in Zermelo (1904; 1908a).

The pragmatists, he explained, because they admit only objects that can be defined in a finite number of words, consider that only one infinite cardinal is possible, the smallest, Aleph-zero, and doubt the existence of Aleph-one. So they do not accept a definition that purports to define a whole set of objects if they cannot also examine the objects in the set, a restriction Cantorians find artificial and even meaningless. This was an issue, said Poincaré, when indirect definition or definition by postulates is concerned, specifically when one defines an object to be related in such and such a way to all the objects of a given set and also to belong to that set (to give an example, the largest member of a set of numbers).[72] The axiom of choice, to take a contentious example, spoke of a collection with one member from each of an infinite family of sets, and asserted that this collection was a set. For Poincaré this was impredicative and to be rejected, but Cantorians disagreed because they did not have to see members of a set individually in order to know the set.

One should perhaps read this debate as one between Poincaré and Hilbert, or between Poincaré and the Hilbertians, of whom, of course, Zermelo was the most forceful proponent, while Hilbert himself had spoken positively of Cantor's theory of the transfinite when placing problems in that subject at the head of the list of problems in his celebrated Paris address of 1900. Behind Poincaré stood the long French tradition of mathematics as a process of discovery; behind Hilbert more and more the forces of abstract mathematical creation. In this spirit, Poincaré bridled at the idea that such postulates can bring mathematical objects into existence. It seemed to him that logic had become creative merely by giving out names.

For Poincaré pragmatists were idealists, who believed that a mathematical object exists only when it is conceived by the mind, and Cantorians were realists who believed in the existence of objects whether known or not. To Poincaré this seemed an untenable position, for while "there is one reality to be known and it is exterior to us, all that we can know of it depends on us, and is then merely a becoming, a kind of stratification of successive conquests. The rest is real but eternally unknowable." He

---

[72] Poincaré's objection is clumsily put; most mathematicians would want to prove the existence of such an object, and still more its membership of the given set. But Hilbert had spoken of defining the real numbers by a system of axioms capable of no further extension.

must have known that the Cantorians in Göttingen were not to be persuaded.[73]

Much later, in the 1920s, Hilbert was to develop a program for talking about infinite sets in finite terms.[74] Very briefly, Hilbert distinguished between mathematics and metamathematics. Formal mathematics is nothing but finite strings of unambiguous marks on paper, which can be manipulated according to certain specified axioms. These axioms, and discussions about content—what the axioms "mean"—and why the axioms should be adopted form metamathematics. So there can be axioms about infinite sets, but everything that is said about these sets is said in finitely many statements each of finite length and therefore, in particular, checkable. Indeed, Hilbert proposed to make the analysis of proofs themselves into a new branch of mathematics—proof theory. On the basis of this distinction he argued in Hilbert (1927) that Poincaré was wrong to deny the possibility of a consistency proof for arithmetic. There was, said Hilbert, an intuitive, contentual induction and formal induction proper "which is based on the induction axiom and through which alone the mathematical variable can begin to play its role in the formal system. Poincaré arrives at his mistaken conviction by not distinguishing between these methods of induction, which are of entirely different kinds." Here Hilbert overstated the merits of his position, as Mancosu, following earlier criticisms of Hilbert argues (Mancosu 1998, 167). All Hilbert can claim is instances of sentences of the form $P(n)$ for each integer n, not the universal claim that "For all n, $P(n)$."

POINCARÉ'S DEFENSES OF SCIENCE

Poincaré had ended his rebuttal of Le Roy's arguments in *VS* (pp. 276, 357) with a peroration that provides an insight into the passions of this

---

[73] Several authors have found Poincaré's position here to foreshadow Brouwer's intuitionism, including Heyting, while Brouwer found that Poincaré had confused construction in mathematics with the role of language—thus misunderstanding Poincaré's position entirely, as Heinzmann and Nabonnand (2008) argue. Folina (1992) also agrees that Poincaré's and Brouwer's positions are very different, because she finds a strongly Kantian direction in Poincaré's thought.

[74] See Hilbert (1925) and, on his finitism or formalism as it is variously called, Mancosu (1998), which also contains translations of a number of valuable sources.

private man:

> It is only through science and art that civilisations are of value. Some have wondered at the formula: science for its own sake; and yet it is as good as life for its own sake, if life is only misery; and even as happiness for its own sake, if we do not believe that all pleasures are of the same quality, if we do not wish to admit that the goal of civilisation is to furnish alcohol to people who love to drink.
>
> Every act should have an purpose. We must suffer, we must work, we must pay for our place in the spectacle, but this is for seeing's sake; or at the very least that others may one day see. All that is not thought is pure nothingness; since we can think only thought and all the words we use to speak of things can express only thoughts; to say there is something other than thought, is therefore an affirmation which can have no meaning.
>
> And yet – strange contradiction for those who believe in time – geological history shows us that life is only a short episode between two eternities of death, and that, even in this episode, conscious thought has lasted and will last only a moment. Thought is only a gleam in the midst of a long night.
>
> But it is this gleam that is everything.

There he had denied, partly out of politeness one suspects, that his target was Le Roy, but it may have been Tolstoy. The immediately preceding paragraph runs:

> We cannot know all the facts and it is necessary to choose those which are worthy of being known. According to Tolstoy, scientists make this choice at random, instead of making it, which would be reasonable, with a view to practical applications. On the contrary, scientists think that certain facts are more interesting than others, because they complete an unfinished harmony, or because they make one foresee a great number of other facts. If they are wrong, if this hierarchy of facts that they implicitly postulate is only an idle illusion, there could be no science for its own sake, and consequently there could be no science. As for me, I believe they are right, and, for example, I have shown above what is the high value of astronomical facts, not because they are capable of practical applications, but because they are the most instructive of all.

The question of what to study occupied Poincaré for some time. He spoke on the choice of facts at a meeting of the University of Paris in 1906. The essay (1906d) begins with a reference to Tolstoy's *On the Significance of Science and Art*, first published in French in 1887. The science Tolstoy was inveighing against in his essay was a Comptean positivism as applied to society at large. He objected both to the claims it made and to its insinuation that, unlike previous philosophies it was "scientific" and therefore beyond reproach. He began by objecting to its claim that it dealt only in facts:

> Contemporary science is also occupied with facts alone: it investigates facts. But what facts? Why precisely these facts, and no others?
>
> The men of contemporary science are very fond of saying, triumphantly and confidently, "We investigate only facts," imagining that these words contain some meaning. It is impossible to investigate facts alone, because the facts which are subject to our investigation are innumerable (in the definite sense of that word), – innumerable. Before we proceed to investigate facts, we must have a theory on the foundation of which these or those facts can be inquired into, i.e., selected from the incalculable quantity.
>
> And this theory exists, and is even very definitely expressed, although many of the workers in contemporary science do not know it, or often pretend that they do not know it. Exactly thus has it always been with all prevailing and guiding doctrines. The foundations of every doctrine are always stated in a theory, and the so-called learned men merely invent further deductions from the foundations once stated. Thus contemporary science is selecting its facts on the foundation of a very definite theory, which it sometimes knows, sometimes refuses to know, and sometimes really does not know; but the theory exists. (p. 4)

Tolstoy continued by saying that it would be better to be guided by our practical and above all our moral needs. Poincaré began by observing that Tolstoy's sense of what is useful was plainly not that of the businessman, still less that of most of their contemporaries. For himself, he said, he was happy with neither ideal, neither that of a greedy and narrow plutocracy, nor that of a virtuous and mediocre democracy led by people without curiosity and who would die of boredom. That being a matter of taste he chose to say no more about it, but to return to the question of what facts to choose. Could it be that one's choice was nothing but a personal whim?

Certainly the triumphs of science that enrich practical men could not have been discovered without the prior efforts of disinterested fools who died poor. And because most people prefer not to think but to let their instincts guide them, it is necessary for those who do like to think to make their thoughts as useful as possible, which is why the most useful laws of science are the most general. There are simple facts, or at least those that appear simple, and simple ideas that seem to allow us to make new discoveries. These facts can be checked regularly, and once their conformity to a simple rule is established they cease to have any interest, and it is the exceptions that become interesting. This, Poincaré noted, is the practice in astronomy or geology: by ranging over great periods of space and time astronomers and geologists seek to find when well-established rules are overturned, the better to understand the much smaller changes happening around us today. And so, Poincaré concluded, the wise man does not choose what facts to observe at random.

There are, he then went on, other aspects to the question. A wise man does not study nature because it is useful, but because studying brings pleasure, and it does so because nature is beautiful. If it were not it would be difficult to study. But this beauty is not the beauty that strikes our eyes, he said, not that there is not such beauty, quite the contrary, but the beauty that only pure intelligence can know. Intellectual beauty suffices in itself and for itself, and perhaps it is for the future of humanity that the wise man condemns himself to long and painful work. Yet, he suggested this sense of intellectual harmony that science provides may not be all that different from a sense of utility. The economy of thought that Mach had praised[75] is both a source of beauty and practical advantage.

Poincaré ended this speech, which had acquired more and more the character of those uplifting utterances that these occasions seem to require, with some speculations on this happy coincidence of the beautiful and the useful. Was it that beautiful things were those adapted to our intelligence and consequently useful. Had evolution bred it into us, by favoring those sensitive to both over those who were not (Greeks over

---

[75] Poincaré's most likely source was Mach (1883).

barbarians)? Poincaré recognized that Tolstoy would reject such a thought with horror, but offered the thought that the disinterested search for truth has its own beauty, and can even make men better. To be sure, he admitted, there had been scientists of a very bad character. Should one then, he concluded, abandon science and study only morals? "And indeed, does one think that the moralists are irreproachable when they come down from their pedestals?"

Tolstoy, as it happened, had already replied to Poincaré's vision of the disinterested scientist:

> I am aware, that, according to its own definition, science ought to be useless, i.e., science for the sake of science; but surely this is an obvious evasion. The province of science is to serve the people.

He then went on to muddle the point, however, adding,

> We have invented telegraphs, telephones, phonographs; but what advances have we effected in the life, in the labor, of the people?

In March 1910 Poincaré gave a lecture at a conference organized by the Catholic journal *Foi et vie* on the topic of "Questions of the present day," later published in their journal as (1910d). He began by contrasting two popular views. The first claimed that science, having constructed mathematics and physics, has now embarked on biology and will tomorrow construct ethics, and we shall all be the wiser. The second claimed that science fouled all it touched, would deprive the universe of its essential mystery and rob the Creator of prestige. Poincaré dismissed both of them. Science described, ethics prescribed, and Poincaré at least, could not derive "ought" from "is." Ethics had to be derived from what we feel: patriotism can hardly be justified by rational arguments, he said, but if France is invaded "our heart will rise up, our eyes will fill with tears, and we shall no longer hear anything."

In particular, science can engender feelings, and

> science keeps us in constant relation with something which is greater than ourselves; it offers us a spectacle which is constantly renewing itself and growing always more vast. Behind the great vision it affords us, it leads us to guess at something greater still; this spectacle is a joy to us, but it is a joy in which we forget ourselves and thus it is morally sound. (1910d, 324)

The scientifically trained mind, he argued, has "a horror of all attempts to deflect the course of experiment [and dreads] as the height of ignominy the censure of having, even innocently, slightly tampered with our results." Is this not, he went on, perhaps with an uncomfortable recollection of his own experiences—as the next chapter indicates—

> the best means of acquiring the rarest, the most difficult of all sincerities, the one which consists in not deceiving oneself?
>
> In our shortcomings, the loftiness of our ideal will sustain us. We may prefer another, but, after all, is not the God of the scientist the greater the farther he withdraws from us? It is true that He is inflexible, and many souls shall be sorry for it; but at least He does not share our pettiness and mean rancor as does too often the God of the theologians. (1910d, 325)

Science cannot only create new feelings, he went on, it can create an edifice based on those feelings by bringing those feelings into harmony and resolving conflicts. But again, he noted, those who fear science fear that it will drive out the good that is concealed in the best of our old traditions. He replied that they have only to fear a hasty and incomplete science. The only point at which Poincaré hesitated, and where he said nothing because he said he had nothing to say, was the vexed issue of determinism. Science, he noted, was deterministic through and through, and this seemed very difficult to reconcile with any system of ethics. So he contented himself with these conclusions:

> There is not now and there will never be scientific ethics in the strict sense of the word; but science can be an aid to ethics in an indirect manner. Science widely understood cannot help but serve it; pseudo-science alone is to be feared. On the other hand, science cannot suffice, because it can see only one part of man, or, if you prefer, it sees everything but it sees everything from the same angle; and secondly, because it is necessary to consider the minds that are not scientific. On the other hand, the fears, like the hopes which are too lofty, seem to me equally chimerical; ethics and science, as they both progress, will surely adapt themselves to each other. (1910d, 329)

Poincaré returned to this theme when he spoke at another conference organized by *Foi et vie* in March 1912, this one being on "The present state of materialism in France." In his lecture, (1912d), he began by saying that he was not sure what was meant by the words "materialist" and "determinist" when applied to the pursuit of science. Certainly most scientists, whatever their philosophical leanings, had a weakness for mechanistic explanations, especially when it came to the "great fortress of mechanism": the kinetic theory of gases. But what was a gas? Poincaré took the opportunity to run through some of the recent advances of science: Perrin's account of Brownian motion, ideas about electrons and their possibly fundamental nature, and most recently Planck's ideas about quanta, which might intervene decisively in the age-old struggle between the atomists and the partisans of continuity. All of this seems to have suggested to Poincaré that the advances of science, while being a victory for humanity, had nothing to say about materialism and determinism in the context of religion.

Quantum Theory

From 30 October to 3 November 1911 Poincaré was in Brussels as one of twenty-one leading physicists at an international conference that had been funded by the Belgian industrial chemist Ernest Solvay to discuss the new quantum ideas in physics. The topic was completely new to him, and he seems to have thrown himself into the task of trying to understand it with enthusiasm, in this way setting to rest the fears of Max Planck, the architect of the quantum approach, who had worried that perhaps only four people had read his papers enough to care about them.[76] Later Planck was to write warmly of Poincaré's response to the new ideas (Planck 1915/1921): "An old man," he said, "will be inclined to ignore the hypothesis, the enthusiast will welcome it uncritically, the skeptic will seek grounds to deny it, the productive man will test it and fructify it. Poincaré, in the profound paper which he dedicated to the quantum theory, proved himself youthful, critical, and productive." The other participants seem to have been equally pleased. At the meeting Poincaré pressed home the questions he had that he was later to write about: the behavior of the quantized resonators, and mechanisms for their interactions. The

---

[76] Famously, Poincaré and Einstein failed to appreciate each other's point of view of relativity theory at this meeting; Poincaré's enthusiasm for the topic of quanta may be part of the explanation.

subject must have fascinated him, for within a month of his return to Paris he had published a six-page paper on the subject (1911d) in which he not only sketched a derivation of Planck's radiation law from the hypothesis that radiation is emitted as quanta, but also a proof of the converse, that the law required a discontinuous hypothesis. He reported more broadly on what had been said at the Solvay conference in a lecture he gave on 11 April 1912 at the Easter meeting of the Société de physique de Paris, published as (1912e). He discussed the topic of magnetons or atoms of magnetism, regarded as a vortex of electrons. He noted recent experimental endorsement of the kinetic theory of gases, Perrin's work on the number of atoms, which led him to proclaim (in keeping with the idealist strain in his philosophy) that "the atom of the chemist is now a reality." The atom, he reported, is complex, "in it...we find electrons (negative) gravitat[ing] around large positive ones." He briefly mentioned Einstein on the photoelectric effect, and claimed to work in the reverse direction. Curiously, there was no mention of the ether, despite the title of the lecture.

In London on 11 May 1912 Poincaré again lectured on the hypothesis of quanta.[77] He reported that at the Solvay conference everyone there had talked at every moment about the new mechanics. This was even newer than Lorentz's mechanics, which, hardly five years ago had seemed the height of boldness, but, he hastened to reassure his audience, this was not in peril. Rather, it was a new stroke of boldness that was proposed, one that would cause the most profound revolution in natural philosophy since the time of Newton and bring discontinuities into natural laws.[78] It arose from ongoing difficulties with the theory of thermodynamics. The distribution of energy in a gas in equilibrium, as determined by the laws of Hamiltonian dynamics, and Rayleigh's law in the theory of a heated body radiating perfectly (a so-called blackbody) do not accord with experiment, and instead Planck had proposed a new law which said that radiation was much less at lower temperatures than Rayleigh's law demanded. Poincaré then indicated that he had been able not only to deduce Planck's radiation law by following Planck's analysis, but to show, as Planck had not, that the new law was necessarily that of a discontinuous process. In his (1912g) he wrote, "A physical system is capable of only

---

[77] The lecture repeated a paper published in February: (1912g).
[78] A view consistent with Poincaré's scant appreciation of Einstein's theory of special relativity.

a finite number of distinct states; it jumps from one of these states to another without going through a continuous series of intermediate states." This discontinuous behavior should apply to any isolated system, he went on, even the universe itself, so there would be a discontinuous variation in time, *the atom of time* (*DP* 125, 86; italics Poincaré's). He ended his lecture by expressing an open mind about what the resolution of these questions would prove to be.

The impact of Poincaré's work on the quantum theory has been analyzed by McCormmach. In his (1967, 52) he concludes that "The enthusiastic conversion of a prominent non-German scientist, Poincaré, to a belief in quanta may in itself have served to direct attention to the new theory. In any case there can be no doubt that his paper was an important factor in the rapid growth of interest in the quantum theory after 1911." In France and in England a continuous stream of publications on the subject starts only after Poincaré had published, and as soon as 1913 his memoir was referred to as "often cited." The physicist Paul Langevin, who had taken Poincaré's courses on Hertzian waves and elasticity when a student at the Sorbonne, wrote after Poincaré's death that the latter's final memoir marked the end of an epoch of three centuries in which natural laws and the language of the infinitesimal calculus were believed to be inseparable. Poincaré, he said in Langevin (1913, 679), had shown with "marvelous clarity the acute conflict" between the older statistical and electromagnetic theories and the new truth that electromagnetic phenomena on the atomic scale could not be the new laws of the atomic world would not be expressed in the mathematics of the continuum; and he believed, with Poincaré, that the discovery of the exact nature of these laws would be the next great task of science.

2

Poincaré's Career

CHILDHOOD, SCHOOLING

Jules Henri Poincaré was born in Nancy on 29 April 1854. His father, Léon, a professor of medicine at the University of Nancy, was then 26 and Henri's mother Eugénie was 24. Henri's sister Aline was born two years later (his cousin Raymond was born in 1860). In 1909 Dr. Toulouse reported that Henri resembled his mother and his maternal grandmother more than his father, and that his maternal grandmother was thought to have a real, but undeveloped gift for mathematics.

The children had a happy childhood. Based on Poincaré's recollections Toulouse recorded that Henri spoke at nine months, and it seems that the only misfortune was a grave bout of diphtheria that afflicted Henri when he was five and which left him with a paralysis of the larynx and an inability to walk. He recovered the use of his legs after two months, but speech did not return for a further six or seven and he communicated through a system of signs that he may have devised himself (Bellivier 1956, 31). His earliest education outside the home was at a family friend's, and he went to school for the first time in October 1862, where he prospered in all subjects. In due course he proceeded to the lycée, and it was there in the 4th class[1] that his talent for mathematics was first recognized (Bellivier 1956, 78). His teacher called at the Poincaré family house to say that Henri would be a mathematician. His mother was not apparently sufficiently surprised or impressed, in part because she liked mathematics, and the teacher went on "I wish to say a great mathematician."

The family liked to travel, and this seems to have encouraged Poincaré's ability with languages. He learned Latin and Greek, then a

---

[1] French classes were numbered in descending order with age.

standard part of the syllabus in the academic schools, well enough to be able to read them in later life, German, and later English, which he was able to practice on a family holiday in Norway, where the Norwegians he met helped him by speaking English slowly and simply (Toulouse 1909, 25).

In 1870 war came to Nancy. The Prussian army invaded and annexed most of Alsace and Lorraine. Metz was taken over and made into a prominent German garrison town, greatly strengthening the western border of the new German state. Nancy was returned to France, but Strasbourg remained German, a matter of sustained national sorrow in France. It was at this time that Poincaré decided to improve his German, because all sources of news in Nancy were written in German.

Poincaré also began to study the two years of *mathématiques speciales* necessary to sit the entrance examinations for the École polytechnique and the École normale supérieure. The course emphasized mathematics and philosophy (which included rhetoric). He passed his baccalauréat ès lettres in August 1871, but when it came to sit the baccalauréat ès sciences in November, Poincaré scored zero on one of his scientific papers. It seems that he arrived late, misheard the question and did not answer the question set. This failure could have prevented his entry to the elite class, but Rollet has argued (Rollet 2010, 64) that Poincaré's reputation had already reached the Faculté des sciences in Nancy, which administered the examinations, and he was allowed to enter the course of elementary mathematics rather than special. There he read Rouché's *La géométrie*, Duhamel's *L'analyse*, and Chasles' *La géométrie supérieure*. Duhamel's book was unusual in advocating that the order in which ideas should be presented was almost always that in which they had been discovered, on the grounds that important ideas are not discovered by chance but in response to what has already been done. Chasles' book also impressed Poincaré. He later called Chasles "the Emperor of geometry" (Bellivier 1956, 101).

At the end of the school year in 1872 Poincaré sat the *Concours général* and graduated second in his class, brought down by poor marks in design despite his high marks in written mathematics and most subjects in the oral part of the examination. He continued to *mathématiques speciales*, where he was taught mathematics by Victor Elliott, a well-regarded professor who was to successfully defend a thesis at the École normale supérieure in 1876 on Abelian functions, and physics by Lefebvre, another but older normalien. Poincaré enjoyed physics, but not chemistry (Appell 1956, 143) perhaps, as Appell speculated, because the teacher was obviously bored. It is from this period that the stories people liked to tell about Poincaré's genius begin. Poincaré did not take notes, and did not give the impression of being a serious student. But when an older pupil asked him to explain a particularly obscure point in the course, Poincaré replied immediately and without pausing for thought, leaving his interlocutor in a state of shock (Bellivier 1956, 109).

One story is indicative of the way Poincaré's mind was always to work (Bellivier 1956, 109). The class was asked to find the points from which a given ellipse subtends a given angle. Poincaré replied at once, "The tangent of the angle will be a ratio. The numerator will contain the first member of the equation for the ellipse, the denominator the first member of the equation for a circle, the position of the vertices of the circumscribed right angles. It only remains to find what exponents and what constant factors enter these polynomials." This makes it clear that Poincaré knew immediately what form the answer would take. It would involve a particular expression for the tangent of the angle and some data determined by the ellipse and the size of the angle. So that was what had now to be calculated. The ability to know what one has to calculate is highly valuable in mathematics. It has also to be accompanied by the technical knowledge and the skill to make that calculation correctly, but knowing the form of the answer in advance is a great help. Poincaré was to demonstrate that ability time and again, often leaving the elaboration of the details to others.

Appell and Poincaré were the only students from Nancy to sit the final exam. Poincaré returned home concerned that he had not finished his final essay, but he need not have worried; a month later Elliott came round to tell the family the result: Poincaré had come top, with Appell second. Rollier, the inspector general who had marked the papers, called Poincaré an extraordinary pupil who will go far. The other examiners, Briot, Bouquet, and Victor Puiseux, observed in their report that Poincaré was distinguished by his knowledge and his aptitude.

The next step was to sit the entrance examinations for the École polytechnique and the École normale supérieure. Elliott was worried that Poincaré's answers would be too cryptic, and indeed there were unexpected problems. Poincaré seems to have failed to satisfy one of the

Figure 2.1. Paul Appell (1855–1930). Source: Ernest Lebon, *Paul Appell: Biographie, bibliographie analytique des écrits* (Gauthier-Villars, 1910).

examiners at the École normale and was ranked only fifth. Appell, the source of this story (Appell 1956, 141) also recalls that the examination asked candidates to produce a working drawing:

> Poincaré, who saw no mathematical interest in drawing lines from memory and making a precise design that bored him, preferred, having put all the data in place, to find the equation of the projection of the curve of intersection by calculation. He therefore found this curve with a degree of perfection that the classical constructions did not attain. But in drawing it on paper he was distracted and reversed it, turning it through 180°. The examiner was intrigued by this solution, which was at once both inexact and perfect.

Things went better at the École polytechnique, which Rollet suggests (2010, 64) was the only place Poincaré intended to go. Appell and Poincaré sat the written part of the examination back home in Nancy, where every window was decked with flags to mark the imminent departure of the German occupiers. Poincaré was excited by the return of the French Army and made a mess of his watercolor, which, Appell wrote, he usually did well, being in a hurry to join his family at the Hôtel de Ville and watch the French troops arrive in Place Stanislas. This was offset by a triumphant oral exam under Tissot, but the miserable watercolor let him down and he scored zero for it. The problem for the École polytechnique was that no one ever before had been admitted who had failed any part of the examination outright, yet his performance was otherwise beyond comparison with the other candidates. To resolve the problem the watercolor was remarked to a score of 1 and Poincaré entered the École polytechnique ranked first. Appell went to the École normale supérieure in the same year.

THE ÉCOLE POLYTECHNIQUE

Once installed in the École polytechnique, Poincaré applied himself to the courses with a significant mathematical content, such as Mallard's course on crystallography. His professors at Polytechnique included Hermite (analysis), Résal (mechanics), Mannheim (geometry), Faye (astronomy), Cornu (physics), Fremy (chemistry), and Zeller (history and literature).[2]

His first publication was his (1874), which established an amusing fact about surfaces that had not been noticed before. Pick a point P on a surface, and draw the perpendicular to the surface at that point. Any plane containing that perpendicular cuts the surface in a curve, and there is a circle that best approximates the curve at P, and this circle has a radius. Do this for every plane containing the perpendicular at P and you obtain a family of curves. Travel out on each of those curves a distance proportional to the square of the corresponding radius of curvature, and what Poincaré proved is that the points you obtain in this fashion lie on a conic (and therefore, in particular, lie in a plane). At a time when the study of surfaces was developing fast, this property of approximations to a surface at a point was worth publishing, but no more than that.[3]

[2] See Bellivier (1956, 127, no. 1).
[3] It belongs with properties of what is called Dupin's indicatrix, which is nicely explained in Coxeter (1961, 364) where, however, this result is attributed to de la Vallée Poussin in 1925, which does at least suggest that it was worth rediscovering.

THE ÉCOLE DES MINES

Poincaré graduated second from the École polytechnique in 1875, just behind Marcel Bonnefoy because once again his marks in drawing (stereotomy) let him down, and with Bonnefoy and Jules Petitdidier, who came third, went to the most prestigious of the specialist engineering schools that the École polytechnique was training students for, the École des mines. However, once at Mines he paid less attention to the more applied courses, if the many doodles in his notebooks from that period are anything to go by. He spent 102 days in Norway and Sweden with Bonnefoy on an educational trip organized by the École des mines, where he also looked for traces of Abel (Bellivier 1956, 188), and graduated in autumn 1878, apparently without a clear idea of what he wanted to do next. He began to study Hermite's work in number theory in some depth, and to think about the theory of differential equations. One might imagine that Poincaré, who was so easily bored and prone to drift off into the world of his own thoughts, would be ill suited to practical life, but he seemed willing to become a mining inspector. There was, however, some uncertainty and concern about where he might go.

One possibility was Clermont-Ferrand, and another was Vesoul, but there was a worrying third possibility, Bône in northern Algeria. Provisionally Bonnefoy was penciled in for Clermont-Ferrand, but he wrote to the ministry to say that if he had the choice he would prefer Vesoul, where, however, an earlier graduate of Mines, Jules Roche, was the likely choice. The prospect of Henri going to Algeria animated his father, who prevailed upon old friends with an influence on the Ministry of Public Works to write to the minister informing him that Poincaré's uncertain health, and his wish to be examined for his doctoral thesis, made an appointment there ill advised. Some days later the minister received a document from the doctor assigned to Mines and Ponts et Chaussées to say the Poincaré suffered from a delicate constitution and had heart trouble, so a long stay in Algeria could have serious consequences for his health.[4] On 1 April 1879 Poincaré was assigned to Vesoul, Bonnefoy to Clermont-Ferrand, and Petitdidier to Angers. Roche went to Nice before joining the trans-Sahara expedition of Colonel Paul Flatters. Of the four,

---

[4] Rollet (2010, 70) notes that this was the first time anyone had raised any doubts about Poincaré's health since his grave attack of diphtheria.

only Poincaré was lucky: Among his contemporaries Bonnefoy was to die in an explosion in a mine in Champagne on 28 May 1881 and Petitdidier, who had poor health and chronic bronchitis, died of his illness on 29 April 1884. Roche also died young, killed by Touaregs in February 1881 on Colonel Flatters's second expedition.

Once arrived in Vesoul, Poincaré was conscientious about his responsibilities. He controlled both mining and railway work, and made several visits. But even so he made time to begin a novel, and to finish his doctoral thesis. His superior, Louis Trautmann, reported that he was satisfactory, although he did not work with sufficient zeal, and that his hopes clearly ran to university life and not toward service (Rollet 2010, 71). That said, on 1 September 1879 Poincaré, was called to investigate the cause of an explosion in a mine at Magny where 16 people had died; rescue efforts lasted for three days. Poincaré considered several ways in which the accident might have occurred before deciding that the cause had been the accidental deterioration of a miner's lamp as the result of being struck by a pick. In his report on 30 November 1879 he listed the human cost as 16 dead, 9 widows and 35 orphans of whom the oldest was seventeen and of whom only 6 were older than twelve.

The same day, 1 December 1879, that the chief engineer wrote accepting every recommendation in Poincaré's report on the disaster at Magny, Poincaré heard that he had been appointed Chargé du cours d'analyse mathématique at the University of Caen. But Poincaré never left the Corps des mines, rising through its ranks to become an inspector general (second class) in 1910. Indeed, for a time he apparently considered pursuing two careers, as an engineer and a mathematician, but mathematics was to prove stronger. He had already been awarded the degree of docteur ès sciences mathématiques from the Sorbonne on 1 August 1879, for a thesis on a delicate problem in the theory of partial differential equations. It was nominally concerned with extending a theorem of Sonya Kovalevskaya's on partial differential equations in the complex domain to understand the nature of the solutions near a singular point of the equation, along the lines of his study of ordinary differential equations in his (1878) (see chap. 7, sec. "Differential Equations" below). He was successful when the singular points belonged only to particular solutions. In this case he showed that the solutions satisfied algebraic equations whose coefficients were holomorphic functions of the variables. But when the singularities were intrinsic to the equation he professed himself defeated. Even in the corresponding case for ordinary differential equations there were difficulties that were to occupy him some years later, in his (1885g) and subsequent papers. His thesis also contained important results on lacunary series and a mild generalization of algebraic functions, which were later to play an important part in his study of complex functions of several variables. The thesis jury was composed of Ossian Bonnet, the director of studies at the École polytechnique, Bouquet, and Darboux from the École normale supérieure. Darboux saw a version in late 1878 and wrote to Poincaré on 4 December, "I continue to believe you have written a good thesis, but it seems to me indispensable that you re-edit it completely and correct all the errors of calculation and changes of notation that make it almost unreadable." Perhaps something was done, but he later wrote of Poincaré's thesis that "it certainly contained enough results to furnish the material for several good theses. But, if one wanted to give a precise idea of the manner in which Poincaré worked, many points in it still needed corrections or explanations" (Poincaré, *Oeuvres* 2, xxi). This Poincaré was never to supply, perhaps because he was settling into his demanding job at Vesoul, perhaps because he was already thinking of other novel approaches to problems about differential equations; his thesis was published for the first time in the first volume of his *Oeuvres* (1928). At all events, 1880 was to be the year Poincaré first did truly original work, and before 1881 was over he had been brought back to Paris as a mathematician.

ACADEMIC LIFE

The move to the University of Caen marks his decisive turn toward his life's work. Almost at once he began to publish. More precisely, he deposited *plis cachétés* on three subjects: the theory of forms and arithmetic, the curves defined by real first-order differential equations, and the theory of linear differential equations in the complex domain. The first of these reveals his interest in the work of Hermite. The second, however, is completely novel, and was to grow into the much more complicated story of celestial mechanics. The third reflects an indirect influence of Hermite, through the mechanism of the prize competitions of the Académie des sciences, that directed Poincaré to the work of the German mathematician Lazarus Fuchs; Poincaré's discoveries of the automorphic functions that he was to call Fuchsian and Kleinian were the ones that made his name among the international community of mathematicians. They will be discussed in chapter 3.

In 1881 he married Mlle Louise Poulain d'Andecy, a relative of Geoffroy Saint-Hilaire (a French naturalist and colleague of Lamarck). The prospect seems to have excited him, for on the final page of a letter from Halphen of 23 February 1881 we can read in Poincaré's hand,

> Letter from M. d'Andecy received Thursday morning. My dear Henri, I received this morning the request for the hand of my daughter that your father put to me in your name. I replied to him with alacrity—and to you at the same time—that we are happy and proud to see you enter our family. My daughter accepts with complete confidence and with unsullied satisfaction the future that you offer her, and I believe that on your side you appreciate more and more the serious qualities of the one you have chosen to accompany.

On Saturday therefore now to receive the paternal accolade.

In October he was appointed a Maître des conférences d'analyse in the Faculté des sciences at the University of Paris. He was proposed for the geometry section of the Académie des sciences, with the strong support of Hermite, and ranked fifth in the election, and he became a member of the Association française pour l'avancement des sciences. His successes with the theory of automorphic functions had come to the attention of Mittag-Leffler, who was in the final stages of creating a new Scandinavian-based journal for mathematics with international ambitions, and his publications in *Acta Mathematica* secured the reputation of the journal from the start, as Mittag-Leffler was always happy to say.[5] By now Hermite had written to Mittag-Leffler, that although "in great fear of being overheard by Madame Hermite" he considered Poincaré to be the most brilliant of his three mathematical stars—the others were Appell and Picard—"and besides, he is a charming young man, who comes from Lorraine, like me, and knows my family very well."[6] In May

---

[5] See, for example, the preface to *Acta Mathematica* (39, 1923). On the creation of *Acta Mathematica* see Domar (1982) and Stubhaug (2010).

[6] See Hermite (1984, vol. 5, p. 150).

1882 Poincaré met Mittag-Leffler for the first time, when Mittag-Leffler passed through Paris on his honeymoon. Mittag-Leffler found Poincaré to be shy and reserved, and his wife to be intelligent and pragmatic and in possession of all the qualities her husband lacked. At that stage, at least, she took charge of Henri's books and letters, and treated him as her beloved and as a child (Stubhaug 2010, 278). Poincaré also met Sonya Kovalevskaya, the leading woman mathematician of the day, for the first time and was impressed by her—perhaps he was charmed by the way "her face lights up with an expression of feminine goodness and of superior intelligence" or by her simple and natural manner, both of which had captivated Mittag-Leffler some years before (Stubhaug 2010, 201). In October he met Sophus Lie for the first time, an associate of Felix Klein who was reshaping the theory of partial differential equations.

In November 1883 Poincaré was appointed répétiteur d'analyse at the École polytechnique, a post he was to hold until 1 March 1897, and he assisted at the Rouen meeting of the Association française pour l'avancement des sciences. He also became a Swedish Knight of the Polar Star in November 1883, at the instigation of Mittag-Leffler, rising to Commandeur First Class in June 1905.[7] Early in 1884, in a letter Hermite wrote to Mittag-Leffler,[8] he remarked that Poincaré wished to assure him of his loyalty to *Acta Mathematica*, adding that he hoped to profit from the papers he has published on celestial mechanics to obtain the chair in astronomy at the Collège de France that was shortly to fall vacant when Serret retired; in the event he was unsuccessful. In another letter, Hermite worried that Poincaré, although "a prodigious force of invention," was working too hard, and had been ill for two weeks with anemia.[9] Also in 1884, the year the Swedish Academy prize was opened, Poincaré ranked fourth in the election to the Académie des sciences, and in 1885 he ranked third. That year he was appointed to the chair of physical and experimental mechanics at the Faculté des sciences in Paris, a fact that remains hard to explain for he had no publications in the subject to his name and between 1882 and 1885 no less than eleven chairs fell vacant at the Sorbonne in mathematics and physics.[10] On 21 December 1885 he

[7] For a full list of Poincaré's many honors, see his *Oeuvres* (11.2, 11–17). The early ones certainly helped him to build his career.
[8] See Hermite (1985, vol. 6, p. 82).
[9] See Hermite (1985, vol. 6, p. 91).
[10] For an account of some of the maneuvering behind the scenes, see Atten (1998).

was awarded the Prix Poncelet of the institut for his mathematical work. In August 1886 he succeeded Lippmann in the chair of mathematical physics and probability at the Faculté des sciences in Paris; thereafter, professionally he was a theoretical physicist, even if his research in many different branches of pure and applied mathematics only continued to expand and deepen, until he became a professional astronomer in 1896. He became president of the SMF, and worked with his father on the Nancy meeting of the Association française pour l'avancement des sciences. This time he ranked second in the election to the Académie des sciences, and finally, in January 1887, at the age of 32, he was elected, in the section on geometry.

His advocates were Hermite and Camille Jordan.[11] Hermite was forthright in his praise for Poincaré. He began with Poincaré's elaboration of a theory of Fuchsian functions, noting the great depth to which Poincaré had penetrated and which he commented on at length. Then he turned to the work on number theory, where his remarks would have carried particular weight. Other work "where the difficulty and importance were worth the attention of the Académie" included Poincaré's papers on Abelian functions and functions of several complex variables, where he answered questions left by Riemann and Weierstrass. That he had overcome such great difficulties, he concluded, testified to the "talent and spirit of adventure of the young geometer."

Jordan had been a member of the Académie des sciences since 1881. His advocacy of Poincaré quickly covered most of his many achievements, starting with number theory, where he remarked that the memoir on cubic ternary forms (1881g, discussed below in chap. 9, sec. "Number Theory") was full of interest, worked out in complete detail, and showed the important role of substitutions. He went on to note the work on complex functions of several variables, the uniformization theorem, and the great generalization of Hermite's theory of the modular function to the theory of Fuchsian functions, which Jordan particularly liked. Then he observed that since the time of Cauchy and Briot and Bouquet the study of differential equations in the complex domain had completely submerged the study of real equations, so he welcomed Poincaré's extremely new and original study, and singled out its topological aspect. He noted the recent work on the three body problem and the shape of

---

[11] Their submissions can be found in Poincaré's dossier in the Académie des sciences.

Figure 2.2. Camille Jordan (1838–1921). Source: Arild Stubhaug, *The Mathematician Sophus Lie* (Springer, 2002).

rotating fluid masses, but it was the conclusion that surely carried the most force: "Poincaré," he said, "was beyond ordinary praise and one could only say, as Jacobi had said of Abel, that he has solved questions nobody before him had dared to imagine."

1888 was no less busy. On 9 November Poincaré was appointed a member of the Comité d'organisation du Congrès international de bibliographique des sciences mathématiques (Exposition universelle internationale de 1889) by the Ministre du commerce et de l'industrie, and on the 16th he was elected president of the committee. On 20 January 1889 it was announced that he had won the Swedish prize competition, with Appell second, and on 4 March he was elected Chevalier de la Légion d'honneur. In July the discovery of a serious error in Poincaré's paper forced Mittag-Leffler to fight hard to avoid a scandal unfolding against this very public background (see chap. 4, sec. "The Prize Competition" below), and it is after this date that a steady stream of international honors came Poincaré's way. Nor did he shirk administrative responsibilities. In July he was elected president of the Congrès international de bibliographique, and president of the Commission permanente internationale du Répertoire bibliographique des sciences mathématiques.[12]

By now Poincaré was a father, his first child, his daughter Jeanne, had been born on 3 June 1887. His second daughter, Yvonne, was born on 9 November 1889, his third, Henriette, was born in March 1891, and his son Léon was born in June 1893.

For a brief while in the early 1890s academic life seems to have become quieter for him. In 1891 he visited Britain to inspect their universities (Parshall 2006, 320). At the end of 1892 he was elected to the board of the Bureau des longitudes, a prestigious meeting place of academics, politicians and high-ranking military men, in recognition not only of his theoretical work (see below, chap. 5) but his capacity for administration, and in July he was appointed ingénieur en chef at Mines. In 1894 he was promoted to Officier de la Légion d'honneur. In November 1896 he succeeded Tisserand in the chair of mathematical astronomy and celestial mechanics in the Faculté des sciences at Paris. The next year he was appointed president of the editorial committee of the *Bulletin astronomique*, published by the Observatoire de Paris, and by 1899 he had published the third volume of his *Méthodes nouvelles*. In June 1900 he was an organizer and president of the second ICM, in Paris, and with the publication of the *International Catalogue of Scientific Literature*, he was elected a member of the Conseil International. He also became the president of the SMF for the second time.

By this time also a public involvement of a different kind had begun for Poincaré, for 4 September 1899 is the date of his first intervention in the Dreyfus affair.

THE DREYFUS AFFAIR

The Dreyfus affair that was to divide France bitterly for many years and to reshape it politically for decades afterwards began in September 1894 when a serving maid working in the German Embassy came into possession of a note, torn into six pieces, that suggested the Germans had a spy in the French Army. Despite the presence of an

---

[12] See Poincaré (1899).

obvious suspect, Charles Ferdinand Esterhazy, a womanizer and gambler with crippling debts who had recently been forced to retire from the army—and who was indeed to turn out to be guilty—Commandant Henry, the counterintelligence officer in charge of the investigation, arrested Captain Alfred Dreyfus, a man of Alsatian Jewish descent. There was no evidence against him except the note, which was said to be in his handwriting, and a closed military tribunal found him guilty in December. He was sentenced to a dishonorable discharge and imprisonment for life on Devil's Island, a French prison in Guyana.

In July 1896 Lieutenant Colonel Picquart looked at the documents that had been presented to the court, and at the irregularities in the trial. He concluded that Dreyfus was innocent, and began a long campaign to clear Dreyfus's name. Despite the army's attempt to silence him by transferring him to Tunisia, he was able through friends to gain the support of influential Senators, and by January 1897 the Dreyfus family, who had always protested Captain Dreyfus's innocence, began to see public opinion move their way. In January 1898 Esterhazy was arrested, tried by a military tribunal—and acquitted. This was the occasion for Émile Zola to publish his famous open letter to Émile Loubet, the president of the Republic: "J'accuse!" Two trials followed at which Zola was fined and sent to prison. To stay at liberty he escaped to London, but as he had hoped, the scandal he had provoked led to another round of arrests. It now emerged that Commandant Henry had forged some evidence that he said had come to light since the trial and further incriminated Dreyfus, and he committed suicide after he was arrested.

Even this did not lead to the release of Dreyfus. In fact, he was convicted again at a retrial in Rennes in September 1899, before being pardoned by Loubet. On 14 December 1900 an amnesty was issued that applied to all related issues, including those involving Picquart and Zola, and eventually, in 1906, Picquart and Dreyfus were allowed to rejoin the army.

Poincaré did not become heavily involved in the Dreyfus affair, unlike some of his colleagues, notably Paul Appell, whose brother Charles had been imprisoned by the Germans on spying charges some years before.[13] It was not his custom to get involved in matters of the day that he felt lay beyond his personal competence, and the only petition he signed

---

[13] This account now follows Rollet (1997; 2002a).

was in January 1898 that called for justice to be done but did not presume that Dreyfus was guilty or innocent. His name appears alongside that of his brother-in-law Émile Boutroux, Gaston Darboux, Paul Janet, Sully Prudhomme, his cousin Raymond Poincaré,[14] Louis Couturat, and André Gide. But a number of mathematicians had become convinced of Dreyfus's innocence by 1897, notably Hadamard and Painlevé. This inspired Painlevé to testify in the trial at Rennes, which he did by reading a letter he had asked Poincaré to write concerning the evidence against Dreyfus. Dreyfus's lawyer in fact called several experts to state that the two handwritings were different, but to no avail. Painlevé's involvement distressed Hermite, who was convinced of Dreyfus's guilt and did not like to see senior members of the army criticized, but when he wrote to Mittag-Leffler (1 September 1898) to ask his opinion Mittag-Leffler replied (9 September 1898) that he shared the universal opinion outside France that Dreyfus was innocent.[15]

Specifically, Painlevé asked Poincaré to give his opinion on the quality of the arguments the Police Chief Alphonse Bertillon had devised to compare Dreyfus's handwriting with that on the note. This was inevitably a matter of comparing stylistic mannerisms and computing probabilities, and Bertillon, who was to become famous for his work in creating a science of police work, was a completely untutored amateur. Poincaré, who had been professor of mathematical physics and probability since 1886, and had just published a lecture course on probability theory in 1896, was the most eminent theorist in the field in France. He had no difficulty in finding that Bertillon's obscure arguments contained elementary errors in computing the relevant probabilities. A significantly rare event in Bertillon's list was said to have a probability of 0.0016 of occurring; Poincaré found the correct figure was 0.7. An allegedly significant number of repetitions, according to Bertillon's scheme for analyzing the handwriting, was found to be almost certain to occur. In a page and a half he dismissed the analysis of Bertillon and his supporters as worthless whereas that of their critics was, he said, exact. He would not pronounce on the guilt or innocence of the accused, but say only that if he is to be condemned it must be on other evidence.

---

[14] Who otherwise kept a safe distance from the Dreyfus affair and refused to take sides.
[15] Other anti-Dreyfusards were Pierre Duhem, Georges Humbert, and Camille Jordan.

Poincaré was called upon again in 1904 as the Dreyfusards' campaign pressed for the now-pardoned man's full restitution. The criminal chamber of the Cour de cassation (the French Supreme Court or Court of Appeal) asked Poincaré, Appell (the dean of the Faculté des sciences in Paris), and Darboux (the permanent secretary of the Académie des sciences) to reach a definitive opinion about the note. They conducted a thorough investigation. They interviewed both Bertillon and the officer commanding the military investigation. Astronomers with precision instruments were called in to examine Bertillon's measurements, and found them to be hopelessly imprecise. On 2 August 1904 the three experts submitted their 100-page report to the Cour de cassation (Poincaré 1908)—in all likelihood it was written by Poincaré.

The report began by distinguishing between two kinds of probability: the probability of a result given certain initial conditions, and the opposite problem, that of finding the initial conditions (the cause) when the result is known. The first concerns a given number of black and white balls in an urn and asks for the likelihood that a ball taken at random is black. The second asks for the distribution of the balls, given that a black one has been removed. The problem in the Dreyfus case is the second, and harder, one of these: Given the note in evidence, is it genuine or a forgery? They found that Bertillon's analysis was flawed throughout. Similarities between Dreyfus's handwriting and the note were treated as evidence of guilt, absence of evidence was treated as a deliberate attempt to conceal further evidence of guilt. They concluded:[16]

> What we are about to say is enough to understand the spirit of Bertillon's "method": it can be summarised in a phrase "when one seeks, one always finds". When a coincidence is noticed, it is overwhelming proof; when one is lacking it is also overwhelming proof that the writer has sought to turn away suspicion. One cannot be surprised by the results obtained by this method. The naivety with which it has unveiled its secrets beggars belief. In short, all these systems are absolutely without any scientific value (1) because the application of probability theory to these matters is not valid; (2) because the reconstruction of the note is false; (3) because the rules of the probability calculus have not been properly applied.

[16] Quoted in Rollet (2002a, 11).

In a word, because the authors have reasoned incorrectly on false documents.

It is hard to know how much weight to attach to Poincaré's interventions. Quite plainly the trial at Rennes was set up to convict Dreyfus, just as the Cour de cassation in 1904 was part of a reconciliation. Already by 1899, and even more by 1904, Poincaré was a major figure in French intellectual life, with positions and international honors that made him an unchallengeable expert. But his refusal to become involved other than as an expert witness, and the care with which he stuck to his brief to comment on Bertillon's analysis of the evidence may have made him a more persuasive witness than some of his more partisan colleagues. There is always an unsatisfactory match between scientific testimony and the procedures of a court of law, just as there is often an uneasy parade of types of expertise and experience (military, police, judicial, scientific, . . .). Perhaps all one can conclude is that Poincaré played his part in the parade of figures who rescued Dreyfus from a great wrong and did something to affirm higher standards for France than the military and political elites would otherwise have been capable of recognizing.

NATIONAL SPOKESMAN

In 1896 Poincaré had become a professor of mathematical astronomy and celestial mechanics at the University of Paris and was awarded the Jean Reynaud Prize of the Paris Académie des sciences—this was a prize of 10,000 francs, roughly a professor's annual salary—for highly original work produced in the previous five years.[17] It was inevitable that Poincaré would be drawn into controversies in physics where his ability to deploy a mathematical theory to get to the heart of the matter made him the ultimate authority. Langevin (1913, 696) observed that for physicists, mathematics was an instrument that was difficult to handle, and learning to play it was as hard as learning Chinese, so it could become impossible to see things beneath a maze of symbols. But Poincaré never got lost, and so the principal difficulty for a physicist was never a problem for him.

---

[17] Jean Reynaud was an ingénieur des mines. The prize, which was to rotate annually between the five académies, was first offered in 1879.

An early call on Poincaré's services came when a Viennese physicist, Gustav Jaumann, put forward a theory that cathode rays were a kind of wave, and not the emission of electrified particles, as was beginning to be the prevailing belief thanks to the work of Perrin and J. J. Thomson. Poincaré refuted Jaumann's analysis, but he kept bouncing back only to be refuted again, with the result that Poincaré's analyses contributed to the success of the emission theory, even though at this stage Poincaré was not personally convinced of the existence of electrons.

The Crémieu and N-Ray Controversies

Poincaré's touch was less sure when it came to two of the controversies in physics that animated the French community and tended to damage it abroad. The first of these concerned Victor Crémieu's attempt to replicate Henry Rowland's experiments from 1876 showing the magnetic effect of a rotating electrified disk.[18] This was a delicate effect predicted by Faraday and also by Maxwell. Crémieu, who had obtained his PhD at the Sorbonne under Lippmann's direction in 1901, became interested in Rowland's work, and when he failed to replicate it began to doubt the existence of the magnetic effect. This conclusion, if confirmed, would have cast grave doubts on Lorentz's particle-based theory of electricity, which was beginning to replace Maxwell's formulation, and the best that could be said for Rowland's work was that it was qualitative. The British physicist Joseph Larmor, however, had produced a particle-based theory of his own which explained Rowland's result, and this was meeting with some success. So Crémieu, who was 29 when the saga began, was playing for high stakes in challenging the prevailing orthodoxy. He needed weighty support, and for a time he obtained it from some of his French colleagues. Lippmann and his colleague Edmond Bouty confirmed Crémieu's results at the Sorbonne; Marcel Brillouin at the Collège de France declared that they did indeed pose a challenge to Maxwell's theory. Others, however, were less convinced. Henri Pellat at the Sorbonne was doubtful, and Alfred Potier, Poincaré's former professor at the École polytechnique said, in a published exchange of letters with Poincaré (1902), that he saw no contradiction between Crémieu's results

---

[18] See the expert account by Walter in Poincaré's correspondence with physicists, (2007), and also Indorato and Masotto (1989).

and Maxwell's theory. But the most consistent support Crémieu gained came from Poincaré.[19]

Poincaré already knew when he wrote the second edition of his *E&O* that Crémieu's results, if confirmed, would refute the ideas of Larmor and Lorentz, and he noted that the contradiction needed to be settled in the laboratory. He repeated this remark in the first edition of his *S&H*, at the end of the essay, commenting that recent agreement among theorists had been disturbed by the experiments of M. Crémieu,

> which have contradicted, or at least have seemed to contradict, the
> results formerly obtained by Rowland. Numerous investigators have
> endeavoured to solve the question, and fresh experiments have been
> undertaken. What result will they give? I shall take care not to risk a
> prophecy which might be falsified between the day this book is ready
> for the press and the day on which it is placed before the public.

Elsewhere he observed with amusement that Crémieu's critics were mostly British and American. Prominent among them was a young American experimentalist, Harold Pender at Johns Hopkins University, who was prompted by the now-bedridden Rowland to conduct fresh experiments. These not only confirmed Rowland's qualitative result, they were sufficiently accurate to give the first quantitative measurements of the effect. He also pointed out some aspects of Crémieu's experimental setup that Poincaré had overlooked. This prompted Poincaré to write to Pender to propose that he and Crémieu work side by side in the same laboratory to settle the matter. With the approval of William Thomson this was arranged, and the two men set up their rival equipment in Bouty's laboratory in the Sorbonne. The Institut de France paid the laboratory, and the Carnegie Institute in Washington paid most of Pender's funds.

The result was decisively in Pender's favor. The Rowland effect was confirmed qualitatively and the discrepancy put down to a dielectric coating on parts of Crémieu's equipment that masked the effect. Poincaré quietly withdrew his remarks about Crémieu's experiments in the second edition of his *S&H*. Crémieu set about trying to restore his damaged reputation by quietly investigating the experimental consequences of the

---

[19] This account follows Walter (2007, 109–113).

field theory of gravity in which gravity is propagated as a wave, as Lorentz and Poincaré had begun to contemplate.

Poincaré's own position was more complicated. In 1901 he had published a long article (1901a), which he reprinted in *S&H*, in which he described the controversy, which had already reached the general public, and explained "at which point the ideas of Crémieu are revolutionary" (see (1901a, 994), a passage omitted from *S&H*). He began with the distinction between an open and a closed current that went back to the work of Ampère. An open current flows in one direction between two charged bodies until their potentials become equal. It is seen in the discharge of a condenser, and for that matter in a bolt of lightning, but Ampère had found them to be too rapid to be a matter for experimental investigation. A closed current is analogous to a circuit $BNAMB$ in which matter goes round and round, loading up with electricity at the point B, then traveling via N to the point A where it deposits the charge and returns empty along AMB, while the charge returns along the wire AMB in the manner of an open current. This is the process in a battery. Ampére called the flow of electricity along BNA a convection current, and the return flow along AMB a conduction current. But all that Ampère could investigate was the effect of one closed current on a portion of another.

Nonetheless, Ampère, and Faraday after him, were able to build up a solid body of experimental results about electricity and magnetism in terms of closed currents and the action of a closed current on a circuit, part of which was fixed and part of which was moving. Since Ampère regarded the moving part of such a circuit as an open current, he believed he had evidence about the action of a closed current on an open one. He now hypothesized how two open currents would interact, and published his (1826) in which he essentially reduced the study of electricity to the action of an inverse square law of repulsion acting between infinitesimal elements of electricity.

Later writers, notably Helmholtz, were able to show that Ampère's theory was inconsistent. It worked very well for closed circuits, but not for open ones. Helmholtz showed, for example, that electrodynamic induction could be deduced from the law of conservation of energy, but only for closed circuits. But Poincaré noted that Helmholtz's theory led to deep problems with the nature of magnetism when open currents were admitted. One solution would be to establish that open currents do not exist, and Poincaré indicated that in effect all the open currents considered by Helmholtz were really closed currents, closed by certain currents analogous to Maxwell's displacement currents:

> Such were the problems raised by the ruling theories when Maxwell appeared and with a stroke of the pen made them all disappear. In his ideas, in fact, there are only closed currents. (1901a, 998, in *Oeuvres* 10, 402)

Twenty years later, Poincaré went on, Maxwell's ideas had been confirmed by Hertz's experiments. The result was that what had seemed to be an open current in the discharge of a condenser was in fact a closed current. But what about a circuit that was half a conduction current and half a convection current? The problem would be solved if the convection current could be shown to be closed, and for this it would be necessary to show that a convection current (a moving conductor) could be made to move the needle on a galvanometer. This is what Rowland was the first to succeed in showing. Since then experiments with cathode rays and currents in solutions had further confirmed the result, and now Lorentz's theory was entirely based on the idea of convection currents and all electrical and magnetic phenomena were to be explained in terms of tiny circulating particles called electrons.[20]

It was this satisfactory theoretical state of affairs that Crémieu's experiments contested, because they suggested that open currents really did exist. Poincaré did not describe Crémieu's experimental setup, which would have taken too long, but confined himself to noting that Rowland's experiments were very delicate and liable to all sorts of misleading results. Rowland was an experienced experimenter, he admitted, and his results had convinced many who had seen them, but the same was now true of Crémieu's contrary results, and, unfortunately, Rowland himself had recently died. It would be very important, said Poincaré, to understand the divergent experimental findings.

Poincaré then turned to the various theoretical opinions that had been expressed. He noted that Crémieu's ideas had alarmed the English, who had therefore applied themselves to showing that they did not have to abandon the conquests of Faraday and Maxwell. Other investigators, Poddington and Wilson, had tried to argue that there was no contradiction between Rowland's results and Crémieu's, but Poincaré

---

[20] The *S&M* chapter ends here with some other remarks.

disputed this. This left the objections raised by Pender, which Poincaré regarded as much more serious, because Pender had not been able to replicate Crémieu's experiment. New experiments, Poincaré concluded, were therefore necessary.

All the experiments to date tried to measure the magnetic effect of a charged, rotating disk, and Poincaré gave a lengthy account of what was involved, with an indication of the advice he had given Crémieu, before he ended this paper by saying,

> Will we find ourselves faced with the paradox of an open current? Will we be obliged to find a new explanation for cathode rays, currents in electrolytes, magnetic polarisation, and the Zeeman effect? That is what we shall soon know. (1901a, 1007, in *Oeuvres* 10, 420)

In view of the final outcome, it is very easy to say that Poincaré should have answered his question with a simple negative, and concluded that it was unlikely that a young French researcher could have discovered a result that would force such a wholesale rewriting of contemporary physics. However, a multiplicity of possibilities was in play. There was surely a patriotic element to Poincaré's involvement, but Crémieu's work came with the endorsement of Lippmann, who was highly respected for his invention of color photography, so it could not be easily dismissed. Poincaré knew as well as anyone how muddled some of the theoretical arguments were, and, more fundamentally, how uncertain and in flux the whole theory of electromagnetism was, and he could see that Rowland's experiments were at best qualitative. He also held no brief for any physical theory in between experiment and pure theory: open currents were hypothetical to him in somewhat the way electrons were. He may even have enjoyed the competitive urge to sort out a mystery. Whatever the mix of factors, the fact remains that Poincaré was rather more on Crémieu's side than Pender's, and to that extent he came out on the wrong side. No shame in that—errors are part of physics—but in such a high-profile controversy, it was a very public mistake.

N-Rays

Nor was it the only one. Physics in the period from 1890 to 1914 was being animated by several unexpected discoveries, of which radioactivity and X-rays were among the most important. The time was right for other claims of this kind, and it fell to the French to host the most famous and embarrassing of them all: N-rays.

N-rays were proclaimed by René Blondlot in 1903.[21] Blondlot was a friend of Poincaré's from Nancy, an accomplished experimentalist who had carefully not got involved in the Crémieu affair. At the end of 1904 he was awarded the Prix Leconte for his life's work in physics by a panel of the Académie des sciences which was chaired by Poincaré; see his (1904f). They were most impressed by Blondlot's careful measurement of the speed of electricity in a wire, but they also mentioned "the curious action of a new form of radiation which he called N-rays," although they found it too early to pronounce upon the significance of this work. Even this cautious opinion covers up part of the story. Nye (1986, 75) has shown that Poincaré rewrote an earlier draft by Becquerel, and suggests that personal loyalties among physicists from the Lorraine blinded them to the extraordinary nature of Blondlot's claims. Lagemann (1977) found no less than 120 physicists, mostly French, in some 300 articles who rapidly claimed to be able to demonstrate the extraordinary effects of N-rays, but they proved to be a collective wave of self-deception.[22]

The N-ray folly was exposed in what has become a classic in the annals of pseudoscience. The American physicist Robert Wood, who had tried and failed to verify the effect himself—few did, outside France—was invited by *Nature* to go to Blondlot's laboratory and see what was going on, because of the growing disquiet internationally concerning Blondlot's work. The experiment involved assessing the brightness of an electric spark in a darkened room under various conditions. Wood secretly removed what was said to be an essential prism, but the observers noticed no change. He replaced a file that was said to emit N-rays with a lump of wood, believed to be inert, and again no one noticed. His report (Wood 1904) put the whole effect down to delusions on behalf of the observers, and quickly N-rays fell into disrepute. It has to be said that Wood was both a good experimental physicist and an enthusiastic hoaxer, and some have disputed his version of events.[23] But hoaxers, and especially

---

[21] The N stood for Nancy, Blondlot's home town.
[22] See http://www.rexresearch.com/blondlot/nrays.htm#lageman.
[23] See Ashmore (1993).

professional magicians, have proved better than practicing scientists at exposing frauds and self-deceptions in the lab. In Blondlot's case it seems that a natural desire to make a discovery and an experimental setup that relied too much on subjective measurements combined with the atmosphere of the time to produce an effect that others could "replicate."

Poincaré's involvement was not limited to preventing his old friend from losing his chance of a major prize (which carried a sum equivalent to five times Blondlot's annual salary). Late in 1904 the *Revue scientifique* carried a series of statements of opinion about the affair. Poincaré (1904d) said that he had been to Blondlot's laboratory in Nancy but had been unable to see the effect.[24] However, there were some photographs that he found precise and convincing, and he exhibited three. Two showed pairs of black spots of different intensity, the one under the influence of N-rays the other not, and Poincaré said that the difference in intensity could not have been due to experimental error. The third photograph showed what the spots looked like when no N-rays were present: the intensity of the spots still differed but by not as much as before, as he admitted but could not explain. On the basis of this evidence he dismissed Wood's criticisms, said that he had the greatest confidence in Blondlot, and called for further experiments to resolve the issue which would, he suggested, be best done with Blondlot's cooperation.

The whole story was a debacle for French experimentalists, who felt that their international reputation had taken a bad blow, and the concurrent award of the Nobel Prize to three Parisian scientists was not reassuring because Pierre Curie had not been educated in the traditional French establishments and Marie Curie was Polish (Nye 1986, 71). By endorsing his friend Blondlot's life's work, and his photographic "evidence," at a time when most of the rest of the scientific world was increasingly doubtful, Poincaré deprived himself of the chance of speaking out against an allegation that was true, damaging, and all the worse because its refutation was an irresistible mixture of scandal and comedy.

---

[24] 29 letters from Blondlot to Poincaré survive, but none in the other direction, and it is impossible to say what Poincaré made of what he was told.

CONTEMPORARY TECHNOLOGY

Poincaré's involvement in the Crémieu and N-ray affairs did him no credit, but several other aspects of his life display a more successful practical side to Poincaré. He was particularly interested in notable technological events of the day, none more so than with the telegraph and later with wireless telegraphy. This grew out of his theoretical work on electricity and optics, of course, but included responses to the issues surrounding long-distance radio transmission, and he went out of his way to listen to and to speak to practitioners, even to the point of writing about the design of the telephone receiver.

The older technology, telegraphy, must count as one of Poincaré's less productive interventions. The successful laying of a transatlantic cable on the advice of William Thomson (later Lord Kelvin) in 1855 was brought about by his shrewd reading of the equation for the propagation of electricity in a long straight wire.

The equation[25] is

$$A \frac{\partial^2 u}{\partial t^2} + B \frac{\partial u}{\partial t} + Cu = \frac{\partial^2 u}{\partial x^2},$$

where $u(x, t)$ is the potential at a point x at time t.

The constants that appear in the equation involve the capacitance K, inductance L, leakance S, and resistance R, per mile of the wire:

$$A = KL, \quad B = RK + SL, \quad C = RS.$$

Capacitance is the ability of a body to hold an electric charge; inductance or self-inductance L in a wire measures the relationship between voltage V and rate of change of current I according to the equation $V = L\frac{dI}{dt}$, which can be derived from Maxwell's laws. If the wire has length ℓ then $S\ell$ and $R\ell$ are respectively called the total capacitance and total resistance of the entire wire.

The equation was obtained for the first time by Kirchhoff in 1857, and in a different way by Oliver Heaviside in 1876. In the ideal case of

[25] For a derivation, see Webster (1966, 44–45), and for its solution Webster (1966, 174–179).

no resistance $(R = 0)$ and no leakage $(K = 0)$ B and C vanish, and the equation reduces to the wave equation. In the case when the inductance is negligible by comparison with the resistance, the constant A may be taken to be zero, and the equation is the one-dimensional heat equation, which is how Thomson regarded it. On this interpretation, the maximum electrical effect at a distance x is inversely proportional to x^2.

Despite the success of the transatlantic cable, the character of the equation is very different when neither resistance nor leakage is negligible. In this case neither B nor C vanish. Poincaré treated the telegraphist's equation in this form in his (1893d) and in his *Cours sur les oscillations électriques* in 1894. In his short paper, Poincaré wrote it in the form

$$A\frac{\partial^2 u}{\partial t^2} + B\frac{\partial u}{\partial t} = \frac{\partial^2 u}{\partial x^2},$$

so he only considered the case where there is no leakage. In units where the speed of light is 1 this reduces to

$$\frac{\partial^2 u}{\partial t^2} + 2\frac{\partial u}{\partial t} = \frac{\partial^2 u}{\partial x^2},$$

and on substituting $u = ve^{-t}$ it becomes

$$\frac{\partial^2 v}{\partial t^2} - v = \frac{\partial^2 v}{\partial x^2}.$$

Poincaré now solved this equation by means of complex Fourier integral methods under the initial conditions that $v(x, 0) = f(x)$ and $\frac{\partial v}{\partial t} = f_1(x)$ when $t = 0$, where $f(x)$ and $f_1(x)$ were given by Fourier integrals (1893d, 1028). By considering particular cases, including that of a simple impulse sent down the wire, and employing Cauchy's theory of residues, Poincaré was led to the general conclusion that the head of the signal moves with a finite speed, and there is no disturbance ahead of it (this is also the case with the transmission of light but not of heat), but the head, once it has passed, leaves behind a disturbance which never vanishes, unlike what happens with the wave equation. This confirmed with rigorous mathematics the existence of a problem that practitioners had already noticed. To deal with this distortion when an attempt is made to transmit a periodic wave down the wire, telegraphists had had to accept a low transmission rate, so as to separate the pulses.

In his short paper, Poincaré did not mention the ingenious discoveries of Heaviside, nor, more importantly, did he in his *Cours* of 1894, where he surveyed a considerable amount of mostly French experimental work, with a view to deciding between the old theory of electromagnetism (due to Kirchhoff) and the modern theories of Maxwell and Hertz. But it had been Heaviside's remarkable discovery in 1887 that the values of the physical constants can be so adjusted that the rate of dispersion is zero, as Yavetz describes in his fascinating book, *From Obscurity to Enigma: The Work of Oliver Heaviside, 1872–1889* (1995). This can be done both mathematically and physically, it merely requires that the leakage be nonzero. What is required is that the signal be transmitted in the form $f(x - ct)g(t)$, which describes a signal that has the shape $f(t)$ at time $t = 0$ moving with velocity c along the wire with that shape and diminishing in size by a factor that depends on the time. For this to happen there must be relations between A, B, and C, and these reduce to $B^2 - 4AC = 0$. This condition is equivalent to

$$(RK + SL)^2 - 4KLRS = 0 \quad \text{or} \quad RK = SL,$$

but this was a possibility that Poincaré excluded when he assumed the leakage was zero, because then it only holds when the resistance also vanishes. However, far from being an inconvenience, this condition is necessary for the production of distortionless telephony. The signal becomes fainter over distances, which can be corrected for by fitting amplifiers. Telephony, which required much higher frequencies, could now be produced, for, with some leakage and a deliberately high self-inductance, it could be undistorted. Long-distance communication was reborn—but the money for the first successful patents went to the American electrical engineer Michael Pupin in 1901, and not to Heaviside. By the same token, in Poincaré's account, experimental work was given a theoretical twist and technological implications were not mentioned.

Poincaré fared better in the dramatic story of the discovery of wireless electric transmission. It had been shown by Faraday in the 1830s that a variable electric field produces a current in a wire, called an induction current, even though the wire is some distance from the source of the field. However, the effect was long regarded as too weak to be of any practical use until in 1888 Hertz created very rapidly alternating electric currents, with a period of between $2 \cdot 10^{-8}$ and $2 \cdot 10^{-9}$ seconds. Because induction currents are sensitive to variation, these rapidly alternating currents produced an effect detectable at several meters, and Hertz was able to demonstrate that the variations in the electric field were transmitted at the speed of light. This confirmed Maxwell's conjecture that light was an electromagnetic phenomenon.[26]

The analogy suggested that these "Hertzian waves," as they were called, could also be used to transmit messages at a distance. Hertz had, however, also showed that the wavelength of the signal was much greater than that of visible light, and this meant that the analogue of lenses and mirrors would have to be of a very great size. If the wavelength is several meters, useful lenses would have to be several kilometers across. The crucial breakthrough that made induced currents easier to detect came with the work of the French physicist Édouard Branly. He built on the discovery that iron filings lining the inside of a tube, which are normally a very poor conductor, become a very good conductor under the influence of a varying electric field. He described his work in a series of articles between 1890 and 1892 that were also translated into English, where they came to the attention of (Sir) Oliver Lodge. The explanation that was offered was that the electric field made the filings cohere, so the device was called a coherer by Lodge in his own work from 1894. Branly and Lodge both discovered that the increased conductivity can be switched off by tapping the tube, and that a combined coherer and decoherer can be made to switch on and off very rapidly, thus enabling the rapidly changing electric field to be picked up.

These laboratory discoveries led to a rush of attempts to make Hertzian waves into a practical means of communication, and this greatly raised the level of public interest. Branly demonstrated the effect over several meters and despite the presence of obstacles in the line of sight (Hertzian waves can turn round corners, as light can, by means of diffraction, but their much greater wavelength makes this process manifest at greater distances, which is why these waves are detectable when there is a hill, but not when the earth itself is in the line of sight from source

---

[26] On Hertz, see Buchwald (1994), which also has a useful comparison of Poincaré and Bertrand.

to detector). Hertz's death in 1894 caught the attention of Gugliemo Marconi, who attached himself to the laboratory of Augusto Righi, who was studying Hertz's work, and Marconi experimented with many ways of extending the efficiency of different combinations of the known components for making and detecting Hertzian waves. In 1895 he had raised the detectable distance to 1.5 km, and failing to raise sponsorship in Italy he traveled to London, where he enlisted the support of William Preece, the chief electrical engineer of the British Post Office, who had independently achieved some of the same results. In the Spring 1897 Marconi transmitted Morse code signals across almost 4 miles of Salisbury plain, and later that year managed some 11 miles across water. Preece energetically publicized Marconi's work, and Marconi refined his apparatus until in 1899 it was capable of ship-to-shore transmissions of over 40 miles. The goal of transatlantic transmission became feasible, and in 1902 Marconi accomplished it.

Much about the physics of radio waves, from the theory to the working of radio transmitters and receivers, was obscure, and naturally Poincaré was drawn in, not only because he was in an excellent position to understand and explain what was going on but because his patriotic sentiments were also engaged. He began with his (1890a), in which he drew attention to an error in one of Hertz's calculations. He had already corresponded with Hertz about this, and Hertz admitted the error which had led to him underestimating the velocity of light in vacuo by a factor of $\sqrt{2}$. He continued with a 22-page report (1894a) comparing the ideas of Maxwell and Hertz for the *Annuaire du Bureau des longitudes*, and concluded with numerous short articles in the *Comptes rendus* that culminate in his 90-page article on the diffraction of Hertzian waves (1910e). In between he wrote a popular article on wireless telegraphy (1902d) for the *Revue scientifique*, articles on Hertzian waves and related subjects for the journals *Éclairage électrique*, and *La lumière électrique*, as well as the book, *Conférences sur la télégraphie sans fil* (1909a) published by *La lumière électrique*. He also published books based on his lecture courses (for example, at the École supérieure de télégraphes): *Les oscillations électriques* (1894b), *La théorie de Maxwell et les oscillations hertziennes* (1899e)—a third edition came out in 1907 (it reprints his (1902e) above)—and the first edition was translated into English and published in 1904 as *Maxwell's Theory and Wireless Telegraphy* with a further text by K. Vreeland (1904). It is a very readable account with lots of experimental details of how rapidly oscillating alternating currents can be created and measured, these being the only ones capable of allowing experiments to decide between the older theories and Maxwell's theories when it came to the electromagnetic theory of optics. Further evidence of Poincaré's practical involvement is provided by his (1907e) on the telephone receiver, which was well regarded in its day.

Throughout these sixteen years Poincaré endeavored to do three things in these books and articles: to convey the latest and most relevant experimental results to his readers; to indicate in broad outline what these results meant; and to fit them to the best theory of electromagnetic phenomena, which he took to be some variant of Maxwell's theory. This last activity brought him up against two problems: One was to decide upon the right idealization of the phenomena to which the mathematics could be applied (Is the earth a sphere? A perfect conductor? Does what lies beyond the earth extend to infinity?) The other was to obtain solutions to Maxwell's equations in a form suited to numerical calculation. And indeed a fourth endeavor might be mentioned, because Poincaré took the opportunity of Fredholm's work on integral equations in 1906 to see if this new technique could assist in the analysis.

None of this was easy. On one occasion Poincaré felt driven to comment,

> It is difficult to speak easily of these facts that are so poorly known and so complicated; ordinary language is insufficient to render their prodigious complexity, and mathematical language, to which one is forced to recourse, pretends in my opinion to a factitious precision that agrees with neither the actual state of our knowledge nor the imperfections in our observational means. The reader must forgive me, and he will take note that all the formulae that follow can only be regarded as gross approximations. (1891c, 610)

And at the end of the long paper (1910e) he noted that he had sometimes been led to contradict himself and to withdraw some of his conclusions, including those in one of his Göttingen lectures. He remarked, "I owe the public an explanation of my successive recantations" and he ascribed them to an erroneous estimate of how many terms were involved when one of his approximate expressions ceased to be reliable. He had originally thought there would be few of these terms, but now he saw they were significant.

The physical issues at stake included:

- the period of a Hertzian oscillator;

- the multiple resonance of Hertzian oscillators, described by two investigators, Sarasin and de la Rive, in 1890 (the occasion for the above remark);

- the propagation and rapid decay of Hertzian waves;

- the diffraction of Hertzian waves and the effect of the curvature of the earth on their propagation.

The first of these was given by a formula due to Thomson, which Poincaré found to be have been derived on dubious grounds: it depended on the flow of current in the wire, which was not fully understood, and on the distribution of electricity in the conductors at the start of each oscillation, which was not known and could not be supposed to be the static state; and it neglected the effects of displacement currents, which could be assumed to be preponderant in rapid oscillations. Moreover, a mathematical analysis of the problem, starting from Maxwell's equations, required that the exciter be regarded as being in a dielectric of indefinite extent (this models the room surrounding the object), but this complicated the mathematics greatly.

The phenomenon of multiple resonance suggested that an exciter emit a continuous spectrum of frequencies, and not a discrete set of harmonics as theory suggested. But, said Poincaré, not only was the theory imperfectly understood, the complicated behavior of the interruption of the electric spark was not understood at all. He went on,

> In the present state of science it could be dangerous to hold to a unique interpretation. It is necessary on the contrary to examine all the possible ones, while looking at new experiments that permit one to decide between them.

This, he said, he had begun to do for multiple resonance in *E&O* (1890) and he returned to the subject in his lectures *Les oscillations électriques* (1894b, 123–129). There he contrasted the explanation of Sarasin and de la Rive to the effect that the measured wavelength of the emitter depended on the resonator that measured it, which selected some of the emissions preferentially, with the idea that the effect was due to damping of both the emitter and the resonator. He found that the second explanation would hold if, as was surely the case, the damping of the emitter greatly exceeded that of the resonator.

The rapid decay of Hertzian waves had been discussed by Hertz himself, but Poincaré was not satisfied with it. Indeed, Hertz did not have a theory governing the behavior of resonators, but only a "justification for the relations that emerged from experiment" (Buchwald 1994, 253). Nor was he alone in this. Oliver Lodge was another, and the experimentalist Blondlot was a third. Poincaré noted that when the exact calculations are done for a spherical exciter, as Lodge had done, the result greatly overestimates the rate of decay, and Poincaré himself was unable to get his field-theoretic explanation to yield some of Hertz's simple formulas. But he did note that Blondlot had eventually succeeded in showing that the velocity of electricity in a wire is almost the same as that of light in a vacuum.

The problem posed by the fact that Hertzian waves can cross the Atlantic was acute once it was shown that diffraction cannot easily account for it, which it cannot because the wavelength is small compared with the radius of the earth. These problems were nicely surveyed by Love in 1915.[27] Love recorded that the prevailing assumptions in the many accounts he looked at were these:

1. The earth is taken to be a homogeneous conductor, surrounded by homogeneous dielectric, the separating surface being a perfect sphere.

2. The sending apparatus is represented by an ideal Hertzian oscillator, or vibrating electric doublet, situated in the dielectric near to the separating surface, and having its axis directed normally to that surface.

3. The waves emitted by the oscillator are taken to be an infinite train of simple harmonic oscillations of a definite frequency.

This problem, he said, had not been solved and had led to discordant results, and further mathematical idealizations had been imposed. H. M. MacDonald (1903) had been the first to give a series representation

[27] As they are from today's point of view historically in Yeang (2003) and mathematically in Pomeau (2006/2010).

for the waves, from which he deduced numerical estimates for the solution, but his arguments were criticized by Rayleigh, who noted that the optical analogy with a metal ball, 1 inch in radius, palpably fails. He attributed MacDonald's error to a neglect of the influence of higher-order terms. Poincaré then showed that MacDonald's series was not uniformly convergent. The result was that MacDonald greatly overestimated the diffraction, and he then reworked his analysis and came to a much smaller estimate.[28] This led him to claim that the departure of the earth from a perfect conductor was negligible. His method, Poincaré's, and another due to the English mathematician J. W. Nicholson agreed that the equation for the problem could be solved by a series expansion which in turn was represented as an integral and then evaluated approximately, but they gave different answers. Poincaré, in a 100-page paper in the *Rendiconti* (1910e) converted the series representing the solution in spherical harmonics to a sum of complex integrals that was dominated by the pole of the spherical harmonic with the smallest imaginary part, and he proved that this contribution decays exponentially, so transmission distance increases with wavelength. But his method left certain crucial quantities undetermined, and could not give quantitative results comparable to experimental data. Nicholson did come up with some quantitative results, but he noted that they did not explain Marconi's success. Love therefore gave a new approach to the summation of the series that led him, after much work, to agree with MacDonald. After surveying several more attempts and looking at the latest experimental work, Love concluded,

> From this critical discussion I draw the inference that there is nothing in the experimental evidence, at present available, to compel us to adopt the view that the diffraction theory fails to account for the facts. On the contrary, that evidence can be interpreted in such a way as to support the view.

The double negative underlines how little confidence he had in these results.

Love did not mention a remark by Poincaré that identified, however tentatively, the weakest of the prevailing assumptions, which was that

---

[28] Hector MacDonald, FRS, was a Cambridge-educated mathematician and the professor of mathematics at Aberdeen. His most important book was to be his *Electric Waves* (1902).

the atmosphere can be modeled as an infinite dielectric. In fact, as the analysis in Watson (1918) was the first to demonstrate, it must be understood as bounded above by the ionosphere, which makes the problem of radio waves a wave-guide problem. As Poincaré put it,

> Perhaps one will be obliged to fall back on another hypothesis that has already been proposed, according to which the upper layers of the atmosphere, having become conductors through ionisation, reflect the waves. What lends support to this opinion is the great difference that exists between long distance transmissions at night and in the day. It is possible, in fact, because the ionisation conditions are not the same at night and in the daytime, that the passage from non-conducting to conducting layers would take place suddenly at night in such a way as to make reflection possible. (1910e, 258)

Another occasion in which Poincaré intervened in a debate about electrotechnology has only very recently been brought to light by Jean-Marc Ginoux, who found papers by Poincaré in the journal *La lumière électrique* that have been forgotten because they were not included in his *Oeuvres*. Very briefly, European street lighting was originally produced by creating a permanent spark between two carbon nodes, but this produced a disagreeable hum, and in London the engineer William Duddell was hired in 1899 to investigate. He discovered that by varying the voltage across the gap he could vary the sounds emitted by the arc; in fact Dr. Simon, a German engineer, had demonstrated this process a year before and dubbed it the "singing arc," and Duddell used it to make a musical instrument controlled by a keyboard.[29] He also discovered that when the current is increased the voltage drops, in contrast to Ohm's law; this phenomenon, which puzzled electrical engineers, was called negative resistance. The spark also emits electromagnetic waves, so it was also useful in wireless telegraphy. What Ginoux has shown is that in November and December 1908 Poincaré gave five lectures to engineers on this subject, in which he not only dealt with the issue of negative resistance but explained how these circuits can produce a sustained oscillation. Most interestingly, he did so by establishing that the differential equation for a singing arc can be studied by his method of first-order differential

---

[29] Plasmatronic sound systems today are based on this process.

equations, and the sustained oscillation derives from the existence of a limit cycle. Poincaré also connected his analysis with his study of periodic solutions in chapter III of his *Méthodes nouvelles*.[30]

INTERNATIONAL REPRESENTATIVE

Among his scientific peers, Poincaré was also known as an ambassador for France on the international stage, having been heavily involved in two successive administrative enterprises in particular: decimalization, and geodesy.[31]

Decimalization

Poincaré was heavily involved in the quixotic campaign for the deci-malization of time.[32] The cause, which embraced the decimalization of the angle, went back to the French Revolution, and was dear to the heart of the Bureau des longitudes, which Poincaré had joined in 1893, becoming president in 1899. An international conference in Washington had been convened in 1884 to raise the question of the location of the zero arc of longitude, and when the French bid was defeated on the grounds that most maps, including all the British and American ones, put the arc through Greenwich, the French were allowed to reopen these questions as a consolation prize. Over a decade later, in 1897, the Bureau des longitudes in France set up a commission to investigate whether France should adopt decimal units, and Poincaré was appointed as secretary to the commission. He was enthusiastically in favor of the change, because he objected to the mixed decimal and sexagesimal units

---

[30] That information seems to have been forgotten until Ginoux's work. When singing arcs were replaced by triode valves their behavior was analyzed along similar lines in van der Pol (1920), and then in 1929 the Soviet mathematician Andronov united Poincaré's theory with the theory of nonlinear oscillations in 1929. For a richly documented account, see a paper by Ginoux and Lozi to appear in *Archive for History of Exact Sciences*.

[31] Poincaré was also for many years the secretary of the *Répertoire bibliographique des sciences mathématiques* created by the SMF in 1885, which, whatever its shortcomings, rapidly became international and ran for 27 years during which 50 mathematicians from 16 countries surveyed some 300 journals and collected systematic information on over 20,000 mathematical works. See Rollet and Nabonnand (2002).

[32] Among other discussions, see Galison (2003).

in the measurements of times and angles, and advocated a system with 400 degrees in a circle.[33] This, alone of the proposals put forward, allowed for the conversion of time to angle, old angle to new angle, and fraction of a circle into degrees in a way that never required dividing by a two-digit number, and Poincaré even produced tables showing the results. But even within France there was opposition. If any of the new proposals were to be implemented, all charts would have to be changed and much equipment on ships and for surveying; watches would show the time wrongly. Physicists objected that their systems of units relied on the old definitions and would also have to be changed. The French Government also discovered, unsurprisingly, that the idea of a change was met with hostility by every foreign government it spoke to about it, and in 1900 the idea was quietly dropped.

The Mission to the Equator

In 1900 Poincaré gave an address (1900c) to a joint meeting of the five académies and wrote a short article (1900b) on geodesy, particularly as it had been practiced by French scientists. He noted that accurate surveying was necessary for the making of accurate maps, but stressed that it spoke to deeper questions: the composition of the interior of the earth, which was no longer the province of Jules Verne alone, and the measurement of the speed of light. But he also noted that from the time of Maupertuis and Clairaut in the Arctic and Bouguer and La Condamine in the Andes, through the Napoleonic era when Delambre and Méchain went from Dunkirk to Barcelona and Arago brought the results home after many adventures including capture by pirates, France stood preeminent if not indeed alone in this science, so much so that "the French ordnance map is a model" (1900b, 565). In 1900, he remarked, French geodesy is in the hands of the army, and "the French colonial empire offers them immense tracts imperfectly explored" (p. 567). In particular, the International Geodetic Association has charged them with making new measurements of the arc of Quito that La Condamine had once attempted, and a new mission has been dispatched under the command of Commandant Bourgeois.

---

[33] See "Rapport sur les résolutions de la commission chargée de l'étude des projets de décimalisation du temps et de la circonférence," *Annuaire du Bureau des longitudes* (1897, 1–12).

There were important intellectual issues at stake: Should the earth be modeled as a stable rotating ellipsoid, such as Poincaré had himself studied earlier (see chap. 5, sec. "Rotating Fluid Masses" below) in which case measures of its flattening near the equator and the poles had to be conducted and the incompatibility of existing measurements explained. Or should one consider modeling the earth as a geoid, an abstract mathematical surface that was an equipotential surface for the earth's gravitational field. This would doubtless express the effect of mountains on the surface and variations in density in parts of the earth below the surface. In the event, however, practical problems made these investigations less than completely successful.

It was a large-scale military and scientific mission that the French organized to measure the shape of the earth in the region of Ecuador and Peru that straddles the equator and which the French had visited with some success in 1735. Poincaré was appointed the reporter for the commission of investigation, and his first report (1902f) conveyed to the Académie des sciences the information brought back by Commandant Bourgeois, who had been in charge of the preliminary expedition and had returned to France in March 1902. A base camp had been established at Riobamba, some 15 miles south-east of the dormant volcano Chimborazo, the highest mountain in Ecuador, and a site for the basic measurements had been chosen. The site was the less convenient of the two considered, being liable to landslides that made positions uncertain, but the alternative was rejected on the grounds that it was too close to Chimborazo, which was capable of distorting the sensitive geodetic measurements. Thereafter, the necessary preliminary work proved more difficult than had been expected. Humidity at night and in the morning softened the wires in the micrometers, dust forced its way into the equipment, security considerations made it impossible to switch observers when repeating observations, which meant that their personal equations had to be measured and taken into account. Very often the best measurements of time were only possible in places where the bulky meridional circle could be set up. Political problems prevented the expedition from extending its measurements into Columbia. Nonetheless a start was made on the network of triangles needed for the determination of latitude and longitude, magnetic observations were taken, and estimates were made of the effect of altitude on the behavior of the scientific pendulum. Plans were accordingly drawn up for the work of the expedition in future years that were, surely inevitably, to prove too ambitious.

Indeed, in April next year Poincaré had to report (1903b), on behalf of Commandant Bourgeois, that it had been found necessary to decide that there would be a six-month delay in completing the work. Personnel were stranded for months at altitudes of 4,000 meters unable to make any observations because the mountain tops were perpetually cloud covered and the camps often shrouded in fog. In places it snowed, and very strong winds were commonplace, making accurate observations impossible. The weather was exceptionally bad, and Poincaré was moved to recall the recent terrible events in Martinique where, on 8 May 1902, Mount Pelée had erupted killing 30,000 people. He noted that the volcanoes surrounding the expedition were more active than expected, and there was unusual seismic activity, which, though it did not affect the work of the expedition directly, might, it was speculated, be responsible for the dreadful weather. Nor did it help that the local Indian population, which had been drafted in as porters, regarded the marking posts erected by the expedition as showing the location of hidden treasure that they then, naturally, dug up. Still, there was some progress to report. The triangulation had been planned out, and would run from Sinigallay at the end of the railway line from Guayaquil in the south to Tulcán in the north, a distance of over 200 miles. Telegraphic measurements, using two different kinds of wire, allowed latitudes and differences of longitude to be measured and found to agree with calculations to at least 10^{-8}. Various other scientific and ethnographic work was done by the expedition's doctor, Dr. Rivet, and once again ambitious plans were drawn up for the next year.

On 25 April 1904 Poincaré had to report (1904g) that the weather conditions had continued to obstruct the expedition, and continued political intervention was necessary to prevent the destruction of the signaling stations. But some work was done and the triangulation extended to include the better surveyed sites at Quito and Guayaquil, near the coast (see fig. 2.3). Only in 1905 could Poincaré report (1905d) that all the bases in the triangulation had been established, and he began his report that year with a further analysis of the problems that had beset the expedition, which now placed some blame on a misunderstanding with the local Indians who, having little interest in long-distance views across the mountains, had not enabled the planners to predict the weather in

Figure 2.3. The arc of Quito. Source: "Rapport sur le projet de révision de l'arc méridien de Quito." *Comptes rendus hebdomadaires de l'Academie des sciences* (1900, 131, p. 220).

a useful way. The members of the expedition had worked very hard, and could in no way be criticized for the lack of progress, but it was necessary either to abandon the expedition, or to do what could be done with the few remaining resources. It was not for the commission to decide, he said, but some remarks could be made about what could be done. To stop now, he said, would be to achieve so little that the honor of France would suffer, and its promises to the International Geodetic Association would not be kept. So one would have to continue as well as one could, and happily Prince Roland Bonaparte had been able to offer them 100,000 francs to continue the original plan. It was therefore necessary to ask the government for a further subvention of 50,000 francs.

The final report, (1907d) of 22 July 1907, opened with the remark that Captain Massenet, who had taken over the expedition from Commandant Bourgeois, had died of typhoid, and so new officers were in charge. In particular, Captain Noirel had made a success of the geodesic measurements at seventy-four stations, and Poincaré reported a happy end to the mission. The botanical and ethnographic findings were valuable, and despite the conditions the geodesic measurements seemed, provisionally, to be as accurate as those taken in France. But no mention of any significant findings in geodesy were made, other than ones concerning the behavior of the equipment. A number of expeditions were mounted by other nations with empires and international political power: the British were similarly occupied in Egypt and South Africa, and these seem to have had a more lasting impact. Certainly, nothing continued afterwards in Peru for many years. The genre of report writing, however, was upheld: bold plans, unexpected difficulties, heroic work on the ground, financial crisis, bailout, and success. One supposes the honor of all involved was saved, but the original scientific goals were quietly forgotten.

THE NOBEL PRIZE

The prize for physics was awarded for the first time in 1901, when it went to Röntgen for his discovery of what are now called X-rays, and in 1902 it was shared by Lorentz and Zeeman, for their theoretical and experimental work on the influence of magnetism on radiation, and Becquerel for his discovery of spontaneous radioactivity. In subsequent years it went to Pierre and Marie Curie, J. W. Strutt (later Lord Rayleigh) partly for his discovery of argon, Philipp Lenard for his work on cathode rays, J. J. Thomson for his work on the conduction of electricity in gases, and Michelson for optical precision implements, until in 1908 it went to the French physicist Gabriel Lippmann for his work on color photography.[34]

It was surely inevitable that Poincaré's supporters would eventually nominate him for the prize, and his career in the first decade of the 20th century continued to garner awards and prestigious honors. On 3 April 1901 he became president of the Société astronomique de France, a

---

[34] The standard account is Crawford (1985).

position to which he was reelected in 1902, when he also became the president of the Société française de physique. In May he became a member of the Conseil de perfectionnement de l'École supérieure des postes et des télégraphes, and on 4 July he was appointed professor of electrical theory in the École professionnelle supérieure des postes et des télégraphes in Paris, in recognition, one supposes, of his new interest in wireless telegraphy. He held this post in addition to the ones already acquired, and it certainly kept him busy. He was required to lecture on different topics as the years went by: on the propagation of electric current (published in 1904), telephony (1907), wireless telegraphy (1908, 1911), and in what turned out to be the last weeks of his life, on the dynamics of the electron (1913).

The first edition of *La science et l'hypothèse* came out in 1902, the same year that Poincaré was elected president of the Société française de physique. In 1903 he was promoted to Commandeur de la Légion d'honneur. In April 1904 he and Appell again reported to the Dreyfus commission, and in September he traveled to the United States to speak at the St. Louis Congress, which he chose over the ICM in Heidelberg. This was to be his only visit. He also became professor of general astronomy at the École polytechnique. In 1905 he won the first Bolyai Prize of the Hungarian Academy of Sciences, which carried a medal and a reward of 10,000 crowns. The citation (see Halsted 1906) began by saying "Henri Poincaré is at the present moment unquestionably the most powerful investigator in the domain of mathematics and mathematical physics. His strongly marked individuality lets us recognize in him the intuitive genius drawing the inspiration for his wide-reaching researches from the exhaustless fountain of geometric and physical intuition, yet capable also of working this out in detail with marvelous logical keenness."[35] He also published *La valeur de la science*, and was elected vice president of the Académie des sciences, becoming president in 1906, which was the year he was elected president of the Commission des finances de l'Association géodésique internationale. In 1907 a rare check to his progress came when he declared an interest in becoming the permanent secretary of the Académie des sciences, but withdrew when it became clear he would not win convincingly.

---

[35] Halsted was well connected with the Hungarians, who respected his advocacy of Bolyai, and he had predicted that the prize would go to Poincaré in his (1905).

On 5 March 1908 he became an immortal, when he was elected to the Académie française on the death of the poet Sully Prudhomme (who had been the recipient of the first Nobel Prize in literature), and on 3 April he agreed to become an honorary professor at the École polytechnique on no pay, because the chair was threatened with closure, a position he retained until the position was abolished some years later. But in April he was gravely ill at the Rome ICM, although he said in an article in *Le Temps* that he had been able to find out what took place by talking to other guests in his hotel.[36] This setback must have been particularly distressing to him because he had gone there to award the first Guccia medal of the Circolo matematico di Palermo (it went to Severi for his work on algebraic geometry; see Poincaré's 1908) and to renew his friendship with Giovan Battista Guccia himself, who was the mainstay of the Circolo and chief editor of its journal, the *Rendiconti*. Poincaré had supported this journal from its inception in 1885, as he had *Acta Mathematica*, and had published several of his most important papers in it. The journal had grown with the dramatic growth of mathematics in Italy since unification, and had rapidly become one of the leading ones in the field, as Poincaré was happy to tell the readers of *Le Temps*.[37] Poincaré recovered, although not completely, and in 1909 he gave six invited lectures in Göttingen, and at the end of the year he became president of the Bureau des longitudes for 1909 and 1910. In October 1910 he went to Budapest to award the second Bolyai Prize to Hilbert.

The Campaign for Poincaré

Support for the idea that Poincaré should be awarded the Nobel Prize in physics grew in the early years of the 20th century and came to a climax in 1910. The idea of the prize seems to have been an immediate success, and the prize was rapidly coveted. After much discussion in the years after Nobel's death in 1895 about how his enormous legacy of 31 million Swedish kronor should be administered, the statutes of the Nobel Foundation were promulgated by King Oscar II in 1900. The Nobel Committee for physics recommends candidates for the prize to the Royal Swedish Academy of Sciences, who then award the prize according to

---

[36] *Le Temps* (21 April 1908, no. 17102).
[37] On Guccia and the Circolo, see Brigaglia and Masotto (1982).

the terms of Nobel's bequest. Up to three candidates may be chosen, and two works. The prize is awarded at a ceremony on Stockholm on 10 December, the anniversary of Nobel's death, so the process of sending invitations to make nominations and campaigns on behalf of candidates starts in January of each year.

In 1902 Poincaré had been very active on behalf of Lorentz, and his letter of nomination (Poincaré 2007, 62.7) was signed by a number of distinguished mathematicians and physicists, including Becquerel, L. Fuchs, Planck, and probably most influentially Röntgen, as well as the principal names in French physics. In it Poincaré noted that after 20 difficult years Maxwell's theory had become accepted by physicists as a result of Hertz's experimental confirmation of its predictions. However, numerous problems remained to do with the nature of light, and the phenomenon of rotatory magnetic polarization that Faraday had discovered still lacked explanation despite all Maxwell's efforts. Lorentz's introduction of the electron had not only been able to explain all that Maxwell's theory had been able to accomplish, it also resolved the outstanding difficulties through his introduction of the ingenious idea of local time ("temps réduit"). His explanation of magnetic polarization was not only simple and natural, it led to novel predictions about the absorption and emission of light that Zeeman had been able to confirm. Lorentz's theory, Poincaré concluded in words that he was to echo in his popular writings, might eventually prove to be false, as so many theories have been shown to be, but it has revealed real and hitherto unknown relationships between apparently disconnected facts, and these are the truths that survive.

In 1904 Darboux, in his capacity as permanent secretary of the Académie des sciences in Paris, and Becquerel, both nominated Poincaré and Lippmann, Poincaré for his book *Électricité et optique* and Lippmann for his discovery of color photography. They each praised Poincaré for his many original theories and fertile views. Becquerel observed that Poincaré's mathematical and philosophical genius surveyed all of physics and was among those that contributed most to human progress by giving researchers a solid basis for their journeys into the unknown. Even these nominations reveal what was to be the greatest problem for those who wanted the Nobel Committee to recognize Poincaré's extraordinary contributions: the absence of a specific discovery, invention, or technique. Nothing in all of his many books and papers quite compared with the discovery of X-rays, the resolution of problems about light and the prediction of the Zeeman effect, or the Curies' work on radioactivity.

The most active of Poincaré's supporters was, once again, Mittag-Leffler. He was well placed in Stockholm to lobby the members of the Nobel Committee and the Royal Swedish Academy of Sciences, and he had the respect of the French mathematical community because of his journal, *Acta Mathematica*. In 1908 the prize had finally gone to Lippmann, who was chosen over Planck, and because Lippmann was French, Mittag-Leffler judged it better to wait a year. In the course of 1908 Mittag-Leffler had come to the opinion that Svanta Arrhenius had been sympathetic to the work of Poincaré, whose name eventually went forward and received five votes. Arrhenius was a physical chemist in Uppsala who had won the Nobel Prize in chemistry in 1903. He was something of a polarizing figure with friends and enemies, but he was influential because he had been involved in the creation of the Nobel Prizes, he was a member of the Royal Swedish Academy of Sciences, and a member of the Nobel Committee for physics. Mittag-Leffler thought that Arrhenius had said to Lippmann that he would support Poincaré's candidacy provided mathematicians were kept out of the picture, and so he went back to Painlevé with that in mind. But when Painlevé began to collect signatures he found that Darboux and Lippmann were already collecting signatures.

The problem was aggravated because, as Mittag-Leffler knew very well, there was a strong feeling among some experimental physicists that the definition of physics should not be broadened to include mathematical physics, even though Nobel himself had specified a broad interpretation of the term. Experimentalists, like J. J. Thomson, argued that, on the one hand, physics was really about discoveries—reliable new phenomena verified in laboratories—and, on the other, that the broader the definition of physics the less likely it was that they would win the prize.[38] This familiar mixture of self-interest and principle was visible on the Nobel Committee, and so Mittag-Leffler devised a ruse to mislead it. He obtained Painlevé's energetic support for the idea that the recent invention of powered flight was worthy of the Nobel Prize in physics, and identified himself with the cause. The attempt failed,

---

[38] See Thomson to Arrhenius, in Crawford (1985, 145).

because the committee found flight was still too dangerous (the prize for 1909 was shared between Marconi and Braun for their work on wireless telegraphy), but it made Mittag-Leffler seem more open to the claims of experimental physics.

In January 1910 the campaign for Poincaré began in earnest. Poincaré had already written an account in 1908 of his principal works in physics (Poincaré 2007, 62.19) which was carefully devised to highlight the experimental implications of his work. He dealt very briefly with his *E&O* but noted that his theoretical refutation of some of Lorentz's ideas had led Abraham to show that the mass of an electron is entirely electromagnetic, and he mentioned his paper on the dynamics of the electron (1908b). He suggested that his theoretical work on polarization by diffraction explained a defect in Fresnel's work that Gouy had discovered experimentally. He observed that his work on Hertzian waves resolved apparent contradictions in the experimental discoveries of Sarasin and de la Rive on multiple resonance. He noted his successful work on the telegraphists' equation and telegraphy, and alluded to his papers on electrotechnology and his popular essays.

Darboux had already amplified this script in January 1909, noting Poincaré's theory of the Zeeman effect, his fundamental work on rigorizing mathematical physics by his study of the Dirichlet problem and partial differential equations, his resolution of the problems left unsolved in Lorentz's theory of the moving electron, and the insights in Poincaré's popular essays. His letter was also signed by eleven other French mathematicians and physicists, among them Lippmann and Picard. In 1910 he repeated this account, stressing first of all the importance of rigorous mathematical physics. Then he praised Poincaré's work on the moving electron and the introduction of Poincaré stresses, noting too Poincaré's remarks about universal gravitation in the same paper. He made it clearer than before that Poincaré had also written on wireless telegraphy. The letter was signed by Appell and Fredholm, and a copy of Lebon's bibliographical analysis of Poincaré's work (Lebon 1909) was attached. Several other French mathematicians and physicists also wrote to the Nobel Committee in support of Darboux's letter, Appell, Boussinesq, Painlevé, Marie Curie, and Pierre Duhem among them. Michelson submitted a short letter citing the importance of Poincaré's papers on the differential equations of mathematical physics, as did Marconi. Volterra wrote at length on the importance of Poincaré's mathematical analyses. Elsewhere

Arrhenius spoke up for the importance of Poincaré's work on the shape of rotating fluid masses. But only equivocal support came from Lorentz and Zeeman, who subscribed to Darboux's nomination of Poincaré but also drew attention to the claims of van der Waals and Kammerlingh Onnes (who indeed was to win the prize in 1913 for his work on low-temperature physics).

Opposition came from the experimentalists, especially those based in England: J. J. Thomson and Ernest Rutherford. But there was a possibly stronger force at work, the nomination initiated by Röntgen in Uppsala for Knut Angstrom (the son of the more famous Anders) for his work on solar radiation. In the event Knut Angstrom died after the nomination period had closed, and although this left him posthumously eligible the prize in 1910 went to van der Waals for his work on the equation of state for liquids and gases. According to Crawford (1985, 147), the only information on the voting figures is that "a handsome number of votes" went to Poincaré. According to the report that Arrhenius then wrote for the Nobel Committee, they were willing to consider his case again in future years, largely because of his work on celestial mechanics which, curiously, had not been much stressed by his nominators. But the campaign in 1911 faltered in the face of yet another attempt by the physicists in Uppsala to secure the prize for one of their own, even though they were unsuccessful (the prize went to Wien), and Mittag-Leffler began to lose interest. With Poincaré's death in 1912 the campaign came to an end.

The Nobel Committee had a thankless task. They were under pressure to maintain the highest intellectual standards while playing fair by the major scientific nations, but could not be immune to personal prejudices and special interest groups. In later years the committee would occasionally give the prize for work done fifty years before, and sometimes they can be criticized for avoiding controversy, but in the early years they tried, surely rightly, to reward recent achievements, and it can at least be said that every one of their choices in the first decade remains a well-respected figure in the world of physics a century later.

Should Poincaré have received the prize? It would have been a departure to choose someone with no deep commitment to discovery, and a recognition of the importance of theory, and so one can ask if Poincaré's theorizing was worth the highest accolade.

Among the winners, only Lorentz can be considered a major theorist; Lenard and Wilhelm Wien understood theory as a guide to their experimental work, as all experimentalists do. But Lorentz's work changed physicists' understanding of electromagnetic theory by introducing the electron and eliminating the generally misunderstood displacement currents in Maxwell's presentation. The mathematics that Lorentz deployed was impressive, but his theory was not exclusively mathematical: it proposed a new constitution for the fundamental objects of the world. It also led to novel, confirmed, predictions. Poincaré's ideas, by comparison, were much more exclusively mathematical, as his nominators implicitly recognized. Moreover, they were not, by the highest standards, innovative. In his work on electricity and magnetism he drew together the existing theories of Helmholtz, Maxwell, Hertz, Larmor, and Lorentz, he identified their fundamental weaknesses and their mutual inconsistencies, and from time to time he improved them, most notably in his paper on the dynamics of the electron. But his mode was more responsive than innovative. Becquerel's advocacy is an attempt to deal with this weakness, but it can be read as saying that the palm belongs to the researchers, not to the man at base camp. No one at the time made the comparison, but if a measure of sheer originality is given by Poincaré's work in mathematics then his work in physics is less outstanding. Add to that the lack of experimental predictions in Poincaré's work by comparison, again, with the example of Lorentz, and one can see why the Nobel Committee took the decision it did. The strange configuration of physics is France at the time may also have played a part. It was heavily weighted to the experimental side, as indeed the French Nobel Prize winners demonstrate. Theoretical physics had prospered in the last years of the 19th century in Germany, but not in France, which is one reason Poincaré played the role he did as the leading theorist in his native land. For all his training at the École polytechnique, for all his lecture courses, books, and papers on many different branches of physics, for all his professorial chairs, he stood out as a mathematician among colleagues building new equipment and grappling daily with invisible (and, as it turned out, dangerous) new forms of radiation. As de Broglie was to say of him in 1954 (Poincaré, *Oeuvres* 11, x), he was more of an analyst than a physicist, and only too obviously so in France in 1910.

1911, 1912

A growing controversy about the reform of the French schools in 1902 caught up with him in 1911 when, along with nearly every member of the Académie française he joined the Ligue pour la culture française, founded by the poet, playwright, and former bohemian Jean Richepin, who had himself been elected to the academy in 1909, at the age of 60. Poincaré's final public speech was "L'union pour l'éducation morale," 26 June 1912, at the opening session of the Ligue. The sentiments he expressed were unexceptional: life is a constant struggle, in ethics there is no reasoning to which we cannot reply; morals must direct our feelings, our energy. This Union should reject no one; we ban only hate. Patriotism, which Poincaré applauded, should be a fight for someone or something and never against others. Finally, we should speak only with respect of those who work alongside us.

For the first three months of 1912 he was director of the Académie française. As *Nature* (16 May 1912, vol. 89, 279) reported, Poincaré gave four lectures in May 1912 at the University of London, South Kensington[39] on mathematical subjects. Three are reprinted in *DP*: (1912c) on 3 May on problems with logic and set theory; (1912f) on 4 May on space and time; and (1912g) on 11 May on quanta. The fourth, on 10 May on number theory, resembles his paper (1912b). He then went to Vienna for a meeting of the Amis du gymnase, which was another movement for educational reform, before being admitted to hospital for a prostate operation. He was worried about his health, and when he wrote to the American astronomer E. O. Lovett on 8 July declining an invitation to the Rice Institute he said that he did not expect to be fully recovered and able to travel until the end of September. It was not to be: after being discharged from the hospital Poincaré died at home from an embolism on 17 July 1912.

Funeral

Poincaré's unexpected death was the occasion for a mingling of private shock and public grief. National mourning at the funeral on 19 July was followed by over a year of writings of all kinds as journals, newspapers,

---

[39] See the *London University Gazette* (1911–1912, XI, 145).

individuals and scientific societies sought to express what they had lost. Internationally, too, people turned to words as if to fill the gap left by the death of one of France's most fluent and popular writers.

His body was taken in an elaborate cortege from his home in rue Claude Bernard where he had died. The pallbearers were the playwright Jules Claretie, the president of the Académie française, Gabriel Guist'hau, the minister for Public Instruction and Fine Arts, General Cornille, commandant of the École polytechnique, René Zeiller, the vice president of the École des mines, the astronomer Guillaume Bigourdan, who represented the Bureau des longitudes, the physicist Gabriel Lippmann, who was president of the Académie des sciences, and the mathematicians Appell, the dean of the Faculté des sciences, and Painlevé, who, like Appell, was a member of the Institut de France. Numerous national and international societies also sent representatives. After the funeral in Saint-Jacques-du-Haut-Pas the body was taken to the cemetery in Montparnasse where eulogies were given and Poincaré's body interred in the family vault.

The man they commemorated, with words at once lofty, conventional, and sincere, was taken to his grave in a ceremony described by one reporter as having "an imposing and moving simplicity."[40] But it was no mere intellectual who had drawn so many to offer condolences to the family, represented that day by Poincaré's son Léon, his brother-in-law Émile Boutroux, and his cousins Lucien Poincaré, the director of Secondary Education, and Raymond Poincaré, president of the Ministerial Council. Poincaré was remembered as a mathematician, a scientist, a public intellectual, as a patriotic servant of France, and as a prominent citizen from Lorraine. Thus Paul Appell, who like Poincaré had cause to remember the annexation of Alsace and Lorraine, ended his address, which had alluded to Poincaré's scientific brilliance, by summoning up Jules Ferry's image of those who understand "how to show behind the blue crest of the Vosges the lamentations of the vanquished." Poincaré, he said, had "worked for truth, for science, and also for his country. May his noble example inspire the youth of France!"

Writers then and later confronted the irresolvable paradox that the achievements of this best-selling author were hard for most writers to convey intelligibly and in a few words. One reporter compared him to

[40] Rollet, *Le Monde Illustré* (27 July 1912).

the Eiffel Tower, which cannot be appreciated by standing at its base but only from afar; a happy comparison that quietly underlined his importance for France.[41] Most fell back on generalities, often drawn from the heroic vocabulary, although Humbert, writing in *La nature*, 27 July 1912 did manage to observe that Poincaré's contemporaries had often had occasion to reproach him for skipping intermediate steps, proposing hasty generalizations, and even making mistakes (Rollet 2002b, 11–12).

The sheer impossibility of capturing another person in words is doubled for those who must write so close to that person's death, and doubled again when the core of the life's achievements are profound, original, and not easily understood. The man who most earnestly grappled with how Poincaré should be remembered was his former advocate, Mittag-Leffler. He had not been able to attend the funeral, but quickly set about organizing an extensive set of essays in Poincaré's memory. They took time to be written, and were eventually published in *Acta Mathematica*, but in a very different world with very different concerns in 1921. They now form the fine set of essays that occupy most of volume 11 of Poincaré's *Oeuvres*. In his short preface, Mittag-Leffler captured the significance of Poincaré in words once said about another of his heroes, Abel: "Where he has been it is impossible to think without him." There is also an interesting collection of essays about Poincaré in volume 21 of the *Revue de métaphysique et de morale* (1913), which contains the first version of Hadamard's account of Poincaré's mathematics (not very different from the one later published in *Acta Mathematica*), but also essays on the physics by Langevin, astronomy by Lebouf, and Poincaré's philosophy by Brunschvicg that were not reprinted in volume 11.

REMEMBERING POINCARÉ

As just noted, Mittag-Leffler's collection of commemorative essays was ready to be printed in *Acta Mathematica* in 1915 but was held over to 1921 because of the First World War. In the same period only one volume (1916, vol. 2,) of Poincaré's *Oeuvres* was edited and published; volume 1, the second to appear, came out in 1928. Picard was in charge, a man Poincaré's wife referred to when talking to Mittag-Leffler in May 1920,

[41] André Beaunier, *Journal des débats* (18 July 1912). Quoted in Rollet (2002b, 10).

as "the enemy," and she correctly predicted that the edition of Poincaré's work would not be finished in Picard's lifetime (Stubhaug 2010, 613).[42] These delays are indicative of a number of factors that made Poincaré's influence problematic. Although there was a doctoral degree in France, the still poorly organized system for graduate training in mathematics meant that there was no "school of Poincaré" to continue his work once he was gone. In fact, it is hard to name other mathematicians of his generation or before in France who had pupils. Neither Jordan nor Darboux among Poincaré's teachers, for example, saw it as their job to train up bright students to continue and extend their work. Nor did Hermite or Bertrand, although they were extremely influential behind the scenes in French mathematical life and established through marriage quite a network of mathematicians, and nor did Poincaré's exact contemporaries, people like Picard, Appell, and Goursat. Nor had he inspired mathematicians or physicists to continue what he had begun. Even Jacques Hadamard, who was closest to Poincaré's interests in dynamical systems and partial differential equations, kept a certain distance from him, remarking in 1937 that Poincaré's works "were admired, but no-one 'dared touch them.'"[43] Élie Cartan was another highly original mathematician who found enough original ideas without feeling a need to feed off Poincaré's inventiveness, nor did good, but not great mathematicians, gravitate around him, happy to pick up, extend, and make precise the brilliant but sometimes vague ideas that Poincaré came up with.

The contrast with Klein and Lie is quite marked in this respect. Both men recruited good mathematicians to write up and extend their work—Poincaré even assisted, by helping Vessiot to study with Lie—and this kept their work alive and, in Lie's case, helped to spread its influence internationally. But Poincaré was no easier to learn from in person than he was in print, and moreover the contrast with the situation in Göttingen could not have been more marked. As Pierre Boutroux observed to Mittag-Leffler,

The opposite of certain researchers, he did not believe that oral communication, the verbal exchange of ideas, could help discovery.

[42] For the series of 11 volumes of Poincaré's *Oeuvres*, see (1956).
[43] Quoted in Maz'ya and Shaposhnikova (1998, 39).

Figure 2.4. Émile Picard (1856–1941). Source: Arild Stubhaug, *The Mathematician Sophus Lie* (Springer, 2002). ©Mathematical Library, University of Oslo, Blindern.

This reserve in my uncle struck me particularly when I spent several months in Göttingen and witnessed totally different customs. One knows what an admirable focus of common thought and collective work is this famous German university. There everything happens in the open. Scarcely has the foreigner disembarked in this small Hannoverian city than he knows already what are important people in the town are working on, where they have reached and what difficulties have stopped them. Ideas, advocated, confronted, discussed, in the course of walks in the forest and in meetings of the mathematical Society, ripen on their own in this fertile milieu, where Klein's always alert curiosity and energy help to sustain an inexhaustible ferment. (P. Boutroux 1921)

This contrast only grew with the rise of David Hilbert, which was so marked that, as Veblen reported to Birkhoff on Christmas Day 1913, "No-one in Germany is interested in anything Hilbert has not worked with."[44]

[44] Quoted in Barrow-Green (2011, 42).

Add to this a tradition that for several decades the best German mathematics departments had cultivated a graduate education, structured around small seminars with a dedicated library of specialist books and journals. There was a role for the supervisor and the doctoral student, and whatever else, it produced a number of students who always spoke warmly of the doctor fathers. These new mathematicians usually began their careers working in the same area as their supervisors, they might even be one of a small group of such students, and while the best of them might have benefited little more from the experience than they would have done in any other setting, and might well move on to other fields, this structured environment was good for keeping the supervisors' ideas alive. It was the German model that E. H. Moore took back to Chicago, and that gradually spread around the world, and it was lacking in France.

The first to build up anything like a coherent group of researchers in Paris was Émile Borel and the people he drew in gravitated to the new analysis, with its inclinations toward set theory, and its predilection for abstraction. Mathematicians such as Baire, Fréchet, Lebesgue, and Montel staked out a territory that was overwhelmingly pure mathematics, and to which Poincaré was little attracted.[45] In mathematics, at least, there was a generational shift underway that moved the subject away from Poincaré's deepest concerns, which were for mathematical physics. Nor, of course, was Poincaré professionally a mathematician. French physicists, on the other hand, were reverential of, rather than inspired by him. It was not until 1911 that Langevin took up special relativity, nor were the French much drawn to quantum mechanics in its earliest years; none of them shared Planck's enthusiasm for the young, enthusiastic, productive 57-year-old and his ideas.

But in the end Poincaré's light dimmed mostly because of the First World War. It was a disaster for the French mathematical community, who sacrificed more of their number in the trenches than did the British or the Germans, and after it was over the German community recovered much faster. In the 1920s Göttingen rose while Paris struggled, and in the 1930s the next generation of mathematicians in France, the Bourbaki

---

[45] As we can see from his work. The famous quote about logic sometimes breeding monsters comes from his paper on the teaching of mathematics in schools (1904c), and makes the point that teaching cannot start there, as logic would seemingly require. It is preceded by sensible comments on how pupils understand the definition of a circle, and argues for intuition before rigor, not against the rigorous study of functions.

generation, opted overwhelmingly to catch up with the Germans in pure mathematics. The field of mathematical physics was sustained by Élie Cartan and Hadamard, and it is possible to see them as Poincaré's successors, keeping alive a vision of mathematics in lively communication with physics, but the disciplinary divide grew, and theoretical physics in France entered a prolonged period of decline.

At the same time, German physics entered another of its glorious periods, and while one should not make too much of the supposed gaps between the mathematics and physics communities in Göttingen and Berlin, and the full story is only being written now, it is clear that German mathematicians were deeply interested in what their colleagues were doing with quantum theory, and Weyl, von Neumann, and van der Waerden made very significant contributions to it.[46] In none of these areas was Poincaré's contribution a live factor any more. Einstein, too, was being marginalized, Lorentz slipping into the comfort of a Nobel Prize (he died in 1928). When functional analysis and integral equations became the staple mathematics of the new physics, there was every reason for this to keep a line open back to the work of Hilbert and his followers from 1904 to 1910, none to send people back to the work of Poincaré which began to be forgotten.

Hilbert's interest in mathematical physics and in logic and the foundations of mathematics deepened in the 1920s, and he began to articulate his disagreements with Poincaré. His disagreement with geometric conventionalism will be looked at below in chapter 11, section "Geometric Conventionalism," and as for logic his much more developed theory of finitistic mathematics with its metamathematical superstructure, whatever its fate at the hands of Gödel may have been, looked a lot closer to the analysis of mathematics in its relation to modern logic than anything Poincaré had managed.

Recently, many factors have worked to change this. There was a great revival of three-dimensional geometry and topology to which Thurston's work was such a notable contribution, a revival in dynamical systems and celestial mechanics with space exploration, a reembrace of mathematical physics. A century after his death, as the subsequent chapters will show, in each of the fields Poincaré worked, his achievements in mathematics, physics, and philosophy are still alive and important today.

---

[46] A German, a Hungarian, and a Dutchman, but all working in Germany.

3

The Prize Competition of 1880

THE COMPETITION

On 10 March 1879 the Académie des sciences in Paris announced the topic for many of its customary essay competitions. Candidates for the Grand Prize for the mathematical sciences were asked

> to improve in some important way the theory of linear differential equations in a single independent variable. (*CR* 1879, 88, 511)

The prize was a medal to the value of 3,000 francs. On 12 April 1880 the academy announced that the panel of judges would comprise Bertrand, Bonnet, Puiseux, and Bouquet, with Hermite as rapporteur. Joseph Bertrand, then 58, was a professor at the École polytechnique who lectured at the Collège de France, and had been a member of the Académie des sciences since 1856. He was influential in scientific circles in Paris, and later Appell and Picard became his nephews.

These competitions were a legacy from the 18th century, when the various learned academies of Paris, St. Petersburg, and Berlin had regularly offered prizes on selected topics. But in the 19th century, when intellectual leadership passed to the universities, the tradition proved steadily harder to maintain. More and more often the question languished, was perhaps readvertised for a further two years, and then the prize was awarded, after all, to someone for their work in the area. Indeed, the previous topics of the Grand Prize for the mathematical sciences had been on algebraic curves and planetary astronomy, and when no essays were submitted these topics were withdrawn.

The Parisian Académie des sciences tried to defy the trend. Care was put into the selection of topics, and doubtless note was taken of the likely entrants. After the Franco-Prussian War topics were selected that connected to the French desire to catch up with the Germans (as was the case with the topic of algebraic curves). In choosing the topic of linear differential equations, Hermite's intention was to inspire young French mathematicians to study the work of Lazarus Fuchs, a German mathematician he respected and who had been writing a series of innovative papers on the subject since 1865, and who most recently had pushed his work in a significant new direction.

Fuchs (1866) had begun by characterizing those linear differential equations that have solutions whose singular points are determined by the equation and which are everywhere meromorphic including at infinity (they have no essential singular points). These equations are the natural generalization of the hypergeometric differential equation:

$$z(z-1)\frac{d^2w}{dz^2} + (c - (a+b+1)z)\frac{dw}{dz} - abw = 0. \tag{3.1}$$

Fuchs showed that such equations have the form

$$\frac{d^nw}{dz^n} + \frac{p_1(z)}{\psi(z)}\frac{d^{n-1}w}{dz^{n-1}} + \cdots + \frac{p_n(z)}{\psi(z)^n} = 0, \tag{3.2}$$

where $p_j(z)$ is a polynomial of degree less than or equal to $j(r-1)$ and

$$\psi(z) = (z-a_1)(z-a_2)\cdots(z-a_r).$$

The singular points of the solutions are precisely the points $z = a_1, a_2, \ldots, a_r$. Equations of this form were later said to be of the Fuchsian class or type in recognition of Fuchs's discovery. In particular, the hypergeometric equation is of this type, with one of its singular points at infinity.

Every solution of the hypergeometric equation can be written as a sum of two independent solutions, say $y_1(x)$ and $y_2(x)$, and an important way to understand the solutions is to look at the behavior of the quotient of these solutions $y(x) = \frac{y_1(x)}{y_2(x)}$ as the variable x is taken on paths in the complex domain. In this way the quotient is revealed to be a many-valued function and the inverse function $x = x(y)$ is consequently a periodic function. Indeed, special cases of the hypergeometric equation are intimately tied to the theory of elliptic functions.[1]

---

[1] See the appendix. The general importance of elliptic functions within 19th century mathematics would take a good part of a book to describe; see, for example, Bottazzini and Gray (2012) or Houzel (2002).

We do not know exactly when Poincaré first began to work on the prize problem of the academy, and certainly 1879 was a busy year for him. Possession of a doctorate enabled Poincaré to take up a position as a lecturer at the nearby University of Caen on 1 December, and he must also have been thinking about real differential equations and their solutions, for on 22 March 1880, he submitted a memoir on this topic to the prize competition. But on 14 June he withdrew it, before the examiners could report on it, and he replaced it with his first account of linear ordinary differential equations in the complex case on 29 May 1880. Like the doctoral thesis, this essay was only published posthumously, in *Acta Mathematica* 39 and the first volume of his *Oeuvres* (1928, 336–373).

The new topic was not only much closer to the work of Fuchs, it was also in line with a French tradition of which Briot and Bouquet had been the leading exponents.[2] Back in the 1850s they had written a series of three papers and then a book (Briot and Bouquet 1856a, b, c; 1859) in which they gave the theory of linear ordinary differential equations in the complex domain, with particular reference to the equations that arise in elliptic function theory. The book was the first in French and one of the first in any language to expound Cauchy's theory of complex functions of a complex variable. Fuchs's work had, however, taken the theory much further.

To understand what Poincaré did, it is necessary to recall the principal results of Fuchs's theory in more detail. Given a second-order linear ordinary differential equation of the Fuchsian class, such as the hypergeometric equation, and a disklike neighborhood in the complex plane containing no singular point of the coefficients of the equation, it is possible to find two distinct solutions of the equation with the property that every solution of the equation is a linear combination of these solutions. These two solutions are called a basis of the solutions, and when these basis solutions are continued analytically around a singular point they will return as a linear combination of the basis solutions. When one of the singular points is at the origin, Fuchs showed that the differential equation takes the form

$$\frac{d^2 y}{dx^2} + \frac{p(x)}{x}\frac{dy}{dx} + \frac{q(x)}{x^2} = 0,$$

---

[2] Briot and Bouquet were regarded as almost inseparable; even Poincaré referred to Bouquet as "le binôme indispensable de Briot"; see Bellivier (1956, 133).

where the functions $p(x)$ and $q(x)$ are holomorphic in a neighborhood of the origin, and the basis solutions are of the form $x^\alpha f(x)$ and $x^\beta g(x)$, where the functions $f(x)$ and $g(x)$ are also holomorphic in a neighborhood of the origin.

When the solution $x^\alpha f(x)$ is analytically continued once in the positive sense around the origin, it returns as $x^{2\pi i \alpha} f(x)$, and the other basis solution returns as $x^{2\pi i \beta} g(x)$. To find the values of α and β one writes $y = x^\alpha f(x)$ and calculates

$$\frac{dy}{dx} = \alpha x^{\alpha-1} f(x) + x^\alpha \frac{df}{dx},$$

and

$$\frac{d^2 y}{dx^2} = \alpha(\alpha - 1)x^{\alpha-2} f(x) + 2\alpha x^{\alpha-1} \frac{df}{dx} + x^\alpha \frac{d^2 f}{dx^2},$$

and substitutes them into the equation. The lowest power of x is $x^{\alpha-2}$ and on letting x tend to 0 one obtains

$$\alpha(\alpha - 1) + p(0)\alpha + q(0) = 0.$$

This is called the indicial equation for the exponents; it has two roots, α and β.

Different cases now arise according to whether the roots differ by a quantity that is not an integer, differ but by an integer, or are the same. The first case is straightforward and the above analysis essentially complete. The holomorphic functions $f(x)$ and $g(x)$ can now be found by the method of undetermined coefficients: one substitutes, for example, $f(x) = \sum_{j=0}^{\infty} a_j x^j$ into the differential equation and calculates the successive coefficients a_j.

In the second case, setting $\alpha = \beta - n$, where n is an integer, the analytic continuation of $x^\alpha f(x)$ and $x^\beta g(x)$ are

$$x^{2\pi i \alpha} f(x) \text{ and } x^{2\pi i \beta} g(x) = x^{2\pi i (\alpha+n)} g(x) = x^{2\pi i \alpha} g(x).$$

In this case, the method of undetermined coefficients finds that $f(x) = g(x)$ but Fuchs showed that an independent solution is obtained that involves a logarithmic term, one of the form $\log(x) f(x)$. In the third case a basis of solutions is given by $u(x) = x^\alpha f(x)$ and $\frac{\partial u(x)}{\partial \alpha}$ evaluated at $\alpha = 0$.

In the essay Poincaré submitted for the prize, he took a second-order linear differential equation of the Fuchsian class, which he wrote in the simplified[3] form $\frac{d^2y}{dx^2} + Q(x)y = 0$, and denoted a basis of solutions by $f(x)$ and $\varphi(x)$. He then focused on the question of when the quotient $z = \frac{f(x)}{\varphi(x)}$ defines, by inversion, a meromorphic function x of z. Fuchs had claimed that for x to be meromorphic on some domain it was necessary and sufficient that the roots of the indicial equation at each singular point, including infinity, differ by an aliquot part of unity (their difference is of the form $\frac{1}{n}$ for some positive integer n). Poincaré found these conditions were neither necessary nor sufficient, because Fuchs had neglected to consider fully the nature of the domain of definition of the inverse function. Poincaré found that if the domain is required to be the whole complex sphere then this condition is necessary, but it is not sufficient.

In fact, the more he looked, the more special cases began to appear, and so he decided to start again with the simplest possible cases. This led him to introduce simple geometrical considerations. When a differential equation has three singular points they may be mapped by a Möbius transformation to the points $z = 0, 1$, and ∞, and the differential equation is transformed to one of the same form as before. These three points are thought of as the vertices of two unusual "triangles": the upper- and lower-half planes. The quotient of the basis of solutions maps a neighborhood of every point of either "triangle" other than a vertex holomorphically onto the image, and when the coefficients of the differential equation are, as Poincaré assumed, real, the sides of the "triangles" are mapped either to straight lines or arcs of circles. The sides of the images meet at angles determined by the exponent differences: an exponent difference of $\frac{1}{n}$ corresponds to an angle of $\frac{\pi}{n}$, as the monodromy consideration shows. Once one image of each half plane has been established all further images are established by analytical continuation of the z variable, which transforms the image triangles by Möbius transformations, as the monodromy equations say. So they too are triangles with the same angles and circular-arc sides, and they fill out a net of such triangles without overlaps (other than exact ones) because the angles are aliquot parts of π. There is no doubt that these intuitive

---

[3] This can always be done, by substituting $\phi(z) = \exp\left(-\frac{1}{2}\int p_1(z)\,dz\right)$ into the general equation $\frac{d^2w}{dz^2} + \frac{p_1(z)}{\psi(z)}\frac{dw}{dz} + \frac{p_2(z)}{\psi(z)^2} = 0$.

geometrical arguments enabled him to sort out the rigorous and the mistaken parts of Fuchs's ideas.

Poincaré began with an example of Fuchs's where the differential equation has two finite singular points and a singular point at infinity, with the corresponding exponent differences $\frac{1}{2}$, $\frac{1}{3}$, and $\frac{1}{6}$. These forced x to be a meromorphic single-valued function of z mapping a parallelogram composed of eight equilateral triangles onto the complex sphere, and $z = \infty$ is its only essential singular point, so x is an elliptic function.[4] The differential equation, Poincaré showed, has in fact an algebraic solution and a nonalgebraic solution. This result agrees with Fuchs's theory.

Poincaré next investigated for several pages when a doubly periodic function can give rise to a second-order linear differential equation, and eventually concluded (p. 79) that Fuchs's theory was correct when one solution of the original differential equation is algebraic, but, for example, extra conditions are needed when the differential equation has four singular points and elliptic functions are involved.

Matters got more interesting, and their implications proved more significant, when it was impossible for the x domain to be the whole z sphere. Poincaré gave an example to show that this could happen even when the differential equation has only two finite singular points. He gave an example where the exponent differences are $\frac{1}{4}$ and $\frac{1}{2}$ at the finite points and $\frac{1}{6}$ at ∞. In this case, when the finite singular points are joined to ∞ by cuts, then as long as x crosses no cuts the quotient $z = \frac{f(x)}{g(x)}$ stays within the quadrilateral $\alpha O \alpha' \gamma$ (see fig. 3.1). The image in the z plane of the upper- and lower-half x planes are triangles that form a quadrilateral joined along the image of the line joining the singular points.

Poincaré now showed that as x is conducted about in its plane, the values of z lie inside the circle HH'. All the images of the upper- and lower-half planes taken together are quadrilaterals that he called "mixtiligne," because their sides are circular arcs meeting the circle HH' at right angles. The same phenomenon happens for a range of similar differential equations: curvilinear polygons are obtained with non-reentrant angles and sides that are circular arcs orthogonal to the boundary circle. They fill out the domain of the inverse function $x = x(z)$

---

[4] To illustrate his argument Poincaré added a sketch of a net of these triangles in which the hexagons are white and the two extra triangles making up the parallelogram are shaded. The shading was done incorrectly, but in what was indicative of a lifelong practice Poincaré submitted it anyway.

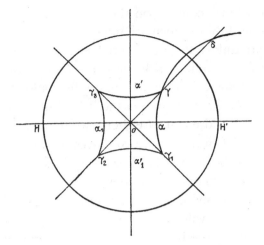

Figure 3.1. A triangle in the disk. Source: Poincaré, *Oeuvres* (1, 365).

in $|z| < OH$, and Poincaré then investigated whether the function x is meromorphic. He reduced this question to showing that, as x is continued analytically, the polygons do not overlap, and indeed they will not overlap if the angles satisfy conditions derived from Fuchs's theory, unless the overlap is in the form of an annular region. However, if the angles are not reentrant, this cannot happen, and so x is a meromorphic function of z.

The Correspondence between Poincaré and Fuchs

Once he had submitted his essay, Poincaré wrote to Fuchs on 29 May 1880 (see 1921b) to raise some of the problems that had occurred to him while reading Fuchs's work (1880). One of his first questions concerned the nature of the inverse function $x = x(z)$. Fuchs had claimed that it is always a meromorphic function, even at points where x is a singular point of the differential equation, and indeed that z is finite at ordinary points and infinite at singular points. Poincaré observed that if, however, x is meromorphic at $x = \infty$ then x is meromorphic on the whole x sphere, and so it must be a rational function of z. This, however, implies that all the solutions of the original differential equation must have been algebraic, which Fuchs had expressly denied.

  Poincaré suggested that the problem was with the domain of definition of the inverse function. There were three kinds of z value: those reached as x traced out a finite contour on the x sphere; those reached on an infinite contour, and those which are never attained. All three situations could occur, and unless the differential equation has only algebraic solutions the last two would, but Fuchs's proof only worked for z values of the first kind. However, Poincaré went on, he could show that $x = x(z)$ was meromorphic in the other cases as well, and he proposed that when all the z values were of the first case then z would be a rational function; if there are values of only the first and second kinds, and x is single valued at values of the second kind, then Fuchs's theorem is still true; but if x is not single valued or if values of the third kind occur and so the domain D of x is only part of the z sphere, then x is single valued on D. In this case the z values of the first kind occur inside D, those of the second kind lie on the boundary of D, and the unattainable values lie outside D. Finally there is a fifth case, when all three kinds of x value occur, but D is annular and values of the first kind fill out the annulus. In this case, said Poincaré, x will not return to its original value on tracing out a closed curve $HHHH$ in D.

Fuchs replied on 5 June (see 1921b). He admitted that his theorem I was imprecisely worded, and proposed to return to the conditions on the exponents at the singular points that he had imposed in his earlier *Göttingen Nachrichten* articles. He added that he excluded paths in which $f(x)$ and $g(x)$ both become infinite, which, he said, ensured that the remaining z values filled out a simply connected region of the z plane with the excluded values on the boundary. Poincaré replied on 12 June to say that some parts of the proof were still obscure. These concerned the limiting behavior as the quotient of the solutions is analytically continued indefinitely and the corresponding question of the domain of the inverse function. He therefore suggested this argument. Join the singular points of the differential equation to ∞ by cuts that do not cross each other; this defines a simply connected region on the x sphere. The image of the region under the map $z = z(x)$ is a connected region F_0. If x crosses the cuts no more than m times, then the values of z fill out a connected region F_m. As m tends to infinity F_m tends to the region Fuchs called F, and Fuchs had claimed that F will be simply connected if F is simply connected for all m. "Now," asked Poincaré, "is that a consequence of your proof? One needs to add some explanation." He agreed that F_m could not form a disk-shaped region with branch points, but it could form an annular region. Yet again, it was the younger man who was gently explaining analytic continuation and the difference between single-valued and unbranched functions to someone who had been an expert for some fifteen years. Fuchs's study was in the tradition of complex analysis that emphasizes local questions about the nature of singular points; Poincaré had in mind global questions about the domains of functions.

Poincaré then added that when there were only two finite singular points the inverse function $x = x(z)$ was a single-valued function, adding,

> That I can prove differently, but it is not obvious in general. In the case where there are only two finite singular points I have found some remarkable properties of the functions you define, and which I intend to publish. I ask your permission to give them the name of Fuchsian functions.

In conclusion, he asked if he might show Fuchs's letter to Hermite.

Fuchs replied on the 16th with a promise to send an extract of his forthcoming complete list of the second-order differential equations of the kind he was considering, and which, he claimed, rendered any further discussion superfluous. He added that he was very pleased about the name, and of course his replies could be shown to Hermite. Poincaré was not to be stopped so easily. In his reply on 19 June he pointed out that when the differential equation was put in the form $\frac{d^2y}{dx^2} + q(x)y = 0$, Fuchs's conditions implied that at all the finite singular points the exponent difference was an aliquot part of 1 and not equal to 1, and there were no more than three finite singular points. When there is precisely one, x is necessarily a rational function of z. When there are two, and the exponent differences are ρ_1, ρ_2, and ρ_3 at infinity, then Poincaré considered two cases. When $\rho_1 + \rho_2 + \rho_3 > 1$, x is rational in z; in this case the triangles are on a sphere and finitely many of them cover the sphere exactly. When $\rho_1 + \rho_2 + \rho_3 = 1$ then x is doubly periodic; in this case the triangles tessellate the plane, but even in this case there were difficulties, as he showed with an example. Finally, if there were three finite singular points, then, as Poincaré showed with another example, x was not a single-valued function of z, so the theorem is wrong.

It is at this point that Poincaré decisively branched out on his own. He now proposed to drop the requirement that Fuchs's functions $x_1 + x_2$ and $x_1 \cdot x_2$ be single valued in u_1 and u_2. As he put it,[5] "this gives a

---

[5] Poincaré to Fuchs, 19 June 1880, *Oeuvres* (11, 22).

much greater class of equations than you have studied, but to which your conclusions apply. Unhappily, my objection requires a more profound study, in that I can only treat two finite singular points." Dropping Fuchs's conditions opens up the possibility that the exponent differences ρ_1, ρ_2, and ρ_3, although still aliquot parts of unity, satisfy $\rho_1 + \rho_2 + \rho_3 < 1$. Now x is neither rational nor doubly periodic, but is still single valued. Poincaré explained,

> It is these functions that I have called Fuchsian, and by the use of this transcendental function and another attached to it I solve differential equations with two singular points...whenever ρ_1, ρ_2 and ρ_3 are commensurable with each other. Fuchsian functions are very like elliptic functions, they only exist inside in a certain circle and are meromorphic inside it.

On the other hand, he concluded, he knew nothing about what happened when there were more than two singular points; his geometrical analysis was confined to triangles.

The process of discovery was recalled by Poincaré twenty-eight years later when he addressed the Société de psychologie in Paris, and is supported by the correspondence with Fuchs as well as what Poincaré went on to do in the summer of 1880. Poincaré explained in the lecture:

> For two weeks I tried to prove that no function could exist analogous to those I have since called the Fuchsian functions: I was then totally ignorant. Every day I sat down at my desk and spent an hour or two there: I tried a great number of combinations and never arrived at any result. One evening I took a cup of coffee, contrary to my habit; I could not get to sleep, the ideas surged up in a crowd, I felt them bump against one another, until two of them hooked onto one another, as one might say, to form a stable combination. In the morning I had established the existence of a class of Fuchsian functions, those which are derived from the hypergeometric series. I had only to write up the results, which just took me a few hours.
>
> I then wanted to represent the functions as a quotient of two series; this idea was perfectly conscious and deliberate; the analogy with elliptic functions guided me. I asked myself what must be the property of these series, if they exist, and came without difficulty to construct the series that I called thetafuchsian. (*S&M* 51, 392)

It is most likely that this account is Poincaré's memory of the period between 29 May and 12 June, when he had discovered a class of functions he wished to call Fuchsian, but it may refer to the period between 12 and 19 June 1880, which would fit with the richer account of the new functions.

Next comes the realization that, more than anything else in this part of his work, was to make Poincaré's name among mathematicians:

> At that moment I left Caen where I then lived, to take part in a geological expedition organized by the École des Mines. The circumstances of the journey made me forget my mathematical work; arrived at Coutances we boarded an omnibus for I don't know what journey. At the moment when I put my foot on the step the idea came to me, without anything in my previous thoughts having prepared me for it; that the transformations I had made use of to define the Fuchsian functions were identical with those of non-Euclidean geometry. I did not verify this, I did not have the time for it, since scarcely had I sat down in the bus than I resumed the conversation already begun, but I was entirely certain at once. On returning to Caen I verified the result at leisure to salve my conscience. (*S&M* 51–52, 392)

When Poincaré wrote to Fuchs on 12 June he said that he had been away from Caen, and went on to speak of the remarkable properties of the new functions, which suggests the bus trip took place before the 12th. In this case the two breakthroughs happened in a rush, which is not quite the impression Poincaré's account suggests. But it could be that the bus journey took place between the 12th and the 19th. Either way, we have two separate breakthroughs. In the first Poincaré found a way to understand the generalized periodicity of the new functions; in the second he made a connection with non-Euclidean geometry that greatly simplified how this new form of periodicity could be analyzed.

The realization on boarding the bus was surely that a simple geometric transformation converts the pictures of a net inside a disk made of triangles with circular-arc sides into a net of triangles inside a disk but with straight sides—and this is the non-Euclidean disk of Beltrami. It follows that the original net of triangles can be regarded as made up of congruent copies of the same triangle, where congruence is to be taken with its non-Euclidean meaning.

The First Supplement

Poincaré now wrote the first of three supplements to his prize essay, which reached the academy on 28 June 1880.[6] It is 80 pages long, and begins by discussing the validity of Fuchs's theorem when there are only two finite singular points, all the exponent differences ρ_1, ρ_2, and ρ_3 are reciprocals of integers, and $\rho_1 + \rho_2 + \rho_3 < 1$. Poincaré showed that the quotient z maps the complex x sphere onto a quadrilateral Q, and analytic continuation in x maps Q onto a neighboring copy of itself obtained by rotating it through an angle of $\frac{2\pi}{\rho_1}$ about an appropriate vertex—the geometric language is new and significant. Another copy is obtained by a rotation through $\frac{2\pi}{\rho_3}$ about another vertex. Poincaré called these rotations M and N, and observed (1997, 31) that the copies of Q obtained by analytic continuation in this fashion fill out a disk, and that each copy of Q can be reached by a succession of crabwise rotations

$$M^{L_1} N^{K_1} M^{L_2} N^{K_2} \cdots .$$

All these motions preserve the boundary circle, and taken together they form a group (1997, 32). In this connection, Poincaré remarked,

> There are close connections with the above considerations and the non-Euclidean geometry of Lobachevskii. In fact, what is a geometry? It is the study of the group of operations formed by the displacements to which one can subject a body without deforming it. In Euclidean geometry the group reduces to the rotations and translations. In the pseudogeometry of Lobachevskii it is more complicated. Indeed, the group of operations formed by means of M and N is isomorphic to a group contained in the pseudogeometric group. To study the group of operations formed by means of M and N is therefore *to do the geometry of Lobachevskii.* Pseudogeometry will consequently provide us with a convenient language for expressing what we will have to say about this group. (1997, 35; italics in original)

Poincaré proceeded to develop the convenient language of non-Euclidean geometry, defining points, lines, angles, and equality of

---

[6] They were omitted from Poincaré's *Oeuvres* and finally published, with a commentary on which this account is based, as (1997).

figures—two figures are equal if one is obtained from another by a non-Euclidean transformation. Since the copies of Q do not overlap, the inverse function $x = x(z)$ is a function "which does not exist outside the circle and which is meromorphic inside this circle." Poincaré continued,

> I propose to call this function a Fuchsian function....The Fuchsian function is to the geometry of Lobachevskii what the doubly periodic function is to that of Euclid. (1997, 37)

Such functions only illuminate the study of differential equations if they can be defined independently of the equations. This Poincaré proceeded to attempt by means of the Fuchsian series he introduced, but he was unable to obtain the full strength of the results he wanted because he could not guarantee convergence of the series involved, and in a note between pages 23 and 24 (1997, 39) he remarked "I have not been able to deduce the results I wanted from the consideration of Fuchsian series; however, I thought I should mention them because I remain convinced that they will find application in the theory of Fuchsian functions."

He then turned to define related classes of functions, including those "we shall call Zetafuchsian functions because they seem to us to be analogous to the Zeta functions one considers in the theory of doubly periodic functions" (1997, 55). These, he said, could be used to solve differential equations with rational exponent differences and two finite singular points. In (1997, 62) he introduced the series, which "I call...the Thetafuchsian series because of its numerous analogies with the functions" (1997, 64) defined as

$$\sum_m H(zK) \left(\frac{dzK}{dz} \right)^m$$

summed over operations K in the corresponding group, where H is a rational function. He proved the series converged when $m > 1$ by a very similar argument to the earlier one, and noted that these series were of two kinds, one holomorphic in the circle if H has no poles inside the circle, and the other meromorphic when H does have poles inside the circle. Moreover, "the quotient of two Thetafuchsian series (corresponding to the same value of m) is a rational function of the Fuchsian function" (1997, 64).

Finally Poincaré summarized his work so far by observing that the new classes of analytic functions solve many kinds of linear differential equation with algebraic coefficients. In particular the new functions allowed one to solve the hypergeometric equation whenever the exponent differences are rational and no logarithmic term appears in the solution.[7]

However, Poincaré remained stuck on the case of the hypergeometric equation until at least 30 July, when he wrote again to Fuchs. He now described his new function this way:

> They present the greatest analogy with elliptic functions, and can be represented as the quotient of two infinite series in infinitely many ways. Amongst these series are those which are entire series playing the role of Theta functions. These converge in a certain circle and do not exist outside it, as thus does the Fuchsian function itself. Besides these functions there are others which play the same role as the Zeta functions in the theory of elliptic functions, and by means of which I solve linear differential equations of arbitrary orders with rational coefficients whenever there are only two finite singular points and the roots of the three determinantal equations are commensurable. I have also thought of functions which are to Fuchsian functions as Abelian functions are to elliptic functions, and by means of which I hope to solve all linear equations when the roots of the determinantal equations are commensurable. In the end, functions precisely analogous to Fuchsian functions will give me, I think, the solutions to a great number of differential equations with irrational coefficients.

This time it was arithmetic that had generated the new idea (see chap. 9, sec. "Number Theory" below), as he recalled in his lecture in 1908:

> I then undertook to study some arithmetical questions without any great result appearing and without expecting that this could have the least connection with my previous researches. Disgusted with my lack of success, I went to spend some days at the sea-side and thought of quite different things. One day, walking along the cliff, the idea came to me, always with the same characteristics of brevity, suddenness, and immediate certainty, that the arithmetical transformations of ternary indefinite quadratic forms were identical with those of non-Euclidean geometry.

---

[7] The term "hypergeometric" was never used by Poincaré in 1880.

Once back at Caen I reflected on this result and drew consequences from it; the example of quadratic forms showed me that
there were Fuchsian groups other than those which correspond to
the hypergeometric series; I saw that I could apply them to the
theory of thetafuchsian series, and that, as a consequence, there
were Fuchsian functions other than those, which derived from the
hypergeometric series, the only ones I knew at that time. I naturally
proposed to construct all these functions; I laid siege systematically
and carried off one after another all the works begun; there was one
however, which still held out and as the chase became involved it
took pride of place. But all my efforts only served to make me know
the difficulty better, which was already something. All this work was
quite conscious. (*S&M* 52–53)

Ternary indefinite quadratic forms (objects of the form $x^2 + y^2 - z^2$)
were exactly what he had been studying in his earliest publications; they
are closely connected to Hermite's number theory.

The Second and Third Supplements

Poincaré now wrote another supplement to his essay, which reached
the academy on 6 September 1880. With his typical honesty, it
begins,

I fear that my first supplement was lacking in clarity, and believe
that it is not pointless, before generalizing the results obtained, to
go over these same results again in order to provide some additional
explanations. (1997, 75)

These additional explanations included a more explicit description
of the non-Euclidean geometry of the disk, defining point, line, angle,
distance between two points, and area. Poincaré also observed that
the maps preserving these quantities (and the boundary circle) are
precisely the maps of the form $z' = \frac{az+b}{cz+d}$, which he called "mouvements
pseudogéométriques," distinguishing between rotations (which, he said,
have two real fixed points) and translations (which have none). The
choice of the word "real" (réel) was unfortunate; he plainly meant "point
inside or outside the circle" as opposed to points on it, which are at
infinity in non-Euclidean geometry.

Then he turned the decomposition of the disk into triangles whose angles are aliquot parts of π and gave a third demonstration (1997, 80), making clear for the first time that every point inside the fundamental circle does lie in some copy of the quadrilateral Q. He now proved it rigorously by showing explicitly how to cover a path from a given point D to the center O with a finite number of copies of Q, using the fact that OD has finite non-Euclidean length.

The first novelty in the supplement was the decomposition of the disk into arbitrary polygons with angles aliquot parts of π. As before, it is necessary to show that the copies of the polygon either overlap completely or not at all. When there are no overlaps, the corresponding function is single valued and continuous on the boundary and takes the same value at corresponding points. Poincaré commented at this point,

> There is always a function that satisfies the conditions stated above. This would not be obvious if we had required our function... to be monogenic [analytic], but we did not do this; in fact, although there are monogenic functions satisfying the stated conditions, as it will be seen later, I have not made this hypothesis because I have no use for it, and because I am not yet in a position to prove the existence of such functions. (1997, 85–86)

This shows that Poincaré did not then know the Riemann mapping theorem, which says that any simply connected domain in the complex plane other than the whole plane itself can be mapped by a holomorphic function onto the interior of the unit disk. The theorem was contentious, but no one disputed that it was other than fundamental if it could be rigorously proved.

Next (1997, 87), Poincaré turned to the connection with the theory of quadratic forms. He supposed that T was a matrix with integer coefficients that preserved an indefinite ternary quadratic form Φ, and that S was a linear substitution sending $\xi^2 + \eta^2 - \zeta^2$ to Φ. Then STS^{-1} maps the quadratic form $\xi^2 + \eta^2 - \zeta^2$ to itself. Suppose that it sends (ξ, η, ζ) to (ξ', η', ζ'). Then the quantities

$$z = \frac{\xi}{\zeta} + \sqrt{-1}\frac{\eta}{\zeta} \text{ and } z' = \frac{\xi'}{\zeta'} + \sqrt{-1}\frac{\eta'}{\zeta'}$$

are related by a transformation $z = zK$ of the non-Euclidean plane for which $\xi^2 + \eta^2 - \zeta^2 < 0$. Poincaré remarked,

> All the points zK are the vertices of a polygonal net obtained by decomposing the pseudogeometrical plane into mutually congruent pseudogeometrical polygons. The substitutions K are those that transform the polygons into each other, or even, as we shall see below, those that reproduce the functions that we are going to define. (1997, 89)

He did not prove these claims, nor that the sheets of the hyperboloid $\xi^2 + \eta^2 - \zeta^2 = 1$ provide a model of non-Euclidean geometry in the z plane—which is quite easy to see—but proceeded at once to generalize his earlier definition of thetafuchsian functions.

After some work showing the utility of these new functions, Poincaré concluded this supplement by saying,

> To every decomposition of the pseudogeometrical plane into mutually congruent pseudogeometrical polygons there corresponds a function, analogous to the Fuchsian functions, and which enables us to solve a second-order linear differential equation with algebraic, but irrational, coefficients.
>
> One sees that there are functions, of which the Fuchsian function is only a particular case, which enable us to integrate linear algebraic differential equations. However, in order to determine whether a given equation is integrable in this way, a long discussion would be required which I do not wish to enter into for the moment, but reserve for later. (p. 23)

The third supplement, received by the academy on 20 December, ran to a mere 12 pages and dealt mostly with a class of differential equations that included Legendre's famous equation for the periods of an elliptic integral as a function of the modulus. At the end Poincaré wrote,

> Besides, I do not doubt that the numerous equations considered by M. Fuchs in his Memoir in volume 71 of *Crelle's Journal*... provide an infinity of transcendents... and that these new functions enable us to integrate all linear differential equations with algebraic coefficients. (1997, 103)

The prize commission reported on 14 March 1881. Six entries had been received, of which, they said, four were worthy of discussion. They unanimously awarded the prize to Georges Halphen for a long memoir on differential invariants that had also taken its inspiration from the work of Fuchs, and accorded the authors of memoirs three and five an honorable mention. The author of memoir number five asked that his name be revealed: it was Henri Poincaré. The commission's report on his essay commented,

> The author successively treated two entirely different questions, of which he made a profound study with a talent that greatly impressed the commission. The second...concerns the beautiful and important researches of M. Fuchs...whose results...presented some lacunae in certain cases that the author has recognized and drawn attention to in thus completing an extremely interesting analytic theory. This theory has suggested to him the origin of transcendents, including in particular elliptic functions, and has permitted him to obtain the solutions to linear equations of the second order in some very general cases. This is a fertile path that the author has not traversed in its entirety, but which manifests an inventive and profound spirit. The commission can only urge him to follow up his research, in drawing to the attention of the Academy the excellent talent of which they give proof. (Poincaré, *Oeuvres* 2, 73)

FUCHS, SCHWARZ, KLEIN, AND AUTOMORPHIC FUNCTIONS

By March Poincaré had already published his first paper (1881e) about his new functions.[8] It had appeared in the *Comptes rendus* (14 February 1881) and began a cascade of ideas. In it he announced the discovery of a large class of new functions, which he proposed to call Fuchsian functions "in honour of M. Fuchs whose works have been very useful to me in these researches," and which, he said, generalized the familiar elliptic functions and could be used to solve differential equations with algebraic coefficients. They were invariant under subgroups of the group $SL(2, \mathbb{Z})$ that leave a "fundamental" circle fixed and that never place a point z

---

[8] This material is also covered in Gray (2000).

and any of its transforms arbitrarily close together. It follows that the "Fuchsian" group, as Poincaré called it, can be studied by looking at how it transforms a polygonal region R bounded by arcs of circles perpendicular to the fundamental circle. He said non-Euclidean geometry would be helpful here, but he did not explain how. Instead he gave some examples of triangular polygons, and asked if any polygon would do. He noted, without further explanation, that any linear substitution group preserving an indefinite ternary form $x^2 + y^2 - z^2$ could also be used; these ideas were later published as a note on discontinuous groups (1882f). Finally he introduced some more new functions, which he called thetafuchsians, by analogy with the theta functions occurring in the theory of elliptic functions, and briefly indicated some of their properties.

The theta functions in elliptic function theory are quasiperiodic. Poincaré's new thetafuchsian functions satisfy the equation

$$\theta \left(\frac{az + b}{cz + d} \right) = (cz + d)^{2m} \theta(z),$$

where m is a positive integer and the matrices $\left(\begin{smallmatrix} a & b \\ c & d \end{smallmatrix} \right)$ are elements of the corresponding "Fuchsian" group.

The very next week (1881b) he pointed out that the quotient of two thetafuchsian functions corresponding to the same Fuchsian group and the same value of m is a Fuchsian function, then said that any two Fuchsian functions corresponding to the same group satisfy an algebraic relation, although he gave no proof, and finally indicated how every Fuchsian function F permitted one to solve a linear differential equation with algebraic coefficients. He gave the hypergeometric equation as an example.

In his third paper (4 April 1881c) he claimed that any two Fuchsian functions corresponding to the same group and whose polygonal region R does not reach to the boundary of the disk are related by an algebraic relation, and that they are all rational functions of any two of them. These properties resemble those of certain families of Abelian integrals, which suggests that Poincaré already knew something of the theory of Abelian integrals, presumably from Hermite, and indeed Poincaré then sketched a relation between the number of generators, $2p + 2$, of the group and the periods of a suitable Abelian integral. This gave him an upper bound, p, for the genus of the algebraic equation connecting the

Fuchsian functions. This connection—as remarkable as it is imprecise—marks the first appearance of the connection between this work and the theory of Riemann surfaces that was to form its most lasting and profound insight.

Poincaré's notes of 18 April and 23 May (1881e, g), dealt in more detail with the role of circular-arc polygons in generating groups. He described under what conditions a Fuchsian group can be said to move a polygonal region around crabwise in the disk. Then he looked for the genus of the equation connecting two functions that correspond to the same group, and which are therefore related algebraically, concentrating on the genus 0 case. He used this discussion to indicate very loosely how a large class of differential equations can be solved, those that have only real singular points, and observed that the hypergeometric equation and the modular function fit into his account. In the next note (30 May, 1881h) he clarified this a little and extended his argument to differential equations with real or complex singular points. Now he outlined a continuity argument to show that if a given differential equation has $2n$ singular points, these can perhaps be pushed onto $2n$ real points and the solutions in this the real case then deformed back to provide those in the general case. He remarked that "If I succeed in showing that these equations always have a real solution I will have shown that all linear equations with algebraic coefficients can be solved by Fuchsian and zeta-Fuchsian transcendents."[9]

Poincaré had a reader he may not have anticipated: Felix Klein. Klein was five years older than Poincaré, and in the autumn of 1880 he had just become professor at the University of Leipzig, where he gave his inaugural address in October on the relationship of the new mathematics to applications. He was ambitious, he had every reason to believe he was the leading German mathematician of his generation and perhaps in the world, and he was beginning to shape himself as the heir of Riemann. In particular he shared Riemann's high opinion of geometry, although his own work was largely in projective geometry whereas Riemann's much deeper work was in differential geometry, and he was beginning to work round to the study of what were already becoming called Riemann surfaces.

[9] Étienne Ghys presented a paper at a conference in Nancy in January 2012 in which he argued that after Poincaré's mistaken counting of constants is corrected, these papers present the first glimpse of the uniformization theorem; I hope he will publish this finding.

Klein had been called to Leipzig to lecture specifically on geometry, but he took the opportunity to deepen his understanding of Riemannian, geometric function theory. He wrote in his autobiography,

> But I did not conceive of the word geometry one-sidedly as the subject of objects in space, but rather as a way of thinking that can be applied with profit in all domains of mathematics. Correspondingly, despite many contradictions, I began my Leipzig professorship with a lecture on geometric function theory [and] began a cycle of lectures on geometry that comprised analytic geometry, projective geometry, and differential geometry. (Klein 1923b, 20)

His second lecture series was to become very well known when slightly reworked and published as *Über Riemanns Theorie der algebraischen Funktionen und ihrer Integrale*. His fundamental idea was to regard an algebraic curve $F(s, z) = 0$ as a closed surface in its own right and a complex function as a pair of flows, with singularities, on the surface. He claimed that this had been Riemann's own approach, and at first Riemann's former student Prym agreed with Klein, but later correspondence between Prym and the Italian mathematician Enrico Betti, who had known Riemann well, makes it clear that Riemann had not thought this way. Klein's description is, rather, an inspired response to reading Riemann. It has the characteristic Kleinian virtues of being visual and intuitive, while the algebraic description of the curve linked it, in Klein's approach, to projective geometry. That said, Klein's work in 1880 had been good but not strikingly original, and he must have been hoping that, now he was in his thirties, he would find the true measure of his worth.

Klein read Poincaré's notes on 11 June 1881 and wrote to him the next day.[10] He observed that he had written about elliptic modular functions in several articles, which "naturally are only a special case of the relations of dependence considered by you, but a closer comparison would show you that I had a very general point of view," and he supplied the necessary references lest Poincaré miss the full weight of his contribution. Then went on to say that in his view the task of modern analysis was to find all functions invariant under linear transformations, of which those invariant under finite groups and certain infinite groups, such as the

---

[10] The Klein–Poincaré correspondence is reprinted in several places, for example (1923), Poincaré, *Oeuvres* (1, 26–65) and Klein, *Abhandlungen* (3).

elliptic modular functions, were examples. However, he had not come to any definite result, and he now wondered if he should not have got in touch earlier with Poincaré or Picard; he hoped his letter would start a correspondence. Finally, he hoped to return to the problem as soon as more pressing duties would permit, and in any case he was to lecture on differential equations in the winter. This was a positive gloss on the state of his research at the time, because in fact he had moved off onto other topics. Now that a rival had entered the field, he was certain to return.

Poincaré's reply (15 June) was characteristically more modest. He was happy to concede priority over certain results to Klein, knowing "how well you are versed in the study of non-Euclidean geometry which is the veritable key to the problem which occupies us." He promised to put this right when he next published his results, meanwhile he would try to find the relevant volumes of the *Mathematische Annalen*, which were not in the library at Caen. But, since that would take time, he asked if Klein could explain some things straight away. The first displays Poincaré's disarming ability to confess to ignorance: Why speak of modular functions in the plural, when there was only one, the square of the modulus as a function of the periods? And had Klein found all the circular-arc polygons which give rise to discontinuous groups and found the corresponding functions?

Klein got the letter on the 18th and replied the next day with reprints of his own articles and a promise to send those of his students. But the burden of his reply was to criticize Poincaré's use of Fuchs's name. The subtext is that Fuchs was a member of the large and dominant school in Berlin, which held a low opinion of Klein and which Klein felt he had to defeat if he was to make it to the top. He argued that all research in this area was based on Riemann's work. His own, he said, was closely related to that of Schwarz (who was even more closely tied to the Berlin leadership), and he urged Poincaré to read it "if you do not already know it." The work of Dedekind (an old friend of Riemann's and much influenced by him) had shown how modular functions could be represented geometrically, whereas Fuchs's work was ungeometric. Even in his work on differential equations Fuchs, said Klein, had made a fundamental mistake which Dedekind had had to correct. He concluded by turning to the subject of polygons and remarking that methods due to Schwarz and Weierstrass showed that it was not necessary to use general Riemannian principles, which apply to functions on any disk-shaped domain whatever. But that

Figure 3.2. Klein's polygon. Source: Poincaré, *Oeuvres* (11.1, 38).

said, some polygons gave rise to discontinuous groups which did not preserve a fixed circle, so "the analogy with non-Euclidean geometry (which is in fact very familiar to me) does not always hold." He gave the polygon shown in figure 3.2 as an example.

Poincaré replied on the 22nd, before the reprints had arrived. He defended the name "Fuchsian function" on the grounds that, even if "the viewpoint of the geometric savant of Heidelberg is completely different from yours and mine, it is also certain that his work served as a point of departure and is the basis for everything that has been done since in that theory. It is therefore only just that his name should remain attached to these functions, which play such an important role there." Klein was not to be so easily persuaded and his brief reply (25 June) directed Poincaré back to Fuchs's original publication in the *Journal für Mathematik* for the purposes of comparison.

Poincaré wrote on the 27th to say that finally the reprints had arrived, having been sent via the Sorbonne and the Collège de France even though they were correctly addressed. He now admitted that "I would have chosen a different name [for the functions] had I known of Schwarz's work, but I only knew of it from your letter after the publication of my results," and he could not change the name now without insulting Fuchs. What Schwarz had done, in his investigation (1871) of when the hypergeometric function is an algebraic function, was to notice one interesting example of when it is transcendental and defines a many-valued function mapping onto the disk, which is summed up in the famous drawing shown in figure 3.3. But Schwarz had missed the connection to non-Euclidean geometry entirely, and when Poincaré began to publish, Schwarz began to complain that Poincaré had not given him any credit. When Poincaré heard of this in a letter from Mittag-Leffler of 18 July 1882, where he said that Fuchs was full of admiration for Poincaré's beautiful discoveries but Schwarz was "almost suffocating

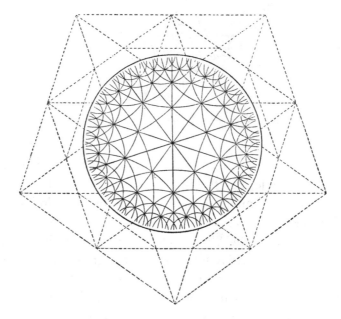

Figure 3.3. Schwarz's tessellation of the disk. Source: Schwarz, *Gesammelte mathematische Werke* (vol. 2, p. 240).

with anger," he wrote back (1999, 100–101) to say that Schwarz had had the chance, and had not taken it, adding that Schwarz was really angry with himself "for having had an important result in his hands and not profiting from it. And I can do nothing about that."

As for the mathematics he was discussing with Klein, he had several questions: Had Klein determined the fundamental polygons for all the principal congruence subgroups? What was the *Geschlecht* in the sense of analysis situs, and was it the same as the *genre* that he, Poincaré, had defined? for he only knew that they both vanished simultaneously. Could he be given a definition of the topological genus, if not in a letter then via a reference to where it could be read? And finally, what did Klein understand by general Riemannian principles?

On 2 July Klein answered these questions as well as he could. "Genus in the sense of the analysis situs" was the maximal number of closed curves which can be drawn without disconnecting the surface, and was *materially the same number* (Klein's italics) as the genus of the algebraic equation representing the surface. Riemann's principles, he went on, do not tell you how to construct a function even when one knows it exists, and so some have decried it as unreliable even though the conclusions that follow from it are correct. Weierstrass and Schwarz, for example, have dealt with the same question by arguments involving convergent series. But Riemannian principles allow one to talk about functions defined on many-leaved surfaces with boundaries in several separated pieces and enable one to prove the existence of a function having prescribed infinities and real parts to their periods. "This theorem," he said, "which by the way I have only completely understood recently, includes, so far as I can see, all the existence proofs of which you speak in your notes as special cases or easy consequences." Finally, the sheer extent of Poincaré's ignorance seems to have dawned on Klein who, for all his own unwillingness to get down to details was a scholar. Referring to another of Poincaré's papers (1881d), he asked why the moduli were presumed to be $4p + 2$ in number when they are really only $3p - 3$. "Haven't you read the relevant passage of Riemann? And is the entire discussion of Brill and Noether... unknown to you?"

When Poincaré replied on 5 July he apologized profusely for asking about the topological genus when it was defined "on the next page of your memoir." Among other matters he replied that the number $4p + 2$ had only been needed as an upper bound and was easy to obtain. He made no mention of his deficient reading. This part of the correspondence closed with Klein's letter of 9 July, in which he pointed out that Riemann's account of Dirichlet's principle was not conclusive, but that one could find stronger proofs in, for example, Schwarz's work.

On 27 June, after he had heard from Klein, Poincaré (1881i) returned to the description of non-Euclidean polygons, this time defined in the upper-half plane (which can be obtained from the disk by an inversion). He sketched how the concepts of non-Euclidean length and angle can be introduced, and polygons defined (with sides equal in pairs and angles that are aliquot parts of 2π) corresponding to a discontinuous group. He then claimed, without a hint of a proof, that every Fuchsian group can be obtained in this way. Turning to Klein's example of an unsymmetric polygon, which does not preserve a fundamental circle, and to his generalization of that to a region bounded by $2n$ circles exterior to one another and possibly touching externally, he said that successive reflections again generated a discontinuous group and so also invariant functions. Making amends for his earlier choice of names, and surely with a twinkle in his eye, Poincaré added,

I propose to call [these functions] Kleinian functions, because it is to M. Klein that one owes the discovery. There will also be theta-kleinian and zetakleinian functions analogous to the thetafuchsian functions.

In his next paper (11 July) entitled "Sur les groupes Kleinéens" (1881j), Poincaré took up the analogy between discontinuous groups and non-Euclidean polygons which Klein had said his example of unsymmetric polygons had broken, and restored it delightfully. He regarded the x–y plane as the boundary of the space $\{(x, y, z) : z > 0\}$ and the polygon as bounded by arcs which were the intersections of hemispheres with their centers on the x–y plane. In this way Klein's groups generated by reflections on the sides are regarded as acting on three-dimensional non-Euclidean space, realized as the space above the x–y plane. More precisely, a pseudogeometric plane is a hemisphere, and a pseudogeometric line is the intersection of two such planes. This paper marks the first time that Poincaré called the new geometry the pseudoge-ometry of Lobachevskii in print; oddly enough, it is three-dimensional, just as the original versions of Lobachevskii and Bolyai were.

The most important paper in the flood from Caen is Poincaré's (1881f), in which he claimed not only that every linear differential equation with algebraic coefficients can be solved by zetafuchsians, but, even more significantly, that the coordinates of points on any algebraic curve can be expressed as Fuchsian functions of an auxiliary variable. This is the first statement of what became known as the uniformization theorem for algebraic curves, which says that if $F(z, w) = 0$ is the equation of an algebraic curve then there are Fuchsian functions $\varphi(x)$ and $\psi(x)$ such that $F(\varphi(x), \psi(x)) = 0$. In other words, the curve can be parameterized by single-valued functions on the x disk which are invariant under an appropriate Fuchsian group. The simplest case is that of the circle $x^2 + y^2 = 1$ with $x = \frac{1-t^2}{1+t^2}$ and $y = \frac{2t}{1+t^2}$. A more significant one is that of cubic curves: the equation for a cubic curve can be reduced to one of the form $w^2 = z^3 - g_2 z + g_3$, where g_2 and g_3 are constants, and the curve is then parameterized by a suitable Weierstrass $\wp$ function on setting $(z, w) = (\wp(z), \wp'(z))$. While the paper was in press in *Acta Mathematica* Mittag-Leffler praised it to his assistant Malmsten (Stubhaug 2010, 292): "Isn't that terrific? In analysis there is no theorem which in its striking simplicity surpasses this," but in fact the search for a valid proof of Poincaré's general uniformization theorem was to draw in several mathematicians and was not to be completed successfully for a further twenty-six years.

Meanwhile Klein had concluded his summer lectures on geometric function theory. On 7 October he sent the manuscript of *Über Riemanns Theorie der algebraischen Funktionen und ihrer Integrale* to Teubner in Leipzig, and was free to return to his old problem. He wrote to Poincaré on 4 December to congratulate him on solving the general differential equation with algebraic coefficients and the uniformization problem, and to ask for an article on this work for the journal he edited, the *Mathematische Annalen*. He explained that he would add a letter to it, connecting it with his own work on modular functions, and that publication would thus acquaint the readership of the *Annalen* with Poincaré's work and explain its relationship to Klein's. One wonders who would gain most by this, but Poincaré, now in Paris, agreed, and the manuscript he sent to Klein came out in the last part of volume 19. Meanwhile, Klein continued to object to the name "Fuchsian functions," on the grounds that Fuchs had published nothing about them whereas Schwarz had. Nor indeed had he, Klein, done anything on Kleinian functions except to bring one special case to Poincaré's attention, but Schottky's name should be mentioned, along with that of Klein's former student Dyck, who had brought out the group-theoretic aspects of the subject in his work. Klein even complained at length to his students in his seminar about the name "Fuchsian," and also about the fact that his journal, *Mathematische Annalen*, was seemingly unknown in France.[11] He warned Poincaré by letter (13 January 1882) that he would protest against the names as much as was in him, and Poincaré indicated that he would like space in the *Annalen* to defend his choice. So, when the paper appeared Poincaré pointed out that it was not ignorance of Schwarz that made him select Fuchs's name, but the impossibility of forgetting the remarkable discoveries of Fuchs, based on the theory of linear differential equations. Furthermore, Schwarz was only a little interested in differential equations, whereas Fuchs had brought forward a new point of view on differential equations "which had become the point of departure for my researches." As for Kleinian functions, the name belonged rightly to the man who had "stressed their

---

[11] See the letter from Georges Brunel to Poincaré, June 1881, where he comments drily that the mathematicians in Berlin are none too sure about that journal either.

principal importance," said Poincaré, turning Klein's observation on
Schottky in his own work (1882a) against him.

Now that Klein had criticized Fuchs publicly, Fuchs came to hear of
the dispute. He was hurt by the allegation that he had published nothing
on invariant functions, and replied in Fuchs (1882) and also privately
in a letter to Poincaré.[12] He pointed to his study (1875) of differential
equations with algebraic solutions, and claimed that Klein himself had
said that these works had led to his own developments in Klein (1876).
This is not entirely true, Klein had there acknowledged the influence
of Schwarz, and Klein repeated this story in his book *Vorlesungen über
das Ikosaeder* (1884); only toward the end of his life did he concede that
Fuchs had been a particular impulse (1922, 257). Klein was determined
to set Poincaré straight. His next letter to Paris (3 April 1882) contains a
lengthy restatement of the historical priorities as viewed from Düsseldorf,
in particular he said that in 1876, when he had solved a problem raised by
Fuchs, contrary to what Fuchs had said: "I did not take the ideas from his
work, rather I showed that his theme must be handled with my ideas."
Klein hoped this letter would close the debate, but he was not to have
the last word. Poincaré replied at once (4 April 1882), to say that he had
not started the debate and he would not prolong it. As for the name
"Kleinian" he had known of no right to the property before Klein's letter.
The debate about names should be over: "Name ist Schall und Rauch"
he wrote in German ("Name is sound and fury" quoting Gretchen in
Goethe's *Faust*).

Poincaré's *Annalen* paper (1882d) much resembles his earlier report in
the Caen memoir (1882c), but it goes into more details on the topics Klein
had requested, although seldom providing proofs. He defined a Fuchsian
group as a group of substitutions of the form $z' = \frac{az+b}{cz+d}$, where a, b, c, d
are all real and $ad - bc = 1$, which move a region R of the upper-half
plane around so that the various copies of R form a sort of web. These
regions are to be non-Euclidean polygons, not necessarily finite in extent,
fitting together along their common edges, and whose angles are aliquot
multiplies of 2π.

Thetafuchsian functions were defined by convergent power series

$$\theta(z) = \sum_j H\left(\frac{\alpha_j z + \beta_j}{\gamma_j z + \delta_j}\right) \frac{1}{(\gamma_j z + \delta_j)^{2m}},$$

[12] The letter is reproduced photographically in Poincaré, *Oeuvres* (11.2, 275–276).

where H is a rational function and m is a positive integer; Poincaré showed that such a function has only finitely many zeros and infinities in any region R, by using the integral of $\frac{\theta'(z)}{\theta(z)}$ taken round the boundary of R to count the number of zeros and poles of the function within R. The quotient of two thetafuchsians is a Fuchsian function; it satisfies $f\left(\frac{az+b}{cz+d}\right) = f(z)$.

If the substitutions of the group have all these properties except that of preserving a half plane or circle, then Poincaré called the group Kleinian, and the corresponding functions thetakleinian and Kleinian, respectively, remarking that their study required processes derived from three-dimensional non-Euclidean geometry. In the *Annalen* he gave no reason for the name, in the Caen memoir he said that Klein had been the first to give an example of such groups.

Poincaré stated four main results for the theory of Fuchsian functions and groups.

- He gave a formula for the genus of the algebraic equation connecting two Fuchsian functions belonging to the same group in terms of the geometry of the corresponding polygon R.

- He gave a complicated theorem concerning the connection between Fuchsian functions and second-order differential equations.

- He sketched a proof of the uniformization theorem derived from the existence of a particular Fuchsian function but which rested on a long numerical calculation, which was not given.

- He claimed that every linear differential equation with algebraic coefficients can be solved by Fuchsian and zetafuchsian functions, or by Kleinian and zetakleinian ones.

As he had promised, Klein's first published response to Poincaré's ideas came in the same volume of the *Annalen*. He confessed that, in order to respond to the younger man, he had felt driven to use methods that he regarded as "to some extent irregular" and to publish his results before he had managed to put them in order. Even so, the contrast between Klein and Poincaré is marked. Poincaré was a visionary, paradoxically helped by the gaps in his mathematical education. His ability was to find the right, hidden generalization, and to push it into unexpected connections. Klein already possessed a unified view of mathematics, based on the importance of projective geometry and the centrality of the group idea. His work on modular functions was almost a new field of study; but also very much in that spirit.

In his first brief paper (1882b), Klein announced without proof a result that has the uniformizatioh of Riemann surfaces of genus greater than 1 as a corollary. He claimed that for each such Riemann surface, there is always a unique function η that is single valued on the simply connected version of the surface, on crossing a cut changes to a function of the form $\frac{a\eta+b}{c\eta+d}$ and so by analytic continuation maps the cut surface without overlaps onto a $2p$-connected region of the η sphere. Klein's description of the generation of the discontinuous group in this context was more general than Poincaré's, who was still using infinite polygons in his discussion of uniformization. As Freudenthal (1954) remarked, the direct route to uniformization is the only point at which Klein surpassed Poincaré. In fact, only Klein invoked the birational classification of surfaces and so established a parameter count that even made it plausible that a correspondence between discontinuous groups and curves might be expected to work.

Klein regarded the analytic continuation of η as if it moved the region that is the image of the cut surface around on the η sphere. In a discussion that reflected his astonishment at Poincaré's mistaken estimate of the number of moduli of a Riemann surface, Klein showed that these movements depend on $3p - 3$ complex parameters (as well as certain inequalities determined by the shape of the fundamental region). He argued that the image of the cut surface may be taken to be a polygon whose $2p$ sides, the images of the cut, are identified in pairs. Each identification, being a fractional linear transformation, contains three parameters, so $3p$ parameters specify the group, but 3 may be chosen arbitrarily since the group does not depend on the initial position of the polygon. But, as he later recognized, this description of the decomposition of a Riemann surface by means of cuts is flawed: $2p$ cuts are needed to render a surface of genus p simply connected, when it becomes a polygon of $4p$ sides. He counted more carefully in his "Neue Beiträge" the next year. There are $2p$ lengths and $4p$ angles in such a polygon, so $6p - 3$ independent real coordinates, and then, using the upper-half plane model, the coefficients of η are real and may be replaced by $\frac{\alpha\eta+\beta}{\gamma\eta+\beta}$ because the first edge cah be put anywhere, so 3 parameters are inessential, and the function η depends on $6p - 6$ real or $3p - 3$ complex parameters. Now, the space of moduli for Riemann surfaces has complex dimension $3p - 3$ and so Klein made the audacious claim that every Riemann surface corresponds according to a unique group.

The next developments were vividly described by Klein himself in a lecture of 1916, reprinted in his *Abhandlungen* (3, 584) and his *Entwicklung* (p. 379). He wrote,

> In Easter 1882 I went to the North Sea to recover my health, and indeed to Nordeney. I wanted to write a second part of my notes on Riemann in peace, in particular to work out the existence proof for algebraic functions on a given Riemann surface in a new form. But I only stayed there for eight days because the life was too miserable, since violent storms made any excursions impossible and I had severe asthma. I decided to go back as soon as possible to my home in Düsseldorf. On the last night, 22nd to 23rd March, when I needed to sit on the sofa because of my asthma, suddenly the "Grenzkreis theorem" appeared before me at about two thirty as it was already quite properly prefigured in the picture of the 14-gon in volume 14 of the *Mathematische Annalen*. On the following morning, in the post wagon that used to go from Norden to Emden in those days, I carefully thought through what I had found once again in every detail. I knew now that I had a great theorem. Arrived in Düsseldorf I wrote all it at once, dated it the 27th of March, sent it to Teubner and allowing for corrections to Poincaré and Schwarz, and for example to Hurwitz.

This was Klein's sleepless night of inspiration.

His student Hurwitz immediately recognized the importance of the theorem, but Poincaré said nothing about it when he replied on 4 April. He did, however, observe in his next note for the *Comptes rendus*, that the results "have been obtained by Klein on other grounds." Schwarz first disputed the theorem, then accepted it, and eventually offered two different ways of proving it.[13]

The *Grenzkreis* theorem (or boundary circle theorem) claims that every Riemann surface of genus greater than 1 can be represented in an essentially unique way by a single-valued function that is initially defined

---

[13] See *Abhandlungen* (3, 584).

on a circular-arc polygon P with $4p$ sides all perpendicular to a common circle C. The function can be extended by analytic continuation to copies of this polygon that fill out a disk-shaped region on the sphere, which Klein called a spherical skullcap bounded by a plane. The circle C is the boundary of the disk, and is a limit that the copies of the polygon approach indefinitely without overlapping. What is novel in Klein's theorem is the idea that the function η can be so chosen that the corresponding image of the Riemann surface on the sphere is the polygon P. This implies the magnificent correspondence between the space of all groups and the space of all curves via the function η.

Riemann's approach had considered a surface as spread out over the complex sphere, rendered the surface simply connected by means of cuts, and then studied functions on the surface in terms of their behavior at singularities and on crossing the cuts; the surface is given intact and sits above the z sphere. In Klein's new approach the surface is given, opened out, and sits underneath the domain of η, which is a disk on the sphere.

Klein's new view was Poincaré's attitude all along. Totally ignorant of Riemann's work (as his ignorance of "general Riemannian principles" makes clear) he was constructing Riemann surfaces from polygons in the disk that were moved around en bloc by discontinuous groups. More surprisingly, Klein, who had done so much to further non-Euclidean geometry in the 1870s, had not noticed its role here or in the related work he had been doing before on elliptic modular functions. This surely derives from his view of geometry as essentially projective, and non-Euclidean geometry as a special case of projective geometry. When he tried to discuss intrinsic metrical ideas on a Riemann surface in his lectures on Riemann's algebraic functions his discussion was clumsy. Differential geometry was not his strength, it was not central to his view of geometry. Poincaré, however, as soon as he had his conformal disk model of non-Euclidean geometry, viewed the transformations as isometries; his ideas were more flexible and better adapted to the problem at hand. Klein's problem was that Poincaré's ideas ran against the current of his own recent work.

In a short note on Fuchsian groups (10 April) mostly concerned with differential equations, Poincaré looked at the fundamental polygonal region with $4p$ sides. Klein was struck by how closely this paper resembled one of his own, and wrote to him on 7 May to say he found the methods interchangeable. He went on,

I prove my theorems by continuity in that I assume the two lemmata: 1. that to any "groupe discontinu" there corresponds a Riemann surface, and 2. that to a single necessarily dissected Riemann surface (under the restrictions of the present theorem) there can correspond only *one* such group (in so far as a group does generally correspond to it. (Klein's italics)

Then he offered Poincaré a new version of his general theorem to accompany the two notes in the *Annalen*, for which he said he had very little time, in support of the conclusion that if the Riemann surface is dissected in a particular way, which he described, then "there is always one and only one analytic function which represents the dissected Riemann surface on the circular arc polygon described in this way." Poincaré replied (12 May) that their methods differed more over details than in the general principle. He had established the first of Klein's results by considerations of power series, he said, and the second "presents no difficulty and it is probable we will establish it in the same way. Once these lemmas are established, and it is in effect there that I begin, like you I employ *continuity*." Two years later, in his paper (1884a, 330–331) Poincaré was to give a more critical response to Klein's argument. Now he said that, if S and S' are two manifolds of the same dimension, and to each s in S there corresponds a unique s' in S', and to each s' in S' there corresponds at most one point in S, then, if this correspondence is analytic, the condition that S is closed (i.e., without boundary) forces the map from S to S' to be such that to every point of S' there corresponds a point of S (the map is an onto map). Therefore, to each s' in S' there always corresponds a unique s in S. But if S has a boundary, the conclusion no longer follows. For Klein's argument to be valid it is necessary to show that the space of groups has no boundary, which said Poincaré, is "not at all evident a priori." The question is made more difficult by the fact that the same group may have many very different fundamental polygons, and the onto nature of the map must be established by a special discussion which Klein had failed to provide. "This is a difficulty one cannot overcome in a few lines" (1884a, 332).

Klein replied by return of post (14 May) with a sketch of an approach that he advanced only in principle, expecting that to carry it out would be a lot of trouble. He added that he had also been to see Schwarz in Göttingen, and offered some of Schwarz's ideas, saying that "this

Schwarzian line of thought is in any case very beautiful." Poincaré agreed (18 May) but said that he had tried and failed to extend Schwarz's ideas and he hoped that Schwarz would have better luck.

By the summer both men were free to write up the first full-length account of their research. Poincaré had been commissioned by Mittag-Leffler to publish his work in the first issue of the new *Acta Mathematica*, and Klein had his own *Mathematische Annalen*. Both men preferred to survey their discoveries to date, perhaps because the uniformization theorem and Kleinian groups were proving to be very difficult.

The high point of Klein's "Neue Beiträge" was an account of why the correspondence in the uniformization theorem was analytic, which Klein rightly admitted constituted only grounds for a proof and not a rigorous proof. As Klein later wrote, the price he had to pay for the work was extraordinarily high. While working on it in autumn 1882, his health broke down completely. Even after a long rest it was his opinion that he was never able to work again at the same high level;[14] elsewhere he spoke of the center of his productive thought being destroyed.

Poincaré's Papers of 1882

Poincaré finished his first long paper for Mittag-Leffler's *Acta Mathematica* in July and it was printed in September; he finished the second in late October and saw it published at the end of November 1882.[15] These papers gave the first detailed account of the theory since the article in *Annalen*, but they say little new.

The first paper (1882i) describes how a discontinuous group of transformations of the upper-half plane $\mathbb{H}$ may be considered geometrically by examining its fundamental polygon. Poincaré noted that the relevant geometry was non-Euclidean, but decided not to employ that terminology in order to avoid any confusion. Poincaré found seven families of polygons, depending on the position of their vertices. A vertex was of the first kind if it lay strictly in $\mathbb{H}$, of the second kind if it was a point of the boundary $\mathbb{R}$, of the third kind if it was a segment of $\mathbb{R}$. Accordingly the seven kinds of polygon were obtained by insisting that either (1) all

---

[14] *Abhandlungen* (2, 258).
[15] There is an English translation of four of the five major papers, with a helpful introduction: Poincaré (1985).

vertices were of the first kind, or (2) all were of the second kind, or (3) all were of the third kind, or (4) all were of the second and third kinds, or (5) all were of the first and third kinds, or (6) all were of the first and second kinds, or (7) were of all kinds. Poincaré claimed that, when the angles at the vertices of the first kind are aliquot parts of 2π the corresponding sides must have the same length (as they evidently will). He then claimed, not entirely correctly, that each polygon determines a unique group.[16]

The action of the group identifies families of vertices in cycles, and this allowed Poincaré to use Euler's formula to compute the genus of the corresponding Riemann surface. For example, if the polygon has $2n$ edges all of the first kind and q closed cycles of vertices, then the genus p, was given by $p = \frac{n+1-q}{2}$, and, if the polygon has n edges of the second kind that implied the genus was $p = n - q$. Finally he added some short historical notes which managed to pay generous tribute to Fuchs while indicating how much he had learned from Klein.

In his second paper (1882b) Poincaré described the construction of the corresponding Fuchsian functions by means of convergent power series. He defined thetafuchsian functions and thetafuchsian series, guided by the analogy with elliptic functions and their theta functions. The fundamental elliptic function in Weierstrass's presentation of the theory is defined as a sum

$$\wp(z) = \frac{1}{z^2} + \sum_{m,n}{}' \left(\frac{1}{(z - 2m\omega_1 - 2n\omega_2)^2} - \frac{1}{(2m\omega_1 - 2n\omega_2)^2} \right),$$

where the sum omits the case $m = 0 = n$. One is free to imagine the fundamental parallelogram with vertices $0, 2\omega_1, 2\omega_2, 2\omega_1 + 2\omega_2$ moved around by the group of all pairs of integers (m, n), as perhaps Poincaré did. Certainly, his new series are carefully constructed sums of the form $\sum_g f(gx)$, where $f(z)$ is a function defined on a fundamental region R that is moved around by the action of a Fuchsian group G and the sum is taken over all the elements $g \in G$. Now, however, the contribution of the different terms is harder to understand and convergence theorems more difficult to establish. He noted that it could be that such a series

---

[16] Nörlund, the editor of vol. 2 of Poincaré's *Oeuvres*, noted (1916, 126) that this is true unless an edge AB has both vertices on the real axis, when the polygon determines an infinity of groups.

converged to the zero function, but he was unable to characterize this awkward case.

Because Poincaré was to return to these questions at intervals in the rest of his life it will be useful to give some more details. Poincaré fixed a particular Fuchsian group G, whose transformations he wrote as

$$g_j(z) = \frac{\alpha_j z + \beta_j}{\gamma_j z + \delta_j}.$$

He then formed the series $\sum_{j=0}^{\infty} \left| \frac{dg_j(z)}{dz} \right|^m$, where m was an integer greater than 1. To prove convergence, Poincaré's early methods made ingenious use of both the non-Euclidean and Euclidean geometry of the disk (as he had done in Poincaré 1997), and took into account whether z was inside or on the fundamental circle. In particular, he was able to estimate the number of points $g_j(z)$ that lie inside any given circle inside the disk with center the origin, and use this estimate to obtain bounds on the partial sums of the series. The result was that the series defined a function holomorphic except at certain points in the disk.

A function satisfying $\theta(g_j(z)) = (\gamma_j z + \delta_j)^{2m} \theta(z)$ he called a thetafuchsian function, and he looked briefly at how the seven types behaved with respect to the fundamental circle. Then he looked for their zeros and poles, noting that it was enough to look at what happens only in the fundamental polygon. Then he looked for Fuchsian functions, $F(z)$, which by definition satisfy $F(g_j(z)) = F(z)$ and have only finitely many zeros and poles in their fundamental polygon. Certainly they arise as quotients of any two distinct thetafuchsian functions with the same value of m. The singularities of such functions are those of the corresponding two thetafuchsian functions.

Poincaré also repeated the proof that a Fuchsian group, and its fundamental polygon, depend on $6p - 6$ real parameters, and hinted that this should imply that there was exactly one Fuchsian group for each algebraic equation in x and y, saying "That is what I shall prove in a later memoir" (para. VI). Then came a discussion of how his new theory connected to the existing theory of Abelian integrals. He took two Abelian integrals of the first kind:

$$G(z) = \int g(z)\frac{dx}{dz}dz = \int g(x, y)dx \quad \text{and}$$

$$G_1(z) = \int g_1(z)\frac{dx}{dz}dz = \int g_1(x, y)dx,$$

defined periods A_j, B_j, A'_j, B'_j as these integrals taken along the appropriate sides of a fundamental polygon, proved that they agreed with the familiar ones, and concluded this brief excursion with the remark "I shall not further emphasis these considerations, which show the extent to which one can benefit from Fuchsian functions in the study of Abelian integrals."

Much of the paper was taken up with exploring the way the seven different families expressed the same story in different ways, and at the end of the paper Poincaré brought it all together. He stressed that the major result of the paper was the theorem that every thetafuchsian can be written as $\left(\frac{dx}{dz}\right) F(x, y)$, where F is a rational function and x and y are two Fuchsian functions in terms of which every other Fuchsian function for the same group can be written rationally and between which there exists an algebraic relation $\Psi(x, y) = 0$. Conversely, every function of the above form can be written as a thetafuchsian function provided it vanishes whenever z is a vertex of R that lies on the fundamental circle. It follows from the theorem that every Fuchsian function can be written infinitely many ways as a quotient of two thetafuchsian functions.

Poincaré's Papers of 1883 and 1884

Poincaré went on to write three more long papers for *Acta Mathematica*: on Kleinian groups (1883a), on the groups associated to linear differential equations (1884c), and on zetafuchsian functions (1884a). Kleinian groups and Kleinian functions were much harder to study, and Poincaré did little more than indicate how much of the analogy goes over, and how three-dimensional non-Euclidean geometry might help. In the Kleinian case the transformations have complex coefficients, and as a result, when they are thought of as transformations mapping a fundamental polygon around in the complex plane there is not a boundary circle that is mapped to itself by the transformations. The question of the boundary of the region that is filled out by the images of the polygon is therefore much harder, and indeed the fundamental polygon itself may now have holes, which (Poincaré had shown earlier) cannot happen in the Fuchsian case.

In both the Fuchsian and Kleinian cases it can happen that there are points where copies of the vertices accumulate. The corresponding function is necessarily singular at these points. In the Fuchsian case these points all lie on the boundary circle. If they occupy it completely the circle is a natural boundary of the function, otherwise the function can be continued analytically outside the circle. In this case the singular points form a nowhere dense perfect set.[17] In the Kleinian case, the situation is much more dramatic. The singular points may still form curves, but they will not be circles, only what are called Jordan curves, and, as Poincaré proved (1883, para. VIII), they never have an osculating circle. He also conjectured that the curve would not have a tangent at any point in the first derived set; this was proved in Fricke and Klein (1897, 1, 415–428). Curves without tangents had been known to Riemann and Weierstrass in the form of graphs of continuous, nowhere differentiable functions, but, as Hadamard said, "Everyone understands the profound difference that exists between a fact obtained in circumstances constructed for the purpose, with no other point and having no other interest than to demonstrate the possibility, some sort of exhibit in a teratological museum, and the same fact encountered in the course of a theory that has all its roots in the most common and the most essential problems of analysis" (Poincaré, *Oeuvres* 11, 162). Even in 1921 Kleinian groups offered him the only example of this behavior of Jordan curves (although in fact other examples were by then known to Fatou and Julia).

The papers on differential equations and zetafuchsians are also substantial. Poincaré raised two questions: given a linear equation with algebraic coefficients, find its monodromy group; and, given a second-order linear equation containing accessory parameters, choose them in such a way that the group is Fuchsian. His conclusions, based on the method of continuity (which he admitted was not at all obviously true) were that every second-order differential equation of the form

$$\frac{d^2v}{dx^2} = \phi(x,y)v,$$

where $\theta(x,y) = 0$ and θ and ϕ are rational in x and y, and the exponent differences of the solutions at the singular points are zero or aliquot parts of a unit is such that, if z is a quotient of two independent solutions, then

[17] Hadamard claimed (Poincaré, *Oeuvres* 11.1, 161) that these sets were discovered by Poincaré before Cantor had published his studies into point-set topology, but this cannot quite be right, because when Poincaré described them he used Cantor's German terms "Punktmenge" and "erste Ableitung" for "point set" and "first derived set."

$x = x(z)$

1. will be a Fuchsian function existing only inside a circle for exactly one choice of the accessory parameters;

2. will be a Fuchsian or Kleinian function existing for all z for exactly one choice of the accessory parameters;

3. will be a Kleinian function existing only in a subregion of the complex plane for infinitely many choices of the accessory parameters.

After 390 pages of *Acta Mathematica*, Poincaré left the subject for others[18], writing,

> This will suffice to make it clear that, in the five memoirs of *Acta Mathematica* which I have dedicated to the study of Fuchsian and Kleinian transcendents I have only skimmed a vast subject which without doubt will furnish geometers with the occasion for numerous important discoveries.

Conclusion

In some ways, Poincaré was lucky. He showed good judgment in separating Fuchs's interest in the quotient of two solutions of a differential equation from the rest of Fuchs's concerns and seeking to understand it. He was right that Fuchs had not fully thought through the intricacies of the problem, and was helped by the simple idea of treating the quotient as a map of a half plane onto a polygon. Under analytic continuation, this polygon fills out a certain net. In the simplest case worth considering, this polygon is a triangle, and he seems not to have known that this elementary geometrical insight had already been exploited by Schwarz. So he was able to exploit it as if it was his own discovery, and undoubtedly it clarified matters. In particular, it makes good sense of the neglected case when the triangle has an angle sum less than π, and it was his great good fortune to see that in this case the triangles can be seen as congruent triangles in the sense of non-Euclidean geometry. This insight allowed

---

[18] Of course, he returned repeatedly, for example in his (1885h, 19) where he proved that the conformal automorphism group of a compact Riemann surface is finite, eight years before Hurwitz's better-known proof.

him to replace the idea of analytic continuation with the more intuitive idea of isometries, albeit in an unfamiliar geometry (a price Poincaré seems to have had no trouble paying).

The second piece of luck came when number-theoretic considerations allowed him to pass from triangles to arbitrary polygons. He was now able to assemble polygons (with the right sorts of angles) until they filled out the non-Euclidean disk, and use that domain as the inverse function to the quotient he had begun with. So he had in play some linear differential equations, and some polygons which could be analyzed geometrically. In particular, the geometry, precisely because it was metrical (in two senses, Euclidean and non-Euclidean) enabled him to generalize the power series arguments used in elliptic function theory to establish convergence arguments for the power series representation of his new functions.

Very quickly he noticed that the sides of a fundamental polygon for a given group were being identified in pairs, and the resulting surface was a Riemann surface. At this point the intervention of Klein led him to relevant mathematics he did not know, but which he quickly mastered. Like Klein, he now had before him a very suggestive way of surveying all Riemann surfaces of a given genus (in the non-Euclidean case, of genus greater than 1). He could give an account of all the meromorphic functions on these surfaces, and even consider complex differential equations defined on these surfaces. The Fuchsian functions resemble the inverse of the quotient of a solution of a differential equation, their inverses in turn (which are the original quotients) are related to the solutions of the differential equations.

But this good luck—and, as is often said, some people make their own luck—was backed up with an enormous amount of energy as the seven different Fuchsian families were analyzed in some depth, and a considerable amount of new, and often delicate, mathematics was discovered. The idea of automorphic functions was essentially new—no one had seen this generalization of the elliptic and modular functions before or even sought one—and so was the geometry of Jordan curves. No one had thought to exploit non-Euclidean geometry in this fashion. No one had treated differential equations in this broad setting before. Any one of these achievements, had they been isolated, would have impressed Poincaré's peers. Taken together, they made his name among mathematicians everywhere.

UNIFORMIZATION, 1882 TO 1907

The uniformization theorem was too good a result to be left unproven by Poincaré, and in his (1883f) he offered a proof that extended it to all many-valued analytic functions, not just the algebraic functions.[19] This suggests that he had seen a much more general argument, or at least no reason why a general argument should not go through, that would show that if y is a many-valued analytic function of x given by an equation of the form $f(x, y) = 0$ then x and y can be expressed as single-valued functions of a complex variable z.

For this it is necessary first to construct a domain for the variable z. In the case of the cubic curves the z domain is the complex plane; for algebraic curves of genus greater than 1 Poincaré believed it would be the non-Euclidean disk. In each of these cases there is, or seems to be, a fundamental domain—a suitable polygon—copies of which fill out the requisite domain. Corresponding points in different copies of the fundamental polygon can be joined by paths, and these paths map down to closed paths on the surface defined by $f(x, y) = 0$. Poincaré therefore considered running this process in reverse, and somehow opening out all the closed paths from a fixed but arbitrary starting point P on $f(x, y) = 0$ and letting their endpoints collectively define the sought-for z domain. A path on the surface will consist of points $(x(t), y(t))$, where t is a real variable and $P = (x(0), y(0))$. Because y is a many-valued function of x it can happen that when $x(t_0) = x(0)$, $t_0 > 0$, nonetheless $y(t_0) \neq y(0)$. Loops in the x plane—closed paths that start and finish at the same point—will not always correspond to loops on the surface $f(x, y) = 0$. It can also happen that even if a loop in the x plane corresponds to a loop on the surface $f(x, y) = 0$, the loop on the surface cannot be shrunk to a point, even though the loop in the x plane can, of course, always be shrunk down to its starting point. It is convenient to use later terminology and to imagine the surface $f(x, y) = 0$ spread out over the complex x plane and then the path $\{(x(t), y(t)) : 0 \leq t \leq t_0\}$ on the surface is called the lift of the path $\{x(t) : 0 \leq t \leq t_0\}$ in the complex x plane.

---

[19] For a thorough account of the uniformization theorem and its history, including the approach taken to it by Poincaré in his (1898b) which connects it to his study of the partial differential equations of mathematical physics, see Saint-Gervais (2010).

The process is complicated, because corresponding to the point $x = 0$ in the x domain there are several, perhaps infinitely many, values of y, say y_j, such that $f(x, y_j) = 0$, and each of these is the starting point for a curve on the surface $(x(t), y_j(t))$ for which $x(t_0) = x(0)$, and one has to look at all the corresponding values of $y_j(t_0)$. Poincaré put into the first class of curves those for which at least one of the functions y_j does not return to its initial value when continued analytically along the loop in x; these lifted paths are not loops on the surface $f(x, y) = 0$. A charitable interpretation of this is that one looks at the list $\{y_j(t_0)\}$ and keeps it if at least one $y_j(t_0) \neq y_j(0)$. Poincaré did not make it clear whether one should distinguish two such lifted paths if they differed for some value of j, but it emerged from his later argument that one should. He formed a second class of paths consisting of those that were loops on $f(x, y) = 0$, so $y_j(t_0) = y_j(0)$ for all j, and divided it into two subclasses: those where the loop C can be shrunk to the point O, and those where it cannot. He left undiscussed the question of what to do with different paths on the surface $f(x, y) = 0$ that start and end at the same list of points.

A case where this process is easy to follow is the parallelograms covering the complex plane that map onto a cubic curve via the Weierstrass parameterization. Here, the paths that are loops in $(x(t), y(t))$ that shrink to a point as the loop in $x(t)$ shrinks to a point are the trivial cases. They can be reduced to paths that do not leave the fundamental parallelogram, and are no use in extending the domain of the z variable. All the others leave that parallelogram, and so play a part in extending the domain. So in his very general setting Poincaré kept these wandering paths, and said that each distinct lift of a loop in the complex x plane defined a point in what would become the z domain. Here, distinct meant distinct lists of endpoints. Poincaré did not consider what might happen if two loops in the x plane with different lifts to the surface nonetheless have the same final list of endpoints $\{y_j(t_0)\}$. The process of following the lift of the first loop in the forward direction (t increasing from 0 to t_0) and then the second lift in the reverse direction (t decreasing from t_0 to 0) would be to trace a loop in the x domain of the second sort— but would it be one to keep or one to discard? Poincaré did not say, and this omission makes this part of his paper very hard to follow. Klein's student Dyck, who was given the task of summarizing this paper for the abstracting journal *Fortschritte*, merely stated the result, which may well mean that he could make no sense of the argument.

Poincaré evidently believed that he had now defined the z domain. It had a point for each noncontractible loop on the surface $f(x, y) = 0$, and in this way had as many copies as necessary of an arbitrary point on that surface. He now let the starting point vary, and let all the copies of the endpoints vary accordingly. The result of doing that, he asserted without proof, is a series of polygonal regions that fit together to form a surface which topologically is a disk. This is the z domain. Poincaré gave no reason to exclude the possibilities that the z domain is a sphere or the complex plane—cases that can arise—or indeed anything else.

Next, Poincaré had to define functions $x(z)$ and $y(z)$ and show that they are analytic. The definition is easy, since each point of the z domain comes labeled with x- and y-coordinates. To establish that these functions are analytic, Poincaré had to show that they were analytic inside any polygonal region (granted that they exist) and also at the vertices, where branch points could be expected to occur. He used a Green's function argument to construct a harmonic function and constructed the analytic functions from the harmonic functions in the way Riemann had pioneered. This leads naturally to a confrontation with the Dirichlet problem.

To find the Green's function Poincaré tacitly regarded the disk as the upper-half plane, to which it is conformally equivalent. He used the elliptic modular function φ, which is holomorphic except at 0, 1, and ∞, and maps the plane onto the upper-half plane, to define a function ψ by the formula

$$\psi(x) = \frac{\left(\varphi\left(\frac{ax+b}{cx+d}\right) - i\right)}{\left(\varphi\left(\frac{ax+b}{cx+d}\right) + i\right)} \tag{3.3}$$

where a, b, c, and d are constants chosen so the function ψ is defined and holomorphic except at the points $x = -\frac{b}{a}$ $x = -\frac{d}{c}$, and $x = \frac{d-b}{a-c}$. He chose b and d so that $\varphi\left(\frac{b}{d}\right) = i$ and $\psi(0) = 0$. He set $y_m = \psi$, noting that one can always do this by, if need be, adding the function ψ to the list of y's. Then he defined a function ζ to be $\log\left|\frac{1}{\psi}\right|$; it is essentially positive, logarithmically infinite at certain points, and otherwise harmonic; that is, it satisfies $\Delta t = 0$ away from those points. This function was introduced to permit crucial convergence results later on.

Poincaré then defined a sequence of functions u_n. He considered a family of circles C_n that are concentric circles with center the point

O—the base point of the opened-out loops— and that gradually fill out the disk S. He then appealed to Schwarz's solution of the Dirichlet problem to say there is a function u_n that is harmonic inside a disk bounded by the circle C_n except for a logarithmic singularity at the point O, and which vanishes on the circle C_n itself. He gave a careful uniform convergence argument to show that the sequence of u_n's tends to a continuous function u, because the sequence of u_n's is constantly increasing but bounded above by the function ζ. Indeed, as he showed, the limit function u is harmonic away from the point O where it is logarithmically infinite because, wherever it is defined, $\zeta > u$.

This argument breaks down at points O_j where the function ζ ceases to be holomorphic, and this is where branch points might be found. Poincaré argued that one could surround the point O_j with a small disk and construct a harmonic function ζ_j that agrees with the function ζ on the boundary of the disk and is holomorphic inside the disk. The function $\zeta_j - u_n$ would then be positive inside the disk including the point O_j and so ζ_j would be the upper bound for u_n even there. He therefore concluded that the function u is harmonic. Therefore it has a harmonic conjugate v making $u + iv$ a holomorphic function (except at the point O). Finally, he showed that the function $z = e^{-(u+iv)}$ is one-to-one everywhere, and therefore the functions y_i are uniformized by an analytic function z, because the surface S was constructed in such a way that they have a unique value at each point of S.

Both halves of this short paper are difficult to understand. The topological considerations—the definition of the z domain and the claim that it is a disk—are little more than assertions. The analytical considerations depend on having a solution to the Dirichlet problem for domains bounded by topological circles, but this lay outside the range of known solutions. And worse, there are infinitely many points where the modular function is not defined, and for which the behavior of the function u has to be sorted out by a different argument. Poincaré wrote two notes at the end of the paper where he argued that these points lay outside the Riemann surface S, but the claim is very weak.

On the other hand, one can respect Poincaré's insight. The separation of the problem into a topological part concerned with the construction of the z domain and an analytical part concerned with the nature of the uniformizing functions, is shrewd. The construction of the domain is ingenious, and with much more care can be made to work. That the uniformizing functions are analytic is plausible, even if the best one can say is that Poincaré merely identified the weakest point in his own argument. The paper also displays what was to become one of Poincaré's hallmarks: a belief that one can pass from finite to infinite problems— in this case algebraic surfaces to analytical ones—with sufficient care and effort.

In the end, the paper did not persuade the community of mathematicians. In 1900, in his Paris address on the mathematical problems facing the 20th century, Hilbert stressed that it was extremely desirable to check that the uniformizing map was indeed surjective. As the prestige of the problems slowly grew, a new generation of mathematicians was brought to consider the question. The German-trained American mathematician William Fogg Osgood (1900), Brodén (1905), and Johansson (1905; 1906), who offered a better majorizing function than Poincaré's, all wrote about it, and in the end Poincaré was brought back to it.

He began his (1907a) with a review of these other earlier attempts, and endorsed Hilbert's criticisms. Then he outlined his new approach. As before, he saw the problem as a Dirichlet problem for a Riemann surface with infinitely many leaves. But now he had a greater command of power series techniques, which had been very important in his work on celestial mechanics, and of the methods of harmonic function theory. He now defined the z domain by using the methods Weierstrass had used to study analytic continuation, which was to form a net of overlapping disks subject to certain equivalence relations. This had the merit of avoiding his earlier discussion of loops on the surface. To do this he successfully extended Weierstrass's notion of an analytic element to include poles and ramification points. Then he constructed a Green's function for this surface using Harnack's theorem (Harnack 1887) and a simplification of the sweeping out method from his (1890c) where he had given his own solution to the Dirichlet problem for an arbitrary domain.

Harnack's theorem says that an increasing sequence of harmonic functions either has no limit or it has a limit which is itself a harmonic function. Poincaré found a suitable majorizing function among the classes of Fuchsian functions he had studied at the start of his career, and this took care of the arbitrariness introduced by the sweeping out method, which involves infinitely many choices. Poincaré dealt explicitly with the complications this can cause. From the harmonic function u obtained in this fashion, Poincaré constructed its harmonic conjugate v and the analytic uniformizing function.

He also mopped up some difficulties he had skipped before. He showed that if the arbitrarily chosen starting point O is varied, the uniformizing functions change by a Möbius transformation. He gave two proofs that the conformal representation was indeed *onto* the interior of the disk, which his paper of 1883 had left obscure, one of his own and one following Osgood (1900). He also raised the issue of when the z domain is not a disk but the whole plane, and showed how these cases could be distinguished.

The uniformization theorem was also rigorously proved independently in 1907 by Paul Koebe, who wrote his doctoral dissertation in 1905 under Schwarz in Berlin. His proof resembled Poincaré's in having a topological and an analytical part, and in distinguishing between the case where the domain of the parameterizing functions is a disk or the complex plane. He then embarked on a lengthy series of papers in which he unashamedly sought to establish himself as the leading complex analyst of his generation. He re-presented his work to show how it compared (and in his view surpassed) papers on the topic by other mathematicians, whom he clearly saw as rivals. In his next paper he noted that he avoided the use of modular functions entirely, unlike Poincaré and others such as Osgood, which he felt led to a notable simplification of the argument (Koebe 1907, 644). The next year he showed that it was possible to avoid Harnack's theorem completely (Koebe 1908). In his (1910) he showed that his ideas lay close to recent ideas of Hilbert's on the Dirichlet principle. All this activity led other authors to excavate short direct proofs of the results buried in the torrent of his lengthy papers,[20] but it made Koebe the authority in the German-speaking world, and it was Koebe, not Poincaré, upon whom Weyl relied in his influential *Die Idee der Riemannschen Fläche* (1913), which marks the start of the first modern account of the theory of Riemann surfaces.[21]

---

[20] For example, Courant (1912) eliminated potential theory in favor of conformal mappings.
[21] See Bottazzini and Gray (2012).

4

The Three Body Problem

POINCARÉ WAS INTERESTED all his life in celestial mechanics, and
made original and lasting contributions in every decade from his
early work on flows on surfaces to his last geometric problem. The
particular case of the three body problem had long been regarded as the
most important; Whittaker in his (1937, 339) estimated that over eight
hundred papers on the subject had been written since 1750. Those which
investigate the stability of the solar system largely assumed that it was
stable, as the majestic long-term behavior of the planets so eloquently
suggests, and that the only risk was collisions. Poincaré's work first offered
a proof of a certain kind of stability (Poisson stability) in which no planet
escapes entirely and therefore almost certainly each one returns infinitely
often to places near where it began. But the implications of his later
work opened up for serious investigation for the first time the idea that
Newton's laws might permit a planet to exit the system altogether.

FLOWS ON SURFACES

The differential equations of celestial mechanics are there to define the
orbits of the planets and other bodies in space. They are real (rather than
complex) ordinary differential equations, and Poincaré's work on such
equations began in 1880, when he announced preliminary results about
the curves defined by a first-order equation of the form

$$\frac{dx}{X} = \frac{dy}{Y},$$

where $X = X(x, y)$ and $Y = Y(x, y)$ are polynomials in x and y. He
emphasized that the coefficients X and Y were real functions of two real
variables whose solutions were real curves. His attention to the real as

Cols   Nodes   and Foci

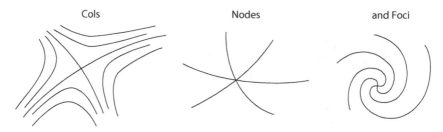

Figure 4.1. Cols, nodes, and foci. Source: drawn by Michele Mayor Angel.

opposed to the complex case reversed the priorities of a generation of pure mathematicians, and he established several wholly new ideas in the subject. From the start, his aim was to study the global behavior of the solution curves, that is to say, the way they fill out their entire domain, which Poincaré initially took to be a sphere (obtained from the x–y plane by stereographic projection). He pointed out that the local study of differential equations showed that except for finitely many points of the sphere each point of the sphere has a unique solution curve (which he called a characteristic) that passes through it. At the singular points there might be either two solution curves, or finitely many solution curves, or an infinite set of spirals that approach the point but never reach it. He called these three types of singularity cols, nodes or centers, and foci, respectively—see figure 4.1.

Poincaré was to stress repeatedly one of his most important novel ideas: to think of the solutions as curves, not as functions, and to consider the global behavior of these curves.[1] Two kinds of topological thinking entered this early work: the algebraic topological ideas of the genus of a surface and the recognition that many surfaces are characterized by their genus alone; and the point-set topological ideas of everywhere dense and perfect sets, which though not original with Poincaré, are put to novel uses.

He began his first major paper on the subject (1881a) with the observations that a complete theory of the functions defined by differential equations would be of great benefit in pure mathematics and mechanics, but that it was evident that most such equations could not be solved by means of known functions. To restrict attention to just those equations that can be solved by definite or indefinite integrals would

[1] See also Barrow-Green (1997) and Gilain (1991).

singularly diminish research and leave most of the questions that arise in applications forever unsolved. It was therefore necessary to study the functions defined by differential equations directly, and to go beyond the study of the solutions to a differential equation in a neighborhood of a point of the plane (for which several methods were known) to study the solutions in the whole plane. He therefore advocated what he called a qualitative theory of differential equations, which studied geometrically the curves defined by the solution functions, and which should be followed by a quantitative part, where the numerical values of the function were found.

He invoked an analogy with the solution of algebraic equations, where one first uses Sturm's theorem to find the number of real roots and then estimates their values by numerical methods. Likewise, he said, one studies an algebraic curve first by the qualitative methods taught in *mathématiques spéciales*, which determine the number of closed branches and infinite branches, and then by quantitative methods that determine a number of points on the curve exactly. He also remarked that this two-pronged approach was required for the resolution of problems such as the three body problem and the long-term structure of the solar system, which is indicative of where he was already planning to take his research.

The first fruits of his global theory were topological: he transferred the flow on the plane to one on the sphere by stereographic projection and claimed that for flows on the sphere the number of nodes plus the number of foci minus the number of cols equals 2. This result was stated in (1880a) and proved in (1881a, 29).

The Russian mathematician Zhukovsky (1876, 180) had drawn pictures of these and three other types of singular points for more complicated flows earlier, in an investigation of the kinematics of fluid bodies. Poincaré did not know of this work, but some appreciation of the simplest types of singular point may well have been common amongst mathematicians working in this area.

He was also unaware that an equivalent result had already been proved by Maxwell in a charming paper entitled "On hills and dales" (1870), who in his turn had proceeded independently of a still earlier discussion (Cayley 1859). Cayley, who was an enthusiastic walker and climber, had described the possible contour lines on a mountainous island. Maxwell invited his readers to consider a level surface rising from below the bottommost depth of the sea. At each moment it cuts the surface of the earth in a contour line. At first it cuts the surface of the earth in a loop that bounds a bowl-shaped region containing the lowest point of the earth. The level surface rises until more bowl-shaped regions are formed in different parts of the ocean, and as the surface rises these regions expand.

The number of these regions will change when two of them merge. There is a precise moment when the level surface cuts the surface of the earth in a figure-eight-shaped loop, each half of which bounds one bowl, and thereafter the corresponding loop bounds only one region. Maxwell called the double point of the figure-eight loop, a bar. He set aside the possibility that more than two regions merge at the same point. The number of regions will also change if a region sends out arms that eventually join to form, one might say, a moat enclosing a region of higher ground. The contour line at the moment the arms merge cuts off two pieces of higher ground from one of lower, and Maxwell called it a pass, and again he set aside singular cases where more than two arms come together at once. These can be handled by gently separating them into several separate sections. Finally the level surface rises and the regions of higher ground shrink to points, called summits, and disappear.

Maxwell now considered the relation between the number of summits and the number of passes. At first, he said, there is only one region and it is above the level surface. For every new region there is a pass, and for every region that reduces to a point there is a summit. Finally the whole earth is below the level surface. So if there are S summits and P passes then $S = P + 1$. Similarly, the number of lowest points or bowl-shaped regions, I and the number of bars, B, are related by the equation $I = B + 1$. Translating this into the theory of functions of two variables led Maxwell to the conclusion that the number of local maxima and minima exceeds by 2 the number of stationary points that are neither maxima nor minima.

Maxwell then considered the lines of slope, as he called them, on the earth. These are the lines of steepest descent, and they cross the level curves at right angles. They run from the summits to the lowest points, with exception of a few that end at a bar or pass. Regions that are joined by lines of slope that start at a common point Maxwell called hills, and regions where the lines of slope end at a common point he called dales. Lines of slope out of a bar or pass that do not end at a summit he called a watercourse, and those that do not end at a lowest point he called a watershed. Special cases aside, which must be counted with multiplicity, there are S summits, I lowest points, W watersheds and the same number of watercourses, and

$$S + I = 2 + W/2.$$

Maxwell now quoted "Listing's rule" (from Listing 1847): $P - L + F - R = 0$, where P is the number of points, L that of lines, F that of faces, and R that of regions (here $R = 2$, because the regions are earth and space). Accordingly,

$$F = L - P + 2.$$

He concluded that if L is the number of lines (watersheds or water-courses), P the number of summits and lowest points together, C the numbers of passes or bars, and F the number of regions formed by a hill and a dale together, then $F = W = P + C - 2$.

To prove his theorem, Poincaré defined several different kinds of curves on a sphere. A simple closed loop he called a cycle, and he remarked that it divided the sphere into two regions: an interior and an exterior. For example, an arc of a solution curve between two points and an arc of an arbitrary curve that only meets the solution curve in the endpoints of the arc is a cycle. Poincaré distinguished two kinds: those for which the rest of the solution curve lay entirely inside the cycle, and those for which it lay entirely outside the cycle. A closed curve with self-intersections (double points) Poincaré called a polycycle. He observed that cycles were analogous to level curves on a geographic map of the earth, and the double points of polycycles were like cols, and also that the points where there was no solution curve were like summits and basins. He then remarked on how these points show up in the theory of functions $z = f(x, y)$.

In general, Poincaré regarded the sphere as defined into two hemi-spheres by the equator, and then studied what happens when singular points do, or do not, occur on the equator or at the poles. This consid-eration of special cases shows how far he was from a truly topological approach which would have started from the observation that it is always possible to rotate the sphere so that nothing singular happens on the new equator or at the poles.

Given a cycle in the northern hemisphere, Poincaré defined a positive direction on it as one that keeps the interior of the cycle on the left, whereas the positive direction on a cycle in the southern hemisphere has the positive direction keeping the interior on the right—an oddly meteorological distinction. He then looked at the value of $\frac{Y}{X}$ as a point traces the cycle once in the positive direction. If there are h points where the expression changes from $-\infty$ to $+\infty$ and k points where it changes from $+\infty$ to $-\infty$ he assigned an index of $i = \frac{h-k}{2}$ to the cycle. He remarked that when a cycle is arbitrarily small and contains no singular point the index is zero, but otherwise it is $+1$ if it contains a col and -1 if it contains a focus. It follows that if a cycle in either hemisphere contains N nodes, F foci, and C cols its index is $C - N - F$, because the cycle can be shrunk down to a sequence of cycles each contains one singular point (a procedure familiar to Poincaré from the theory of contour integration in complex function theory). Poincaré had then to calculate the index of the equator, and to break up cycles that cross the equator into two cycles, and then he could conclude thus (1881a, 29):

> The total number of nodes and foci equals the number of cols plus 2.
> The total number of singular points is a multiple of 4 plus 2.

To describe how the solution curves filled out the entire sphere, Poincaré next turned to how they relate to closed curves. A point where a solution curve was tangent to a cycle he called a contact point on a cycle, and if the cycle is defined by an algebraic curve it will have at most finitely many contact points. He was particularly interested in cycles with no contact points, because such cycles mark out regions where either all the solution curves enter or they all leave, and this determines the singular points they can contain.

He then looked at what can happen to a solution curve as it is followed in one direction. If it never meets a singular point it can be followed forever. If it meets a node, it stops there. But if it meets a col, he said that one then follows either the left-pointing or the right-pointing direction, but never straight ahead. With this convention, it is possible to follow solution curves that start at nearby points. The points are supposed to lie on an arc having no contact points. It can happen that the solution curves starting at these points meet the arc again. He called this approach the method of consequents; it later became known as the first return map. In this case, either at least one of the solution curves is closed, or they all form spirals. With a little more work, Poincaré deduced that any solution curve that does not end in a node is either a cycle or a spiral.

Another possibility that interested Poincaré was that solution curves might spiral outwards to a closed curve, or away from such a curve. These solutions Poincaré called asymptotic, and he distinguished between those which approach the closed orbit as $t \to \infty$, and those which approach it as $t \to -\infty$. Today these are called the forward and backward asymptotes. The realization of the existence and importance of asymptotic solutions is one of the first fruits of Poincaré's new point of view, and it seems to be original to him. The closed curves that the asymptotic solutions approach or depart from he called limit cycles. They are solution curves, and Poincaré remarked that these closed paths with no singular points could never be revealed by a purely local analysis. The importance of limit cycles in his (1881/82) is that they permitted Poincaré to divide the sphere into regions having, none, one, or perhaps more than one limit cycle, and so to give a global description of the flow. Later, he observed that it is the closed solutions, the periodic ones, that enable one to understand at least some other solution curves, those which spiral toward or away from them. He proved that it was always possible to exhibit a system of cycles without contact, polycycles without contact, and limit cycles that filled out the sphere, and whose cols, nodes, and foci were those of a given equation. The theory of consequents and the idea of limit cycles were to prove very significant in his analysis of the three body problem, which he began to write about in his (1883b) and (1884b).

In December 1881 Poincaré stated the index theorem for the general case of a flow on a surface of arbitrary genus (1881c). This is a remarkable application of topological ideas to the theory of differential equations, and without precedent in the literature. To study flows on an arbitrary surface, Poincaré interpreted a differential equation of the form $F(x, y, y') = 0$ where $y' = \frac{dy}{dx}$ as defining a flow on the surface given by the equation $F(x, y, z) = 0$. He rapidly saw that the qualitative nature of a flow is determined by the topology and specifically by the genus of this surface. His general index theorem then asserted that on a surface of genus p the number of nodes plus the number of foci minus the number of cols is equal to $2 - 2p$. It follows that the torus, for which $p = 1$, is the only surface upon which there could be a flow with no singularities,

and partly for that reason, and partly because it is the first case where
the genus is not 0, he devoted a long section of the paper to nonsingular
flows on the torus.

Poincaré took the angular variables ϕ and ω as coordinates of position
on the torus, and accordingly wrote the differential equations as

$$\frac{d\omega}{dt} = \Omega, \quad \frac{d\phi}{dt} = \Phi,$$

where Ω and Φ are functions of cosines and sines of ϕ and ω. The simple
case

$$\frac{d\omega}{dt} = a, \quad \frac{d\phi}{dt} = b,$$

with $\frac{a}{b}$ irrational, known today as irrational flow on the torus, has the
solution curves $\frac{\omega}{\phi} = \frac{a}{b}$, which is an example of a flow with no limit cycles.
Every flow line is dense in the torus, and so, in accord with the general
theory he was to develop in 1885, all the orbits are stable.

To study more general flows on the torus, he invoked the first return
map. He considered the case where Φ and Ω are always positive and
there are no singular points. These conditions imply that the solution
curves are always moving forwards and, one might say, clockwise, so no
solution curve is ever tangent to the meridian curve μ with equation $\phi = 0$
and every flow line that meets it crosses it. The first return map sends
a point M on the meridian μ to the first point on the solution curve
through M where it meets the meridian again. This makes it possible
to reduce the study of a flow on a surface to an iterative process on a
curve. Poincaré now asked whether some points could have stable orbits
and others unstable ones. He considered the set P of all the images of a
point M on μ under iterations of the first return map. Following Cantor,
Poincaré defined the derived set of P, denoted P'. Poincaré showed that
P' is what Cantor called a perfect set. Poincaré could show that for any
choice of starting point, M, either the set P is a subset of the set P' or else
they have no point in common; in the former case the orbit of M is stable,
otherwise unstable. However, even after a long and delicate topological
analysis, Poincaré was unable to determine whether some points could
have stable orbits and others unstable ones.

In what he regarded as the third and final part of his investigation into
curves defined by differential equations, his (1885f), Poincaré proved the
general index theorem and went on to discuss the stability of orbits for

the first time. Finally in 1886 he wrote a paper (1886b) in which second-order differential equations are discussed, thus bringing the equations of celestial mechanics into his new framework.

Poincaré began the paper (1885f) with the remark,

> One cannot read the previous two parts of this memoir without being struck by the resemblance between the various questions treated there and the great astronomical problem of the stability of the solar system [where, however] one meets a new difficulty in this problem, essentially different from those that we have overcome in the study of first order equations, and I intend to bring it out if not in this third part then at least in later work.

Evidently these words were picked up attentively by Hermite and Weierstrass.

Poincaré then gave explicit examples of what can happen with equations of the form

$$\frac{dx}{X} = \frac{dy}{Y}, \quad \text{or, equivalently,} \quad \frac{dx}{dt} = X, \quad \frac{dy}{dt} = Y,$$

to bring out the question of the stability of the solutions. Solutions might be closed curves, which are undoubtedly stable. Or they might be given in polar coordinates by an expression of the form

$$\rho = 2 + \sin\left(\frac{\omega}{h} + C\right),$$

where C is a constant of integration. If h is not commensurable with 2π then the solution curves are not closed, but they display a certain kind of stability: draw an arbitrarily small disk around any starting point on any solution curve and that curve will return infinitely often to that disk. Indeed the solutions are confined to the annulus $1 \le \rho \le 3$ and fill it entirely. Any plane region inside the annulus contains points of the trajectory; "the Germans," said Poincaré, with Cantor in mind, "say that the point set [*Punktmenge*] formed by the different points of the trajectory, is everywhere dense [*überalldicht*] in the interior of the annulus." This appreciation of the emerging subject of point-set topology was characteristic of Poincaré; it marks him out from the generation of his teachers. Poincaré also gave other examples of everywhere dense curves that fill out the plane, and gave the logarithmic spiral as an example of a curve that was in no sense stable: it never returned to a neighborhood of an initial position, nor did it approach a limit cycle.

In chapter XI of his (1885f) he looked at differential equations of the first order and degree, and investigated when the origin can be considered as a center of a family of orbits. A center is like a summit geographically, which means that solution curves reach it, and so the trajectory of a mobile point is stable. Poincaré was led to contemplate an infinity of conditions that have to be satisfied by any power series representing a solution curve. He found that the method of undetermined coefficients produces series that formally satisfy the differential equation but in which the angular variable occurs both within trigonometric terms and on its own (*Oeuvres* 1, 109), and commented,

> It is impossible not to be struck by the analogy between the terms thus introduced and the terms astronomers call *secular*. There is, however, an essential difference that it is important to note. When one meets a secular term, in the usual methods of celestial mechanics, one is not allowed on that account to conclude that the orbit is unstable; for it can happen either that the series is divergent, or that the term thus obtained is only the first term of a development of which the sum always remains finite. It could perhaps be the first term in the development
>
> $$\sin \alpha\omega = \alpha\omega - \frac{\alpha^3\omega^3}{6} + \frac{\alpha^5\omega^5}{120} + \cdots .$$
>
> It is not the same in the case that concerns us and with the method that I am going to expound. If, in the sequence of calculations, one comes across a secular term, one can immediately infer instability.

Next Poincaré turned his attention to differential equations of the form

$$F\left(x, y, \frac{dy}{dx}\right) = 0.$$

He first recalled how these can be thought of as defining curves on the surface $F(x, y, z) = 0$, and observed that these curves belong to one of four kinds:

- cycles or closed curves;

- trajectories that stop at a node;

- trajectories that spiral into a focus;

- trajectories that continue indefinitely and show none of the above behavior;

and he gave a proof of his index theorem for flows on these surfaces by using the idea of the index of a cycle to divide up the surface. Made precise, this is the content of the Poincaré–Bendixson theorem, which asserts that if a solution curve is not closed, and if it does not tend toward a single point, it tends toward a closed orbit (see Bendixson 1901, chap. 1, theorem 2). Although it is in (1881a) and (1882a) that Poincaré is at his most topological, it was Bendixson's great contribution to rework this theory in what is surely its intended setting, that of the continuously once-differentiable functions.

Only now did Poincaré seek to go beyond the content of his earlier papers. His account of motion on a torus is the most important aspect of the new work; it is singled out by the index theorem as the only surface that admits a flow with no singular points. He took the torus with equation

$$z^2 + \left(R - \sqrt{x^2 + y^2}\right)^2 = r^2$$

that is obtained in standard fashion as a surface of revolution, and took latitude and longitude coordinates on it, ω and φ, while remarking that his conclusions would apply to any surface of genus 1, which were all the same from the viewpoint of the "geometry of situation." He reminded his readers of the simple equations that give rise to closed orbits and those that give rise to a dense family of orbits, and considered equations for which ω and φ are always increasing. This allowed him to look at how they cross the closed curve C with equation $\varphi = 0$, which, as he pointed out, is a closed curve without contact.

If $M(0)$ is a point of C then the solution curve through it meets C again at $M(1)$, and he introduced Cantor's terminology:[2] the points $M(1), M(2), \ldots, M(j)$ are the successive consequents, and the points $M(-1), M(-2), \ldots, M(-j)$ the successive antecedents of $M(0)$. The set of points $M(j)$ for all integers j Poincaré denoted P, and its derived

---

[2] Cantor's work, he observed, had been published in German in *Mathematische Annalen* (vols 15, 17, 20, 21), and in French in *Acta Mathematica* (2); Poincaré evidently found that these articles spoke directly to his own concerns.

set—the set of points any neighborhood of which contains infinitely many points of P—he denoted P'. Poincaré now asked what could be said about the set P and its successive derived sets; and in what order the points $M(j)$ are distributed on C as C is traversed in the direction of ω increasing.

First he showed that either $P \cap P' = \emptyset$ or $P \cap P' = P$. Closed trajectories aside, the first case corresponds to an unstable orbit, because if $M(0)$ does not belong to P' then the moving point never returns close to its starting point. Next he took two starting points $M(0)$ and $N(0)$ and showed that if for some value of j both $M(j)$ and $N(j)$ lie in the interval $M(0), N(0)$ then there is a limit cycle—this is clear on looking at successive jth iterates $M(nj)$ and $N(nj), n = 1, 2, \ldots$.

The circular order in which the iterates appear on C was analyzed by Poincaré in terms of the limiting value as n increases of the quantity

$$\frac{\alpha_j + \alpha_{j+1} + \cdots + \alpha_{j+n}}{n},$$

where α_j is the length of the arc $M(j)M(j+1)$. He showed that this limit existed, was independent of the value of j, and was incommensurable with $2\pi r$, the length of C, and he wrote it in the form $\frac{2\pi r}{\mu}$. He showed that the order of the successive consequents of $M(0)$ was determined by μ. In other words, the long-term behavior of the iterates more and more approximates the effect of giving the cycle C a rotation through an irrational angle $\frac{2\pi r}{\mu}$. In later language, this says that the successive iterates are uniformly distributed on C.

Poincaré deduced that in this case $P' \cap P'' = P'$, which means, he said that the set P' is what Cantor had called a perfect set, and Cantor, he said, had distinguished between perfect sets that contain no intervals and those that do. This gave Poincaré three cases to consider:

- All points of C belong to P';

- Some arcs on C belong to P but some do not; and

- No arc on C belongs entirely to P'.

However, the second of these cases can immediately be rejected; it contradicts the uniform nature of the long-term behavior. The first case happens, as Poincaré noted, for the irrational flow on the torus. The third case requires that for some trajectories $P \cap P' = P$ and for some others $P \cap P' = \emptyset$. This does not contradict uniformity, but at this juncture Poincaré insisted that the function that determines the position of the consequent be holomorphic, and he said he had been unable to find such a flow or to show that one did not exist. So the paper ended inconclusively, or should one say with some open questions, and with some further remarks about celestial mechanics that indicate where Poincaré's ideas were taking him. But his insight that the rotation number, μ, was important was profound, and his study of its geometrical implications has become a major part of the modern study of dynamical systems.

STABILITY QUESTIONS

Further evidence of Poincaré's long-standing interest in planetary astronomy is provided by a string of papers he communicated on the subject of trigonometric series in the early 1880s. These are series of the form

$$\sum_{j=0}^{\infty} \left(a_j \sin \alpha_j t + b_j \sin \beta_j t \right),$$

and they had been introduced into mathematics by Riemann in the 1850s as a source of functions with counterintuitive properties. But they also arise naturally in celestial mechanics, and had in particular been studied by the Swedish astronomer Anders Lindstedt. In his (1882g) October 1882 Poincaré took the example $\sum_j a_j \sin \alpha_j t$, where a_j and α_j are positive and $\frac{1}{a_j}$ and α_j tend to zero as j increases, and showed in a page that the sum can be arbitrarily large. Then he drew out the implication for celestial mechanics that certain tempting deductions about convergence of these series could be false. In December 1883 Poincaré (1883e) turned to a recent paper by Lindstedt on the three body problem in which he had used trigonometric series to represent the motions, and commented that his earlier paper showed that one could not deduce the stability of the solar system from Lindstedt's series. Poincaré remarked that Lindstedt had explicitly assumed that the series converged for all values of t, or at least that they did so for some interval in t. So he then asked three questions:

1. If the series converges for some interval in t does it do so for all t?

2. Is it certain that one can find values for the constants so that the series always converges?

3. Even if the series do not converge can they still yield indefinitely good approximate solutions?

He showed that if the series converges absolutely for some interval in t then it does so for all values of t; this was a qualified yes to the first question. The other two questions likewise had affirmative answers.

In the first issue of the *Bulletin astronomique* Poincaré returned to the third of those questions (1884b). He opened with the remark that it was Chebychev who had said that a divergent series in the time variable may be useful if the error committed in truncating the series at a certain term remains sufficiently small for a certain interval of time. This, he said, is why the series commonly used can indeed be used, although almost all of them are divergent. He then extended his earlier analysis until he could show that: if a trigonometric series converges absolutely it does so for all values of the time variable, but if the series is only asymptotically convergent this asymptotic convergence will generally only hold for an interval of time. Moreover, if the convergence is absolute and uniform the series represents a function that is bounded, but if the convergence is not uniform the function can grow indefinitely.

POINCARÉ'S ESSAY AND ITS SUPPLEMENTS

The circumstances surrounding the award of the prize have already been described. As Barrow-Green has uncovered, the events leading up to it, most of which Mittag-Leffler was at pains to suppress, were almost as dramatic.

It is not clear if the idea of a mathematical prize competition came originally from King Oscar II or Mittag-Leffler, but inevitably the task of running the competition and making it a success devolved onto the professional mathematician. We learn from a letter he wrote to Kovalevskaya (Barrow-Green 1997, 53) that at an early stage there was a move to have a German judge (probably Weierstrass), a French or Belgian one (Hermite), an English or American one (Cayley's and Sylvester's names were mentioned) and a Russian or Italian one (perhaps Chebychev). The prize was to be offered on a regular basis (in the event, this did not happen). Kovalevskaya's opinion was that Weierstrass, Hermite, Cayley, and Chebychev would never agree on the merits of a memoir and each of them would probably refuse to serve on a jury containing the other three. Doubtless it was for reasons such as these that the jury eventually consisted only of Hermite, Weierstrass, and Mittag-Leffler himself.

In due course they announced that a prize of a gold medal and a sum of 2,500 Swedish crowns would be awarded for the best attempt at answering one of four questions, or, failing that, for the best essay received on any topic. The topics were as follows:

1. A system being given of a number whatever of particles attracting one another mutually according to Newton's law, it is proposed, on the assumption that there never takes place an impact of two particles to expand the coordinates of each particle in a series proceeding according to some known functions of time and converging uniformly for any space of time.

2. Mr. Fuchs has demonstrated in several of his memoirs that there exist uniform functions of two variables which, by their mode of generation, are connected with the ultra-elliptical functions, but are more general than these, and which would probably acquire great importance for analysis, if their theory were further developed. It is proposed to obtain in an explicit form those functions whose existence has been proved by Mr. Fuchs, in a sufficiently general case, so as to allow of an insight into and study of their most essential properties.

3. A study of the functions defined by a sufficiently general differential equation of the first order, the first member of which is a rational integral function with respect to the variable, the function, and its first differential coefficient. Mr. Briot and Mr. Bouquet have opened the way for such a study by their memoir on this subject (*Journal de l'École polytechnique*, cahier 36, pp. 133–198). But mathematicians acquainted with the results attained by these authors know also that their work has not by any means exhausted the difficult and important subject which they have first treated. It seems probable that, if fresh inquiries were to be undertaken in the same direction, they might lead to theorems of high interest for analysis.

4. It is well known how much light has been thrown on the general
   theory of algebraic equations by the study of the special
   functions to which the division of the circle into equal parts and
   the division of the argument of the elliptic functions by a whole
   number lead up. That remarkable transcendant which is
   obtained by expressing the module of an elliptic function by the
   quotient of the periods leads likewise to the modulary equations,
   that have been the origin of entirely new notions and highly
   important results, as the solution of equations in the fifth degree.
   But this transcendant is but the first term, a particular case and
   that the simplest one of an infinite series of new functions
   introduced into science by Mr. Poincaré under the name of
   "fonctions fuchsiennes," and *successfully* applied by him to the
   integration of linear differential equations of any order. These
   functions, which accordingly have a role of manifest importance
   in analysis, have not as yet been considered from an algebraical
   point of view as the transcendant of the theory of elliptic
   functions of which they are the generalisation. It is proposed to
   fill up this gap and to arrive at new equations analogous to the
   modulary equations by studying, though it were only in a
   particular case, the formation and properties of the algebraic
   relations that connect two "fonctions fuchsiennes" when they
   have a group in common.

One can hardly imagine a set of questions better contrived to attract
Poincaré: all four questions could have been tackled by him. Hermite
even admitted that his question had been chosen with Poincaré in mind,
Weierstrass might also have been interested to see what Poincaré would
do with the topic of celestial mechanics given the chance, and Mittag-
Leffler was already known as a champion of Poincaré. Curiously, no one
complained about this.

But there was trouble immediately. Leopold Kronecker, Weierstrass's
younger colleague in Berlin and a man with whom he had a deteriorating
relationship, complained that the whole competition was set up to
promote *Acta Mathematica*, the journal Mittag-Leffler had founded and
edited and which would publish the prize-winning memoir. Barrow-Green
speculates that Kronecker's ire was aroused by being passed over for the
jury in favor of Weierstrass, and that may well be, but, as she also notes,

he also complained that the fourth of the proposed questions had already been investigated by him some years ago and shown to be impossible. He even threatened to tell the King about this if something was not done. Because the question had been proposed by Hermite, this put Mittag-Leffler in a very difficult position, but he pleaded ignorance, smothered Kronecker with flattery, and the incident passed over.

Worse was to come. In August 1885, and again in October 1886, Kronecker complained to Mittag-Leffler that Weierstrass's preamble to question 1, which drew on remarks made by Dirichlet, misrepresented those remarks, and moreover Kronecker was in a position to know because the remarks in question had been made to him. He eventually made these criticisms publicly at a meeting of the Berlin Academy of Sciences in 1888. The commission did not know how to respond. Hermite refused to enter the fray, and claimed it was his patriotic duty to stay out of this entirely German affair. Weierstrass indicated that he would have no difficulty replying, but understandably wanted any reply to come from the whole commission. They resolved to postpone the matter until after the competition was over.

This competition, as was the practice at the time, called for entries to be submitted anonymously in a sealed envelope that was linked by an epigraph to another envelope containing the author's name, but Poincaré had failed to follow the rules and had written and signed a covering letter as well as sending a personal note to Mittag-Leffler. Of course, once the envelopes were opened there was no possibility of disguising which was Poincaré's; everything from his inimitable manner of exposition, and the links to his earlier work, to his handwriting gave it away. It took Mittag-Leffler and Weierstrass only the month of August, which they spent together, to decide that Poincaré should win, and they wrote to Hermite to say so. But it was to take them much longer to understand what Poincaré had written.

They recognized this was a typical problem when dealing with Poincaré:[3]

> But it must be acknowledged, in this work as in almost all his researches, Poincaré shows the way and gives the signs, but leaves much to be done to fill the gaps and complete his work. Picard

---

[3] Quoted in Barrow-Green (1997, 65).

has often asked him for enlightenment and explanations on very important points in his articles in the *Comptes Rendus*, without being able to obtain anything except the statement: "it is so, it is like that," so that he seems like a seer to whom truths appear in a bright light, but mostly to him alone.

Mittag-Leffler, who was concerned about how the final essay would look in *Acta Mathematica*, started to write to Poincaré requesting clarification of various points. This was in defiance of the rules, and Weierstrass asked him to conceal the fact that he knew Poincaré had entered until the end of the competition. Poincaré complied with a sequence of notes that added 93 pages to his memoir. He also produced a suitable sealed envelope so that the announcement of the prizewinner and the revelation of the author's name could proceed according to plan. Even so, the ceremony lacked Weierstrass's signature on the official report; he had been too ill to sign it.

The report itself was short and direct. It might have been too short, because the astronomer Hugo Gyldén, a member of the editorial board of *Acta Mathematica* concluded from it that he had found some of Poincaré's important results two years earlier, and he reported this opinion to the Swedish Academy of Sciences. The King called for a reply, and Mittag-Leffler had to ask Poincaré to sort the matter out, which he swiftly did, but it left a rift between Mittag-Leffler and Gyldén that was to worry him later.

The Original Essay

Poincaré opened his essay (1889b) with a typical disclaimer that can only have worried the commission. He announced that although he was responding to the first of the four questions proposed, his results were so incomplete that he would have hesitated to publish them had it not been for the importance and difficulty of the problem. He then surveyed the field, writing of his earlier work in the third person, before announcing that he intended to bring together three lines of investigation: the traditional methods of celestial mechanics with their emphasis on trigonometric series, the rigorous calculus (a nod here to Weierstrass) for the theory of convergence, and Poincaré's geometric methods for a study of the stability question.

He also restricted his attention to a version of the three body problem for which, he said, the English term was "two degrees of freedom." Three bodies were studied. One has considerable mass, one only a little, and the third an infinitesimal amount. The first two orbit in circles in a common plane around their center of gravity, and the orbit of the third is in the common plane. The problem of determining the orbit of the third body (sometimes called the planetoid) has become known as the restricted three body problem. Poincaré said that he could prove rigorously that the orbit was stable, in the sense that he could give precise bounds on the maximum distance the planetoid escaped from the other two, and he hinted that he had been able to avoid the obstacle to convergence posed by the existence of small divisors by exploiting the use of periodic solutions pioneered by Hill.

The opening chapter of the memoir was devoted to the theory of differential equations of the form

$$\frac{dx_1}{dt} = X_1, \quad \frac{dx_2}{dt} = X_2, \ldots, \quad \frac{dx_n}{dt} = X_n, \tag{4.1}$$

where t is the time variable, $x_1, x_2, \ldots, x_n$ are unknown functions, and the expressions $X_1, X_2, \ldots, X_n$ stand for given functions of $x_1, x_2, \ldots, x_n$. If

$$(x_1, x_2, \ldots, x_n) = (\varphi_1(t), \varphi_2(t), \ldots, \varphi_n(t))$$

is a solution, one can look for other solutions of the form $(\varphi_1 + \xi_1, \varphi_2 + \xi_2, \ldots, \varphi_n + \xi_n)$ and which are supposed to differ very little from the original solution. In these circumstances one can ignore the squares and higher powers of the ξ terms, and the equations for them become the equations of variation:

$$\frac{d\xi_j}{dt} = \frac{dX_j}{dx_1}\xi_1 + \frac{dX_j}{dx_2}\xi_2 + \cdots + \frac{dX_j}{dx_n}\xi_n, \quad 1 \le j \le n, \tag{4.2}$$

where the expressions $\frac{dX_j}{dx_k}$ are to be understood as known functions of time by replacing each x_k with the corresponding $\varphi_k(t)$. This is a linear system of equations.

Poincaré said that equations (4.1) are canonical when the x variables divide into two families $x_1, x_2, \ldots, x_p$ and $y_1, y_2, \ldots, y_p$ and the equations can be written as

$$\frac{dx_j}{dt} = \frac{\partial F}{dy_j}, \quad \frac{dy_j}{dt} = -\frac{\partial F}{dx_j}. \tag{4.3}$$

In this case the system is said to have p degrees of freedom. The method for solving differential equations of the form of (4.3) was to look for what he called invariant integrals, functions of the x_j and y_j that are constant. One such is the kinetic energy, and the system was said to be solved when p such invariants have been found. Another example was taken from hydrodynamics: the equations of motion of a fluid are

$$\frac{dx}{dt} = X, \quad \frac{dy}{dt} = Y, \quad \frac{dz}{dt} = Z,$$

where $\frac{dX}{dx} + \frac{dY}{dy} + \frac{dZ}{dz} = 0$ when the fluid is incompressible. In this case the volume $\iiint dx\, dy\, dz$ is invariant.

Poincaré now indicated his general approach. He would investigate the restricted three body problem by writing the differential equations that define it in canonical form. He would assume a solution was known, and indeed that it defined a periodic orbit for the planetoid starting from a position P, and investigate what happened had the planetoid started from a nearby position instead. He would indeed suppose that the nearby positions were strung out on an arc through P, and he would seek to track this arc as it swept out a surface in space by looking for invariant integrals: functions of the x and y variables that have the same value on the arc for all values of t. The existence of an invariant integral would allow him to show that the planetoid was confined to a bounded region of space, and from this it would follow that it would return to an arbitrarily small region of space infinitely often. With some more work it could be shown that at least some of the nearby orbits were also periodic—at least, that was the claim in the original memoir.

The great success of this part of the memoir was what has become known as the Poincaré recurrence theorem (para. 4).[4] This says that if the motion of a point is constrained to lie in a bounded region of three-dimensional space and the volume of any region of space is invariant under the motion then almost every point returns arbitrarily

---

[4] Poincaré called this Poisson stability only in the published version (1890c, para. 8).

close infinitely often to any region of the space that it ever visits. His argument is very simple. Consider an initial position of the point P and time τ_0, and an arbitrarily small ball around it. If the ball never overlaps any of its previous positions at successive times $\tau_1, \tau_2, \ldots$, then it must sweep out a region of increasing volume as time goes by, but the point is moving in a space of finite volume, so the ball must start to overlap some of its earlier positions. If these overlapping positions occur at times τ_p and τ_q, $p < q$, then it is clear there was also an overlap of the balls at time τ_0 and time τ_{q-p}. Now consider successively all the forward and backward iterates of all the points in the common region that return to this common region, and look at the points that belong to all of these iterates. This is a set of points that returns infinitely often arbitrarily close to the initial point P, as required. This result is the closest Poincaré came to establishing the stability of planetary orbits and thus to addressing the tacit question raised by the prize competition.

The proof that some solutions near to a periodic solution are also periodic was to prove invalid. This will be discussed below, when it can be compared with the corrected version, but we now follow the original memoir into a discussion of the existence of the initial period solution (this forms chap. 3 of pt. I). Poincaré proposed to argue as follows. Consider the simplest case of the restricted three body problem, where instead of the two masses of the major bodies being $1 - \mu$ and μ they are 1 and 0. If it can be shown that there is a periodic motion of the planetoid in this case, then, he said, there will also be periodic orbits for the planetoid for some range of values of μ greater than zero. Strictly speaking, this means that Poincaré considered not one restricted three body problem but a range of them as μ varied.

Poincaré examined what might happen to a periodic solution as μ increased from zero by looking at how the equations of variation altered as μ altered. He found that varying μ could only cause a periodic solution to vanish if it merged with another one, and so periodic solutions disappear in pairs, like the real roots of an algebraic equation. But it could be that there were infinitely many periodic solutions, and on analyzing this case he found that periodic solutions might still exist as μ increased, but the length of the period might also vary.

Next he looked at the equations of variation for solutions near a periodic one. In general, the solutions of these equation, $\xi_1, \xi_2, \ldots, \xi_n$ take the form of a sum of n terms, each one an exponential $e^{\alpha t}$ multiplying a trigonometric term, and if all the characteristic exponents α are distinct, real, and negative then nearby solutions must approach the periodic one, so Poincaré said that in this case the periodic solution was stable, and otherwise unstable.[5] His question therefore became one about the $n \times n$ matrix of the α's, and its dependence on $\sqrt{\mu}$. His conclusion was that if there is a periodic solution when $\mu = 0$ for which none of the characteristic exponents vanish, then there will be a periodic solution for small values of μ.

Poincaré looked in particular at the special case where the equations of motions are canonical, and the function F depends in a real analytic fashion on μ:

$$F = F_0 + \mu F_1 + \mu^2 F_2 + \cdots .$$

A long and delicate analysis led him to a differential equation previously studied, he said, by Hermite "that one meets quite often in celestial mechanics, and that several mathematicians have already looked at":

$$\frac{d^2\rho}{dt^2} + n^2\rho + m\rho^3 = \mu R(\rho, t),$$

where m and n are constants, μ is a very small parameter, and R is a function of ρ and t that can be developed as a power series in ρ that is periodic as a function of t. He quickly showed that this can be written as an equation in canonical form, and wrote down a basis of solutions for it in terms of elliptic functions.

For the canonical equation with three degrees of freedom, the characteristic exponents for a periodic solution are all zero. Poincaré investigated what happens as μ increases from zero, and showed that they could be written as power series in $\sqrt{\mu}$. This eventually allowed him to show that for small values of μ there was at least one stable and one unstable periodic solution and precisely the same number of each. This part of the paper concluded with some remarks about asymptotic solutions.

In part II of the memoir Poincaré turned to the n body problem in the case where there are two degrees of freedom.[6] In this case the canonical

---

[5] Poincaré also discussed the case where some of the exponents are equal and the above remarks about the solutions fail, and when two characteristic exponents vanish.

[6] If there is only one degree of freedom the equations are straightforwardly integrable.

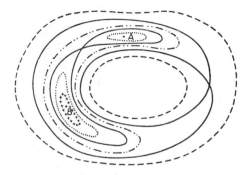

Figure 4.2. The first approximation to a section of the asymptotic surface. Source: Poincaré, *Oeuvres* (7, 419).

equations are, as Tisserand had expressed them in his own work,

$$\frac{dx_1}{dt} = \frac{dF}{dy_1}, \qquad\qquad \frac{dx_2}{dt} = \frac{dF}{dy_2},$$

$$\frac{dy_1}{dt} = -\frac{dF}{dx_1}, \qquad\qquad \frac{dy_2}{dt} = -\frac{dF}{dx_2}, \qquad\qquad (4.4)$$

where Poincaré thought of x_1 and x_2 as linear variables and y_1 and y_2 as angular variables. The kinetic energy, as he called it, $F(x_1, x_2, y_1, y_2)$ is an invariant integral.[7]

It followed from what he had established before that the equations (4.4) have an infinity of periodic unstable solutions. These give rise to asymptotic solutions, and these solutions give rise to asymptotic surfaces that depend on $\sqrt{\mu}$ and exist at least for small values of μ. Poincaré analyzed these surfaces in two stages. First, he looked only at the $\sqrt{\mu}$ term itself.

He found that, in a particular example, the corresponding surface cut out a figure-eight-shaped curve with a double point on a transverse section. He therefore concluded that, to a first approximation, the asymptotic surfaces were closed (see fig. 4.2). Next he looked at the way the surface depended to any finite degree on $\sqrt{\mu}$. His conclusion

---

[7] It is true that $\frac{dF}{dt} = 0$, which makes F an invariant integral, but calling it the kinetic energy of the planetoid is merely traditional in this branch of dynamics—after all, the planetoid has no mass.

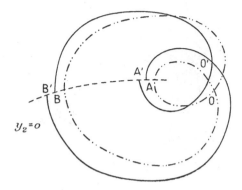

Figure 4.3. The second approximation to a section of the asymptotic surface. Source: Poincaré, *Oeuvres* (7, 438).

now, based on the erroneous result in part I, was that the power series expansions for the surface converged, and therefore the asymptotic surface was closed and divided space into three parts (see fig. 4.3).

The memoir ended with a summary of positive and negative results. Among the positives were the existence of infinitely many stable and infinitely many unstable periodic orbits. The unstable ones gave rise to infinitely many asymptotic orbits that escaped from one closed orbit and converged on another. The ones corresponding to a given unstable periodic orbit filled out a surface that divided space (i.e., the three-dimensional subspace of (x_1, x_2, y_1, y_2) space defined by the only invariant integral) into three regions, and each of these regions contained infinitely many unstable closed orbits. But since it was impossible to leave one region and enter another, the (Poisson) stability of the orbits was rigorously proved. The orbit of the planetoid could therefore be one of three types: closed, asymptotic, or none of the above. Finally, it was highly likely that every region of space was traversed by infinitely many closed orbits.

Among the negative results were the nonexistence of any other invariant integral that was single valued and analytic. This followed, as Poincaré quickly sketched, from the complicated nature of the asymptotic surfaces, which contradicted the presumed analytic nature of any surface defined by an analytic invariant integral. This confirmed a result established by Bruns in a different way, and showed that the series Lindstedt had used in his work on celestial mechanics could not be convergent (else they would lead to an invariant integral just shown to be impossible).

Hermite, Weierstrass, and Mittag-Leffler can be forgiven for finding Poincaré's memoir difficult to understand. It made many claims that were weakly established, if at all. In the end Mittag-Leffler wrote several times to Poincaré for more information, and received nine notes that were placed at the end of the memoir when it was prepared for printing. The first and the seventh expanded on the observation that Lindstedt's series cannot converge. The second translated the main results into the language of astronomy and connected them to earlier work of Hill and Bohlin. The third did a better job with the theory of invariant integrals, the fifth brought out a connection with Poincaré's still-unpublished thesis and related work of Kovalevskaya on partial differential equations, the sixth gave an analytic rather than a geometric account of the theory of asymptotic surfaces. The others were more technical.

None of these eminent men found the major mistake in the memoir, the one that vitiated some of its most important findings. The honor for that seems to rest with Poincaré himself, after some prodding from Phragmén, who Mittag-Leffler had charged with editing the memoir for publication in *Acta Mathematica* in accordance with the conditions of the prize. He found some obscure passages—not an unusual experience in the circumstances— and turned to Mittag-Leffler for help, who in turn wrote to Poincaré on 16 July 1889. The crucial observation was this one about the asymptotic surfaces:

You suppose that x_1 and x_2 are developed in powers of $\sqrt{\mu}$, and you write

$$x_1 = x_1^0 + x_1^1 \sqrt{\mu} + x_1^2 \mu, \quad x_2 = x_2^0 + x_2^1 \sqrt{\mu} + x_2^2 \mu.$$

How can one know that these developments are possible? In fact, it seems that in the asymptotic solutions (pp. 91–92) $\sqrt{\mu}$ is introduced into the denominators because of the divisors

$$\gamma\sqrt{-1} + \sum \alpha\beta + \alpha_i$$

where γ is zero, and it is not very easy to see how one can avoid this. I will be very grateful to you, my dear friend, if you explain to me how to overcome this little difficulty.

Poincaré first replied with the above-mentioned note on asymptotic solutions, in which he showed how the negative powers of μ disappeared.

But something seems to have made him rethink the whole subject of asymptotic surfaces, and on 1 December 1889 he wrote to Mittag-Leffler:

> I have written this morning to M. Phragmén to tell him of an error I have made and doubtless he has shown you my letter. But the consequences of this error are more serious than I first thought. It is not true that the asymptotic surfaces are closed, at least in the sense which I originally intended. What is true is that if both sides of this surface are considered (which I still believe are *connected* to each other) they intersect along an infinite number of asymptotic trajectories (and moreover that their distance becomes infinitely small of order higher than μ^p however great the order of p).
>
> I had thought that *all* these asymptotic curves having moved away from a closed curve representing a periodic solution, would then asymptotically approach the *same* closed curve. What is true, is that there are an infinity which enjoy this property.
>
> I will not conceal from you the distress this discovery has caused me. In the first place, I do not know if you will still think that the results which remain, namely the existence of periodic solutions, the asymptotic solutions, the theory of characteristic exponents, the nonexistence of single-valued integrals, and the divergence of Lindstedt's series, deserve the great reward you have given them.
>
> On the other hand, many changes have become necessary and I do not know if you can begin to print the memoir; I have telegraphed Phragmén. In any case, I can do no more than to confess my confusion to a friend as loyal as you. I will write to you at length when I can see things more clearly. (italics in original)

The news threatened Mittag-Leffler's standing with the King as well as his reputation among mathematicians. Gyldén and Kronecker could be expected to renew their criticisms of the competition, and the future of *Acta Mathematica* would be at risk when people would begin to wonder why the issue with the memoir was delayed. Moreover, there was the potential embarrassment for Hermite and Weierstrass in having awarded the prize to a memoir marked by a grievous error. Mittag-Leffler wrote back to Poincaré on 4 December to say how perplexed he was by the news. He did not doubt, he said, that the memoir would in any case be regarded as a work of genius by most mathematicians, nor did he think that the prize had been other than rightly awarded. "But here is the great misfortune: your letter arrived too late and the memoir has already been distributed."

He asked Poincaré to write a letter to *Acta Mathematica* explaining the nature of the error that had been discovered in the course of correspondence with Phragmén, outlining the new discovery, and promising a new work indicating the necessary modifications to the memoir. He concluded that he knew the adversaries he had acquired through the success of *Acta Mathematica* would cause a scandal, but he regarded that with great tranquillity "because there is no shame in having been deceived at the same time as you and because I am firmly convinced that it is you who will one day unveil the most hidden mysteries in this difficult question."

The next day he wrote again to Poincaré asking him to write a new memoir incorporating material from the old notes and any new material as he saw fit, with an introduction in which he was to explain that in reworking the memoir for publication and expanding on points only hinted at in the original memoir an error had come to light that had been passed over in the earlier work and which was now corrected. It would be necessary, he said, for Poincaré to pay for the costs of the new printing, because he could not ask again for the subvention from the Scandinavian academies that supported *Acta Mathematica*, and he asked that Poincaré keep everything secret until the publication of the new memoir. "Happily," he said, "Kronecker has not received a copy. As for Gyldén's and Lindstedt's copies I shall try to get them back without arousing their suspicions."

Poincaré paid the costs without demur. At 3,500 crowns they were half as much again as the value of the prize. With his finances in order and no risk of a leak there, Mittag-Leffler set about, with whatever tranquillity he could muster, reclaiming all the copies that had been distributed. These were the prepublication copies sent to the editorial board, to Hermite and Weierstrass, and about a dozen others. He explained everything to Hermite, but a week later Hermite replied that Poincaré had let him down: he had not come up with a summary of the results in the memoir and so he (Hermite) had been unable to report on the work to the Académie des sciences. Mittag-Leffler tried to play down the damage with Weierstrass, telling him only that the delay in publishing was due to the correction of some minor errors. This left him unable to refute the rumors that Gyldén and Wolf brought to Berlin in February 1890, and he wrote to Mittag-Leffler to demand an explanation. Mittag-Leffler again tried to smooth things over, but this only left Weierstrass worried lest the published report no longer match the eventual memoir (in the event, it did).

As for the other copies, as Barrow-Green records (1997, 67, no. 142) there is one in the library of the Institute Mittag-Leffler, inscribed in Mittag-Leffler's hand with the phrase which in Swedish reads "whole edition destroyed." When Poincaré sent the revised memoir to Phragmén at the beginning of January, Mittag-Leffler could start to relax. Printing began in April, but was delayed by a backlog of other work, and volume 13 only came out in November 1890. It carried Poincaré's memoir and Appell's, Hermite's report on Appell's work, but only a promise of a report by Weierstrass on what Poincaré had accomplished that was never to be kept. This may be partly because of his age, Weierstrass was 75, but Barrow-Green also notes that Weierstrass felt under a moral obligation to write about Poincaré's error, and Mittag-Leffler was determined it should not be revealed; very likely he took no steps to get the report from his former mentor.

The Published Memoir

The revised version (1890c) took Poincaré only a month to produce, despite the extensive changes he had to make and the novel findings he had to describe. It is in many places unchanged from the one that won the prize. In others it includes the material first submitted as one of the notes, and in others, where the original was in error, it is completely new.

The memoir now opens with a chapter in which Poincaré set up his definitions and his notation, explained at greater length than before the use of the calculus of limits and its application to the theory of partial differential equations, and then discussed the theory of linear ordinary differential equations with periodic coefficients. The second chapter, on the theory of invariant integrals, was left largely unchanged, until it came to the error itself. This has been carefully analyzed by Barrow-Green. Poincaré considered a flow in a space of three or more dimensions that maps a transverse section to itself, and proved that when there is a positive integral invariant no such section is closed (*Oeuvres* 7, 323, theorem II). This generalizes the result familiar from physics that a volume-preserving flow cannot have a closed invariant section, for if it did material would either accumulate inside the surface or dissipate from it. He then showed that if a point on the flow meets the section first at (x_0, y_0, z_0) and next at (x_1, y_1, z_1) then provided that $x_1 - x_0$, $y_1 - y_0$, and $z_1 - z_0$ are sufficiently small they can be expanded as power series. This gave him a hold on how the distance between two points on the section was related to the distance between their iterates. He now considered what happened to an arc on the transverse section. He took two points on it, A and B and looked for their iterates A_1 and B_1. He supposed that the distances AA_1 and BB_1 tend to zero with μ, and that the arcs AA_1 and BB_1 had finite curvature. But whereas the original memoir had erroneously concluded that this forced the arc to be part of a closed curve, Poincaré now deduced correctly (*Oeuvres* 7, 328, theorem III) that the arcs AA_1 and BB_1 merely crossed.

The consequences were considerable. In the restricted three body problem there are four variables x_1, x_2, y_1, and y_2 and asymptotic solutions can be written in the form

$$x_1 = \varphi_1(t, Ae^{\alpha t}), \quad x_2 = \varphi_2(t, Ae^{\alpha t}),$$

$$y_1 = n_1 t + \varphi_3(t, Ae^{\alpha t}), \quad y_2 = n_2 t + \varphi_4(t, Ae^{\alpha t}),$$

where A is an arbitrary constant and the φ's, as functions of the first variable alone, are functions of period T, and $n_1 T$ and $n_2 T$ are multiples of 2π. The variables t and $Ae^{\alpha t}$ can be eliminated, giving

$$x_1 = f_1(y_1, y_2), \quad x_2 = f_2(y_1, y_2)$$

as the equations of the asymptotic surfaces. But it now turned out that when the φ's are expanded in powers of $\sqrt{\mu}$ the series expansions no longer converge, but are merely asymptotic for suitably small values of μ.

To analyze this problem, Poincaré developed the solution in three stages: in terms of $\sqrt{\mu}$, as polynomials in $\sqrt{\mu}$, and finally as asymptotic series in $\sqrt{\mu}$. He found now that because an invariant arc and its iterates crossed (see fig. 4.4), there was a doubly asymptotic trajectory, and as a result two such trajectories, and finally he deduced that there will be an infinity of doubly asymptotic trajectories (p. 227).

LES MÉTHODES NOUVELLES DE LA MÉCANIQUE CÉLESTE

The first volume of the *Méthodes nouvelles* (1892c) is a work of almost 400 pages, which Poincaré must have begun to write when the ups and downs of his work for King Oscar's prize were still vivid in his mind, and it and

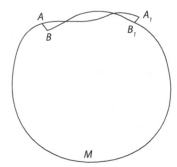

Figure 4.4. The correct behavior. Source: Barrow-Green, *Poincaré and the Three Body Problem*, HMath (11, p. 91).

the two volumes that followed became the mature statement of his ideas that spoke to mathematicians for the next century and more. It was, he said, only the absolute rigor of his paper in *Acta Mathematica* that gave his results on periodic and asymptotic solutions their value, because it offered solid ground upon which one could build with confidence.

The first volume described the formulation of the three body problem, the use of power series methods in solving it, the importance of periodic solutions of the first kind, the method of characteristic exponents, the nonexistence of uniform solutions, the perturbation function, and, finally, the existence of asymptotic solutions. In the introduction Poincaré claimed that the final goal of the theory was to solve the great question: Do Newton's laws explain on their own all astronomical phenomena? He noted that the methods of Lagrange, Laplace, and more recently Le Verrier had solved this question for all practical purposes, although it would always be necessary to refresh their predictions from time to time. But it was the theoretician's job to march ahead of the calculator of tables of ephemerides, perhaps by many years. These calculational methods were based on power series expansions in the known functions that enter the equations of motion of the planets (such as their masses), and gave answers in the form of power series. They had thrown up the existence of what are called secular terms in which the time t appears in an infinite series of sines and cosines, and as a result the convergence of these series was doubtful for large values of t. But these secular terms, said Poincaré, do not belong to the nature of the problem but only to the method used, as is shown by the distinction between the function $\sin \alpha m t$ and any of its polynomial approximations. This disparity had long given rise to the hope among astronomers that another method could be found which avoided this artificial problem, and he suggested that the study of secular terms by means of a system of linear differential equations could be regarded as more of a new method than an old one in this sense.

However, he regarded Hill's work, albeit unfinished, as the contribution that contained the germ of most of the work done since, and Gyldén as "beyond doubt" the man who had done most to advance the new astronomy. His work had then been simplified in certain cases by Lindstedt, and the problem of the secular terms could be regarded as solved (again, for all practical purposes). But all was not yet done. The power series they used did not in fact converge, in the sense mathematicians meant by the term, even though the leading terms gave a very satisfactory approximation, and so these series would eventually give poor predictions. They were therefore no use in discussing the long-term stability of the solar system, for example. Here too one saw the need to go beyond the need to calculate astronomical tables to the proper investigation of Newton's laws.

Poincaré began with a quick summary of Jacobi's formulation of dynamics that leads to the canonical or Hamiltonian equations of motion. There are two series of p variables: $x_1, x_2, \ldots, x_p$ and $y_1, y_2, \ldots, y_p$ and a function F in these $2p$ variables, all connected by p equations,

$$\frac{dx_j}{dt} = \frac{dF}{dy_j}, \quad \frac{dy_j}{dt} = -\frac{dF}{dx_j}, \quad j = 1, 2, \ldots, p. \tag{4.5}$$

Typically the x's are position coordinates and the y's are velocities of a system of p point masses, in which case the jth point mass is said to have mass m_j, position coordinates $x_{3j-2}, x_{3j-1}, x_{3j}$, and velocity coordinates $y_{3j-2}, y_{3j-1}, y_{3j}$. The function F is now supposed to depend only on the positions of the masses, in keeping with Newton's inverse square law of gravity, and when energy is conserved there is a potential function V of the x-coordinates such that

$$F_j = \frac{dV}{dx_j}.$$

The kinetic energy of the system depends only on the velocities:

$$T = \sum_j \frac{m_j}{2} \left(y_{3j-2}^2 + y_{3j-1}^2 + y_{3j}^2 \right),$$

and conservation of energy says that $T - V = $ constant. In this setting the functions y are determined by the functions x, so one says there are $3p$ degrees of freedom and $3p$ independent variables. The prudent thing to do now is to arrange for as many as possible of these variables to be constant, and ideally zero, thus making some of the equations trivial.

The three body problem of course involves just three point masses, and therefore has nine degrees of freedom. If their orbits are supposed to lie in a plane then the problem reduces to one of only six dimensions. To make progress, even further simplifications must be imposed. If one of the point masses has negligible mass—meaning that its orbit is affected by but does not affect the other two—then the two massive bodies will orbit each other in ellipses around their common center of gravity. If furthermore their motion is circular then one may choose coordinates centered on the center of gravity, which cannot move, and rotating with the line joining the two massive bodies, which passes through their common center of gravity. This, Poincaré showed, produces a problem having only two degrees of freedom, which can be taken to be the position coordinates of the massless third body. These considerations led Poincaré to propose the general problem (p. 32) of solving the equations (4.5) when

$$F = F_0 + \mu F_1 + \mu^2 F_2 + \cdots ,$$

where F_0 depends only on the x-coordinates and $F_1, F_2, \ldots$ are periodic functions with period 2π in the y's.

Even the problem with two degrees of freedom is too difficult to admit of a complete answer, as Poincaré had demonstrated in his prize-winning essay. He regarded Hill's breakthrough in his study of the motion of the moon as the attention to periodic orbits. "What makes these periodic solutions so precious," he wrote (p. 82) "is that they are, so to speak, the only breach by which we can penetrate a fortress hitherto considered inaccessible."

Hill had examined the motion of the moon with respect to a system of uniformly rotating axes that have the earth at their center and the earth–sun line as the x-axis. He obtained two homogeneous second-order differential equations for the x- and y-coordinates of the moon, in which a parameter m appears[8] and the period of the moon is 2π in suitable

---

[8] This parameter is the ratio of the synodic lunar period to the sidereal period of the earth.

units of time. There is an invariant integral $-\frac{1}{2}(\dot{x}^2 + \dot{y}^2) + \frac{1}{r} + \frac{3}{2}m^2x^2 = C$, where C depends on m, and for each value of m there is a unique periodic solution of the differential equations. The most interesting of these has two cusps where the moon is exactly on the y-axis, and Hill seems to have thought that these would be the last of their kind, a conclusion Poincaré was to refute in his *Méthodes nouvelles*, as Hill was later to acknowledge.[9] But more importantly, Hill's approach had placed the earth in a circular orbit around the sun, thus formulating the problem as a restricted three body problem.[10]

Inspired by the conclusions Hill had been able to draw from the existence of a periodic orbit, Poincaré set himself the problem (p. 81) of considering a system of differential equations of the form

$$\frac{dx_j}{dt} = X_j, \tag{4.6}$$

where the X_j are given functions of $x_1, \ldots, x_n$ that depend on a parameter μ. He supposed that the equations were solvable when $\mu = 0$ and that among the solutions $x_j = \varphi_j$, $1 \leq j \leq n$ were some that were periodic: $\varphi_j(t) = \varphi_j(t + T)$, $1 \leq j \leq n$. He then asked,

> Under what conditions has one the right to conclude that there will still be periodic solutions for small values of μ?

For example, in the three body problem with masses $\alpha_1, \alpha_2\mu$, and $\alpha_3\mu$, and for which μ is very small, the explicit solutions when $\mu = 0$ have the two small bodies orbiting the large one in Keplerian ellipses, and there is an infinity of periodic solutions. He now promised to show that when μ is sufficiently small the problem for each value of μ still has infinitely many periodic solutions.

It might seem, he said, that periodic solutions were no use, because the odds against the initial conditions being such as to give rise to a periodic orbit were almost zero, but even so they could be used as a starting point in the investigation of other, more probable, orbits. Indeed, he conjectured, but admitted he could not rigorously prove, that given a solution to equations (4.6) there would always be a periodic solution,

---

[9] On this, see Nauenberg (2006/2010, 121–126).

[10] Hill reduced the problem to a single second-order differential equation $\frac{d^2w}{dt^2} + \Theta w = 0$, where Θ can be written as a Fourier series, and solved by treating it as one involving a doubly infinite sequence of algebraic equations—another topic congenial to Poincaré. Hill also introduced complex coordinates.

perhaps with a very long period, with the property that the difference
between the two solutions was as small as one wants for as long as one
wants.[11] Moreover, he could also show (pp. 152–160) that the trick of
comparing a real orbit with a close approximation that is periodic was
central to Hill's success with the motion of the moon, and with Laplace's
account of the motion of the satellites of Jupiter.

The lesson professional astronomers drew from this part of Poincaré's
work was that the method of periodic orbits might be one way to reduce
the sheer tedium involved in computing orbits based on the method
of perturbing simple solutions, which required lengthy convergence
arguments and tended to give series solutions that converged too slowly
to be any use when studying long-term stability questions.[12] Much of the
work of G. H. Darwin in England and F. R. Moulton in the United States
drew directly on the inspiration they took from Poincaré and Hill, as
Whittaker (1899) describes.

In his investigation of how the periodic solutions depend on the
parameter μ Poincaré found that they disappear in pairs like the real root
of algebraic equations. He argued as follows: Suppose that in fact when
$\mu = 0$ all the solution functions φ_j, $1 \leq j \leq n$ have period 2π, and that
when μ is very small and $t = 0$ then $x_j(0) = \varphi(0) + \beta_j$. Suppose also that
$x_j(2\pi) = \varphi(0) + \beta_j + \psi_j$. So the functions ψ_j are functions of $\beta_1, \ldots, \beta_n$ and
μ, and a solution will be periodic if

$$\psi_1 = 0 = \psi_2 = \cdots = \psi_n. \tag{4.7}$$

Poincaré then appealed to the implicit function theorem, which he had
proved in the previous chapter, to argue that if the Jacobian determinant
$\left(\frac{\partial \psi_j}{\partial \beta_k}\right)$ is not zero then the β's can be written as functions of μ, $\beta_j = \theta_j(\mu)$,
which all vanish when $\mu = 0$. Moreover, the functions θ_j can be written as
power series in μ, by one of those convenient assumptions that Poincaré
was prone to make, and so he concluded that for small values of μ there
are values of the β's such that the ψ's vanish and there are still periodic
solutions. Tacitly, he argued that the values of the β's for which the ψ's all
vanish depend smoothly on μ and cannot disappear.

---

[11] This was eventually proved in Hopf (1930).
[12] I owe this observation to Tatiana Roque, who is preparing a full-length study of the topic.

If, however, the Jacobian vanishes, then equation (4.7) has $\beta_1 = \cdots = \beta_n = 0$ as one of, say, m solutions, and Poincaré now argued that if equation (4.7) has m_1 solutions for μ small and positive and m_2 solutions for μ small and negative, then by earlier lemmas that rested ultimately on results in his (still unpublished) doctoral thesis the numbers m, m_1, m_2 are all of the same parity and so periodic solutions can only appear and disappear in pairs.

The *Méthodes nouvelles* goes much further into the general theory of the three body problem than does the prize-winning memoir. In the general problem there are 18 variables—the position and velocity coordinates of the three bodies—but they are connected by a number of invariants. These are the equations that say that the center of gravity of the system is fixed, that the total angular momentum of the system is conserved, and that the total energy of the system is also conserved. This reduces the number of effective coordinates from 18 to 8. Each of these equations is algebraic in the relevant coordinates, and so interest had attached to the search for other algebraic invariants which would, if they existed, further reduce the number of variables involved in the problem, but in his paper of 1887, Heinrich Bruns, a former student of Weierstrass, had shown that there are no other algebraic invariants for fixed values of the masses.[13] In the prize-winning memoir Poincaré had shown that the restricted three body problem admits no new transcendental invariants that are analytic functions of the masses, and in the *Méthodes nouvelles* he reworked and extended this proof to cover the general three body problem. This involved him in the notorious problem in the subject, the so-called small divisors.

When the mean motions of the bodies are incommensurable, the bodies will sometimes be unusually close to each other. This shows up in the power series expansions of the functions describing their orbits in the form of coefficients, which are of the form

$$\sum_{j,k} \frac{A_{jk}}{(jn_1 + kn_2)} \sin\left((jn_1 + kn_2)t\right),$$

where n_1 and n_2 are the mean periods of the two planets and the A_{jk} are constants that rapidly tend to zero as j and k increase. Because n_1 and

---

[13] For a detailed account of Bruns' work, see Whittaker (1937, 358–379).

n_2 are incommensurable, there are values of j and k that make $jn_1 + kn_2$ arbitrarily small. These terms therefore contribute disproportionately to the function. Famously, Laplace had encountered the problem of small denominators in investigating the way the orbits of Jupiter and Saturn deviate from Keplerian ellipses. The ratio of the mean orbital periods is, to a good approximation, $5 : 2$, which means that the term where $j = -2$, $k = 5$ contributes significantly to the motion, and in 1785 Laplace had been able to show that as a result the mean longitude of the system rotates with a period $2\pi/(-2n_1 + 5n_2)$, or about 900 years.

However, in the problem that Poincaré confronted all that is required is to know when small denominators occur, and Poincaré was able to extend some ideas of Darboux to analyze the matter and circumvent the difficulty. But, as so often in his work, the conclusions he reached were valid only for small values of his parameter μ, and Aurel Wintner was to point out that both his result and that of Bruns are only valid for some unspecified value of the masses and are therefore inapplicable in any given situation (see Barrow-Green 1997, 155, no. 253).

The second volume of the *Méthodes nouvelles* appeared in 1893 and was devoted to explaining the methods used by professional mathematical astronomers. As is quite frequently the case when mathematicians communicate with specialists in other mathematical fields—today, for example, in quantum field theory—it is not a simple matter of acquainting each side with the tricks of the other side's trade. In this case it was not so much that the astronomers had not proved that the power series expansions on which they relied converged, as of observing that the series they worked with demonstrably did *not* converge, even though they gave increasingly useful results. The mathematicians looked at the behavior of a power series as more and more of its terms were added in; astronomers contented themselves with looking at perhaps the first twenty terms and seeing if they decreased sufficiently rapidly. If they did, the astronomers pronounced the series convergent. Therefore, said Poincaré on page 1,

> To take a simple example, let us consider the two series which have as their general terms
>
> $$\frac{1000^n}{1 \cdot 2 \cdot 3 \cdots n} \quad \text{and} \quad \frac{1 \cdot 2 \cdot 3 \cdots n}{1000^n}.$$

Mathematicians will say that the first series converges and even that it converges rapidly, because the millionth term is much less than $10^{-1000000}$, but they regard the second as divergent because the general term increases without limit.

Astronomers, on the contrary, regard the first series as divergent because the first 1000 terms increase, and the second as convergent because the first 1000 terms decrease and this decrease is initially very fast.

But, he went on, both rules are legitimate, the first in theoretical work and the second in numerical calculation. His task was to reconcile the two approaches, and his preferred example was that of Stirling's series. Cauchy had shown (1843) that the terms of this series for $n!$ initially decrease and then increase, but if one takes the sum only as far as the last term in the decreasing part a strikingly accurate approximation to $n!$ is obtained. Poincaré had gone on to find other series of this kind and develop a method for handling them that, in the context of astronomy, could make rigorous mathematical sense of the astronomers' seemingly rough and ready methods. In this spirit he then analyzed the methods of Hugo Gyldén, which he called the "most perfect"; Lindstedt, who had offered some important simplifications of Gyldén's difficult work; Delaunay; and Bohlin. In 1901 Poincaré returned to the subject of Gyldén's work and now found some faults with it. As a result, in 1904 and 1905 Poincaré was challenged by Gyldén's successor, Victor Bäcklund, who believed that it was Poincaré who had erred, and Poincaré this time came to a more negative conclusion. He found some of the papers obscure, some parts were correct but could have been done better, a great number of the results were false, and most were so poorly presented it was impossible to decide if they were true or false.

The third volume of the *Méthodes nouvelles* came out in 1899, and has proved to be the one that has spoken most deeply to mathematicians. It opens with a treatment of the theory of invariant integrals that greatly improves the account in the prize-winning memoir, then Poincaré moved on to discuss periodic solutions "of the second class"—those that make more than one orbit around the primary. Here he compared his theoretical results with the numerical solutions produced by the English astronomer G. H. Darwin. He found a large measure of agreement, and was able to show in the one case where they disagreed that Darwin's results about the stability of a certain family of solutions were in error. The volume ends with an account of singly and doubly asymptotic solutions. The one advance he made on the memoir of ten years before was to consider not merely the homoclinic solutions, which arise from the doubly asymptotic solutions associated to a single unstable periodic orbit, but a new class of what he called heteroclinic solutions. These arise from the asymptotic solutions associated to two distinct unstable periodic orbits, and as a result are much more complicated.

Poincaré was able to show that, in either case, the existence of one hetero- or homoclinic solution implies the existence of infinitely many more, but otherwise, the complexity defeated him. As he put it,

> When one tries to depict the figure formed by these two curves and their infinity of intersections, each corresponding to a doubly asymptotic solution, these intersections form a kind of net, web, or infinitely tight mesh; neither of the two curves can ever intersect itself, but must fold back on itself in a very complex way in order to intersect all the links of the mesh infinitely often.
>
> One is struck by the complexity of this figure that I am not even attempting to draw. Nothing can give us a better idea of the complexity of the three-body problem and of all the problems of dynamics in general where there is no single-valued integral and Bohlin's series diverge. (*Méthodes nouvelles* 3, 389)

As Barrow-Green commented,

> The difficulties that Poincaré had encountered in trying to understand these doubly asymptotic solutions is evident from the fact that almost ten years had elapsed since he had first introduced the idea..., and despite the time interval he had added relatively little to his original discussion, apart from the important inclusion of heteroclinic solutions. (1997, 162)

Such was the complexity, indeed, that for some time mathematicians were unable to do more than echo his words. But these tangles of solution curves, with the marked unpredictability of the solutions and their sensitive dependence on the initial conditions, are the hallmarks of chaos, the modern study of which has become a central feature in the study of dynamical systems. They have gone from being indicative of why the three body problem is so difficult to a gateway to much elegant mathematics.[14]

---

[14] For a good account of how this happened and many of its implications to the present day; see Diacu and Holmes (1996).

POINCARÉ RETURNS

Geodesics on Convex Surfaces, 1905

Such were the complexities of any version of the three body problem that Poincaré spent some time looking for a related but simpler problem that could be helpful by having only the difficulties that were in the essence of the three body problem but had none of what Poincaré called its secondary difficulties. In 1904 he decided that the subject of geodesics on a closed convex surface was the problem he was looking for.[15] It was, he explained, a problem in dynamics, but it was the simplest of all, having only two degrees of freedom, and it shed the difficulty that arises in the three body problem of points where the moving particle stops. Indeed, in the new problem one could assume that the geodesics were traced at a constant speed that could be regarded as one of the givens of the problem.

He acknowledged that a paper (1898) by Hadamard on geodesics on surfaces of negative curvature had been an influence, but claimed that convex surfaces posed a problem closer to the astronomical one and, unfortunately, harder.[16] He had, he said, been able to come to some results on the existence of closed geodesics on a convex surface; such geodesics were the analogue of periodic solutions in the three body problem. He first presented his results at the meeting of the American Mathematical Society in St. Louis; he then published them in the *Transactions of the AMS* (1905g).

A convex surface is one whose principal curvatures at each point are always positive and bounded; an ellipsoid is a useful special case, and a sphere is another example. On such a surface one may consider the geodesics through an arbitrary point. On the sphere these geodesics very nearly describe a system of polar coordinates covering the whole sphere. The origin is put at the north pole, and one measures length along a geodesic (a line of longitude) from the origin. Different geodesics make different angles with the zero meridian, and so almost every point has a unique coordinate obtained by specifying the unique geodesic on which

---

[15] For an interesting comparison of Poincaré's work in celestial mechanics and on geodesics on spheroids that brings out Poincaré's uses of geometry, see Robadey (2004).

[16] In Poincaré's problem, but not Hadamard's, the long-term behavior of the solutions is much harder to study.

it lies and how far along it lies. This system has one disadvantage: these geodesics all meet again at the south pole. On a general convex surface, matters are more complicated. Poincaré considered two geodesics emanating from a point O in almost the same directions, and looked for the first point M where they meet again, F_1, and all the subsequent such points where they do this (here he referred to another paper: Hadamard 1897). These points are called foci. Poincaré called the collection of all foci of a point O the caustic associated to O; it is the envelope of geodesics emanating from O.

Poincaré did not pause to explain the optical language he used, but it comes about by considering light emanating from a point on the surface and constrained to propagate in the surface. Initially the light will spread out like a wave, and the wave front will be approximately spherical. But gradually it will distort and parts of it may converge. It is natural to call a point where they converge a focus. The term caustic is used because it applies to the envelope of a bundle of light rays, such as one sees on reflecting sunlight in a cup or glass. In his lectures on dynamics, edited by Clebsch, and in unpublished work edited and published in the last volumes of his *Gesammelte Werke* (1891) Jacobi conjectured that when a general point is taken on an ellipsoid its antipodal point will be surrounded by a figure with four cusps.[17]

Arguments involving the curvature on the spheroid allowed Poincaré to deduce that the point O is surrounded by nested, nonintersecting ovals composed of arcs of caustics. In this respect the sphere is, as so often in this context, a misleading example to have in mind, and a better one is given above. By speaking of ovals and not mentioning cusps it would seem that Poincaré was not aware of Jacobi's ideas, although they had been published only recently. Nor did he recall the way cusps had appeared in his own analysis of Hill's work.

Poincaré now switched to the alternative way of thinking about geodesics on a surface, not as curves of shortest length between their points but as curves of least energy. If one imagines stretching an elastic band between two points on a convex surface, the shape it will take up minimizes the energy in the band—and it is a geodesic between those points. He considered spheroids, surfaces that differ only slightly (in ways

---

[17] See the beautiful figure by Knill, available at http://www.math.harvard.edu/~knill/caustic/exhibits/ellipsoid/index.html.

he did not specify) from a sphere. He therefore began with a sphere. Geodesics on the sphere are great circles, and movement along a great circle can be specified as follows: give the coordinates of the poles of the great circle, specify a uniform speed on the great circle, and the longitude of the moving point with respect to a fixed but arbitrary point on the great circle.

Poincaré then wrote down the expression for the energy, T_0, involved in traveling along an arbitrary curve on the sphere, which is an expression closely related to the metric on the sphere. He then imagined that the sphere was perturbed into a spheroid. The spheroid has a slightly different metric, and the energy expression is therefore slightly different; Poincaré showed that it could be written as $T = T_0 + \mu T_1$, where μ is very small and T_1 has the same form as T_0: a quadratic in two variables U and V. He now set about deriving the Jacobi equations for geodesics on the spheroid. His method was plainly taken from his work on celestial mechanics. Not only did he set up the situation when $\mu = 0$ and then increase μ slightly, he borrowed technical terms from Laplace's astronomy. He imagined that as μ increased from zero each point on the sphere moved to its corresponding point on the spheroid, where corresponding points had parallel tangent planes. This gave him a good way to relate the maps of the sphere and the spheroid on a plane.

He could now investigate when a closed geodesic on the sphere (a great circle) remains a geodesic on the spheroid. He obtained a way of measuring how close two infinitesimally close points are on the spheroid in terms of the separation of the corresponding points on the sphere, and from that defined a function R of the coordinates on the spheroid whose derivatives vanished on a curve that was a geodesic. The corresponding function R_0 can be defined on the sphere itself, and now its derivatives, which vanish at the maxima, minima, and saddle points of R_0, are well understood. In fact, by his paper on curves on surfaces (1881a) the number of maxima, m_+ say, plus the number of minima, m_-, and minus the number of saddle points, s, is 2: $m_+ + m_- - s = 1$. The number 2 obtained topologically has been halved, because Poincaré's method of counting distinguishes the direction of the geodesics. So the number of closed geodesics is

$$m_+ + m_- = 1 + 2s,$$

which is odd and nonzero. This number depends continuously on μ and so it cannot vanish for small μ, which shows that for a spheroid there is at least one closed geodesic. Or rather, and here Poincaré corrected himself, it only suggests that the result might be true, for this calculation is only approximate and valid to the first order in μ alone. Poincaré now asked himself if a rigorous proof valid to all orders in μ could be extracted from this approximate proof by means of the methods he had used in *Méthodes nouvelles* (1899, vol. 3, chap. 1), and he gave two answers. In the first, he assumed that the spheroids from $\mu = 0$ to $\mu = 1$ were defined by an analytic equation $F(x, y, z, t) = 0$, where t is a parameter and $0 \le t \le 1$. Four coordinates x_0, y_0, x_0', y_0' specify a geodesic, and they satisfy some analytic conditions when this geodesic is closed. So, taking account of the parameter t, a closed geodesic is specified by suitably related points in some five-dimensional space. This turns out to be an analytic curve, and Poincaré deduced that as a result geodesics could only appear or disappear in pairs. Since the process starts with at least one, it remains odd and therefore nonzero.

Poincaré next had to be sure he had a geodesic that does not self-intersect, as any nonclosed geodesic will if conducted far enough, so he had to investigate the number of self-intersection points of a closed curve on a spheroid and consider what might happen if the closed curve is also a geodesic. A consideration of what geodesics can and cannot do allowed Poincaré to conclude that there is always an odd number of closed geodesics without self-intersections on a spheroid.

In fact, Poincaré's argument was not entirely conclusive. He did not discuss the possibility that there might be infinitely many geodesics on the spheroids, as there are on the sphere, although he did mention it in the analogous section of the *Méthodes nouvelles* (1892, vol. 1, sect. 37) as an exceptional case. Nor did he consider that the geodesics might grow too rapidly in length; it was shown much later that the hypothesis on the curvature of the spheroid rules this out (see Anantharaman 2006/2010, 158).

Poincaré now turned to the question of the stability of geodesics close to a closed geodesic on a convex surface. He said a closed geodesic was stable if nearby (nonclosed) geodesics stay close by, and unstable if nearby geodesics moved away, and he showed that the number of stable geodesics exceeds the number of unstable ones by a constant amount. In the case of the ellipsoid, the shortest and the longest of the three closed geodesics are stable, the middle one is unstable, so this number is 1, and so Poincaré concluded that there was always one stable geodesic on a convex surface.

Then Poincaré gave a second proof of the existence of a closed geodesic on a convex surface, which is in some ways more direct and intuitive. It has proved to be the proof that has drawn the most attention, and it is also a charming illustration of his ingenuity and of the difficulties that mathematics has to be developed to solve.[18] Any simple closed curve on the sphere divides it into two regions. If the curve is a geodesic, the so-called Gauss–Bonnet theorem says that the integral of the curvature over either region is 2π. Poincaré therefore considered the space of all simple closed curves on the surface that have the property that the integral of the curvature over either region is 2π. He observed that these curves cannot be arbitrarily short, or the integral on one region could be made arbitrarily small. So it was possible to speak of a curve of this kind of the shortest possible length, and, he gave an argument to show that this curve would be a geodesic.

"Let us," he said, "examine the objections that one can make to this incomplete reasoning." One must check that there is a set of points realizing this minimum and that it is indeed a simple closed curve. Then, is it a sufficiently smooth curve for the methods of differential geometry to apply, so that it can be proved to be a geodesic? Can it be shown that it does not, for example, have a cusp? Poincaré regarded the first set of problems as belonging to analysis, and dispatched them with the remark that one should consult recent work by Hilbert.[19]

The other problems he regarded, intriguingly, as physical, and he proceeded to deal with them. He assumed he had a curve of minimal length, and sought to show it had to be a geodesic. The problem is to show that the curve is simple. On the face of it, the curve could overlap itself as if fitted with a zip, but Poincaré was able to show that such a configuration would not be stable, and the overlap would undo itself. But this does not show that the curve is of minimal length, only that it is an extremal of length. The second variation has to be considered, and this Poincaré did by following a geometrical argument until he had found some more necessary conditions which he then had to show were sufficient. He now

---

[18] For a fuller account see Anantharaman (2006/2010) and the literature cited there.

[19] Poincaré did not specify which. Most likely he had in mind Hilbert's Paris address on mathematical problems; it is unlikely he had seen Hilbert (1905b).

concluded the paper with the words "I need not treat this question here, which the ordinary methods of the calculus of variations permit one to resolve."

The Last Geometric Problem, 1912

On 9 December Poincaré wrote to Guccia to say that he was thinking of submitting a paper that he was worried he would never finish. He began the resulting paper, which appeared in the *Rendiconti*, by saying that he had never before presented the public with an unfinished work. He then reminded his readers of his long-standing interest in the existence of periodic solutions in the three body problem, and said that in reflecting upon the problem he had been led to believe that in dynamical problems with two degrees of freedom the existence of a periodic solution depended on a very simple geometric theorem. But, in tackling this problem he had been led to consider too many special cases. Those he had analyzed fully supported his belief that the theorem was true in general, but for two years he had been unable to provide a general proof. He had therefore put it aside for several years, thinking to pick it up again one day, but feeling that at his age he could no longer guarantee that he would be able to do so, and in view of the great importance of the result, he judged it best to share his partial results in the hope that someone would be more successful than him.

The theorem that had thus defeated him he then stated as follows. One is given an annulus formed by two concentric circles of radii $b < a$ as in figure 4.5, and a continuous invertible map T of this annulus to itself that maps the inner circle to itself and the outer circle to itself. The map T satisfies two conditions: First, T maps every point on the inner circle in one direction although by varying amounts, and every point on the outer circle in the opposite direction, again by varying amounts. Second, the transformation is area preserving (the area integral is an invariant integral, in the language of his work on celestial mechanics). Then there will always be two points in the interior of the annulus that are each mapped to themselves by the map T.

He noted that the theorem was equivalent to claiming that if the first condition holds but there is no point that is mapped to itself, then the transformation T cannot be area preserving, and also that, by his

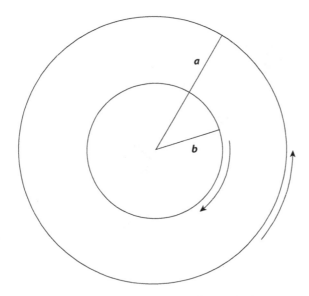

Figure 4.5. The annulus. Source: drawn by Michele Mayor Angel.

earlier work, the number of points mapped to themselves by T must be even.[20]

To see that some condition such as area preservation must enter the theorem, consider the situation in figure 4.6.

The annulus is filled with curves that unwind from the inner circle, change direction and spiral out to the outer circle. A flow along such curves defines a one-parameter family of maps from the annulus to itself, but for any value of that parameter at some point a thin annulus B will be mapped onto a longer annulus A of the same width, and so area is not preserved for this flow.

Poincaré first explained the implications of the theorem for dynamics by means of two examples. One concerned geodesic flow on a convex surface, the subject of his (1905x), the other the restricted three body problem. In each case a periodic solution (a closed geodesic or a closed orbit) was represented by a closed curve C_0 in a certain abstract three-dimensional space, which Poincaré supposed was spanned by a surface, A. Poincaré supposed that all the other flow lines that met the surface A did so by crossing in and none were tangent to it. The first return map

---

[20] Either transplant the annulus to an equatorial region on the sphere and complete it with two spiral flows out of the polar caps, or double the annulus to form a torus.

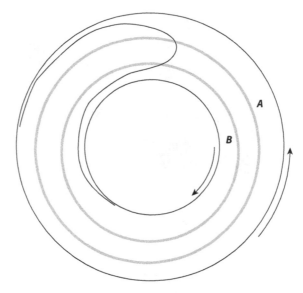

Figure 4.6. A flow that does not preserve area. Source: drawn by Michele Mayor Angel.

then defines a map from the surface A to itself, and after several pages of work that drew on ideas from his earlier papers and books Poincaré was able to show that the theorem he was struggling to prove would apply to this surface provided it contained a region inside it in which there was a flow in the opposite direction to the flow along C_0.

In order to prove the theorem Poincaré took polar coordinates x and y in the annulus, so each point is expressible in the form xe^{iy}, and then looked at the corresponding strip in the x–y plane, which is bounded by $x = b$ and $x = a$. So the expression $xe^{i(y+2k\pi)}$ represents one point in the annulus but infinitely many in the x–y plane. This gave him the problem of distinguishing between closed curves in the annulus according to the number of times that they wrap around the inner circle, and which therefore open out differently when mapped to the strip. The same complication enters into the depiction of the images of curves in the annulus or strip under the area-preserving transformation.

He now attempted to prove the theorem by constructing a loop that would either wholly contain or be wholly contained in its image, thus contradicting the area-preservation property. To that end, he defined a curve in the annulus whose existence would imply the truth of the theorem, and sketched a deformation process for curves in the annulus which he hoped would show that any initial curve can be deformed into the required curve. This process would be followed by looking at pictures in the strip, and, unusually for him, he provided several pictures to illustrate how the process worked in particular cases. But, as he had admitted, he was unable to find a general argument.

The challenge to prove Poincaré's last geometric theorem was taken up by several mathematicians, but the acknowledged winner was the young American mathematician George David Birkhoff. He had already been interested in Poincaré's work on dynamics, and he was looking for a major problem with which to make his reputation in Europe, but even so it took him several months of work to come up with his elegant and convincing solution (Birkhoff 1913). But it did everything he wanted: it greatly impressed the French, and it established him, very soon after Poincaré's death, as the new leader in the field. The subject of dynamical systems was to prosper in his hands for a further thirty years.[21]

At the end of his own paper Birkhoff commented on why he might have succeeded when Poincaré had not. In his opinion, Poincaré had been too naive in his interpretation of the area-preserving nature of the map T:

> It is interesting to notice that Poincaré uses only a single property consequent on the existence of an invariant integral, namely that no continuum on [the annulus] R can be transformed into part of itself by the transformation T. It seems improbable that this condition is equivalent to the condition that there exists an invariant integral. The existence of such an invariant integral is a fact which enters more intimately into the proof which I have given above. (Birkhoff 1913, 22)

It has turned out that Poincaré's argument was nearer to success than either he or Birkhoff realized[22]—one can only wonder at the effect on the history of mathematics had Poincaré actually succeeded.

---

[21] For a detailed study of Birkhoff's work on the last geometric theorem and some of his later ideas about the important problems in dynamical systems, see a forthcoming paper by Barrow-Green.

[22] For a good account of what Poincaré did and how a suitable loop can be constructed and a successful proof obtained, see Golé and Hall (1992).

5

Cosmogony

ROTATING FLUID MASSES

There are many reasons for caring about the motion of a rotating fluid mass, most notably that we live on one, but the most spectacular sight of the problem is Saturn's rings. They had first been detected by Galileo in 1610, who saw them as handles on either side of the planet; in 1655 Huygens, with an improved telescope of his own design, saw them as a disk surrounding it. Robert Hooke was able to see shadows cast on them, and in 1675 Domenico Cassini was the first to see gaps in the disk and therefore to see them as rings. In 1787 Laplace suggested that they might be formed of a large number of solid ringlets. This idea prevailed until Maxwell, in an essay that won the Adams Prize for 1856 at the University of Cambridge, showed that solid rings would not be stable and would be broken apart. This raises one of the most interesting problems in dynamics: not so much the existence of this or that orbit or configuration, but the possibility of stable ones, ones that in some sense resist the influence of their surroundings and remain for long periods of time with the same shape.[1]

Maxwell noted that if a ring is supposed to be solid then three aspects of the motion must be noted: the gravitational pull of Saturn on the ring, the centrifugal force of each part of the ring as it rotates, and the gravitational attraction of the ring to its own center of gravity. The third of these is the one that presented mathematicians with genuine difficulties. Laplace had shown that a solid uniform ring cannot be in orbit around a planet, because any displacement of the center of the ring

---

[1] See Harman (1992), who discusses the interesting speculation that problems in calculating the hopelessly complicated motion of particles in the rings led Maxwell to develop the statistical techniques that he went on to deploy in his theory of gases.

from the center of the planet would end in the ring falling into the planet. He therefore proposed that the rings be irregular, nonuniform solids. Maxwell examined this hypothesis, and found that it might be possible if the ring was loaded at one point "with a heavy satellite about $4\frac{1}{2}$ times the weight of the ring" (1890, 295), but rejected this as too artificial a solution because small variations in the weight would again make the ring unstable, and in any case the idea was inconsistent with the appearance of the rings. This left Maxwell to consider the idea of "a ring of independent satellites, or a fluid ring."

Maxwell next investigated the idea that variations in the ring might accumulate and eventually tear it apart, and he noted that this very hypothesis was "one of the leading doctrines of the 'nebular theory' of the formation of planetary systems," so it could not be idly dismissed. Indeed, he said, he could show that this destructive tendency exists, but the motion of the ring converts it into a force for stability. A further analysis of this dynamical tension showed that on the one hand Saturn must be at least 42.5 times denser than the ring, and on the other hand that it cannot be more than 1.3 times as dense if the inner and outer parts of the ring are to orbit with the same angular velocity. Maxwell then investigated and rejected the idea of a single ring made of parts of a broken ring, and therefore concluded that the ring system was made up of several rings of satellites.

In 1874 Sonya Kovalevskaya wrote a paper as part of her doctoral thesis that addressed a different question that had also been considered by Laplace: the shape of the rings. It was published in the *Astronomische Nachrichten* only in 1885 however, which is when Poincaré came to hear of it, and by then he had already taken up the subject of rotating fluid masses. He was much impressed by her work, and frequently referred to it in his own papers. Two aspects caught his attention. First, she had expressed the shape of the ring as a Fourier series, and so her computation of the potential of the ring depended on the infinitely many coefficients of the series. This led her, via some impressive juggling, to confront a system of infinitely many equations in infinitely many unknowns, a topic that was to interest him later. She sketched a way of tackling such equations, but could not establish any theorems to justify her approach. Second, the juggling itself introduced some elliptic integrals, and he was intrigued by their use. Cooke, in his study of Kovalevskaya's work (Cooke 1984, 80) points out that two reasons for the inconclusive nature of her account are its "Brobdingnagian" complexity and her awareness that Maxwell had already shown that it was "doubtful whether Laplace's view of the structure of the rings of Saturn is acceptable."

Another stimulus for Poincaré was a series of unproved statements in Thomson and Tait's major work, their *Treatise on Natural Philosophy*, that occupies the second part of volume I, paragraph 778. As they put it in 1879,

> During the fifteen years which have passed since the publication of our first edition we have never abandoned the problem of the equilibrium of a finite mass of rotating incompressible fluid. Year after year, questions of the multiplicity of possible figures of equilibrium have been almost incessantly before us, and yet it is only now, under the compulsion of finishing this second edition of the second part of our first volume, with hope for a second volume abandoned, that we have succeeded in finding anything approaching to full light on the subject.

Poincaré published a series of five short notes on the subject between February and July 1885 in a variety of journals. In his first note, (1885b), Poincaré sought to establish rigorously this claim of Thomson and Tait's:

> If the condition of being a figure of revolution is imposed, without the condition of being an ellipsoid, there is, for large enough moment of momentum, an annular figure of equilibrium which is stable, and an ellipsoidal figure which is unstable. It is probable, that for moment of momentum greater than one definite limit and less than another, there is just one annular figure of equilibrium, consisting of a single ring....The condition of being a figure of revolution being still imposed, the single-ring figure, when annular equilibrium is possible at all, is probably stable. It is certainly stable for very large values of the moment of momentum.

Poincaré tackled this by first claiming that it was easy to prove that if an arbitrary system is stable under the action of certain forces, and is then subjected to some infinitesimally small perturbing forces, it will occupy a new stable equilibrium position infinitely close to the original one. Then, without realizing that he was following Kovalevskaya's approach, he assumed there was a solution in the shape of the ring, which he represented by a Fourier series, and computed the potential of the ring by first ignoring all but the first term of the Fourier series and using a power series method that appealed to the theory of elliptic integrals, and then putting the other terms back in on the assumption that their coefficients are arbitrarily small.

In his next note, (1885c), Poincaré observed that in problems of this kind one cannot appeal to the method of successive approximations to prove the existence of solutions, but must appeal instead to a principle of continuity, such as he had already used in his early studies of the three body problem. He now indicated how this could be done. In his third note, (1885d), he observed that his methods were close to Kovalevskaya's, which is perhaps why he published this note in the *Bulletin astronomique*. A note followed on new types of equilibrium solids of revolution that were not ellipsoids, and a further note observing that almost all of these new figures were not stable. These notes served as publicity for his major paper (1885e) which appeared in *Acta Mathematica* in September 1885. Poincaré began by listing no less than eleven results stated, mostly without proof, by Thomson and Tait that he proposed to tackle. He also promised to discuss some new figures of equilibrium.

First of all Poincaré considered a highly simplified version of the actual problem, in which there is a holomorphic function F of n variables $x_1, x_2, \ldots, x_n$ and a parameter y with the property that the values of the x's that correspond to equilibrium positions depend on the parameter y. He argued that there might be several distinct equilibrium positions given by

$$x_1 = \varphi_{j1}(y), \ldots, x_n = \varphi_{jn}(y), \ 1 \leq j \leq k$$

that vary smoothly with y. For some values of y, some of the x values of φ_{jl} may coincide; what happens then, he asked? Mathematically, this condition is expressed by the vanishing of the Hessian determinant Δ whose entries are $\frac{\partial^2 F}{\partial x_j \partial x_l}$, and it can be studied as a function of y.

Poincaré gave several examples to show what can happen before proceeding to the general case, and here he made an excursion into topology. He supposed he had two variables x_1 and x_2 and a single parameter y, and regarded x_1, x_2, y as the coordinates of a point in space. The equilibrium conditions are

$$\frac{\partial F}{\partial x_1} = 0, \quad \frac{\partial F}{\partial x_2} = 0,$$

which represent two surfaces S_1 and S_2 meeting along a curve C. He further supposed that the equations

$$x_1 = \varphi_1(y), \quad x_2 = \varphi_2(y)$$

represent a branch B of C, and that at a point M of B where $y = \beta$, say, Δ has one sign at P with y-coordinate $\beta - \varepsilon$ and another at Q with y-coordinate $\beta + \varepsilon$. He then deduced that the curve C branched at M.

To prove this, at each point of the arc PQ he attached a circle of radius r parallel to the x_1–x_2 plane and with its center on PQ, forming a curved cylinder, Σ. By construction, no part of the branch B lies on Σ, and the question is: How many points of the curve C lie on Σ? To answer this, Poincaré invoked the Kronecker characteristic (Kronecker 1869) of the surfaces Σ, S_1, and S_2. This gives the number of points of C on Σ satisfying a certain condition minus the number of points of C on Σ not satisfying that condition as an integral evaluated on the circles at the ends of the cylinder, taken in appropriate directions. In the present case Poincaré, without offering any explanation, said that the integrals are either $+1$ or -1 in the four regions defined by the signs of the functions $\frac{\partial F}{\partial x_1}$ and $\frac{\partial F}{\partial x_2}$, and their sum was 2. So he could deduce that at least two points of C lay on the cylinder and no points of B did, so the curve was branched at M. The same argument about counting the number of points where a curve meets a surface was to arise a few years later when Poincaré produced his duality theorem for manifolds (see chap. 8, sec. "Poincare's Work, 1895 to 1905" below).

In the original context, because multiple roots occur if and only if the Hessian determinant vanishes, it follows that a position of equilibrium for which $\Delta \neq 0$ remains stable under infinitesimal variations in the parameter y. The vanishing of Δ can signal either that some of the x values will become imaginary when y is varied, or that the same equilibrium values belong to two sets of values φ_{jk}—that is, for two distinct values of j. The former case corresponds to a limiting form, the latter to a bifurcation.

Poincaré therefore directed his attention to the quadratic form $\sum \frac{\partial^2 F}{\partial x_j \partial x_k} X_j X_k$. The equilibria will be stable if and only if this form is negative definite, and this form can be written, after a suitable change of coordinates, as a sum of squares $\sum \alpha_j Y_j^2$, and Poincaré called the coefficients α_j the coefficients of stability. Some, say v, of them will be positive and the remaining $n - v$ will be negative; Δ will be positive if

$n - v$ is even, negative if $n - v$ is odd, and will vanish if any of the α vanish. Poincaré was able to show that the form will correspond to a bifurcation if both Δ and one of the coefficients of stability change sign; he called this the exchange of stabilities, and we shall see (in this section) that it was to be crucially called into question.

Poincaré then turned to a more realistic problem, in which the function F depends on infinitely many variables, and was able to show that the preceding considerations extended to the new case, provided it could be assumed that if a stable equilibrium configuration is perturbed infinitesimally then a new stable equilibrium configuration is adopted. He then deduced, as he had done in an earlier paper, that Thomson and Tait had been right to claim that there was a ring-shaped figure of revolution.

Poincaré regarded the book by Thomson and Tait as the only account worth having of stability relative to a set of rotating axes, and in (1885, para. 7) he explained their approach. For a finite system of particles, the differential equations of motion can be approximated by ignoring all but first-order terms, a process known as linearization that is often reflected in the way the original mathematical description is formulated, and which amounts to giving a description of the instantaneous motion of the system. The resulting equations are easy to solve, and the motion is given by a sum of exponential functions $e^{\lambda_j t}$. The motion is therefore stable if all the λ's are purely imaginary. There can still be stability if the λ's have their real part zero or negative, but otherwise the motion is unstable. So everything depends on the equation for the λ's, which is a polynomial of degree $2n$. For a rotating body one can consider the forces which act on it (the mutual attraction of the particles), and the kinetic and potential energies of the system, denoted T and U, respectively, of which T is a quadratic form in the velocities of the particles and U is a quadratic form in the coordinates of the particles. For an absolute equilibrium—one in which the rotation is zero—stability is assured when U is positive definite. This condition remains sufficient, but is no longer necessary, when the particles describe a rotating fluid body; as Poincaré had already shown, stability can happen when all the coefficients of stability are negative. When viscosity can be ignored, and if in these circumstances the relative equilibrium is stable, then there is also ordinary stability. If the equilibrium remains stable when viscosity is taken into account the stability is said to be secular. Poincaré observed that one can have ordinary stability without secular stability; this can happen when the viscosity is very weak and the system persists in one configuration for a long time before suddenly falling apart. For secular stability it is necessary and sufficient that the quadratic form U be positive definite, which is to say that all the coefficients of stability be negative. But the conditions for ordinary stability are much more complicated, and Poincaré restricted himself to a simple case before observing that all of these conclusions that Thomson and Tait had reached required that the work done by the viscosity was negative (and not merely zero), and that this condition was not always met.

Poincaré's enthusiasm for the work of Thomson and Tait is appropriate, but it is also odd. The line of descent in these problems is from Newton, who argued that rotating fluids, if ellipsoidal, would be spheroidal and slightly flattened at the poles, and was pleased to note that Jupiter was indeed this shape. After some controversy, careful measurements under the direction of Maupertuis had confirmed that the earth also has this shape in 1738 (see Terrall 2002). In 1740, and in more detail in 1742, Colin MacLaurin had argued for the stability of Newton's ellipsoids of revolution, but his calculation of the eccentricity of the rotating ellipse did not agree with observations. The problem was clarified by d'Alembert in 1773 when he showed that a given mass of fluid with a given angular velocity ω would have one of two eccentricities, one large and one small, provided that ω was less than a certain critical velocity. If ω equals this critical velocity the two ellipsoids have the same shape, and after that one of these ellipsoids ceases to exist. Laplace, in his *Mécanique céleste*, explained this by arguing that as the angular velocity of the ellipsoid with large eccentricity increases so does its moment of inertia, and this happens in such a way that the angular momentum tends to infinity, so there is only one MacLaurin ellipsoid for any given value of the angular momentum.

Lagrange also argued, less successfully, that rotating ellipsoids must be spheroids. But in 1834 Jacobi wrote to the Académie des sciences in Paris to announce his discovery of new equilibrium ellipsoids with three distinct axes; such an ellipsoid, although rotating, is not an ellipsoid of revolution. He accompanied his letter with some disparaging remarks about the "pitiable state" to which celestial mechanics had sunk since Laplace's time.[2] This was taken by Liouville as a challenge to French

---

[2] See Lützen (1990, 479) upon which these paragraphs are based.

mathematics, and he quickly derived Jacobi's results from material in the *Mécanique céleste*. But he remained interested in the topic, and the more he worked on it the more he lost confidence in Laplace and the more his appreciation for Jacobi grew. Several other mathematicians joined in, and by 1843 Liouville was able to use the techniques of Jacobi's student Meyer to analyze the shape of a Jacobi ellipsoid in terms of its angular momentum. Meyer had already shown that there is precisely one Jacobi ellipsoid for every value of the angular velocity below a certain critical velocity. For that angular velocity the Jacobi ellipsoid is also a MacLaurin ellipsoid, and thereafter no Jacobi ellipsoid exists. Liouville showed that the angular velocity decreases as the angular momentum increases, and as the angular momentum increases from a minimum, the ellipsoid changes its shape from a spheroid (at the minimum) to a cigar shape.

Liouville went on in unpublished work (see Lützen 1990) to claim that for low values of the angular momentum there is only one stable ellipsoid, which is the MacLaurin one, and above a certain transitional value only the corresponding Jacobi ellipsoid is stable. The problem was reanalyzed by Riemann in his (1861), who took it up from an account by Dirichlet.[3] He supposed that the positions of the individual particles of the rotating fluid ellipsoid were linear functions of their original positions, and showed that under these hypotheses the Jacobi ellipsoid is always stable, but the MacLaurin ellipsoid is only stable when its eccentricity is very small.

Seen in this light, Poincaré's apparent ignorance of the work of Riemann, which was shared by Thomson and Tait, is disappointing, and rather hard to explain.[4] One possibility is that Poincaré had dispensed with the requirement that the rotating fluid be some sort of ellipsoid at all. It would indeed be preferable to have either a demonstration that the shape must be an ellipsoid that was based on less restrictive assumptions, or the discovery of nonellipsoidal rotating bodies of fluid in stable equilibrium, and this latter task was the one that Poincaré

[3] Chandrasekhar (1989) comments, "In many ways, the most curious aspect of the subject has been the almost total neglect of the fundamental papers of Dirichlet, Dedekind, and Riemann." He also noted and corrected several points in Riemann's analysis.

[4] Contrary to Lützen (1990, 488) Poincaré did mention Liouville's published work on the stability of rotating ellipsoids in his (1885), for example on p. 41 (p. 77 in *Oeuvres* 7), where he explicitly adopted Liouville's notation.

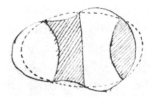

Figure 5.1. A conjectural pear-shaped figure of equilibrium. Source: Poincaré, *Oeuvres* (11.2, 283).

was most pleased to have accomplished. However, before getting there Poincaré chose to rework the old material on ellipsoids according to his new methods.

First, he summarized the old results in this way in his (1885e, para. 8), using the less explanatory concept of angular velocity as the key parameter. For $\omega^2 < 4\pi \times 0.093$ there are four ellipsoids, two MacLaurin ones and two Jacobi ones. When $\omega^2 = 4\pi \times 0.093$ the two Jacobi ellipsoids and one of the MacLaurin ellipsoids coincide, and for $\omega^2 > 4\pi \times 0.093$ until $\omega^2 < 4\pi \times 0.112$ there are only two ellipsoids of revolution. When $\omega^2 = 4\pi \times 0.112$ these ellipsoids coincide and thereafter there is only one. Then Poincaré derived expressions for the coefficients of stability in terms of the shape of the ellipsoid, and was led in paragraph 12 to consider the Jacobi ellipsoids. He deduced that one of these ellipsoids was a bifurcation ellipsoid by showing that one of its coefficients of stability vanished, and so gave rise to a new series of equilibrium figures. He investigated further, and concluded that it was pear-shaped. One of his drawings is shown in figure 5.1.

Then, to study the stability of the ellipsoids, Poincaré found it necessary to reconsider the criteria he had laid down in paragraph 7, which were that all the coefficients of stability must be negative. This is not the case for the Jacobi ellipsoids, but Poincaré argued that that criterion had been derived under an assumption that no longer applied. Instead, energy considerations led to a new criterion, one that was both necessary and sufficient for secular stability, and which allowed Poincaré to confirm the conclusions of Thomson and Tait. However, if the condition of being ellipsoidal is no longer imposed then only the ellipsoids of revolution and the Jacobi ellipsoids that are (in each case) nearly spheroidal are the only ones to enjoy secular stability. Most importantly, Poincaré also declared that the same analysis also showed, after a lot of work, that his pear-shaped figures were also secularly stable.

The first person to respond to Poincaré's work was the Russian mathematician Aleksandr Mikhailovich Liapunov. He had studied mathematics at St. Petersburg, where he was greatly influenced by Chebychev, and on his advice he began to write a master's thesis on nonellipsoidal figures of equilibrium (Mawhin 2005). When he found that the method of successive approximations, which Chebychev had expected to succeed, was blocked by insuperable complications, he wrote his thesis instead on "The stability of ellipsoidal forms of equilibrium of a rotating liquid." It was published in Russian in the *Mémoires de l'Université de Kharkow* in 1884, where Liapunov also obtained a teaching position, and was later translated into French in 1904.

Poincaré's first notes on the subject chimed in strikingly with his own findings. In particular he too had found new surfaces infinitely close to certain algebraic surfaces that, in the first approximation, were in equilibrium. He wrote to Poincaré on 9 October 1885 to say,[5]

> The algebraic surfaces of which I spoke are precisely those that belong to the linear series of equilibrium figures that you have found. But I do not see how one can prove that these figures really satisfy the equilibrium condition. The method of successive approximations is not, I believe, suitable for giving such a demonstration.

This was indeed the weak point of Poincaré's argument, and he replied to Liapunov within the month. He was, he said, unable to read Russian, but he had asked the astronomer Rodolphe Radau to provide him with a summary of his memoir, and he could see, he said, that Liapunov was ahead of him at a certain number of points. He agreed that the method of successive approximations was not applicable "because the second approximation, which in any case tells us no more than the first, is accompanied with almost inextricable difficulties." So he explained he had taken a new approach drawn from his work on the three body problem, which, "to tell the truth, is not a complete proof" because it introduced a new principle "which might, perhaps, be difficult to prove but which, I believe, cannot seriously be doubted." This was the idea that

---

[5] The correspondence was published in Smirnov and Youschkevich (1987, see p. 3).

a rotating fluid in equilibrium and given a suitably small perturbation will adopt a new equilibrium figure:

> This theorem will be evident when it concerns a discrete system of points, but when applied to a fluid mass it raises a certain number of difficulties concerning the use of infinity that are perhaps more interesting to analysts than applied mathematicians.
>
> I don't doubt that one can overcome them some day, but I have not done so yet. (Smirnov and Youschkevich 1987, 4)

There matters rested between the two, apart from the publication of a note in the *Bulletin astronomique* (vol. 2, pp. 522–525) on Liapunov's work that Poincaré may have instigated, until Liapunov found the time to read Poincaré's long memoir in *Acta Mathematica*. Then, toward the end of 1886 he wrote to Poincaré again. He once more regretted the absence of rigor, but he also pointed out that their aims overlapped in some places but not others. In particular he pointed out that he was less interested in the equilibrium figures per se and more in the general concept of stability. He had based his discussion of stability on the idea of Thomson and Tait that it corresponded to a minimum of the total energy, but admitted that he had not been able to come up with a definition of stability that was entirely rigorous.

Poincaré replied to ask if an extract of the letter could be published (again in the *Bulletin astronomique*—Liapunov agreed) but also to ask for more information than he had been able to glean from Radau about Liapunov's idea of stability. In his reply, Liapunov said that the main difficulty he had encountered was defining the notion of "infinitely close" configurations. For example, a deformation of a sphere can be measured by looking at the normals from each point of the sphere to the new surface, but he freely admitted that this approach was highly arbitrary. Or one could consider the volume between the equilibrium surface and the perturbed surface. Deformations that kept this volume arbitrarily small while allowing the largest normal to grow indefinitely he said he called "projections." Then it was also necessary to constrain the notion of an initial perturbation, but this had proved too difficult and so he had instead modified the definition of stability and could now conclude that "a figure of equilibrium could be regarded as stable even in the case when, after infinitely small perturbations the figure can depart more and more from the equilibrium figure *if this departure is only made by means of deformations associated with infinitely small projections.*"[6] One should imagine that Liapunov's projections have the form of small protuberances, such as lead to the formation of a drop.

Poincaré replied in December to say that one could, for example, imagine a deformation of a sphere corresponding to a small increase in energy. Then after a time it could happen that all this energy was converted to kinetic energy and concentrated in a small volume of the fluid. If the mass of this piece was very small it could acquire a very high velocity, even though its kinetic energy was very small, and it would then escape the sphere completely and go to infinity. Therefore the energy theorem could not be used rigorously to show that a very small initial perturbation of a sphere would not alter a sphere in a noticeable way. But otherwise, Poincaré said that Liapunov's explanations had satisfied him completely—so completely, it was to turn out, that the correspondence between them ceased.

Liapunov was indeed the analyst to Poincaré's applied mathematician. In an article in 1905 he returned to Poincaré's new principle, and observed that it paid a price in reasoning that was less than rigorous:

> It is not a proof, rather it is a generalisation by analogy and Poincaré himself recognised this when he wrote "There will certainly be objections to make, but one cannot demand of mechanics the same rigour as in analysis when it comes to the infinite".
>
> But I am not of this opinion...one cannot behave this way when it is necessary to solve a precise problem (in mechanics or physics) posed in a mathematically rigorous fashion. The problem then becomes a purely analytical one and must be resolved as such.
>
> As a result, for all their subtleties, I cannot consider that Poincaré's studies contain the solution of the problem. (Smirnov and Youschkevich 1987, 12–13)

Liapunov's most famous work, "The general problem of the stability of motion" (1892), was his doctoral dissertation at Kharkov. He regarded Poincaré's "Sur les courbes définies par les équations différentielles" as the only work to have investigated this question without resorting to a linearization of the equations, a process that is restricted to the first

---

[6] See Smirnov and Youschkevich (1987, 9). As they note, this prefigures ideas Liapunov was to publish in his (1907).

approximation, and he felt that the methods put forward in that paper were capable of considerable generalization. As he put it,

> In spite of the fact that Poincaré restricted himself in his research to very special cases, the methods he uses allow of much more general applications and they can still lead to many new results. We shall see that in what follows, for in most of my research I have been guided by ideas developed in the memoir just mentioned. (1907, 205)

Unlike Poincaré, he regarded a solution as stable if, informally, nearby solutions remain close for all time. He gave two methods for discussing stability. In the first, he defined the characteristic numbers associated to a linearized system and proved that if these numbers are all positive the unperturbed motion is stable and all nearby motions tend asymptotically toward it. In the second he introduced the concept of a positive definite function $V(\mathbf{x}, t)$ which generalized the role played by a potential function in some simple systems previously analyzed by Dirichlet; a positive definite function $V(\mathbf{x}, t)$ is one bounded below by a continuous, increasing function $\varphi(||\mathbf{x}||)$ that vanishes[7] at 0. Liapunov then showed that there is a stable solution to the system of equations

$$\frac{d\mathbf{x}}{dt} = \mathbf{X}(x_1, x_2, \ldots, x_n, t)$$

if there is a (positive or negative) definite function whose derivative along the solution curves is such that

$$\langle V_{\mathbf{x}}(\mathbf{x}, t), X(\mathbf{x}, t) \rangle + V_t(\mathbf{x}, t)$$

either has a fixed sign opposite to that of V or is identically zero in a neighborhood of the origin.

The influence of Poincaré is apparent from the opening reference, and it extends from the topic to many of the methods. Poincaré's prize essay, and the first volume of *Méthodes nouvelles* were just coming out when Liapunov wrote his memoir, as he remarked, while carefully claiming independence for himself. Mawhin (2005, 672) summarized the influence

---

[7] The vector notation is not in Liapunov's paper.

as being that of a geometer who wrote wide-ranging memoirs on an analyst who preferred well-organized presentations.[8]

Cosmogonic Implications

In the essay (1892b) Poincaré set out what he saw as the implications of his discovery. It was, he said, generally agreed that the stars were solids that had originally been liquid or gaseous and that those which no longer retained the primitive fluidity nonetheless revealed the shape that they had taken when still fluid.[9] That being the case, the astronomer's task was to explain the shapes of rotating fluid masses, but the problem was as important as it was difficult, and until recently only the ellipsoids named after MacLaurin and Jacobi were known. Later, Thomson and Tait proposed some new solutions, after which Kovalevskaya and Poincaré himself had shown the existence of some new equilibrium figures, and then Liapunov had done the most important new work on the subject.

Poincaré then briefly described how the MacLaurin and Jacobi ellipsoids change their shapes as their angular momentum increases, before turning to the question of the existence of altogether new shapes. He asked his readers to imagine a rotating ellipsoid divided into $n + 1$ zones by n parallels, and into $2p$ segments by p equally separated meridians. These lines of latitude and longitude divide the ellipsoid into a number of regions, which can be colored alternately black and white like a chess board, and one can imagine that in the black ones the surface rises while in the white ones it sinks, in the manner of spherical harmonics. Thus new equilibrium figures arise that do not differ much from an ellipsoid. Now one can imagine that these peaks and troughs increase in size and the lines separating them deform more and more. For this to happen, he added, the initial shape of the ellipse and the choice of the latitudes is not arbitrary.

What can happen for small values of n and p? The case $n = 0 = p$ is obviously that of an ellipsoid. The cases $(n = 0, p = 1)$, $(n = 1, p = 0)$, and $(n = 1, p = 1)$ are essentially those of the MacLaurin ellipsoids, and

---

[8] For the subsequent history of attempts to establish a rigorous stability criterion, see Smirnov and Youshkevitch (1987) and Roque (2010). Picard discussed periodic solutions, asymptotic solutions, and instability in the manner of Liapunov (1897) in 1908 in the second edition of his *Traité* (chap. VIII).

[9] There was at this time no idea of the sources of stellar energy.

the case ($n = 0$, $p = 2$) is a Jacobi ellipsoid. But one can start with a Jacobi ellipsoid and divide it along its lines of curvature into $n + 1$ zones, and as soon as $n = 3$, new equilibrium figures arise. Indeed, the case ($n = 3$, $p = 0$) in which the zones produce a mountain, a valley, a mountain, and a valley is the pear-shaped figure. That said, the deeper question concerns the stability of these figures. What is required is what is called relative stability, in which the body seems to be at rest with respect to an observer using axes that rotate with the body. Now, if a body is rotating very fast it can be relatively stable without its energy being a minimum, as the sleeping top demonstrates. However, the existence of friction, however slight, will always destroy this relative equilibrium after a time, as Thomson had shown (said Poincaré) and as every disappointed child learns. So one must distinguish between ordinary stability, which friction eventually destroys, and the more interesting case of secular stability, which friction cannot overcome. The secularly stable ellipsoids are the nearly spheroidal MacLaurin and Jacobi ellipsoids, said Poincaré, and the newly discovered pear-shaped figure.

The stability of the pear-shaped figure was, however, contested by Karl Schwarzschild, the German theoretical astronomer, in his inaugural thesis of 1898, who argued that Poincaré's principle of the exchange of stabilities did not apply in this case. This provoked one of Poincaré's English admirers, G. H. Darwin, to take up the problem. He wrote to Poincaré to express his interest and they divided the problem between them. In his (1901) Darwin calculated the axes of the critical Jacobian ellipsoid. Poincaré then established new criteria for stability in his (1901d), on the basis of which Darwin then concluded (Darwin 1902) that the pear-shaped figure was stable. But in 1905 Liapunov gave a counterargument that showed that the pear-shaped figure was unstable. Jeans resolved the question in his (1916–1917), where he showed that Darwin had been inconsistent in his decisions about what terms to include or ignore and that a consistent treatment led to the conclusion that the shape was indeed unstable. In 1928 Élie Cartan proved that the pear-shaped figure will break up into at least two pieces, and becomes unstable in both the secular and the ordinary sense of the term, so it will fly apart and not form a double system with small eccentricity as might have been supposed. Chandrasekhar, who revived the subject with a penetrating modern analysis in 1969, commented (see his 1989, 842) that after the excitement generated by Poincaré's ideas and the fission theory, as it was

Figure 5.2. Sir George Darwin (1845–1912). Source: http://en.wikipedia.org/wiki/File:George_Darwin_sepia_tone.jpg, photo by J. Russell & Sons.

known, of the origin of double stars, Cartan's discovery was so effective that "at this point the subject quietly went into a coma."[10]

As for the stable ellipsoids, Poincaré argued in (1892b) that the implications for astronomy can be deduced by supposing that the stellar bodies cool very slowly and uniformly. The total angular momentum will remain constant in this process, and since the moment of inertia will decrease the angular velocity must increase. Now, if the initial angular velocity is small the result is that the ellipsoid will become flatter and flatter and eventually cease to be secularly stable. If moreover the cooling

---

[10] From which it has awakened recently with work on such topics as the nonradial oscillations of stars.

process is very slow, the ellipsoid will be a Jacobian one for a while, but even this form is not stable and eventually the rotating mass will split into two that orbit around their common center of gravity, and each is slightly pear-shaped because of the gravitational effect of the other. Eventually these bodies settle into steady orbits and, if sufficiently separated, assume spheroidal shapes. In this way the sun might have given rise to the separate planets, said Poincaré, although he conceded that this mathematical description was surely a considerable simplification of the actual process.

How might the rings of Saturn fit into this picture? Maxwell had shown that they could not be solid, that if they were a fluid the mean density of the fluid must be much less than that of Saturn, and finally, by a less than rigorous argument, that if they were made of rubble the mass of the ring must be very much less than that of the planet. To explain these results Poincaré invoked an analogy of lines of force in electromagnetism to eliminate a slew of abstract and complicated mathematical formulas, and was able to conclude that the mean density of a *fluid* ring that was at most 1/16th that of Saturn itself, which was not only close to Maxwell's estimate but sufficiently implausible for the third hypothesis to be the only one left viable. And indeed, said Poincaré, Trouvelot's recent observations make this conclusion all the more likely.

In 1911 Poincaré lectured at the Sorbonne on cosmology (1911a) (and also published the article (1911e) in *Revue du mois*). He reviewed the various hypotheses then current for the origin of the solar system, and indeed of the stars and the galaxy we see as the Milky Way. Prominent among these was Laplace's nebular hypothesis, which suggested that the sun and the planets condensed out of a rotating cloud of gas. The sun formed at the center, and the rest of the gas settled under gravity into the equatorial plane, then into rings, and then into planets. This explained why the planets all orbit the sun in the same direction and lie more or less in the same plane. Poincaré subjected Laplace's account to a careful mathematical analysis, and this led him, via the topic of ring-shaped orbits, back to the rings of Saturn, the work of Maxwell, and his own work of 1885. Poincaré now observed that Laplace's gaseous ring would also break up under the effect of Kepler's third law, which says that the orbital period τ varies with the radius r of a circular orbit as $\tau = r^{3/2}$. The result would be small planetary nebulas in orbit around the nascent sun, and these nebulas would in turn settle into a planet and its satellites. Stability considerations led to an estimate of how far away a satellite could be from its parent body, and, like many of the estimates in Laplace's theory the agreement with observation was crude but satisfactory.

Poincaré next looked at objections that had been raised against the nebular hypothesis. It had been calculated, he said, that the process would take 150 million years, and that was too long given the limitations that thermodynamics imposed on any account of the age of the universe. (This objection, raised most acutely by Lord Kelvin, was unanswerable before the implications of radioactivity were understood and any adequate theory of how the stars operate had been proposed.) Additionally, stability considerations led to an estimate of how far away a satellite could be from its parent body, and exceptions had been found: one of the satellites of Mars, and the inner satellites of Saturn were too close.[11]

In their day, these lectures were well regarded. Hadamard said of them (Poincaré, *Oeuvres* 11, 227) that "it took a successor of Laplace who was at the same time a successor of Clausius and Boltzmann to write the *Leçons sur les hypothèses cosmogoniques*." Today they may look quaint, but while we have all manner of photographic evidence and even some chemical samples, we still have no better way to investigate the origin of the solar system, with its many planets and their multitude of moons, than to rely on Newton's laws and to try to decipher their consequences.

A shorter, but equally flawed, contribution that Poincaré based on Kelvin's work has, however, turned out to be more lasting. In his (1910f) he took up the question of whether the earth has a solid or a liquid core. His arguments were based on how the two cores would contribute to the nutation and precession of the earth, and he agreed with Kelvin that there was no significant difference and one might as well assume the core is solid. Even though it was shown in the 1920s that there is a solid inner core surrounded by a liquid outer core, Poincaré's mathematical approach has become the basis for modern discussions of nonviscous flow in the core, and the so-called Poincaré flow of uniform vorticity is a profitable way to study how the motion of mantle and core interact (see Noir 2000).

---

[11] The modern objection to Laplace's theory is that it cannot account for the fact that the planets have almost all the angular momentum of the solar system; Poincaré did not mention this.

6

Physics

THEORIES OF ELECTRICITY BEFORE POINCARÉ: MAXWELL

The study of magnetism and electricity in the second half of the 19th century was marked by disagreement at every level. There were at least two theories popular in continental Europe, Wilhelm Weber's and the later theory of Hermann von Helmholtz. They were followed by the much more sophisticated theory of James Clerk Maxwell, which was the first fully mathematical theory of the static and dynamic aspects of the complex interrelations between electricity and magnetism. It also held out the prospect of explaining light by treating it as an electromagnetic wave. For reasons to be discussed, it was never fully understood or accepted on the Continent, but it was energetically embellished by the British Maxwellians. In the late 1880s Heinrich Hertz, working with Helmholtz's theory, which also predicted electromagnetic radiation but at different velocities in air and in wires, came to conclusions that he at first thought incompatible with Maxwell's ideas. Other experimenters, Sarasin and de la Rive, came to results that made Hertz change his mind, and ultimately Hertz both developed a theory that predicted the existence of electromagnetic radiation and conducted suitable experiments that confirmed Maxwell's suggestion and refuted the work of Helmholtz. Then in the 1890s came the theories put forward by Hendrik Antoon Lorentz. None of these theories agreed with each other, none were wholly satisfactory on their own terms, and in particular some were fundamentally different at the level of physical processes, to the point of resting on an explanatory void. One consequence of this confusion, which Poincaré entered in the late 1880s and worked with for the rest of his life, was his marked conventionalism about the interpretation of experimental results and theoretical terms.

The most significant difference between the British and their Continental peers concerns current. Continental scientists argued that when an electrical circuit is closed and current flows along a wire, something travels in the wire. There were one-fluid theories and two-fluid theories, and neither type of theorist had much idea of what an electric fluid might be, but they agreed that the flow of something was being turned on and off as a circuit was closed and then opened. Maxwell argued that, be that as it may, what one actually deals with is fields, the electric field and the magnetic field. It is fields that enter the equations, fields that are the fundamental objects, and fields that have to be understood. It happens that they may be discontinuous in certain situations, and this jump in their strength manifests itself as current. Historians of science seem agreed that Continental theorists could never quite accept this analysis, which ran counter to their entire way of thinking about what causes electricity and magnetism to behave as they do.

A further difference between theorists grew as the 19th century came to an end. Most notably with the work of Lorentz it became more usual to think of electricity as the motion of charged objects, perhaps little particles, whose nature was somehow responsible for what was observed although there was no hope of observing an individual such particle. These particles, in later theory, became known as electrons, and clarity on this point was decisive for the situation by the time of Einstein— which is not to say that the theory of how the particles produced the fields was in a fully satisfactory state. But, as the title of the best book on this development has it, there was a shift from Maxwell to microphysics.

Since this dynamic confusion is exactly what Poincaré undertook to explain to himself, his students at the Sorbonne, and the wider world of French physicists, it might seem possible to let him take over at this point, but because he shared the Continental failure to fully understand Maxwell it seems better to start with Maxwell. In this way we can also recall the basic laboratory results that any theory of electricity and magnetism had to explain, although to do that we must be more naively Continental with our language.

We are all familiar with the idea of a magnetic field. The earth's magnetic field is detected by any small magnet, such as a compass needle, and in the simplest cases we talk comfortably about lines of force going from one pole to the other of a bar magnet (or from the earth's north

Figure 6.1. Heinrich Hertz (1857–1894). Source: http://commons.wikimedia.org/wiki/File:PSM_V45_D304_Heinrich_Hertz.jpg, *Popular Science Monthly* (May–Oct. 1894, vol. 45).

magnetic pole to its south magnetic pole). It is a simple experiment to show that current flowing in a wire exerts an effect on a sensitive magnet, and a little more delicate to observe that the magnet also exerts an effect on a current-carrying wire (else conservation of momentum would fail and a perpetual motion machine would be possible). Current-carrying wires in parallel will attract each other when the current is going in the same direction and repel when the directions are opposite. It was Faraday who discovered that a changing magnetic field will produce a current in a wire, whereas a static magnetic field does not.

How these observations, or rather their quantified versions, were turned into mathematical expressions, involved a certain amount of idealization. After Coulomb's very careful experiments, it became possible to speak of a certain amount of electricity, say e, confined in one small body, another amount, e' confined in another body, and the repelling force that each body exerts on the other when they are a considerable distance r apart. It was found[1] to be proportional to ee'/r^2. This was very similar to the force of gravity, which obeys an attractive inverse square law, and between the 1820s and the 1840s George Green in England and Gauss in Germany developed the mathematics to deal with this.

In each case (electromagnetism and gravity) the force is a vector; it has three components at each point in space. Green's idea, which went back to Lagrange and was independently available to Gauss, who also used it, was to replace the force by a function, which is a much simpler object. This function, handily called the potential function by Green, has the property that its partial derivatives at each point define the force. Potential theory—the mathematical theory of potential functions and the forces they encode—formed a large part of the generalization of the calculus to functions of two and three variables in the mid-19th century. It proved ideal for organizing the theory of electric and magnetic fields independently of the physicists' ideas about what magnetism and electricity might be, and Maxwell took it over in order to put Faraday's discoveries into mathematical form.

Maxwell was concerned to explain the very different electrical properties of different media. He took from Faraday the term "dielectric" to refer to those media which do not immediately conduct electricity. He argued that it follows from Coulomb's work that if there is more electricity in one place than another and these places are joined by a wire then electricity will flow along the wire, and he said this motion was a manifestation of an electromotive force, noting prudently that this force cannot always be derived from a potential function. In a dielectric, however, it is necessary to distinguish, he said, between the electromotive intensity at each point and what he called the electric polarization at each point produced by the field, and these vectorial quantities will point in different directions. He deduced that an electric field places a stress on a dielectric and moves the electricity in the direction of the electromotive force. Therefore, "the variations of electric displacement evidently constitute electric currents" (*Treatise*, para. 60).

---

[1] For an engaging account of what is involved in replicating this classic experiment, see Martinez (2011, 128–146), "Coulomb's Impossible Experiment."

He then summarized his theory as follows:

62.] The peculiar features of the theory are :-

That the energy of electrification resides in the dielectric medium, whether that medium be solid, liquid, or gaseous, dense or rare, or even what is called a vacuum, provided it be still capable of transmitting electrical action.

That the energy in any part of the medium is stored up in the form of a state of constraint called electric polarization, the amount of which depends on the resultant electromotive intensity at the place.

That electromotive force acting on a dielectric produces what we have called electric displacement, the relation between the intensity and the displacement being in the most general case of a kind to be afterwards investigated in treating of conduction.

He then argued for an analogy, and for multiple explanations in physics:

Whatever electricity may be, and whatever we may understand by the movement of electricity, ... in every case the motion of electricity is subject to the same condition as that of an incompressible fluid, namely, that at every instant as much must flow out of any given closed surface as flows into it.

It follows from this that every electric current must form a closed circuit. The importance of this result will be seen when we investigate the laws of electro-magnetism.

Since, as we have seen, the theory of direct action at a distance is mathematically identical with that of action by means of a medium, the actual phenomena may be explained by the one theory as well as by the other, provided suitable hypotheses be introduced when any difficulty occurs.

At the end of this passage Maxwell also remarked,

In many parts of physical science, equations of the same form are found applicable to phenomena which are certainly of quite different natures, as, for instance, electric induction through dielectrics, conduction through conductors, and magnetic induction. In all these cases the relation between the intensity and the effect produced is expressed by a set of equations of the same kind, so that when a problem in one of these subjects is solved, the problem and its solution may be translated into the language of the other subjects and the results in their new form will still be true.

Buchwald, in his helpful analysis of how to read Maxwell and compare it with modern theory (Buchwald 1985, 28–29), quoted a passage from Maxwell's *Treatise* that is worth repeating here as indicative of how Maxwell used his theoretical terms:

> If the medium is not a perfect insulator, the state of constraint, which we call electric polarization, is continually giving way. The medium yields to the electromotive force, the electric stress is relaxed, and the potential energy of the state of constraint is converted into heat. The rate at which this decay of the state of polarization takes place depends on the nature of the medium. In some kinds of glass, days or years may elapse before the polarization sinks to half its original value. In copper, a similar change is effected in less than the billionth of a second.
>
> We have supposed the medium after being polarized to be simply left to itself. In the phenomenon called the electric current the constant passage of electricity through the medium tends to restore the state of polarization as fast as the conductivity of the medium allows it to decay. Thus the external agency which maintains the current is always doing work in restoring the polarization of the medium, which is continually becoming relaxed, and the potential energy of this polarization is continually becoming transformed into heat, so that the final result of the energy expended in maintaining the current is to gradually raise the temperature of the conductor, till as much heat is lost by conduction and radiation from its surface as is generated in the same time by the electric current.

This passage is instructively obscure. Maxwell saw all types of matter, and all states (solid, liquid, or gas) on a continuum as far as the conduction of electricity was concerned. Conductors and insulators were both dielectrics; they differed in whether the motion of electricity through them generates a lot of heat, as it does in conductors, or only negligible amounts, as it does in insulators. (This is not the later, and modern, view, which concerns how fast electricity might travel through them.)

The appearance of equal and opposite, positive and negative charges on opposite sides of every particle of a dielectric was, for him, the manifestation of a single phenomenon—electric polarization. A shift of some incompressible substance generates a certain amount of potential energy which is converted into heat, a process that in conductors, where the flow happens very fast, is regarded as the passage of electric current. This displacement, **D**, is related to the electromotive intensity **E** by the simple equation $\mathbf{D} = \frac{K}{4\pi}\mathbf{E}$, where K is a constant determined by the dielectric and what Maxwell called the "dielectric inductive capacity" of the medium. So the driver is polarization, which can be brought about in a number of ways. It generates an electromotive force with a certain intensity. We shall return to this point when Poincaré's treatment is considered.

To return to paragraph 60 of the *Treatise*, a great deal of mathematics was then introduced, along with a wealth of attention to the different circumstances in which electricity and magnetism manifest themselves and the ways in which they can be measured. Static phenomena are described, and then dynamic ones. In part IV, paragraph 786—well toward the end of a book of one thousand pages—an electromagnetic theory of light is introduced. Maxwell considered the nature of the medium through which the disturbance propagates, and concluded (para. 782) that it was a material medium in which energy was transmitted by the action of its contiguous parts. The theory of undulatory phenomena required that

> this energy is supposed to be partly potential and partly kinetic. The potential energy is supposed to be due to the distortion of the elementary portions of the medium. We must therefore regard the medium as elastic. The kinetic energy is supposed to be due to the vibratory motion of the medium. We must therefore regard the medium as having a finite density.

Thus he baptized the luminiferous ether.

He now considered what his equations had to say about motion in a nonconductor, and was able to reduce them (para. 784) to three equations for F, G, and H, the components of the electromagnetic momentum, of the form

$$K\mu\frac{d^2F}{dt^2} + \nabla^2 F = 0,$$

(with similar equations for G and H). He recognized these as similar to the equations of an incompressible elastic solid (thereby adding another property to the ether) and as such solved by Poisson and applied to the theory of diffraction by Stokes. He immediately deduced that they define a disturbance propagating with velocity $V = \frac{1}{\sqrt{K\mu}}$, and remarked (para. 786) that if "light is an electro-magnetic disturbance, propagated in the same medium through which other electro-magnetic actions are transmitted" then these disturbances must travel at the speed of light, and agreement or disagreement on this point would test the electromagnetic theory of light.

He noted that if the medium is air and electrostatic units are used then $K = 1$ and $\mu = \frac{1}{v^2}$ where v is the ratio of the electrostatic to the electromagnetic unit. His prediction was that the speed of an electromagnetic disturbance must equal the ratio of the electrostatic to the electromagnetic unit. He then cited estimates of the velocity of light by Fizeau, Foucault, and of the ratio of the units by Weber, William Thomson, and himself which broadly confirmed an agreement, and remarked,

> It is manifest that the velocity of light and the ratio of the units are quantities of the same order of magnitude. Neither of them can be said to be determined as yet with such a degree of accuracy as to enable us to assert that the one is greater or less than the other. It is to be hoped that, by further experiment, the relation between the magnitudes of the two quantities may be more accurately determined. (para. 787)

It is this prediction that was verified later by Hertz. That this important experiment was not carried out by a British follower of Maxwell such as Fitzgerald, the leading Maxwellian and the one who might be expected to have a head start on understanding the theory, is partly explained by Hunt (1991, 34–47). He observed that Maxwell's theory is not always equivalent to action at a distance, but that Fitzgerald mistakenly thought so for some time, on the basis of a nonetheless intelligent reading of Maxwell's *Treatise*, and therefore believed that the electric production of waves like those of light was impossible. When he came to correct his error, in 1883, he could not see how to detect such waves, thus leaving the way open to Hertz. Hunt summarized the situation by saying,

"Extracting a consistent and comprehensive theory from Maxwell's book was among the greatest and most difficult of the Maxwellians' tasks" (1991, 34).

As every historian of science who writes on the subject remarks, what we know as Maxwell's equations are not to be found in Maxwell's *Treatise*. Instead, in chapter IX of book II, entitled "General equations of the electro-magnetic field," Maxwell gave a series of equations, first in coordinate form and later in a quaternionic form, that were rewritten by Heaviside in the mid-1880s in the now familiar vector form. These equations became generally accepted by the mid-1890s.[2] This little remark conceals numerous stories. On the extraordinary character and career of Oliver Heaviside, one can consult Yavetz (1995). On Maxwell's liking for quaternions and the way they were split into their scalar and vector parts, one can consult Crowe (1967). The story of the international movement to introduce vector notation in mathematics and physics has not been fully told anywhere to my knowledge. Most importantly, what a modern physicist understands by Maxwell's equations and what Maxwell, Heaviside, or Poincaré understood by them, are different.

The question of what Maxwell's theory says—how we are to summarize or understand a book of a thousand pages—depends on who is asking it. An experimental physicist would start with, and perhaps care most for, the descriptions of how electromagnetic phenomena can be measured in the laboratory. A certain kind of mathematician would care for the use of potential theory. These are clear ways in which the book is a success. But many readers might have hoped that in the end one would find out what electricity and magnetism actually *are*; most professional physicists have a hard job withholding existence from the objects in their theories and the language of discovery is heavily biased that way. If demonstrable things happen—compass needles move, wires heat up, current (whatever it might be) flows in a circuit and lights turn on, if more grandly electric effects can produce magnetic effects and magnetic effects electric ones—then it seems natural to say that some microscopic, even invisible processes are at work that, in aggregate, cause these visible changes. It is somehow less satisfactory to say that all of this is to be handled by mathematics that rests on nothing more than some tentative, and forever unverifiable, hypotheses.

[2] See the appendix in Hunt (1991), "From Maxwell's Equations to 'Maxwell's Equations.'"

Maxwell did not, however, conclude his book with a rousing account of discovery. The penultimate chapter is devoted to an explanation of magnetism. His conclusion is that the action of magnets on each other "can be accurately represented by the attractions and repulsions of an imaginary substance called 'magnetic matter,' " and that this is a molecular phenomenon. So a magnet is a multitude of molecules, within each of which circulates a system of electric currents. These, he noted, were the views of fifty years before, proposed by Poisson and Ampère. That said, "Still, however, we have arrived at no explanation of the nature of a magnetic molecule, that is, we have not recognized its likeness to any other thing of which we know more" (para. 833). Maxwell therefore advocated "extending our mathematical vision into the interior of the molecules," and accordingly the gains he obtained were in the form of simplifications of his theory. The *Treatise* then ends with a rather downbeat chapter on action at a distance and the medium through which this action is propagated. Maxwell looked at the ideas of Gauss, Weber, Helmholtz, and then the more recent and more mathematical ideas of Neumann, Riemann, and Betti, and concluded,

> Hence all these theories lead to the conception of a medium in which the propagation takes place, and if we admit this medium as an hypothesis, I think it ought to occupy a prominent place in our investigations, and that we ought to endeavour to construct a mental representation of all the details of its action, and this has been my constant aim in this treatise. (para. 866)

Nor can it be said that Maxwell's book resolved all of the issues it addressed: it left many problems for its readers quite apart from the problematic nature of current. Energy was one. Helmholtz had introduced the concept of conservation of energy into physics in 1847 and it began to be generally accepted by physicists by about 1860.[3] The British naturally attributed the concept to Joule and his series of careful experimental measurements, and in the preface to their *Treatise on Natural Philosophy* (1879, vi) Thomson and Tait could write, "One object which we have constantly kept in view is the grand principle of the *Conservation of Energy*. According to modern experimental results,

---

[3] See Elkana (1974) for the much more complicated story that I am summarizing here too crudely by attributing everything to Helmholtz. My purpose is merely to point out the role of Helmholtz in asserting such a law.

especially those of JOULE, Energy is as real and as indestructible as Matter."

But if there was an impressive consensus by the 1870s that energy was a fundamental concept in physics, it was much less clear what energy actually was. In particular, it was not clear how it could move from place to place, a problem that was first resolved in the mysterious case of electromagnetic theory by the British physicist John Henry Poynting in 1884.[4] By transplanting Rayleigh's analysis of energy flow as sound waves propagate in air to Maxwell's theory, which involved some considerable ingenuity, Poynting showed that the flow of energy in the presence of an electric field **E** and a magnetic field **H** can be represented by the vector **E** × **H**, which is today called the Poynting vector; conveniently, energy flows in the direction in which Poynting's vector points. Poynting deduced from this that when a conduction current flows along a wire, in which case the electric field is parallel to the wire and the magnetic field flows in loops around it, the energy does not flow along the wire but flows radially into it at each point of the wire, where it is transformed into heat.[5]

If the British found Maxwell difficult, even misleading, all the more so did Continental readers. His theory attracted attention after Hertz's success in 1888 with electromagnetic waves. French readers could learn it from Poincaré's *E&O* (1890) and German readers could learn it from Boltzmann's (1891) as well as later books by Föppl, Helmholtz, and other authors. Both groups read Maxwell's theory with spectacles of a Helmholtzian shade, in which electricity was a substance, not a consequence of some displacement, because the previous theory of electromagnetism was the influential account in Helmholtz (1870). Buchwald (1985) analyzed this work carefully, and showed that a critical term in Helmholtz's theory is the continuity equation, which says that electricity, whatever it may be, is conserved, and accumulations of free charge derive from inhomogeneities in the current density. This shows that free charges play a role in the theory for which there is no analogy in Maxwell's theory. However, the equations that Helmholtz found are in many respects similar to those of Maxwell, and the fit can be made even better by assuming that a parameter, which expresses the so-called polarizability

---

[4] For a full account of how Poynting came to his discovery, and the importance of his training in Cambridge, see Warwick (2003).

[5] For a discussion of how this works in Maxwell's theory, see Buchwald (1985, 41–49).

of the ether, is infinite. Helmholtz himself had argued (1870, 125), in responding to Maxwell (1865), that Maxwell's theory could be obtained from his by letting this parameter tend to infinity, so he called this (and another similar condition) Maxwell's limiting conditions, thus making Maxwell's theory a special case of his own. But, as Buchwald has shown, this is false: the theories are incompatible and "simply put, taking the limit does not lead to Maxwell's theory; it only deprives Helmholtz's of physical significance" (Buchwald 1985, 182).[6]

POINCARÉ'S *ÉLECTRICITÉ ET OPTIQUE*, 1890

Poincaré tried hard to keep two concepts of electricity in play in his account of Maxwell's *Treatise*. He rightly emphasized that Maxwell's theory was characterized by the central role of dielectrics, which, he said—rather less certainly—Maxwell considered were occupied by a hypothetical elastic fluid. This fluid was analogous to the ether and, said Poincaré, Maxwell called it electricity. Poincaré, reasonably enough, found this confusing and called it a *fluide inducteur* (inductive fluid or fluid inductor), reserving the word "electricity" for its usual Continental meaning.[7] An electric displacement is a change in position of a molecule of this inductive fluid. If the displacement is given by (f, g, h) then

$$(f, g, h) = -\frac{K}{4\pi}\nabla\psi,$$

where ψ is the potential due to the distribution of electricity. Poincaré now investigated what happens when a closed surface filled with a homogeneous dielectric containing two charged conductors in equilibrium is given a slight increase in charge. When equilibrium is restored the inductive fluid will be in a different place. Poincaré applied Gauss's law

---

[6] Buchwald, in a private communication, writes that "actually the procedure bears a certain resemblance to the technique of renormalization in quantum electrodynamics, which works perfectly well mathematically (and experimentally) but whose physical interpretation is complex."

[7] Even so he was criticized by Fitzgerald (1891) on the grounds that in Maxwell's theory there is no *fluide inducteur* with in any case contradictory properties, but a medium.

to investigate how many molecules had crossed the closed surface, and found that

> the quantity of inductive fluid that leaves the surface is equal to the quantity of electricity that enters. Everything happens as if the electricity chases the inductive fluid, or, in other terms, as if the inductive fluid and the electricity are two incompressible fluids. (p. 17)

He then added that $\nabla \cdot (f, g, h) = 0$, so indeed the inductive fluid is incompressible.

On Poincaré's reading of Maxwell's theory, when electricity moves, so does the inductive fluid, and indeed what the inductive fluid vacates electricity occupies, so as far as Poincaré was concerned Maxwell advocated a two-fluid theory of electricity. In keeping with the two types of electricity there were also two types of current, at least in the ordinary (Continental) theory: closed currents, generally permanent, and open ones, generally instantaneous. But, said Poincaré, in Maxwell's theory there would only be closed currents. However, there would be conduction currents, that flow in a circuit, and displacement currents, which are caused by a displacement of the inductive fluid. If both currents occur together the result is what, in another theory, is called an open circuit. If there is only a displacement current, electricity displays its connection to the theory of light. These two types of current obey different laws: conduction currents obey Ohm's law, Joule's law, and Ampère's law, but displacement currents obey only Ampère's law.

Energy considerations came next. In chapter II Poincaré showed how potential energy can be defined in Maxwell's theory, but in chapter IV he showed that there were problems with the way Maxwell tackled the way electricity acts on matter. He argued that considerations of conservation of energy meant that electric energy could be neither consistently potential nor kinetic, and that the forces between conductors were not understood.

In chapter V, on electrokinetics, Poincaré denoted the velocity of electric displacement by (u, v, w) and observed that continuity of the electric fluid requires that $\nabla \cdot (u, v, w) = 0$. So displacement currents depend on the magnitude of the displacement, but conduction currents on the velocity of the displacement.

Magnetism is considered as a force, as it was by Maxwell. Poincaré denoted it by (α, β, γ) for which there is a potential Ω such that

$$(\alpha, \beta, \gamma) = -\nabla\Omega.$$

Maxwell (*Treatise*, para. 383) defined the magnetic moment of a bar magnet to be the product of its length and the strength of its pole, and the intensity of magnetization of a magnetic particle to be its magnetic moment divided by its volume. Poincaré, following Maxwell, investigated the magnetic force inside a magnet by considering excavating a cavity in the magnet, and deduced (para. 101) that the magnetic induction at a point is given by (a, b, c), where

$$(a, b, c) = (\alpha, \beta, \gamma) + 4\pi(A, B, C),$$

and (A, B, C) denotes the magnetization.

Unlike Maxwell, Poincaré now turned at once to electromagnetism. By considering the force exerted on a closed circuit by a magnetic pole, and making his habitual use of Green's theorem, Poincaré deduced (para. 118) that

$$(u, v, w) = \frac{1}{4\pi}\nabla \times (\alpha, \beta, \gamma).$$

The electromagnetic moment (F, G, H) was then introduced and provisionally defined as

$$\nabla \times (F, G, H) = (\alpha, \beta, \gamma).$$

However, this supposed that a current does not flow through a magnet, but when Poincaré now looked at the effect of a current flowing through a magnet he found that it was, on the face of it, uncertain whether one should use (a, b, c) or (α, β, γ). He resolved the matter by using the fact that $(\alpha, \beta, \gamma) = \nabla \times (F, G, H)$ to show that $\nabla \cdot (\alpha, \beta, \gamma) = 0$, which he had shown was not the case for the magnetic force inside a magnet, and so he adopted the definition

$$\nabla \times (F, G, H) = (a, b, c).$$

Poincaré also noted that in order to solve for (F, G, H) uniquely, Maxwell had further assumed that $\nabla \cdot (F, G, H) = 0$, which was not strictly necessary, and he indicated the form the equations would take if this assumption was rejected.

There followed a chapter on induction, in which Poincaré defined a suitable Lagrangian in terms of the position of a molecule of the inductive fluid in terms of an expression for the kinetic energies of the material molecules and the molecules of the inductive fluid. He then deduced that if (P, Q, R) denotes the electromotive force of induction then

$$(P, Q, R) = (a, b, c) \times \frac{d}{dt}(x, y, z) = \frac{d}{dt}(F, G, H) - \nabla \psi,$$

where ψ is an arbitrary function, and for the moment assumed that $\nabla(F, G, H) = 0$.

Then came chapter X, "The equations of a magnetic field," which can be regarded as Poincaré's version of Maxwell's *Treatise*, "General equations of the electromagnetic field," (vol. 2, chap. IX) in which the previous results are pulled together. This was followed by a chapter on the new electromagnetic theory of light, which drew in part on some of Poincaré's earlier lecture courses, and a final chapter on more recent developments, such as Rowland's work and the Kerr effect, in which Poincaré noted some problems raised by Maxwell's theory.

In 1879 the American physicist Edwin Hall, who was studying in Henry Rowland's laboratory in Johns Hopkins University, Baltimore, had become interested in the effect of a magnetic field applied at right angles to an electric current. He found two effects. In one, contrary to his original expectation, the current is moved in a direction perpendicular to the magnetic field and the original direction of the current. In the other, the plane of polarization of a ray of light is altered on reflection at the surface of a magnet. He published his findings in the *American Journal of Mathematics* (vol. 2), which was edited in Johns Hopkins, and in the *Philosophical Magazine*.[8] Hall's interpretation of his discovery was thoroughly Maxwellian, but it did not convince the British Maxwellians, and in 1881 the Cambridge physicist Richard Glazebrook, working in Maxwell's laboratory, the Cavendish, found a way to derive the Hall effect within Maxwell's theory. Continental physicists such as Lorentz in 1884 and Boltzmann in 1886, who had a different theory of electric

---

[8] See Buchwald (1985, chap. 10) for a full account of Hall's work.

current, naturally came to other interpretations still. Rowland himself believed that Hall's discovery required an alteration in Maxwell's laws, and connected the Hall effect, as it was becoming called, with a delicate optical analysis made by the Scottish physicist John Kerr into the way an electric field alters the refractive index of a material.[9] A few years later Fitzgerald gave an account of Kerr's results that drew on an interpretation of the Hall effect that was not, however, orthodox enough for the Maxwellians, and in 1891 Basset rewrote it in a more acceptable way that, however, J. J. Thomson was to criticize and rework in 1893.

Many of these conflicting arguments and others were drawn together in the concluding pages of the first edition of E&O under the heading "Difficulties raised by Maxwell's theory." Here Poincaré came down in favor of the resolution offered by Alfred Potier in 1880, who had offered an explanation involving the magnetic behavior of molecules, while noting that Lorentz had offered an explanation in terms of their electric charge.[10]

In all of this account it is clear that Poincaré resonated with the bells in Maxwell's theoretical belfry, as described in Maxwell's review of Thomson and Tait's *Treatise on Natural Philosophy* (1879, 783–784) (see chap. 1, sec. "Paris Celebrates the New Century" above).

Poincaré cast these bells in the bronze of the Lagrangian approach to dynamics, in which arbitrary coordinates $q_1, \ldots, q_n$ are chosen—Poincaré suggested these would be the measurable quantities—and these and their derivatives, $q'_1, \ldots, q'_n$ in the case of kinetic energy, contribute to the expression for the potential energy U and kinetic energy T of the system. The Lagrangian equations of motion of the system are of the form

$$\frac{d}{dt}\frac{dT}{dq'_j} - \frac{dT}{dq_j} + \frac{dU}{dq_j} = 0,$$

and if the theory is a good one the solutions of these equations describe the motion of the system correctly.

---

[9] See Buchwald (1985, chap. 13).

[10] Alfred Potier had become a répétiteur in physics at the École polytechnique in 1867 and had taught Poincaré there. His interests included Fresnel's theories of light and he was one of the first French physicists to take up the ideas of Maxwell, whose *Treatise* he translated into French. In his obituary of Potier, Poincaré praised him for the substantial nature of his courses.

However, any expressions for U and T will do provided they agree with the values of the potential and kinetic energy when converted to the familiar coordinates of position, so, said Poincaré in italics in the introduction,

> If a phenomenon has a complete mechanical explanation it will have an infinity of others that give an equally good account of all the particularities revealed by experiment.

Therefore, he went on a few lines later, one can now easily understand what Maxwell's fundamental idea is:

> To show that it is possible to give a mechanical explanation of electricity, we do not have to concern ourselves with finding the explanation itself, it is enough to know the expression for two functions T and U which are the two parts of energy, to write down the Lagrange equations in terms of these two functions and then to compare these equations with the experimental laws.

Furthermore, when it comes to choosing between all the possible explanations, "Perhaps one day physicists will lose interest in these questions, which are inaccessible to positivist methods and abandon them to the metaphysicians. This day has not yet come. But man cannot resign himself easily to eternal ignorance of the essence of things." So the choice must, for now, be guided by personal taste, in large part, there being some solutions that everyone rejects as too bizarre and others that everyone prefers on account of their simplicity. Maxwell, not without trying in some of his other writings, made no choice in his *Treatise*.

One can wonder to what extent Poincaré's approach makes him a physicist. Bertrand's surprise at Poincaré's agnosticism occupied the first half of his review of the book in the *Journal des savants* for 1891. He marveled at the way in a single sentence Poincaré could mix praise and criticism (Maxwell excited both unease and admiration), but he could not accept that physics must abandon the quest for the real causes of effects to metaphysics, and he cited ongoing debates between those who held the Fresnel and Neumann theories of light (to be discussed below) as evidence that physicists would want something more substantial than Poincaré's faith in Lagrangian dynamics.[11] On the other hand, German physicists appreciated Poincaré's account highly (in the German translation 1891a), and preferred it to Maxwell's (see Darrigol 2000, 354).

Throughout the first (Maxwellian) part of his *Électricité et optique* Poincaré was, accordingly, as neutral as he could be about the "real" underlying physical processes. So the attitude ushered in, in chapter IX, is significant. There Poincaré wrote about the theory of induction, first as it had been given by Helmholtz when he wrote his memoir on the conservation of energy, this being something like the form Poincaré preferred, then as it had been developed by Lord Kelvin and then by Maxwell "in a way that is different and more complete in some respects." This refers to the adoption of the Lagrangian approach.[12]

As Poincaré put it in paragraph 152:

In the preceding chapters we have been led to conclude that the hypotheses proposed by the English scientist were only provisional, and that, though they are more satisfying than the two-fluids hypothesis, they do not have, even in their author's eyes, any more objective reality. *On the contrary, we touch here, I believe, on the true thought of Maxwell.*

At the start of his theory, Maxwell made the two following hypotheses:

1. The coordinates of the molecules of the imponderable fluid depend on the coordinates of the material molecules of the body subject to the electric phenomenon and also on the coordinates of the material molecules of the hypothetical fluids (positive electricity and negative electricity) in the ordinary theory of electricity; but we are completely ignorant of the law of this dependence;
2. The electrodynamic potential of a system of currents is nothing other than the "demi-force vive" of the fluid of Maxwell; it is therefore the kinetic energy.

[11] Bertrand also found some of Poincaré's mathematics less than rigorous, thus supporting the French tradition of rigor that Poincaré had said made it difficult for French students to read Maxwell in the first place.

[12] See Maxwell, *Treatise* (vol. 2, chap. VII), "Theory of electric circuits."

The flexibility of the Lagrangian approach was intended to make redundant any appeal to physical objects and processes that constitute electricity.

Poincaré's account of Maxwell's theory was intended to do three things. One was to acquaint a French audience with Maxwell's theory per se, insofar as a French mathematical sensibility could do so without passing on Maxwell's errors. The second was to acquaint the same readers with the best ways of thinking about the fundamental physics of electricity, magnetism, and light. The third was to insert a particular philosophical position into the debate about the nature of mathematical physics.

His account of Maxwell's ideas missed its deepest features, as did every Continental writer's. This is evident in Poincaré's early comments that an electric displacement is a change in position of a molecule of the inductive fluid, and his consequent deduction that Maxwell advocated a two-fluid theory of electricity. Whereas, to quote Buchwald (1985, 187), the field-theoretic image of "charge" is "nothing more than a discontinuity in the displacement vector due entirely to the alteration in conductivity at the boundary." However, Maxwell, it must be said, left all but his most assiduous readers with a confused impression about the nature of "charge." It is quite easy to find him writing about it as a substance, for example, the initial account of the electric field (para. 44) is about the force exerted on a body carrying a "charge" e. In paragraph 45 Maxwell wrote about how a charged body moves. In paragraph 49 he wrote about the tendency of electricity to flow along a wire connected to two electrodes at different potentials. In paragraph 52, he explained how a dielectric may be charged without immediately distributing their electricity (it is the very definition of a dielectric that it can do this) and therefore acts as an accumulator holding a charge.

To give one final example, lest it be thought that Maxwell withdrew this language in the later parts of his *Treatise*, in paragraph 617 he wrote, "We may therefore adopt, as a definition of **A**, that it is the vector potential of the electric current, standing in the same relation to the electric current that the scalar potential stands to the matter of which it is the potential." Here it is almost inevitable that any but a well-trained Maxwellian would gloss "current" as the motion of a "charge"—but it is not so in Maxwell's theory.[13] In Maxwell's approach any medium carries a vector field **J** with

[13] This account, including the introduction of λ, follows Buchwald (1985, 24).

a zero divergence, $\nabla \cdot \mathbf{B} = 0$. So we may think of $\mathbf{J}$ as representing the flow per unit area of a conserved quantity. This vector field may vary with time, and Maxwell introduced the vector field λ with the property that $\mathbf{J} = \frac{\partial}{\partial t}\lambda$. So λ represents the shift in location of a conserved substance. It is connected to the Maxwellian displacement $\mathbf{D}$ which creates the shift in λ but is not identical with it. For example, a charged sphere embedded in a dielectric will, in Maxwell's theory, create a displacement where the dielectric touches the sphere, thus giving the surface of the dielectric a positive charge, whereas there is no displacement in the sphere, so its surface is uncharged. This differs markedly from any modern theory.[14] In fact, for Maxwell, charge is produced by an electric field which is carried by an ether. Charge is discontinuity in the displacement associated with a field. This displacement produces a strain in the dielectric that depends on the conductivity of the medium, which causes the medium to breakdown and dissipate the energy in the form of heat. The field then restores the strain, and the process repeats, too rapidly to be detected other than as an average of buildup and discharge that we call electric current.

But Maxwell's theory was soon to be abandoned in favor of a microphysical theory of electrons, so it can be argued that Poincaré's misunderstanding was a fortunate one, because it promoted a more viable theory. To what extent, then, was it a satisfactory theory of electromagnetism in its own right? The excitement around the topic of electromagnetism was centered on the theory of optics that followed from it. This was a weakness in Hertz's theory and a strength of Lorentz's, as Poincaré was to discuss. One indication of that was Poincaré's public disenchantment with another theory that was losing ground: Fresnel's wave theory of light. In 1891 the advocates of Fresnel's theory believed they had found an experimental result with which to ward off the advances of Maxwell's electrodynamics, which involved a delicate study of how polarization is affected by diffraction. Poincaré gave reasons in his (1892d) to suggest that the experiment did not really bear the interpretation that was given,

---

[14] Intriguingly, Bertrand (1891, 745–6) quotes an unnamed French physicist "most familiar with Maxwell's work" (Potier?) as saying to Poincaré that he understood everything in his book except what is an electrified sphere, only to be told by Poincaré that he had thought of removing from the concept of electric displacement the sort of indeterminacy that is the cause of the obscurity, but "I did not want, in making the thoughts of the original author precise, to go beyond them and so to betray them."

but then pushed for an electromagnetic interpretation of the phenomena. The mathematics is complicated, and must deal with idealizations of the necessarily imperfect diffraction gratings used in experiments, and he had scarcely dared, he said, to publish such incomplete results had he not been encouraged by Fizeau's remark, "one can always hope that in applying the calculus to some of the theoretically simplest cases that one will arrive at rigorous deductions that can clarify the question." He had found it easier, he said, to use the language of electromagnetism rather than that of wave theory, but he insisted that there were no facts that the one theory could explain and the other not.

His tentative conclusion was that incident and reflected light is polarized perpendicular to the plane of diffraction. However, a number of observational factors, such as the color of the diffracted rays, remained unaccounted for, and Poincaré spent a number of pages in the second half of this memoir investigating the extent to which they might be due to the differences between real gratings and his simplified, mathematical ones. He then set the problem aside until he was brought back to it by a paper that Arnold Sommerfeld published in 1896 that was essentially his Habilitation. In Poincaré's opinion, what Sommerfeld did was to rediscover and complete his results by an extremely ingenious method. Sommerfeld indeed claimed that Poincaré's method lost track of the domain of validity of the approximation formula he used, and produced some spurious infinities that his own method could explain, and moreover his method could confirm the validity of Poincaré's formulas. Sommerfeld's method, as befitted a student of Klein's, made considerable use of complex function theory, and this seems to have inspired Poincaré to do the same in his second memoir, (1897b). And, as Walter has pointed out in (2007), Poincaré also devoted part of his lectures on elasticity and optics to an exposition of Sommerfeld's method, a fact for which we only have Langevin's notes on the course.

LARMOR AND LORENTZ: THE ELECTRON AND THE ETHER

In the years between 1890 and 1904 physical theory changed significantly, and the study of electromagnetism became based on the idea of microphysical particles, called ions or electrons. This brought to an end Maxwell's particular field-theoretic approach, although it was still a theory of electric and magnetic fields that were now supposed to be caused by the behavior of these ions, and numerous problems with the wave theory of light were left unresolved. A quick appreciation of the merits of wave theory will not only bring out one of the principal experimental results upon which Poincaré relied, it will help explain why the ether theory lasted so long in the face of the problems that it raised, and upon which Poincaré insisted so forcefully.

A vivid picture of the state of the art by 1904 is provided by the introduction to the survey (Lorentz 1904a) of the theory of electricity and magnetism that Lorentz handed over to the editors of the *EMW* in December 1903. It was in two parts: an 80-page essay on Maxwell's theory, and one of 140 pages on extensions of the theory, subtitled electron theory. In the first part he observed that it was the job of an essay in the encyclopedia to give as simple as possible an overview of the whole field. Therefore he would start with the equations governing the theory, and then indicate how they accounted for the experimental facts; deriving the equations first was not the point. That would come later, using whatever assumptions to account for appearances that seem appropriate. This was the right thing to do, he wrote, because the field equations, at least in the simplest cases, seem to be more certain than the hypotheses by means of which one has sought, with more or less luck, to ground them. There could scarcely be a better admission of the radical state of flux into which the fundamentals of electromagnetic theory had been plunged.

Lorentz then introduced Maxwell's equations in the form presented by Hertz and Heaviside, using the new vector notation, which he carefully explained. He then deployed them to account for a variety of physical phenomena, but when he came to Fizeau's experiment he called it a difficulty that Hertz's theory could not overcome. This was not the only problem. The success of the wave theory of light, which drove out the former corpuscular theory as the 19th century progressed, brought with it an increasingly famous problem, that of explaining the medium in which the waves oscillated, the mysterious all-pervasive ether. But the wave theory required rethinking a number of other optical phenomena, for example stellar aberration. At its simplest, stellar aberration says that the relative motion of the earth with respect to a distant star means that a telescope pointing directly at the star does not give its true position, but its apparent position. If light travels through space with a velocity c and

Figure 6.2. Hendrik Lorentz (1853–1928). Source: http://osfundamentosdafisi ca.blogspot.com/2011/07/especial-de-sabado.html.   ©Museum   Boerhaave Leiden.

hits an observer moving through space with velocity **v** then that observer would surely record the light as having a relative velocity of **c** − **v**. The English astronomer James Bradley had reported in 1728 that observation of several stars showed that the velocity of light in space was the same for all stars. In his (1810) the French theorist François Arago praised this result, but recognized that it posed a problem for the corpuscular theory of light, because the velocity should depend on the relative velocity of the source and the emitter. In 1818 he turned to Augustin Fresnel, a leading exponent of the wave theory of light, to see if he could explain aberration that way.

Fresnel replied that in the wave theory, light traveled with a speed determined by the medium it was in, but not by the velocity of the source. If light traveled through the ether with a speed v and entered a body with refractive index N it would travel with a speed of $v(1 - 1/N^2)$, which he explained by saying that the moving body dragged some of the ether with it. So the velocity of light in the body, $\mathbf{c}'$, was given by

$$\mathbf{c}' = \mathbf{c} + \left(1 - 1/N^2\right)\mathbf{v}.$$

For motion in the ether, for which $N = 1$, one has $\mathbf{c}' = \mathbf{c}$. In 1851 Hippolyte Fizeau verified Fresnel's argument by determining experimentally the velocity of light in water, a result that was confirmed more accurately by Michelson and Morley in 1886, and which Poincaré regarded as one of the best, most reliable experimental results in physics.

But was the ether at rest, or in motion? How could it be utterly at rest when massive objects like the earth moved through it? In a paper of 1886 Lorentz was led to suggest that as a transparent body moves through the ether its constituent molecules are affected by the ether, which remains at rest. His theory therefore explained stellar aberration in two ways, depending on whether the source or the observer was taken to be at rest.

The challenge Lorentz set himself in his (1892) was to derive the electromagnetic properties of matter in motion from his version of Maxwell's equations. His first goal was to derive Fresnel's coefficient of entrainment $(1 - 1/N^2)$, but now on the assumption that the ether was absolutely at rest. Moreover, the electric and magnetic fields were obtained by averaging over the behavior of charged vibrating particles, so the approach was entirely microphysical. He succeeded in deriving a theory that was accurate to first order in v/c, as he had in 1886 but now in a more satisfactory way, and deducing the expected value for Fresnel's coefficient. That said, he was willing to pay a high price to get there. He began with Maxwell's equations expressed in space and time coordinates of an ether-based frame of reference, and converted them to a system of equations with respect to a frame moving with uniform velocity with respect to the ether in the standard way one switched from one reference frame to another. This did not change the time coordinate. But this coordinate transformation did not convert the wave equation in the ether frame into an equation in the moving reference frame which had the right form to be a wave equation. So Lorentz applied a further coordinate transformation that resulted in the final transformation of the original equation into another equation that had the form of a wave equation. This second coordinate transformation did affect the time coordinate, and the new t variable Lorentz called the "local time" of the moving body. He regarded the second transformation as a mathematical trick, and not only attached no physical significance to the local time, he seemed not to bother that the new theory could not be an ether-based theory of light.

By 1892 the null result of the Michelson–Morley experiment of 1887 had been worrying Lorentz for five years, because it contradicted Fresnel's theory. Lorentz finally came to propose that matter might contract along the direction of its motion. As Miller writes,

> Thus, the hypothesis of contraction was proposed to explain a single experiment; it was obtained from reasoning unconnected with the electromagnetic theory; its connection with the electromagnetic theory was at best tenuous—in short, the hypothesis of contradiction was ad hoc. (1981, 31)

The sole attractions of the hypothesis for Lorentz, one supposes, were that it reconciled his theory with the Michelson–Morley results and allowed his theory to be accurate to second order in v/c.

Lorentz rederived these results in what has become his best-known and most highly regarded paper, his *Versuch* (1895).[15] He again addressed the problem of the aberration of light by assuming an ether at rest with respect to ponderable matter, and based his derivation of the fundamental equations of electromagnetic theory on the properties of hypothetical ions. In a section of chapter II called "Application to electro-statics," he outlined a program for solving problems in electrodynamics by reducing them to problems in electrostatics. Built into this process were two assumptions: that the mathematical transformations are what are called Galilean ($x' = x + vt$, $t' = t$ when motion is in the x direction with velocity v) and that ions are rigid. This results in a modified Poisson's equation that can be replaced by the usual one on stretching the x-axis. Later, in the discussion on applications to optics, he again used his device of local time, so the theory was far from the unified one he would have liked.

Although Lorentz's ideas were the ones that most impressed Poincaré, he came back to electromagnetic theory with a series of papers in which he responded to the work of Larmor. Joseph Larmor was a Cambridge

---

[15] This account follows Miller (1981).

physicist who had been led, with some prodding by Fitzgerald, to propose a microphysical theory of electromagnetism. This was a complete departure from Maxwell's electromagnetic theory, and Larmor struggled for some time to reformulate the new theory, but by 18 April 1895 "Larmor had finally abandoned Maxwellian theory."[16] The new theory naturally retained some features of the old one, but it had to be created on the basis of a completely new idea of electric current, which was now attributed to the motion of electrons.

Poincaré's response to Larmor was a series of articles which began in April 1895 with his (1895b). He began by reminding his readers of the many different theories then in circulation, and the problems they entailed. Poincaré took up this challenge in the next part of the paper (18 May 1895). The equations of electrodynamics had been stated by Hertz, so the task was to translate them into a modified version of Fresnel's theory and then in Larmor's theory. But before doing that it was necessary to consider whether Hertz's equations, which only rested on some bold inductive leaps, could be accepted unaltered. Poincaré thought this was very doubtful. In a page of analysis he showed that Hertz's theory implied that a plane wave of light would be entrained completely by a dielectric in motion, but this was absolutely contradicted by Fizeau's experiment. So Hertz's theory had to be modified, but now there was an almost embarrassing range of choices in the literature, none of them entirely satisfactory. Poincaré thought that Fresnel's theory was the best among them, so he set about seeing if it could be transcribed into a modern electrodynamical theory. One way would be to follow an approach Helmholtz had developed in response to Maxwell's work, but Poincaré showed that this violated the law of conservation of electricity.

Another, and in some respects more satisfying, approach, said Poincaré, was due to Lorentz, and he summarized it this way:

The fundamental hypothesis is the following:

A very large number of small particles carrying charges with which they are indissolubly connected, are spread through the volume of conductors and dielectrics. They run through the conductors in

---

[16] See the detailed description of his journey and what it entailed for the theory of electromagnetism in Buchwald (1985, quote on p. 171).

every direction at very great speeds. In dielectrics, on the contrary, they can only undergo small displacements and the action of nearby particles tends to restore them to the equilibrium position that they have left.

This theory automatically takes care of the principle of conservation of electricity, but it violates Newton's third law: there is no equality of action and reaction. For, as Poincaré explained, the ether acts on a charged body but there can be no reaction of the charged body on the ether because the ether is supposed to be absolutely at rest.[17] This was not a problem with Hertz's theory, so each theory was seemingly flawed. In reflecting on this, Poincaré said he had been led to consider how the entrainment of light could be compatible with Newton's third law. This was a problem tackled unsuccessfully by others, such as J. J. Thomson.

In the third part of the article (October 1895) Poincaré again reminded his readers that the task was to find a consistent theory that took account of Fizeau's experiments, obeyed the laws of conservation of electricity and magnetism, and preserved Newton's third law. He was then able to show that Hertz's theory was the only one that satisfied the last two requirements, and so no theory satisfied all three:

> Consequently, we can only hope to escape this difficulty by profoundly modifying generally admitted ideas. . . .It is therefore necessary to give up developing a perfectly satisfying theory and to hold provisionally to the least defective of them all, which appears to be that of Lorentz. This will be enough for my purpose, which is to deepen the discussion of Larmor's ideas.

At this point Poincaré renounced whatever allegiance he may have had to Maxwell's account of electric current, and returned to a one- or two-fluid theory of current, perhaps as a circulation of microscopic charged particles. But, he concluded,

> Experiments have revealed a slew of facts that can be summarised in the following formula: it is impossible to demonstrate the absolute motion of matter, or better the relative movement of ponderable matter with respect to the ether; all that can be exhibited

---

[17] The equality of action and reaction guarantees the conservation of linear momentum, but the principle was still interpreted at this time as applying only to moving substances.

is the motion of ponderable matter with respect to ponderable matter.

The theories proposed take good account of this law, but only under two conditions:

1. It is necessary to neglect dispersions and various other secondary phenomena of the same kind;
2. It is necessary to neglect the square of the aberration.

Now, that is not satisfactory; the law seems to be true without these restrictions, as a recent experiment of M. Michelson has shown. There is therefore a gap which is not perhaps unrelated with what the present article is intended to draw attention to.

As for a possible way forward, Poincaré concluded this part of his paper with this remark:

And in fact the impossibility of exhibiting the motion of matter relative to the ether, and the equality which without holds between action and reaction without taking account of the action of matter on the ether, are two facts whose connection seems evident. Perhaps these two gaps can be filled at the same time.

This series of essays came to its inconclusive end with some further reflections (November 1895), with the problem of finding a theory that met all three of Poincaré's aims simultaneously, unsolved. But the problem was at least clearly stated, and on the way Poincaré, like Larmor, had abandoned Maxwell's physics for a theory based on the movement of charged particles, such as Lorentz had proposed in his *Versuch*.

Poincaré then left the subject of electromagnetic theory for over a year, before returning to it with two papers on how Lorentz's theory explained the Zeeman effect. Lorentz's former student Pieter Zeeman had discovered in 1896 that the spectral lines of sodium are split by a magnetic field. Lorentz immediately explained this effect in terms of the motion of charged ions relative to the fixed part of a molecule of which they were a part. From the observed frequencies Lorentz was able to calculate a charge to mass ratio for the mobile ions, and compare it with an already determined charge to mass ratio for the hydrogen ion in electrolysis. He found they were comparable, and published his findings in 1898. The next year J. J. Thomson determined the charge on a cathode ray, and found it to agree with the charge on a hydrogen ion in electrolysis. This made it almost certain that Lorentz's elementary charged particles were, in fact, electrons.[18] Poincaré's first paper (1897) compared the theories of Lorentz and Potier, and demonstrated the close link between Zeeman's discovery and the ideas of Faraday. His second paper (1899) showed how to generalize Lorentz's theory so that it covered more recent experimental investigations of the Zeeman effect, which had found subtler spectral patterns that the original account could not explain.

POINCARÉ ON HERTZ AND LORENTZ

Poincaré's lectures on electromagnetic theory are more important than these last two papers. He gave them at the Sorbonne in 1899 and they were published in 1901 in the second edition of his *E&O*. Here he offered a detailed, comparative account of the theories of Hertz and Lorentz, and concluded by reprinting his four-part response to Larmor. Hertz's theory may be regarded as a Continental Maxwellian theory, Lorentz's, as we have seen, as a microphysical one. The fundamental difference between them is that in Lorentz's theory there is no true magnetism: everything is fundamentally electrical in nature. So where Hertz described a magnetic force, this is reduced in Lorentz's theory to a particular manifestation of the electric force. Moreover, Lorentz's theory is a theory of charged particles at rest and in motion, so electric current is analyzed differently. Both theories agree that electricity is conserved, and that energy is conserved; as Poincaré again explained, Newton's third law is broken in Lorentz's theory, but not in that of Hertz.

Lorentz's microphysical theory explains the existence of the dynamical forces of electricity and magnetism in two stages. First, it observes that Maxwell's equations are more than adequate for many purposes, but leave it impossible to discuss how electrical phenomena depend on the state and chemical constitution of different materials. Therefore, said Lorentz (1895, 8), one introduces positive and negative electrons and discusses electricity and magnetism in terms of the distribution and motion of these extremely small particles. Second, it attempts to derive Maxwell's equations from hypotheses about the electrons and the (fixed)

[18] On the complicated history of the discovery of the electron, see Martinez (2011, chap. 8).

ether.[19] The ether permeates even the electrons, which have a small but finite size, and the charge, ρ, on an electron was taken by Lorentz to be concentrated in the electron and falling rapidly to zero in a thin boundary of the electron. Because charge varies continuously there is no need for surface discontinuities to enter the theory. It is supposed that the charge associated with an electron does not alter as the electron moves, but one is free to speculate about its shape. So Lorentz wrote down Maxwell's equations at the level of an electron, in which ρ appears, and showed that at the macroscopic level, where one can set $\rho = 0$, they reduce to the familiar Maxwell's equations. It was then possible for Lorentz to apply his new theory to situations where the old Maxwell's equations are too crude.

Poincaré began the second part of his *E&O* with a quick review of Hertz's ideas about the different physical theories under consideration at the time. Some were theories of force as action at a distance; this, Hertz noted, requires at least two bodies. One could turn to potential theory, which can handle single bodies as well. One can imagine a dielectric full of "cellules" which can become electrified and polarized, as Poisson had done. In this theory, action at a distance plays only a modest role and attention focuses on the behavior of neighboring cellules; this had been Helmholtz's approach. One could even suppress ideas of action at a distance entirely and consider only the polarization of a dielectric; this was Maxwell's field-theoretic approach. But Hertz had found this theory lacking in clarity because the word "electricity" was used imprecisely (in its popular meaning, as an impressible fluid, and so on) and more importantly, because Maxwell's *Treatise* was a ragbag of theories. Hertz had therefore dropped everything from Maxwell except his equations. Since his presentation of the theory from that starting point considered first bodies at rest and second bodies in motion, Poincaré proposed to divide his account of Hertz's theory in the same way.

Poincaré carefully sketched Hertz's ideas, which were not, in fact, simply Maxwell's equations but rather a system of similar equations with somewhat different physical interpretations and connections to experiment. It was therefore also necessary to translate between the two theories, and Poincaré showed how to do this. Hertz's electromagnetic theory for bodies at rest, he concluded, was based on the four concepts of

---

[19] Rigorous derivation had to wait until the 1960s.

true and free magnetism and true and free electricity, which he defined, only to comment,

> One therefore sees that for Hertz what one calls electricity and magnetism is not a fluid or anything material but a purely analytical expression: an integral; what really exists is the electric and magnetic force. (para. 299)

He therefore completed the purely analytical account, and from it showed that the principles of conservation of magnetism and electricity were satisfied.

The question of conservation of energy was more interesting. Maxwell's and Hertz's ideas of energy differed only trivially over the definition of electric energy, but significantly over magnetic energy. Maxwell, Poincaré observed, had in any case given two inconsistent definitions of magnetic energy, one when speaking of magnetism without electric currents and one when speaking of currents. So they could not be admitted. Hertz's definition of magnetic energy was different again and reduced to expressions that agreed with Maxwell's no-current analysis if all magnetism is permanent, and with his current-based analysis only on the assumption that there was no permanent magnetism. Poincaré now showed that on Hertz's definition magnetic energy was conserved and so (para. 305) Hertz's definition was the only acceptable one.

Poincaré now turned to Hertz's theory of the electrodynamics of moving bodies. An entirely mathematical derivation led him to compare Hertz's and Maxwell's equations, and he was able to show that Maxwell's and Hertz's equations were mathematically equivalent only if magnetism is permanent. But, he noted, this was not always the case: magnetism can be destroyed by heat.

Once again it was necessary to check on the fundamental conservation laws, and Poincaré showed that electricity and magnetism were indeed conserved. Moreover, Hertz's equations took the same form whether referred to fixed or moving axes, and so kept the same form whether motion was relative or absolute. This had one happy and one worrying consequence. Happily, Newton's law of the equality of action and reaction was preserved in Hertz's theory. Worryingly, Hertz's theory could not account for Fizeau's experiment which implied the partial entrainment of light. Poincaré therefore looked to see if he could slightly modify Hertz's theory to make it compatible with Fizeau's result and the theory of optics.

He found that an otherwise indefensible ad hoc modification could do so, although it was vulnerable to experiments on magnetic induction. He did not pursue this idea, however, because, he said, it was the impulse for Lorentz's theory and would be dealt with in due course.

A much lengthier analysis (paras 322–328) was needed for Poincaré to show that Hertz's theory of electrodynamics conserved energy, and then Poincaré showed (para. 329) that the principle of the equality of action and reaction was also conserved. He argued that Hertz's equations took the same form in relative and absolute motion, as did the expression for the energy, J. So Poincaré compared the rate of change of energy under an absolute and a relative motion, and checked that the difference vanished for a translation. In this way Poincaré concluded his settling of accounts with Hertz, and took up the evolving ideas of Lorentz.

Poincaré's account of Lorentz's work was, if not unsympathetic then in some sense indicative of a temperamental divide. He opened it by saying that while Hertz's account, although perfectly coherent, failed to explain crucial results in optics, Lorentz's account, which did deal with the optical question, failed to keep the law of equality of action and reaction. One way to see this separation is to register it as Poincaré's dislike of Lorentz's hypothesis that electrical phenomena are the result of minute electrical particles, called ions or electrons, which pervade all of matter. Lorentz, he said, used this idea to explain magnetism by reducing it entirely to electric currents after the fashion of Ampère. It followed that in his theory an electric current in a circuit really was the motion of these electrons. In dielectrics the electrons cannot move more than a small amount, and as a result the action of an electric field is to polarize a dielectric.

Poincaré suggested two ways to analyze electrical phenomena. One, available to beings with very subtle, highly developed senses, would be to see the individual electrons as they move; such beings would never see magnets. Observers with cruder senses—like ourselves, he commented— would never see these ions. They would only see aggregate effects, and would believe in magnets. One might well wonder if the first kind of observer was one Poincaré could ever truly believe in.

At all events, such an observer, he explained, could analyze electric current, and find it was a mixture of displacement currents and Rowland currents (currents caused by moving electrons). Poincaré now cast the theory into a Lagrangian formalism, attributing some energy to matter and some to the ether. Some of the coordinates q give the position of the ions (Poincaré allowed Lorentz's argument that any rotation they might have could be neglected), and some refer to the position of the ether. Poincaré now set about separating the two kinds of variable, until he had obtained Lorentz's equations for the state of the ether, which relate P, Q, R and α, β, γ, where

$$(P, Q, R) = \frac{4\pi}{K_0}(f, g, h),$$

K_0 is a constant, and (α, β, γ) are the components of the magnetic force. Here (f, g, h) satisfies

$$\frac{d}{dt}(f, g, h) = (u, v, w) + \rho(\xi, \eta, \zeta),$$

where (u, v, w) represents the total current, and is regarded as the sum of displacement current and a Rowland or convection current, (ξ, η, ζ) is the velocity of a point P, and $\nabla \cdot (f, g, h) = \rho$. He found (para. 338) that the ether obeyed the equations

$$\nabla \times (P, Q, R) = \frac{d}{dt}(\alpha, \beta, \gamma).$$

Poincaré now compared Lorentz's equations with those of Hertz, and found that to do so he had to reintroduce the effects of magnetic induction, noting that Lorentz's equations permitted this in a way he would sketch later. A further analysis of the Lagrangian equations, this time those that do not involve the ether, allowed Poincaré to compare Lorentz's theory with Hertz's in every respect. Lorentz's theory had no magnetic force. The electric force in Lorentz's theory included effects due to magnetic induction and electrostatics, and so agreed with Hertz's in this respect.

Poincaré now showed that electricity was conserved in Lorentz's theory, as was energy (trivially, because the theory was a Lagrangian one, where energy conservation is built-in). But, as he had signaled in advance, the Newtonian principle of the equality of action and reaction was lost. An electromagnetic variation acts on a charged particle by producing a force that moves the particle, but the particle exerts no reactive force; to underline the point, Poincaré explicitly calculated the magnitude of the motion. He noted that Liénard had attempted to explain away the disparity, but gave an argument using Poynting's vector (here he referred to *E&O* p. 27). He then investigated whether Newton's law could be recovered by looking at the motion of the ether, and showed that his earlier calculation implied an impossible relationship between the strength of the field and the velocity of the ether.

Poincaré did not reject Lorentz's theory on this account, much though he found it unwelcome. He now wrote down Lorentz's equations (they fill para. 353) and set about integrating them. To allow for the propagation of electromagnetic disturbances at the speed of light he introduced the method of retarded potentials. He introduced the expression

$$\Box U = \triangle U - K_0 \frac{d^2 U}{dt^2},$$

and showed Poisson's equation can now be written as $\Box V' = -4\pi f$, where f is a given function and V' is unknown. Then, "provided one assumes everything starts from rest" (an unclear phrase) and all the functions vanish at infinity, V' is determined by f, and the equation $\Box U = \Box V$ implies that $U = V$.

Poincaré now turned to Lorentz's world as a crude observer like ourselves would see it. The great success of Lorentz's theory was its account of optics. Poincaré set the scene this way:

> The most important of these [optical] phenomena is astronomical aberration. This phenomenon gives evidence of the motion relative to the ether of a ponderable medium that it penetrates. . . .
>
> Lorentz's theory, as we shall see, explains these facts very well. It is necessary, however, to make one hypothesis: *if one requires that the optical phenomena are not influenced by the motion of the earth it necessary that one neglects in the formulae terms of the order of the square of the aberration* (that is to say, of the order of 10^{-8}). . . .
>
> In almost all experiments these terms are in fact negligible; however, there is an exception for Michelson's experiment, which shows that the motion of the earth has no influence on optical phenomena that one observes on its surface and where he finds that the terms of the order of the square of the aberration are not negligible. (pp. 516–518; Poincaré's italics)

Lorentz's theory of aberration made use of his theory of local time, and Poincaré placed this in the context of a theorem (para. 412, pp. 523–525) that said: if one neglects the squares of ξ, η, ζ, which are the components of the velocity of the earth, the motion of the earth has no influence on optical phenomena. To prove it, Poincaré changed coordinates from a frame at rest with respect to the ether to one moving with respect to the ether but where time is local time. On neglecting the squares of ξ, η, ζ the new equations for the motion of light have the same form as the old ones. It followed that the most sensitive optical experiments, those involving interference fringes, would not detect any motion: light and dark patterns would agree completely. Also, the difference between actual time and local time, being of the order of $\frac{1}{3 \times 10^9}$ would be too small to detect. A more delicate analysis then showed that the intensity of the interference fringes would also be the same in the two frames.

This good agreement nonetheless left Michelson's experiment unaccounted for. One could resolve the problem, said Poincaré, by supposing the lengths of moving objects contracted suitably:

> This strange property would be a veritable push [*coup de pouce*] given by nature to avoid having the motion of the earth reveal itself through optical phenomena. This would not satisfy me, and I believe I must say here what I feel: I regard it as extremely probable that the optical phenomena only depend on the motion relative to solid bodies that are present, light sources and optical apparatuses, and *that not on quantities of the order of the square or the cube of the aberration but rigorously.* To the extent that experiments become more exact this principle will be verified with greater precision.
>
> Will there be a need for a new *coup de pouce*, a new hypothesis for each approximation? Evidently not: a properly constructed theory will lead to a demonstration of the principle at a stroke and with complete rigour. Lorentz's theory cannot do this yet. Of all those that have been proposed, it is the best able to do it. One can therefore hope to make it perfectly satisfactory in this respect without modifying it too profoundly. (p. 536; Poincaré's italics)

On 11 December 1900 the University of Leiden celebrated the 25th anniversary of the award of a doctorate to one of its most distinguished faculty members, Hendrik Lorentz, with a conference in his honor. Poincaré gave a paper, which he introduced by admitting that it might seem strange that at a meeting in honor of Lorentz he should raise objections to his theory, but he would not plead that he had come to reduce his objections. Rather, he had an excuse a hundred times better: "Good theories are supple," and, in line with his remarks at the recent ICP in Paris, he proposed to bend Lorentz's ideas so that they overcome some serious objections. In triumphing, he suggested, they would also be transformed.

Poincaré now spoke of electrons, but, more reasonably, called electromagnetic energy a fictitious fluid that obeys Poynting's laws. He then considered the laws of a system composed of matter, electromagnetic energy, and energy that is not electrical. His aim was to rescue Newton's third law, and his strategy was to impute properties to the fictitious fluid, such as its inertia, and then to demonstrate that "if some apparatus, after having produced some electromagnetic energy radiates it in a certain direction, then this apparatus must recoil like the cannon that fires a projectile" (*Oeuvres* 9, 471). Indeed, he added, all the theories under consideration predict the same recoil. This, however, was odd, because Hertz's theory accommodated Newton's third law, and Poincaré explained this by examining how the two theories dealt with energy in a dielectric.

He then returned to the importance of Newton's third law, and observed that without it perpetual motion would be possible. Would that still be the case with the new theory? Poincaré answered yes, because it was a consequence of two other principles: conservation of energy and the relativity of motion, "which imposes itself imperiously on the mind, when applied to an isolated system" (p. 482). But, he then said, in the new theory we do not have an isolated system, because the whole ether has been invoked. Confronted with the issue of the motion of the earth through the ether the principle of relativity makes so feeble a claim that it has been necessary to conduct experiments to find evidence of this motion; experiments that, to be sure, had found no evidence—but that had been rather surprising.

Lorentz's theory did explain those negative results, and so the principle of relativity was no longer a priori but rescued a posteriori, but only by a compensation of effects: one must neglect terms of the order of v^2; use the theory of local time; reconcile the motion of real and apparent energy; and allow that the relative motion of a body producing electromagnetic energy is subject to an apparent force that vanishes when the motion is taken to be absolute. Apropos the second point, Poincaré interpreted local time as the time measured by two observers correlating their clocks by light signals transmitted back and forth on the assumption that the signal travels with equal speed in each direction. Poincaré concluded his paper by arguing that these difficulties are all overcome by developing the theory of real and apparent forces, or, otherwise put, of considering matter and ether simultaneously, never singly. Most pointedly, he observed that without such a development it was the case that Fizeau's experiment was incompatible with Newton's third law.

On 20 January Lorentz wrote to Poincaré to thank him for his contribution to the event, and to say how touched he was by how many scientists had expressed an interest in his work despite its imperfections. He was, he said, particularly grateful for Poincaré's paper, but he had to confess that he did not see any way to modify his theory so that the fundamental difficulty would disappear. Instead, he was willing to contemplate abandoning the principle of action and reaction by using another of Poincaré's formulas to attribute yet another property to the ether, which might, he went on, lead to a theory in which matter is regarded as a modified kind of ether. But he was not fully happy with this reply, and in several later papers came back to the issue that Poincaré had raised so forcefully.

At the end of his second *EMW* article Lorentz surveyed the state of the theory at the end of 1903. He offered a defense of his contraction hypothesis as plausible behavior of matter, noting on the one hand that it offered an explanation of the null result of the Trouton and Noble experiment, and on the other that it presently took no account of the motion of molecules within a body. He admitted that Poincaré (in his 1901e, 22–23) had rightly criticized the theory for introducing a hypothesis solely to explain Michelson's result[20]:

> Poincaré has rightly criticized the theory (although we should be permitted the excuse, that we are entering into this domain cautiously groping forward) on the grounds that it has introduced

---

[20] Translation from Miller (1981, 70).

an hypothesis contrived for the purpose of explaining just the experiment of Michelson, and that perhaps similar artifices would be necessary to account for new experimental results.

In his paper of 1904, published in English, Lorentz surveyed the accumulating negative evidence for the earth's motion through the ether and addressed Poincaré's criticism:

> Surely this course of inventing special hypotheses for each new experimental result is somewhat artificial. It would be more satisfactory if it were possible to show by means of certain fundamental assumptions and without neglecting terms of one order of magnitude or another, that many electromagnetic actions are entirely independent of the motion of the system. Some years ago, I already sought to frame a theory of this kind. I believe it is now possible to treat the subject with a better result. The only restriction as regards the velocity will be that it be less than that of light. (1904b, 13)

He now supposed that the equations of electrodynamics written with respect to a frame fixed in the ether could be modified to fit a frame moving with velocity v, by allowing for a contraction along the direction of motion (taken as the x direction) that also affected the time t' measured in that frame (the local time). The corresponding equations were

$$x' = \beta l x, \quad y' = l y, \quad z' = l z, \quad t' = \frac{l}{\beta} t - \beta l \frac{v}{c^2} x,$$

where the coefficient l is a function of v that takes the value $l = 1$ when $v = 0$ and otherwise differs from unity "by no more than a quantity of the second order." He noted that this concept of local time reduces to his earlier one when $\beta = 1, l = 1$.

He then made a new assumption:

> Thus far we have used only the fundamental equations without any new assumptions. I shall now suppose that the electrons, which I take to be spheres of radius R in the state of rest, have their dimensions changed by the effect of a translation, the dimensions in the direction of motion becoming βl times and those in perpendicular directions l times smaller.

In this deformation, which may be represented by $\left(\frac{1}{\beta l}, \frac{1}{l}, \frac{1}{l}\right)$ each element of volume is understood to preserve its charge. (para. 8)

So Lorentz's electrons have a size, and a distribution of charge upon them, and he considered the change in their shape—they become flattened ellipsoids—in a way that was independent of any concept of time.

Lorentz now compared the moving system to a hypothetical static one whose coordinates agreed with the primed coordinates, noting that the equations of motion were not the same in the two cases. For such a hypothetical state to correspond to a real one, he showed that l must be a constant, and therefore equal to 1. This simplified the coordinate transformation to

$$x' = \beta x, \quad y' = y, \quad z' = z, \quad t' = \frac{1}{\beta}t - \beta\frac{v}{c^2}x.$$

Lorentz now showed that interference patterns in the two systems will be the same:

> It appears that those parts which are dark while the system is at rest, will remain so after it has been put in motion. It will therefore be impossible to detect an influence of the Earth's motion on any optical experiment, made with a terrestrial source of light, in which the geometrical distribution of light and darkness is observed. (1904b, 29)

This is because if the relevant variables vanish in one frame they vanish in the other. As he observed, "Many experiments on interference and diffraction belong to this class," including Michelson's and the Trouton and Noble experiments.

ST. LOUIS, 1904

Poincaré addressed his audience in St. Louis with these opening words:

> What is the present state of mathematical physics? What are the problems that it has been led to pose? What is its future? (VS 170, 290)

These words may have been intended to remind his audience of Hilbert's address to the ICM in Paris in 1900, but the metaphor Poincaré proceeded to adopt was that of the prudent doctor, as he put it, unwilling to offer a prognosis but perhaps a diagnosis: there were indications of a serious crisis, as of an impending transformation. However, we should not worry, it would not be fatal, there would certainly be a recovery—the history of the subject guaranteed it.

As Poincaré saw that history, there had been a prolonged period where physical processes were explained in terms of laws of attraction: Newtonian gravity, Laplace's theory of capillarity. Only Fourier's theory of heat departed from that model. However, there had come a time when people recognized that the universe was analogous to a machine whose inner mechanism, the cogs of the machine, would be forever hidden from us. Scientists then had recourse to general principles to illuminate the input and output of the machine, such as the law of conservation of energy. Poincaré enumerated five further principles: Carnot's principle (the degradation of energy); Newton's third law; the principle of relativity; conservation of mass; and the law of least action. This second phase of physics was successful, different from the first but impossible without it, and the transition from one to the other the grounds for optimism in the face of the impending crisis.

Would there have to be a new phase? Carnot's principle raised the uncomfortable paradox that the processes of dynamics were time-reversible, but most day-to-day processes were not, in conformity with the theory of thermodynamics. A drop of wine in a glass of water diffuses and never separates, as Boltzmann, Maxwell, and more recently Gibbs had sought to explain in the language of statistical mechanics. However, the phenomenon of Brownian motion showed that it would be difficult to believe simultaneously in Carnot's principle and conservation of energy. "And there already one of our principles is in peril."

Next, relativity. Two charged bodies, at rest with respect to the earth, are carried along by the motion of the earth, and as such exert an attractive force on each other, but when this force is measured, it reveals the speed of the earth, not with respect to the sun and the fixed stars, but absolutely. Nor can one appeal to the ether, which was remaining firmly undetectable, as Lorentz had explained, at the cost of accumulating hypotheses. The most ingenious of these, said Poincaré, was that of local time. Two observers synchronize their clocks by exchanging light signals, allowing, of course, for the time the light takes to travel between them, which they assume is the same in each direction. This is easy to do when the observers are at rest, but when they are moving relative to one another, even with constant velocity, their clocks can only reflect a local time, because the time the signal takes to go from one to the other is affected by the motion. However, when allowance is made for this, the observers find their estimates of time are still slightly erroneous, unless it is further assumed that lengths contract slightly along the line of travel by an amount determined by the relative velocities of the observers. And it is not only distances that are affected, but the measurement of forces—all because the observers are moving relative to one another with constant velocity. Poincaré conceded that the principle of relativity had been valiantly defended, "but even the energy of the defence shows how serious was the attack."

And so it went with the other principles. Newton's third law was confronted with the issue of radiation pressure; the conservation of mass with the velocity-dependent mass of the electron, and if all matter was composed of negative and positive electrons then all of mechanics was incorrect. This, said Poincaré, was why the experiments on cathode rays could justify Lorentz's doubts about Newton's third law:

> From all these results, if confirmed, would an entirely new mechanics be deduced that would be characterised, above all, by the fact that no speed could exceed that of light?

Here Poincaré added a footnote: "Because bodies impose an increasing inertia to the causes that tend to accelerate their motion, and this inertia becomes infinite as one approaches the speed of light."[21] He then went on to explain that mass has two aspects: it is the coefficient of inertia and an attractive mass that enters as a factor into the Newtonian law of gravity; but if the coefficient of inertia is not constant then what is the attractive mass? Finally, conservation of energy was threatened by recent discoveries involving radiation, and only the vaguest of the six principles, that of least action, was left.

What to do, surrounded by all these ruins? One could note that all these disturbing experimental results concerned extremes: very small

---

[21] In the original publication and in *VS* (197, 305).

objects, very high speeds, very rare elements. Could one perhaps defer to the experimental physicists and get on with theoretical work in the large domain that was surely not imperiled by their recent findings? No, theorists had contributed to the present crisis, and the new ideas had also to be subjected to further theoretical analysis before they could be allowed to overturn the old principles.[22] One might, for example, be able to replace Lorentz's theory of the contraction of lengths with some further hypothesis about the behavior of the ether. In any case, a better analysis of stellar aberration was required. If one returned to earth, the dynamics of the electron raised many problems, among them the recently discovered Zeeman effect.

The Zeeman effect, which concerns the splitting of the spectral lines caused by a magnetic field, was one that physicists had measured very precisely, but theoretical explanations were still lacking. In St. Louis, Poincaré mentioned Lindemann's, which he found laudable but unconvincing; in *VS* he chose instead to mention the theory of the Japanese physicist Nagaoka, without fully endorsing it either.[23]

Then Poincaré turned to consider what would happen if all efforts to support the old principles failed. What then would be the implications for his conventionalism? It would seem that experiments had, after all, brought down principles that, as conventions, albeit arbitrary, had been elevated beyond all risk. "Well," said Poincaré, "I was right then and I am not wrong today." Suppose, he went on, that, as had been suggested, one rescues the principle of conservation of energy by imagining a new sort of radiation that fills space and which only interacts with radium. How convenient this would be! It is unverifiable, and irrefutable. It could rescue the principle of conservation of energy from any attack. In this sense the principle is immune to any experimental evidence. But this *coup de pouce* has rendered the principle useless, and when a principle ceases to be fertile it dies, without ever having been contradicted.

---

[22] Here Poincaré offered a paragraph of remarks about statistical mechanics that he dropped when republishing the essay in *VS*.

[23] Nagaoka Hantaro was a professor of physics in Tokyo, and had spoken at the ICP in Paris. In 1904 he proposed a Saturnian model of the atom in which electrons are modeled as rings surrounding the nucleus and bound by electrostatic forces. Aspects of this theory were confirmed by Rutherford's experiments in 1911, and Nagaoka used it to explain the Zeeman effect, but, as he recognized, such an atom is unstable, and he abandoned this theory in 1908.

So one must rebut the challenge in another way. It was not necessary to conclude that science was the work of Penelope, undoing at night what she did in the day. Science had survived a crisis before, it would do so again, perhaps on the model of the kinetic theory of gases, in which the physical laws rest on the mathematical laws of chance. Or it might be that we have glimpsed a new mechanics in which the speed of light is a bound that cannot be exceeded, and ordinary mechanics would be a simple first approximation. Then we would not have to regret having believed in the old principles, which would remain useful in their place, for all but the highest speeds. "We are not there yet," Poincaré concluded, "but nothing has yet proved that we will not return from the struggle victorious and intact."

Langevin and Poincaré traveled to America together as part of the French delegation. In his obituary of Poincaré (1913, 702) Langevin has left us this vivid picture of Poincaré in America:

> In the course of our conversations, during the week he gave me the pleasure of travelling alone with him across the vast plains of North America, returning from the St. Louis Congress, I had the opportunity to see the passionate interest with which Henri Poincaré followed every phase of the revolution that he saw coming about in our most fundamental ideas. He saw with little disquiet the shake up, thanks to instruments he himself had forged, of the old edifice of Newtonian dynamics that he had so recently crowned with his remarkable work on the three body problem and the equilibrium forms of celestial bodies. But if his enthusiasm was more reflective than mine he was, like all of us, driven by a fever to enter an entirely new world.

Paul Langevin's address in St. Louis makes an interesting comparison with Poincaré's.[24] His address placed great emphasis on the new experimental fact of the "discontinuous corpuscular structure of electrical charges." He praised Faraday, Maxwell, and Hertz for giving us a "precise knowledge of the properties of the electromagnetic ether," and Lorentz for his theory of the electron, which gave the connection between ether and matter, before going on to give a thorough survey of the evidence for the existence of electrons. The famous problem of detecting motion

[24] See *Congress of Arts and Science* (7, 121–156).

relative to the ether was, he said, solved by Lorentz. Halfway through his paper he came to the problem he called the dynamics of the electron: what would the motion of a free electron be in a given electromagnetic field?[25] He recognized that the issue of radiation pressure, the recoil of matter as it emits light, is one that violates Newton's third law, but this did not seem to bother him. In the paper's only reference to Poincaré, Langevin noted that Poincaré had made some proposals for restoring the third law, and that Max Abraham had been able to carry out the necessary calculations for electrons with small acceleration (the quasi-stationary case). However, Langevin preferred to formulate the problem differently, by regarding ether as more fundamental than matter, and he took his paper into a discussion of the nature of electrons and radioactivity.

All in all, this exposition of the electromagnetic world view was much more reassuring than Poincaré's survey of physics. It reads like many popularizations of science before and since: strong on experimental results, keen to suggest the arrival of something new and important which, though undoubtedly conceptual, is only hinted at, and weak on analysis. Poincaré typically provided the opposite: experimental results were referred to as if the reader would know them from elsewhere, there was a genuine problem posed in understanding the new material, and much analytical reflection was offered. The former mode of exposition stands or falls on its selection of experimental facts, and even when prescient may fade with the passing of time. Langevin's enthusiasm for the electron is surely swamped by his positive attitude to the ether, making it all the more ironic that it was to be Langevin, and not Poincaré who was the first to speak up for Einstein's theory of special relativity in France.[26] On the day, in 1904, it was Poincaré the mathematician, and not the younger physicist Langevin, who caught the sense of a world of physics about to change.

THE DYNAMICS OF THE ELECTRON

In 1905, 1906, and 1908 Poincaré published three important papers on the dynamics of the electron. Of these, the first is a note in the *Comptes*

[25] Here he described the motion of the electron in what later became called the theory of Langevin waves.

[26] Langevin was one of the first to discuss the twin paradox, in 1911.

rendus that summarizes the second paper, in which Poincaré introduced the Lorentz group, and the third (1908a) is a review of the mutually contradictory state of the fundamental principles of physics, as Poincaré took them to be.

The (1905e) begins by observing that attempts to detect the motion of the earth relative to the ether that could have detected effects measurable in $\frac{v^2}{c^2}$ had failed to do so. Lorentz, in his *Electromagnetic Phenomena* (1904b) had explained this null result by supposing that bodies contracted in the direction of motion of the earth, thus making any attempt to detect motion relative to the ether impossible. This paper, said Poincaré, was so important that he had been moved to take up the question, and found that his results agreed in all important respects with those of Lorentz, which he now proposed to modify and complete in some points of detail.

Without discussing how or why Lorentz had argued as he had, Poincaré took from him the observation that the equations of the electromagnetic field were unaltered by transformations of the form

$$x' = kl(x + \varepsilon t), \quad y' = ly, \quad z' = lz, \quad t' = kl(t + \varepsilon x),$$

where x, y, z are the coordinates and t the time before the transformation and x', y', z' and t' afterwards. Furthermore, ε is a constant that depends on the transformation, and $k = \frac{1}{\sqrt{1-\varepsilon^2}}$. These transformations, together with those that describe rotations, must form a group, and so, said Poincaré, it turns out that necessarily $l = 1$, a conclusion, he observed, that Lorentz had reached in another way. Poincaré's derivation tied the result directly to the fundamental principle of relativity, that two observers in constant relative velocity must make measurements that differ in a way that only takes note of their different situations (positions and velocities).[27]

To derive equations for how the density of an electron and the (electromotive) force are transformed, Poincaré argued that the first set follow immediately from the equations just given, and the spherical electron becomes an ellipsoid, but the second depend on a further property of the electron. If it retains its mass, one set of equations results, if it retains its volume (as Langevin had suggested) another set of

[27] Poincaré had first explained this argument to Lorentz in a letter in May (?) 1905; see Miller (1981, 40–41), where a facsimile of the letter is given.

equations is deduced. Both agree with Kaufmann's experiments, but, said Poincaré, the hypothesis of constant volume did not allow one to deduce the impossibility of showing motion relative to the ether.[28] Therefore Poincaré took Lorentz's side, and argued that it is enough to assume that $l = 1$.

Poincaré now noted that it was also possible to explain the contraction of the electron by supposing it was subject to some sort of constant external pressure doing work in proportion to the volume of the electron. On the assumption that this pressure was totally electromagnetic in origin and that inertia is likewise solely electromagnetic then, said Poincaré, the principle of least action implied that the ether would be forever unobservable. Furthermore, one can suppose, as Lorentz had done, that all forces, including gravity, transform in the way just described. This can be done by assuming that gravity propagates at the speed of light, and then, said Poincaré, a Lorentz-invariant theory of gravity will be the usual one, at least when the speed of the massive bodies are not too great and the square of v/c can be neglected. One can therefore suppose that no astronomical measurements would be capable of testing the discrepancy.

The long paper (1906g) delivers the details. The impossibility of detecting motion with respect to the ether is raised to the status of a law of nature and dignified with the name "the postulate of relativity." However, the original hypothesis of contraction proposed by Lorentz and Fitzgerald had proved insufficient to deliver his postulate, and so Lorentz modified it in his 1904 paper. These Lorentz transformations do imply the postulate of relativity, said Poincaré, but the new Lorentz contraction must then be explained, which shrinks lengths along the direction of motion while preserving them in orthogonal directions.

But the problem is not merely one of explaining something, it involves knowing what has to be explained. Imagine, said Poincaré, an astronomer before Copernicus, who notices that every planet moves on one of two circles, an epicycle or a deferent, that is traced out in the same time. This could not be by chance; it would have to be seen as a mysterious connection between the planets. But Copernicus, simply by making a

---

[28] Later experiments showed that Kaufmann was right in concluding that the mass of the electron depends on its velocity, but wrong in concluding that Abraham's theory of the electron was correct.

new choice of the fixed axes, made this appearance vanish, because each planet now traveled on one circle with no apparent connection between the lengths of their orbits (until Kepler made his discoveries).[29] It is the same here, said Poincaré. If we admit the principle of relativity, the same number—the speed of light—crops up in many branches of physics. This might be because everything in the world is electromagnetic in origin, or it might derive from the way we measure everything, which is based on the assumption that two lengths are equal if it takes light the same time to traverse them. It would be enough to renounce this definition for Lorentz's theory to be as completely overturned as Ptolemy's was by Copernicus, and if that happened one day, it would not prove that Lorentz's work had been a waste of time, because Ptolemy's work had been useful to Copernicus. That said, Poincaré also apologized for bringing forward partial results at a time when the whole theory might be in danger from fresh discoveries in the study of cathode rays (i.e., fast-moving electrons).

To study the effect of a Lorentz transformation on electromagnetic phenomena, Poincaré wrote Maxwell's equations in the potential form, and in units in which $c = 1$. He then observed (p. 499) that if in one frame the equation of a moving sphere is

$$(x - \xi t)^2 + (y - \eta t)^2 + (z - \zeta t)^2 = r^2,$$

then the equation in transformed coordinates described a moving ellipsoid, the shape of which depends on the velocity of the sphere. He next obtained the components of the electric and magnetic fields in the new coordinate frame, noting that his results agreed with those of Lorentz, and observed that Maxwell's equations were still satisfied. Finally he deduced that there was a problem with understanding the stress on an electron, which was regarded by everyone as having a very small but finite size and shape; the stress ensures that the electron is in equilibrium.[30] He also wrote down the new addition law for velocities (p. 500): if two electrons have velocities ξ and ξ' relative to an observer, one has a velocity of $\frac{\xi + \xi'}{1 + \xi \xi'}$ relative to the other.

---

[29] Poincaré's point survives his misrepresentation of Copernicus's theory.
[30] This issue had been raised by Abraham in 1904; see Miller (1973, 230–231) and Miller (1981, 55 et seq), who notes that Poincaré's general Lagrangian framework enabled him to deal with different hypotheses about the shape of a moving electron.

In paragraph 2 of his paper Poincaré stated a version of the principle of least action and used it, as Lorentz had done, to deduce a formula for the pressure on an electron. In paragraph 3 he then showed that a Lorentz transformation leaves the action unaltered, and so reobtained the invariance of Maxwell's equations under Lorentz transformations.

The short paragraph 4 in which Poincaré showed that the Lorentz transformations form a group provided that $l = 1$ is notable for using Lie's method of infinitesimal generators. Poincaré began by considering transformations of the form

$$x' = kl(x + \varepsilon t), \quad y' = ly, \quad z' = lz, \quad t' = kl(t + \varepsilon x), \tag{6.1}$$

where $k^{-2} = 1 - \varepsilon^2$. He showed how they compose, and then that the infinitesimal transformations that generate the group can be found as follows. When $l = 1$, so ε is infinitely small, there is a transformation

$$\delta x = \varepsilon t, \quad \delta y = 0 = \delta z, \quad \delta t = \varepsilon x,$$

which he called T_1. This can be written in Lie's fashion as

$$T_1 = t\frac{\partial \varphi}{\partial x} + x\frac{\partial \varphi}{\partial t}.$$

The transformations T_2 and T_3 are found similarly on replacing x by y and x by z. There is also the transformation found by setting $\varepsilon = 0$ and $l = 1 + \delta l$, when

$$\delta x = x\delta l, \quad \delta y = y\delta l, \quad \delta z = z\delta l, \quad \delta t = t\delta l,$$

which can be written as

$$T_0 = x\frac{\partial \varphi}{\partial x} + y\frac{\partial \varphi}{\partial y} + z\frac{\partial \varphi}{\partial z} + t\frac{\partial \varphi}{\partial t}.$$

These infinitesimal generators give rise to transformations that can all be written as composites of ones of the form

$$x' = lx, \quad y' = ly, \quad z' = lz, \quad t' = lt,$$

and those that preserve the quadratic form $x^2 + y^2 + z^2 - t^2$. For these to form a group, in which l is a function of ε, it follows from a consideration of a rotation of π around the y-axis that equation (6.1) implies that $l(\varepsilon) = l(-\varepsilon)$. But this transformation is its own inverse, and so $l = \frac{1}{l}$, and so $l = 1$.

In paragraph 5 Poincaré used his theory of Lorentz transformations to rederive the Langevin waves that describe the electromagnetic field produced by a single moving electron. In paragraph 6 he considered the much-discussed topic of the Lorentz contraction of electrons. He surveyed various hypotheses: those of Abraham (spherical, undeformable electrons), Langevin (constant volume electrons), and Lorentz (electrons contracted along the direction of motion). He used a Lorentz transformation to pass to the case of a motionless electron, and came up with a theory of stresses on an electron with a variable parameter that left the theory compatible with each of the three hypotheses. Feynman (1966, 28-4) discussed the Lorentz and Poincaré theories, and observed that a purely electromagnetic theory of mass gave inconsistent estimates according as momentum or energy considerations were pursued. The forces involved were handled by calculations based on conservation laws, and Poincaré argued that the forces holding the electron together must be included in these calculations, and in this way a consistent theory is returned—but is no longer purely electromagnetic in character.[31]

In paragraph 7 Poincaré returned to the question of whether the contraction hypothesis makes it impossible to detect absolute motion. He gave a two-stage argument: first the quasi-stationary case when the motion of the electron is slow; second, the general case. In the first argument he recalled Abraham's discussion of transverse and longitudinal mass, and showed that Lorentz's hypothesis was the only one compatible with the impossibility of detecting absolute motion, but offered what he called the true reason: the Lorentz transformations must form a group and for this necessarily $l = 1$. In paragraph 8 he extended this argument to cover the general case, by using the principle of least action. This left the possibility of detecting absolute motion to those phenomena that were not of electromagnetic origin, such as gravitation. Lorentz had, of course, argued that gravitation behaved under Lorentz transformations as the electromagnetic forces do. Poincaré examined this hypothesis by

---

[31] Feynman made an interesting distinction (1966, 28-4) between classical mechanics, which he regarded as a consistent theory that does not agree with experience, and classical electrodynamics, which "is an unsatisfactory theory all by itself." Indeed, he went on, the difficulties remain even after the introduction of quantum mechanics.

considering the effect of a Lorentz transformation on any function of time, position, and velocity such as would arise in any analysis of time. He further assumed that any suitable law of attraction would reduce to Newton's law for bodies at rest, and would not disagree with astronomical observations of slowly moving objects. He looked for invariants under the action of the Lorentz group, and found that if speeds faster than light are allowed then time can pass negatively. He excluded this possibility, and deduced that he was left with the proposition that gravity would travel at the same speed as light, but that further investigations were called for.

The reception of Poincaré's ideas, and indeed Einstein's, on the subject of relativity is strange to modern eyes and has been much discussed by historians.[32] Einstein's ideas were not an immediate success, and only in 1908, with experimental confirmation, did it begin to seem preferable to any of the alternative theories in the field.[33] After Minkowski's account of space and time was published in 1908[34] German theorists such as Sommerfeld swung behind the new theory, and this influenced Langevin in France, who used Sommerfeld's four-dimensional vectors in his lectures at the Collège de France in 1910–11. French mathematicians such as Borel and Cartan then made their way to the new theory. From the boldness of the Lorentz group Poincaré retreated to a Galilean position in keeping with his views about measurement, and eventually lost the enthusiasm for the subject that came to animate the next generation. This is discussed further below in section "Poincaré and Einstein."

POINCARÉ AND EINSTEIN

[Poincaré] left to Einstein the glory of seeing all the consequences of relativity and, in particular, . . . , the true physical character of the relationship the principle of relativity establishes between space and time. (De Broglie in Poincaré, *Oeuvres* 11.2, 66)

[32] See, for example, Darrigol (2004; 2006) and Walter (2009b) and the literature cited there. It is worth noting that there never was a Nobel Prize for the discovery of special relativity.
[33] See Darrigol (2000) for the situation in 1905.
[34] There was also a French translation in 1909.

Poincaré undoubtedly discovered many of the ideas that now form our mental picture of the theory of special relativity and associate with the name of Einstein. In his analysis of the relativistic nature of investigations into space and above all, time (1905h), he discussed how different observers can compare time measurements by exchanging light signals,[35] he called for a new physics in which the speed of light is an impassable limit, and he came up with the Lorentz group—all of this independently of Einstein and mostly before him. He was indeed well ahead of Einstein in speculating about a truly relativistic theory of gravity. Yet in the seven years from 1905 until Poincaré's death the two men showed little sign of recognizing the importance of each other's ideas, let alone any common significance.[36] They did not even agree or disagree about questions of priority, and they did not build on each others' work. It would seem that the two men best placed to know, and who otherwise show relatively little competitive spirit, thought that they had done very different things. How is this disparity to be understood?

It is possible to be partisan in these matters, and it is possible to supply any amount of speculation about influences.[37] Writers about Einstein can, if they wish, marginalize or even ignore Poincaré's papers; the pro-Poincaré faction can behave like lawyers and cut and paste their man's words into a fairly impressive list of insights. It would be more useful to respect a few historical facts. First, and most important, the publication of Einstein's paper on special relativity did not cause a sensation overnight. It did not convert its readers on the spot, there was no instant recognition of its insights: it did not immediately change the way physics was done. Appreciation of its merits began slowly in 1906 with comments by Max Planck, although the first paper on the subject in the French literature came out in 1909 (when Borel translated Minkowski's "Raum und Zeit"). Second, one should take note of the issues as they were formulated by all those involved in the crucial years; not just Poincaré and Einstein, but Lorentz, Abraham, Langevin, and others. The dynamics of the

---

[35] This is how Einstein analyzed the measurement of time in 1905, but Miller does not list this essay by Poincaré among the sources he is confident Einstein was familiar with by the time he wrote his relativity paper.

[36] Poincaré did recommend Einstein in November 1911 for the chair in theoretical physics at the ETH, Zürich, which he took up; see Walter (2008, 232).

[37] For a measured assessment, and a refutation of some of the wilder proposals, see Darrigol (2004).

electron was a major problem for them, with its own store of insights, confusions, ad hoc devices, and mathematics of various kinds. Third, any reconstruction of Poincaré's ideas must somehow take note of the passing of the years: he may have shifted his positions, failed to draw conclusions we see as obvious, looked in other directions.

Physicists proceed by accepting, perhaps provisionally, certain observations as facts. In the early 1900s, these facts included the failure of the most delicate experiments to detect the motion of the earth relative to the ether to the order of $(v/c)^2$. This "fact" is worth taking apart. It is not a brute fact. The relevant brute facts are the nonappearance of interference fringe shifts in certain elaborate and delicate experiments. The fact just cited is an interpretation of those brute facts according to some theory, in which the nature of light and the measurement of time are somehow formalized, and the mysterious ether is somehow to be entwined with the motion of matter. Indeed, by this time the nature of matter, and the idea that it might not be a primitive concept but reducible to electromagnetic phenomena, has also to be entered into the debate. In a confusing subject every investigator has also to consider which facts to accept; perhaps some will prove to be false, or to be capable of other interpretations: Kaufmann's experiments of 1902 and discussed above are a case in point. It seems, for example, that Einstein set little store by the Michelson–Morley experiment.

Poincaré had three professional reasons to care about the measurement of time: his work on geodesy and the determination of longitude; the study of electricity, magnetism, and optics; and his philosophical reflections on the practice of science. From the first he drew the simple observation that when synchronizing clocks in different locations by exchanging signals along a wire it is necessary to allow for the time the signals take to travel. Even if this is the source of his idea that Lorentz's theory of local time could be interpreted that way, it does not break with the idea that there is a sense of time on which all observers can agree, and crucially it does not involve the idea that observers might be in a state of relative motion. His more philosophical reflections capture more of the problems with the concept of time. He pointed out that we cannot compare the duration of two separate intervals of time: an hour today and an hour yesterday, that we cannot naively speak of the simultaneity of distant events, there is no absolute time, and, famously, that all motion is relative. Before we concede that these remarks take us into the domain of special relativity it is worth asking how many of them Newton would have disagreed with.[38]

Newton, in his *Principia*, had proposed a relativistic theory of mechanics, in the sense that his mathematics referred to measured quantities whose motions can only be considered absolute by reference to some body regarded as immovable (and he even admitted that such a body may be fictional).[39] So although he spoke of an absolute space and an absolute time (the one that flows equitably), he distinguished them from the relative spaces and times that we actually measure. He argued that constant relative velocity was equivalent to a state of rest, but acceleration was absolute, and as for time,[40]

> Absolute, true, and mathematical time, in and of itself and of its own nature, without reference to anything external, flows uniformly and by another name is called duration. Relative, apparent, and common time is any sensible and external measure (precise or imprecise) of duration by means of motion; such a measure—for example, an hour, a day, a month, a year—is commonly used instead of true time.

He even allowed: "It is possible that there is no uniform motion by which time may have an exact measure. All motions can be accelerated and retarded, but the flow of absolute time cannot be changed" (p. 410).

Without descending into philosophy, any physicist in any branch of physics must wrestle at some stage with the same conundrum. It is easy to invoke some mathematical abstractions, anchor them loosely in some observations, and draw some useful conclusions: mass is the quantity of matter in an object, time is measured by this sensitive watch, distances by this contraption, here are some selected experiments— please observe the tidy agreement between the experimental data and the formulas involving m, t, and s. Work on heat or electricity is handled similarly. Several of Poincaré's essays begin by observing that these sorts of arguments are too glib, and the relation between abstract concept and a measurable concept is problematic. But they do not end in despair at the impossibility of ever knowing about the "true" values of the

---

[38] The analysis below agrees with DiSalle (2002, 33–56), and Rynasiewicz (2008).
[39] Newton, *Principia* (p. 411).
[40] Newton, *Principia* (bk. I, scholium to definition 6).

physical quantities: instead Poincaré drew attention to the conventional element involved in allowing us to pass to talk about mass, time, distance, temperature, charge, and so forth.

What is the difference between saying time is absolute, but we only have measurements of duration obtained from instruments that may be flawed, and saying time is not absolute but may proceed as if it is by subscribing to a convention? Generations of heated debate notwithstanding, the answer for physics (as opposed to metaphysics or theology) may be: very little indeed. Newton recognized that some assumption was needed to pass from his mathematics to the observations he so diligently demanded from professional astronomers; Poincaré did the same. Newton, with his absolute concept of time, could speak of distant events happening at the same time, and the equal length of different hours (which is what he meant by it flowing equitably). But in practice this was made possible by the existence of excellent clocks whose measurements fitted so closely to his mathematics. Poincaré, when discussing Roemer's analysis of the motion of the moons of Jupiter, likewise remarked that the astronomer's way of resolving these philosophical issues is the simplest and the only one with any merit. Where there may be a valuable distinction is when we leave Newton's era and come to the early 1900s, when there is a genuine issue about the nature of time, but the difference may only be a pragmatic or a psychological one. Someone with a belief in absolute time but confronted with a serious failure to match experiment and theory might be less willing to settle for very different measurements of time than someone who firmly believes there is no absolute time anyway. But a lot would depend on how tightly such a person would adhere to the idea that some convention will secure the passage back to a concept of time that we can all not only share but determine.

The conclusion is that there is not such a large philosophical gap between Newton's ideas and those of Poincaré. When Poincaré said that we cannot compare the duration of two separate intervals of time, or speak of the simultaneity of distant events, he was not saying that statements of that kind were simply nonsense. He was saying that we have to recognize that when we make these remarks we do so by means of a convention amply supported by experience and which could, just possibly, be revised. When he said there is no absolute time he certainly wanted to say precisely that, and he also wanted to alert his readers to what he felt was a pervasive problem in what he sometimes called modern metaphysics: the dangers of talking about a concept without explaining how it is measured. It follows that Poincaré's ideas belong to a tradition of reflecting on the practice of science, and in particular how its fundamental concepts are measured on the one hand, and formulated mathematically on the other. His conventionalism was formulated before fast-moving electrons posed their deepest challenges to contemporary physics, and there is no reason to label his ideas about time "proto-Einsteinian."

Nonetheless, they are the ideas that he brought to the many problems posed by the electron. Poincaré was a late convert to the idea of electrons, or any other physical embodiment of electricity, as the long quotation from Maxwell at the front of *E&O* attests. He was not keen on explanations based on the supposed fundamental properties of objects it was impossible, or almost impossible, to detect, and preferred to argue from certain explicit generalities. This temperamental difference with Lorentz runs deeply into debates about how physics is done. Certainly, to count as a theory of *physics*, any theory must make some statements about the physical world which it will take for granted, some of which may turn out to be the explanations for others. But it is perfectly possible to build up special relativity with a minimum of statements about light (it travels in straight lines at a speed independent of source and observer) and to deduce time dilation, length contraction, the dependence of mass on velocity, and $e = mc^2$, and to wind up saying that objects with mass cannot travel as fast as light because their mass would then become infinite and photons do travel at the speed of light because they have no mass. Or, as some physicists do, to take the dependence of mass on velocity as a fundamental physical fact, add other facts of that kind, and spin the story of special relativity out of them.

Poincaré chose to aim for a rigorous derivation of Maxwell's laws of electrodynamics that would fit Lorentz's explanation of electricity and magnetism in terms of electrons, which he thought of as having a shape and to be spherical when at rest, and which would invoke the ether and its interactions with matter. He wanted a theory that was thoroughly relativistic in a sense Newton would have recognized, one that saw no fundamental distinction between rest and any uniform motion. It would presumably be necessary to ascribe some properties to electrons in motion, and some properties to the ether. Lorentz's theory of 1904 had taken the eponymous contraction to be real, and to apply to electrons, and relied on his theory of local time. Poincaré now regarded the principle of relativity as implying that it would always be impossible to detect motion with respect to the ether, and showed that Maxwell's equations were invariant under a group of transformations that could be made to form a group (later called the Lorentz group). To get from the electrons to an electromagnetic field it is necessary to know the density of the electrons in the body, which depends on their speed, because of the Lorentz contraction and its effect on the shape of the electron. Lorentz and Abraham agreed; the only dispute was about the details.

The Lorentz group is an inspired discovery. It eliminates the dubious mathematics Lorentz used, it turns Lorentz's mathematical contrivance into a plausible definition of local time by making it the time on an moving observer's clock, it makes it possible to ignore the ether. It is in this spirit that Lorentz in his (1915/1921, 251) commented that he had not thought to describe the phenomena "in the system x', y', z', t' in *exactly the same manner* as in the system x, y, z, t" and that as a result he had not found what Poincaré did: "the perfect invariance of the equations of electrodynamics" (p. 252).

But it may not have signified to Poincaré what it does to anyone familiar with special relativity today. To see this, consider what, if anything, is the difference between saying

1. the length of this moving rod is such and such, and

2. the length of this moving rod appears to be, or we measure it to be, such and such.

Poincaré might have argued that there is no real distinction between these two remarks, because the meaning of the first statement is that "if you measure the length, you will find it to be such and such." But the discussion of the shape of the electron is consistent with a more naive reading: the moving rod shrinks. This confusion is at the heart of many of the familiar paradoxes in special relativity that later authors delighted in finding, such as the very fast moving train heading for a gap in the rails. The train driver sees the gap greatly shrunk by the phenomenon of length contraction, and predicts a smooth ride. But a man standing by the track sees a greatly contracted train, seemingly much smaller than the gap, heading for disaster: who is right? The train driver, happily, inasmuch as there is no accident, because the man by the track has incorrectly understood what it is for the front and back of the train to be over the gap at the same time. The paradox works so strongly upon us because it appeals to our sense that the man by the track is "really" at rest, and his observations and predictions are the true ones.

That Poincaré may have been prey to this confusion is suggested by several remarks. For example in (1905e, 491) he wrote,

> But with Lorentz's hypothesis the agreement between the formulae is not all; one obtains at the same time a possible explanation of the contraction of the electron, by supposing that *the deformable and compressible electron is subjected to a sort of constant exterior pressure whose work is proportional to the changes in volume*. (Poincaré's italics)

He indicated that he could derive this result from the principle of least action if certain assumptions about the nature of electromagnetism were made, and went on, "One thus has the explanation of the impossibility of showing absolute motion and the contraction of all bodies in the direction of the motion of the Earth." In the next paragraph Poincaré discussed how Lorentz went on to claim that all forces are effected by a uniform motion in the same way. In the major paper (1906g, 496), Poincaré said that "in order to preserve Lorentz's theory and free it of some intolerable contradictions, it is necessary to suppose a special force" that explains the contraction, and when Poincaré came to discuss the contraction of electrons (in para. 6) he argued that a moving electron could be studied by using a Lorentz transformation to obtain an equivalent ideal immobile electron. Now, he said, "because the energy is principally located in the parts of the electron closest to the centre," the electron can no longer be considered as a point, so while the ideal immobile electron is a sphere, moving electrons are deformed into ellipsoids. This led him to consider what forces hold the electron together.

Poincaré's language is not thoroughly relativistic; he seems to want to refer back to the observer at rest, who is a privileged observer. The Lorentz contraction is real, and it explains why the Michelson–Morley experiment gave a null result. It is as if the poor moving electrons are doing work battling through the ether, which appears as some sort of a head wind.

The comparison with Einstein's approach at this point is instructive. One of his objections to the old theory was that it offered asymmetrical explanations of what happens in situations that ought, on relativistic grounds, to be symmetrical. His paper of 1905 begins:

> It is known that Maxwell's electrodynamics—as usually understood
> at the present time—when applied to moving bodies, leads to
> asymmetries which do not appear to be inherent in the phenomena.
> Take, for example, the reciprocal electrodynamic action of a magnet
> and a conductor. The observable phenomenon here depends only
> on the relative motion of the conductor and the magnet, whereas
> the customary view draws a sharp distinction between the two
> cases in which either the one or the other of these bodies is in
> motion. For if the magnet is in motion and the conductor at rest,
> there arises in the neighbourhood of the magnet an electric field
> with a certain definite energy, producing a current at the places
> where parts of the conductor are situated. But if the magnet is
> stationary and the conductor in motion, no electric field arises in the
> neighbourhood of the magnet. In the conductor, however, we find
> an electromotive force, to which in itself there is no corresponding
> energy, but which gives rise—assuming equality of relative motion
> in the two cases discussed—to electric currents of the same path
> and intensity as those produced by the electric forces in the former
> case.

This led Einstein to derive a physics that raised what he called the
principle of relativity to the status of a postulate, and to make explicit the
assumption that "light is always propagated in empty space with a definite
velocity c which is independent of the state of motion of the emitting
body" (1905, 279). He then established the coordinate transformations
between what he called a stationary and a moving system—but did he
not make explicit that they form a group (he only considered observers
moving in the same direction, and noted that these transformations
form a group). But, significantly, he established these transformations
on entirely kinematic, not dynamical, grounds. He also noted that two
observers moving with uniform relative velocity will agree that a point
source of light emits a spherical wave front that satisfies

$$x^2 + y^2 + z^2 = c^2 t^2$$

in each coordinate frame, and remarked that this showed that his two
fundamental principles were in agreement. He also added a footnote

to say that the equations of the Lorentz transformation may be derived more simply from this invariance than the method he used in the text. Next Einstein considered the physical meaning of his equations, deduced that a moving clock will appear to run slow, and gave the addition law for velocities in the same direction that Poincaré had done.

Only now did Einstein take up electrodynamics and show that Maxwell's equations were invariant under Lorentz transformations. Then he dealt with aberration, radiation pressure, deduced that all problems in optics may be reduced to the optics of stationary bodies, and concluded with a discussion of the dependence of mass on velocity.

Einstein observed that a clock moving with constant speed and taken on a closed path that returns to its starting point will show a different time to a clock kept at the starting point. Later writers souped this up to the famous twin paradox. Poincaré seems never to have thought of it, although it is implied by the Lorentz transformations, and it leads one to ask how he regarded Lorentz's local time.

An answer comes from his lectures at the Sorbonne, which were given in 1906–07 and published as *Les limites de la loi de Newton* (1953) from notes taken by Henri Vergne. The penultimate chapter is on the dynamics of the electron, and this is the only place in Poincaré's writings where he enters into the formulas for special relativity. Here he once again explains that two observers in uniform relative motion can only keep their local times. Then he said that Lorentz hypothesized a contraction of bodies along the direction of their movement, and asked, since every physical body obeys this law, how can it be observed? He defined a unit of length in terms of the time it would take light to travel along it. Then he produced an argument to show that a moving and a stationary observer will agree on the shape of a spherical wave front, because the Lorentz contraction of length is exactly compensated for by a dilation in time, and so we will be forever unable to detect motion with respect to the ether.

The crucial point is that Poincaré always compared true time with apparent time, and never took the step that Einstein did to regard time on a clock as the only time there is, with all the paradoxes which that entails. It has been suggested that one reason Poincaré may have failed to appreciate Einstein's work is that he was always looking for a dynamical explanation for the Lorentz contraction, whereas Einstein gave it a kinematical one. Damour quotes the following anecdote:

Poincaré and Einstein met only once, at the Solvay meeting of 1911. Maurice de Broglie (who was one of the secretaries of this first Solvay meeting) wrote... the following: "I remember that one day in Brussels, as Einstein was expounding his ideas [on the new, that is to say, relativistic, mechanics], Poincaré asked him: 'What mechanics do you adopt in your arguments?' Einstein replied 'No mechanics,' which seemed to surprise his questioner." This conversation on "relativistic mechanics" (which contrasts the "dynamical approach" of Poincaré to the "kinematical" one of Einstein) is not reported in the official proceedings of the 1911 Solvay meeting which concerned the (old) theory of quanta. (Damour 2004, 156)

On the other hand, in 1905 at least it is clear that Poincaré saw significant implications for the theory of gravity that Einstein did not: gravity would propagate at the speed of light, and if all matter is electromagnetic in origin there were quantifiable effects on the orbit of Mercury.

In his address in 1954, de Broglie said that it was "without doubt his a little too hypercritical turn of mind, due perhaps to his having first been a pure mathematician, that was the cause" of Poincaré not taking the decisive step that Einstein did (in Poincaré, *Oeuvres* 11.2, 66). But the evidence would seem to point to Poincaré's deep-seated interest in philosophy. For him the nature of space and time was not a matter of science alone, but of epistemology, with its attendant element of conventionalism. In this connection, an interesting perspective is afforded by an observation drawn from the discovery of non-Euclidean geometry. It is difficult to think about space, especially three-dimensional space, without familiar Euclidean ideas entering one's mind. But it is not so hard to think about x, y, and z, what formulas for distances between points might look like, and what novel trigonometric formulas might look like: one can put one's trust in neutral letters and only later impose a geometric meaning on them. In some measure, this is what Bolyai and Lobachevskii did. In the case of space and time it is what Einstein did in his paper of 1905. Minkowski, however, when he reworked those ideas in 1908 was much more explicit that the nature of time and space had changed: "Henceforth space by itself, and time by itself, are doomed to fade away into mere shadows, and only a kind of union of the two will preserve an independent reality." For Poincaré, who was committed to the epistemological aspects of physics, to have grasped the full implications of special relativity he would have to have been not Einstein, but Minkowski. That is surely asking too much of anyone in 1905.

EARLY QUANTUM THEORY

In the early 1900s Planck was interested in attempts to specify the radiation emitted by a body as a function of its temperature.[41] This was a surprisingly difficult topic that drew heavily on electromagnetic theory, thermodynamics, and statistical mechanics. To obtain a law that fitted experimental results over a large range of temperatures, Planck had been driven to suppose that a heated body must be considered as a system of Hertzian resonators that emit energy in discrete amounts— later to be regarded as multiples of the basic quantum of energy. This quantum varies with the resonators, and is inversely proportional to the wavelength of the emitted light. This led Poincaré in his (1912g), to explain how energy is exchanged between these resonators. It cannot be done directly, because the wavelengths of the two resonators will in general be incommensurable. It can be done by radiation or by collision, and Poincaré found that for the recently discovered Wien's law to be true the quanta involved when energy is exchanged by collision must be inversely proportional to the wavelength, which he said was a rather mysterious property of the resonators. Moreover, this aspect of the theory, which Planck had left unanalyzed, seemed to imply there should be a minimum possible intensity of light, but light less intense than this had been observed. Later in the lecture, Poincaré noted that Planck had modified his initial account. There he had supposed that emission of quanta was discontinuous, and so was absorption, but this ran into problems with the varied nature of the resonators. Now he proposed that absorption was continuous, but Poincaré found this to be confusing.

As was his custom, Poincaré based this lecture on a more detailed analysis that he published elsewhere (1912e). In this paper he began by considering the exchange of energy between two resonators with different

---

[41] See Kuhn (1978).

periods. The state of the system will be represented by a point in the plane where a certain probability measure has to be defined. Poincaré showed how to do this in principle, starting from an analysis of the behavior of each resonator separately, but observed that there was no point in taking the analysis of such an absurdly unrealistic system too far. He then embarked on a lengthy analysis of the case of a large number of resonators. For it to be possible that states of thermodynamic equilibrium can occur, he remarked, it must be the case that a probability measure is obtained that is independent of the number of resonators, and the distribution of their periods. This he found to be unlikely, but he could show that the measure was independent of the ratio of the numbers of resonators of the two kinds.

Now Poincaré took up Planck's hypothesis that the energy of a resonator must be a multiple of the quantum of energy, and deduced Planck's law. This left him dissatisfied, because the nature of the partition law still rested on an ad hoc hypothesis, and it was still possible that Planck's law could be derived from other hypotheses. To explore further, Poincaré turned to the methods of Fourier integrals, and by expressing the probability as a Fourier integral and then inverting the integral he was able to conclude that it was only the quantum hypothesis that led to Planck's law, and also that if the total radiant energy is finite only a discontinuous probability law would work (1912i, in *Oeuvres* 9, 649). Here Poincaré observed that he had had to rely on classical electrodynamics at one point in his argument, which was, of course, unsatisfactory and indicative of the trouble the subject was in.[42]

The underlying mathematics shows how radical Poincaré was prepared to be. If you strip out the meaning Planck attached to the symbols in his theory of blackbody radiation, you are left with a system of n variables $x_1, x_2, \ldots, x_n$ evolving in time according to a system of n equations

$$\frac{dx_j}{dt} = X_j(x_1, x_2, \ldots, x_n). \qquad (6.2)$$

Such equations are ubiquitous in applied mathematics; for example, the Hamiltonian theory of dynamics, which Poincaré had used in his study

---

[42] For an elucidation, see Lorentz (1915/1921, 256).

of celestial mechanics, is of this form. As he knew very well, there is an extensive theory of such things. In particular, Jacobi had introduced a function in 1844 called the last multiplier, whose role is to prove the existence of an invariant integral of the system. Indeed, for Hamiltonian systems, but not a general system of the form of (6.2), the last multiplier turns out to be a constant, and this implies a theorem due to Liouville which says that a suitable expression in $\sum_j dp_j\, dq_j$ is invariant, which means that the area of any region is unaltered by the flow. There was also a theory of when and how an arbitrary system of the form (6.2) can be reduced to Hamiltonian form, in which a nonconstant last multiplier plays a role.

The last multiplier for the system (6.2) is a function $W(x_1, x_2, \ldots, x_n)$ which satisfies the partial differential equation

$$\sum_j \frac{\partial(WX_j)}{\partial x_j} = 0. \tag{6.3}$$

So, if it turns out that W is a constant then the continuity or volume-preserving condition holds for the given flow. Now, (6.3) is also the equation which says that the expression $Wdx_1 \cdots dx_n$ is a measure of the probability that the system is in a region of $(x_1, x_2, \ldots, x_n)$ space. Poincaré, at this point in his essay (1912g, 121–122), gave this long quotation from Planck:

> The probability of a continuous variable is obtained by considering independent elementary domains of equal probability. . . .In classical dynamics, in order to find these elementary domains, the theorem is used which affirms that two physical states, in which one is the necessary effect of the other, are equally probable. In a physical system, if one of the generalized coordinates is represented by q and the corresponding moment by p, according to Liouville's theorem the domain $\int\!\int dpdq$ taken at any instant whatever is an invariant with respect to time, if q and p vary in conformity with Hamilton's equations. Moreover, p and q can, at a given instant, assume all the possible values independently of one another. From which it follows that the elementary domain of probability is infinitely small, of the size of $dpdq$. The new hypothesis must have as its objective to limit the variability of p and q, in such a way that these variables no

longer vary except by jumps, or that they are considered as linked in part to one another. We thus succeed in reducing the number of elementary domains of probability so that the range of each of them is increased. The hypothesis of the quanta of action consists in supposing that these domains, all equal to each other, are no longer infinitely small but finite and that for each of them $\iint dpdq = h$, h being a constant.

He then commented, "The theory does not yet satisfy the mind. It is necessary besides to explain why the quantum of a resonator is in inverse ratio to the wave length. And this is what induced Mr. Planck to modify the method of setting forth his ideas." He therefore set out to interpret the function W, and succeeded in showing that for the system to describe blackbody radiation the function W is necessarily discontinuous, and so the system can only be a number of discrete states separated by jumps.[43]

---

[43] See the succinct account in Dugas (1988, 622–626), where the quotation is, however, mistakenly ascribed to Poincaré.

7

Theory of Functions and Mathematical Physics

FUNCTION THEORY OF A SINGLE VARIABLE

The core discipline in French mathematical education was that of analysis, particularly the theory of ordinary and partial differential equations, but also of Fourier series and the evaluation of integrals. It covered everything from the rigorous foundations of the calculus after the manner of Cauchy, to applications of analysis to geometry and mechanics. Originally analysis meant the study of real functions of a real variable with only modest excursions into functions of a complex variable, but complex or analytic function theory was one of the great growth areas of mathematics in the 19th century. Indeed, it grew much faster, and more systematically, than did real function theory, contrary to the preferences of 20th-century mathematicians. It was given a boost by Abel's and Jacobi's discoveries of elliptic functions in the 1820s. These complex functions are rich generalizations of the familiar trigonometric functions, and on the one hand they proved full of uses in many branches of mathematics, from number theory to mechanics, while on the other hand they called out for a grounding in a theory of complex functions, one that was supplied in the 1840s and 1850s. In the 1820s, and still more in the late 1840s and 1850s Cauchy developed the foundations of complex function theory and gave it many of the theorems that still shape the subject today. The subject was further developed by Riemann in the 1850s and early 1860s and, after his death, by Weierstrass, who began his long professorship in Berlin in his forties. In France, Hermite was the acknowledged expert in this branch of mathematics and its implications for Gaussian number theory, as was his counterpart in Berlin, Leopold Kronecker.

For the French, the core of this theory was the idea that a complex function of a complex variable is analytic in a domain if its real and imaginary parts satisfy the "Cauchy–Riemann" equations. Cauchy had proved that an analytic function has a convergent power series expansion on any disk within that domain. The converse is straightforward, so the result was that complex analytic functions and convergent power series were more or less identified. For those in the Weierstrass empire that came to dominate the subject, the very definition of an analytic function was as a convergent power series on a disk, but by Cauchy's theorem the higher reaches of the theory were similar on either side of the Rhine.

Weierstrass endlessly reworked his approach to the subject for the start of his two-year, four-semester lecture course, and this involved him in thinking through ever more precisely and algebraically what the key features of the subject should be. Among his important contributions was a sharp distinction between a so-called finite pole of a function and an essential singularity. The function $f(z)$ has a pole of order n at a point z_0 if $\lim_{z \to z_0}(z - z_0)^n f(z) = 0$ but $\lim_{z \to z_0}(z - z_0)^{n-1} f(z)$ is infinite. The function has an essential singularity there if no such n can be found. He showed that in the first case the function may be said to tend to infinity as z tends to z_0, but in the second case the function gets arbitrarily close to any value at all as z tends to z_0. He was the first to show that there is an analytic function with any prescribed set of points where it takes the value zero, provided only that that set of points is suitably scattered (precisely: that it has no accumulation point). He also gave what became the standard account of elliptic functions. Central to his investigation was the idea that a function may be extended beyond its initial disk of convergence by a process called analytic continuation onto a family of overlapping disks until its entire domain of definition is established and the points beyond which it cannot be continued are defined; these form what he called the natural boundary of the function.

It can happen that some or all of the boundary of the disk of convergence is part of the natural boundary of the function, in a way that in principle is determined by the coefficients of the original power series. It might seem that this would have been a natural topic for mathematicians in Weierstrass's orbit to take up, but in fact French mathematicians were the first to do so. They took their lead from Edmond Laguerre, an examiner at the École polytechnique who, in 1883, had also become a professor of mathematical physics at the Collège de France, and they brought to the subject another concern that turned out to be related, which asked for estimates of how fast a complex function grew as $|z|$ increased. This in turn connected with another mysterious topic, one that, as it happened, Poincaré left alone: the Riemann zeta function.

Laguerre, who had a liking for small, precisely focused papers, had noticed that some transcendental functions grew little faster than polynomials, so much so that the theorems of Rolle and Descartes about the location of their zeros applied to them. For example, Riemann's ξ function, which is closely related to the zeta function, is of this kind.[1] Laguerre looked at Weierstrass's representation of a function with arbitrary zeros, and more particularly at their so-called primary factors, which are functions of the form

$$E(u, 0) = 1 - u, \quad E(u, n) = (1 - u) \exp\left(u + \frac{u^2}{2} + \cdots + \frac{u^n}{n}\right),$$

and in three short notes on the matter in 1882 (and one in 1884) he singled out the primary factors of the form $(1 - \frac{x}{a})e^{x/a}$ which he called of "genre" 1, those of genre 0 being functions with no exponential factor. He also showed how one might determine the "genre" of any given entire function (one with no poles).

Poincaré now took up the subject. He defined a function to be of genre n if its primary factors were of the form $e^{P(x)}\left(1 - \frac{x}{a}\right)$ where $P(x)$ was a polynomial of degree n (Poincaré 1882h). He investigated the rate of growth of functions of genre 0, and showed that if F is of genre 0 and $e^{\alpha x}$ tends to 0 along a ray, then it tends to 0 more strongly than F tends to infinity; or, even more shortly, that $e^{\alpha x}$ dominates F. But he noted with some regret that this and some other properties he presented did not characterize functions of genre 0, and had to content himself with observing that if F was of genre n, then $\exp\left(\alpha x^{n+1}\right)$ dominated F.

Worse, as he went on to observe in Poincaré (1883a), it seemed very difficult to establish such basic results as:

1. the sum of two functions of genre n is also of genre n;

2. the derivative of a function of genre n is also of genre n.

Indeed, he said, one could not be sure that the results were true. His instinct was sound, for in 1902 Pierre Boutroux showed that pairs of certain types of function of genre n had a sum of genre $n + 1$.

---

[1] Hadamard (1893) gave the first rigorous proof that Riemann's function ξ is indeed of genre 0 as a function of z^2. Riemann had offered only an insufficient proof of this assertion.

Blocked in this direction Poincaré turned aside, publishing in April on another topic pioneered by Weierstrass, the theory of lacunary spaces, and in May on the uniformization theorem. Functions with lacunary spaces had only recently been discovered by Weierstrass. They are functions analytic on a domain and whose natural boundaries surround a hole, which means that the function cannot be continued analytically into the hole. Poincaré had encountered them in his doctoral thesis and in his work on Fuchsian and Kleinian functions, many of which cannot be analytically continued beyond the unit disk. They are not hard to construct; elsewhere Poincaré gave the example $\sum_{j=1}^{\infty} \frac{a_j}{z-b_j}$, where the a_j are chosen so that $\sum |a_j|$ converges and the b_j are chosen to be a dense set of points on an arbitrary closed curve. The sum defines a function inside and outside the region bounded by the curve, but has this curve as a natural boundary. It was natural to wonder if in some sense two functions, one defined on one side of such a boundary and one on the other, were not somehow the analytical continuation of the other across the boundary, and Weierstrass had shown that this was not necessarily the case.

In this paper (1883c), which Hermite sent on Poincaré's behalf to Mittag-Leffler for publication after impressing on him that he already considered Poincaré to be a veritable genius,[2] Poincaré considered the situation where one has two functions f and g, where f is defined only on the upper-half plane and g only on the lower-half plane, so the domain of definition for the one is a lacunary region for the other and each has the real axis as a natural boundary. In a highly counterintuitive move, he showed that it was possible to define two new functions φ and ψ on the whole plane such that in the upper-half plane $\varphi(z) + \psi(z) = f(z)$, in the lower-half plane $\varphi(z) + \psi(z) = g(z)$, the function φ is single valued in any domain not containing the line segment $(-1, 1)$, and the function ψ is single valued in any domain not containing the line segments $(-\infty, -1)$ and $(1, \infty)$. Moreover, and Poincaré said he wished to insist on this point, you might imagine that when functions φ and ψ exist that are defined on the whole plane and satisfy $\varphi(z) + \psi(z) = f(z)$ in the upper-half plane then they must also satisfy $\varphi(z) + \psi(z) = g(z)$ in the lower-half plane—but this was not true. The theorem holds even when the function f cannot be analytically continued below the real axis and the function g is entirely arbitrary.

[2] See Hermite (1984, vol. 5, p. 110).

The Poincaré–Volterra Theorem

Many-valued functions were, and to some extent remain, a particular feature of complex analysis. They are exemplified by the square and nth root functions, $f(z) = \sqrt{z}$ and $f(z) = \sqrt[n]{z}$, and the logarithm function $\log(z)$, which takes infinitely many values, differing among each other by integer multiples of $2\pi i$. In Weierstrass's theory of analytic continuation it is natural that a chain of disks may overlap the starting point. When this happens the question arises of whether the continued function takes the same value as the original one at the same point, and if the function is many valued the answer can be negative. This leads on to the question, how many values can it take at a point? Weierstrass himself raised this question in a letter to his former pupil and would-be acolyte Schwarz on 14 March 1885 when he asked: "Given an arbitrary analytic function $f(u)$ could one prove that for each value of u that the corresponding values of $f(u)$ form a countable set so that they could be arranged in a series? No doubt the answer is affirmative," he added. He also claimed to be able to give a proof that drew on the theory of minimal surfaces that he and Schwarz had created some twenty years before. In a rare flash of appreciation for geometry, which he increasingly distrusted, he told Schwarz that by means of such surfaces he had made the whole family of the pairs $(z, f(z))$ "much more intuitive to my mind than through the Riemann's surface"; but he immediately observed that it would "certainly be possible to prove the theorem without such help," for example by assuming that Poincaré's uniformization theorem of 1883 was correct.

In saying that a countable set of values might be possible, but not an uncountable one, Weierstrass was appealing to the theory of infinite sets of another of his former pupils, Georg Cantor. Indeed the result had first been communicated by Cantor to Weierstrass "many years before," as Cantor explained in 1888 in a letter to the Italian mathematician Vivanti.[3] Some months later, on 26 August, Cantor repeated the claim in a letter he wrote to another Italian mathematician, Vito Volterra, who was professor of mechanics at Pisa.[4] The occasion of the letter was Volterra's recent publication of the result, albeit with an incorrect proof. Vivanti published his proof in his (1888a, b), saying (p. 150) that the theorem

---

[3] Cantor's letter is published in Ullrich (2000, 381–382).
[4] Published in Israel and Nurzia (1984, 176).

"had recently been communicated" to him by Cantor who had also invited him to produce a proof. His argument rested on the assumption that the sheets of the Riemann surface of a monogenic function that meet at a branch point are countably many in number and then an appeal to "the basis of the principles of set theory." This connected the theorem with Poincaré's uniformization theorem, and both Poincaré and Volterra responded quickly. Poincaré's letter to the editor of the *Rendiconti* (1888), was published in the same volume of the journal as Vivanti's paper, and gave a proof based on Weierstrass's definition of analytic functions by means of power series and analytic continuation through disks. Volterra independently gave a similar argument that led to a detailed, rigorous proof of the theorem.[5] This explains why a theorem that ought to be attributed to Cantor is known today as the Poincaré–Volterra theorem.

Asymptotic Analysis

Poincaré had already noticed in 1884 (see chap. 4, sec. "Stability Questions" above) that divergent series can be surprisingly meaningful in astronomy. The classic example of this phenomenon that is invoked to study what is today called the theory of asymptotic series, arises in the study of the Gamma function, which extends the factorial function to a complex function: $\Gamma(n) = (n-1)!$. Stirling's series for $\log \Gamma(x)$ is a divergent series for all values of x that nonetheless provides excellent approximations to the value of the Gamma function for large values of x. It is

$$\log \Gamma(x) = \frac{1}{2} \log(2\pi) + \left(x - \frac{1}{2}\right) \log x - x + \frac{B_1}{1 \cdot 2x} - \frac{B_2}{3 \cdot 4x^3} + \frac{B_3}{5 \cdot 6x^5} - \cdots,$$

where the B_k are the Bernoulli numbers. This series diverges, yet for each n, if S_n denotes the sum of the series up to and including the term in $\frac{1}{x^n}$, the expression $x^{n+1}(\log \Gamma(x+1) - S_n)$ tends to 0 as x increases.

Poincaré found such series occurring as the formal solutions in a neighborhood of $z = \infty$ of linear ordinary differential equations that were not of the Fuchsian type. A simple example is provided by

---

[5] For a detailed account of the story see Israel and Nurzia (1984) , which includes the relevant letters between Volterra, Vivanti, and Cantor.

the differential equation $x^2\frac{dy}{dx} + y = 0$, which formally has the solution $y = e^{1/x}$, a function with an essential singular point at the origin. He decided to define a divergent series $A_0 + \sum_j \frac{A_j}{z^j}$ as representing a function $J(z)$ asymptotically, or as an asymptotic series, if S_n, the sum of the first $n+1$ terms, was such that $z^n(J - S_n)$ tends to 0 as z increases. It follows that $J - S_{n-1} = \frac{A_n + \varepsilon}{z^n}$, where A_n is finite and ε very small. He showed (1886c) that such series can be added and multiplied together according to the usual rules for convergent series, and that they can validly be integrated term by term. He then applied these ideas to linear differential equations, and gave conditions for these asymptotic series to represent the solutions asymptotically. He showed that if an asymptotic representation is valid for all values of the argument of z it is in fact convergent, and that for any given fixed value of the argument of z the asymptotic series representing a given function is unique if it exists at all.

Poincaré's paper was criticized by Thomé, who claimed in his (1887) that Poincaré's results were scarcely of interest, arguing by means of an example that the exponents in the formal divergent expansions could be completely different from the ones in the actual expansions in series. Poincaré replied (1887b, 334) that "when facing two equally unsolvable problems it is not useless to show that they reduce to each other."

Whatever its merits in the theory of differential equations, Poincaré's intervention seems to have turned existing ad hoc techniques into a theory of asymptotic expansions, and this helped the topic of divergent series to be dealt with using rigorous mathematics, as Borel's (1901) demonstrates. It is also true, as Lord Rayleigh and others pointed out, that the study of optics, and especially diffraction, continued to generate delicate problems involving the estimation of integrals involving a large parameter for which it has been argued (Berry 1989) that Poincaré's methods are less suitable than those of Stokes. In the early 20th century the idea of asymptotic expansions was enriched with the theory of complex integration (see Debye 1909) and applied to diffraction problems, and it has been valuable ever since.

Differential Equations

Poincaré's interest in ordinary differential equations was an early feature of his career, and it led to his first two spectacular successes, where he introduced a new field of study in the real variable case and a greatly enriched perspective in the complex case. His interest in the more straightforward approach of complex function theory was however intermittent and less successful.

In what was only his second publication (1878), the only one for 1878, he discussed the solution of a first-order differential equation of the form $x^m \frac{dy}{dx} = f(x, y)$ in the manner of Briot and Bouquet. They had observed that if the coefficients of the equation are holomorphic in the neighborhood of a point then so are the solutions, and had then considered what happens when the coefficients are singular. In these cases the solution may be a holomorphic function y of x or of $x^{1/n}$, or matters may be more complicated. More precisely, if the singular point is at the origin, it might be that the equation can be written as

$$x \frac{dy}{dx} = f(x, y),$$

where $f(0, 0) \neq 0$, or where $f(0, 0) = 0$, or that the equation can be written as

$$x^m \frac{dy}{dx} = f(x, y),$$

for some integer m. They concentrated on the first case, which they argued was the general one, and showed that much depended on the value, λ, of $\frac{df}{dy}$ at the origin, or rather, upon its nature as a positive integer, or as a positive rational, or as a complex number with positive or negative real part. For example, if λ is not a positive integer then they showed that there is a holomorphic solution that vanishes at the origin.

They did not study the nonholomorphic solutions that arise, and when Poincaré looked at them he found that when λ is not a positive integer but does have a positive real part then the solutions of the equation $x \frac{dy}{dx} = f(x, y)$ are holomorphic in x and x^λ, and when λ is a positive integer the solutions are holomorphic in x and $x \log x$. The proof he gave is a good if unremarkable exercise in manipulating power series and demonstrating convergence, and it shows only that in 1878 Poincaré was still largely unaware of what German mathematicians were doing.

In 1888 the Académie des sciences elected Hermite, Jordan, Poincaré, Darboux, and Picard to administer their Grand Prize for 1890, and they decided to try to revive interest in an old paper of Darboux's on first-order, first-degree differential equations. In 1878 Darboux had investigated algebraic equations of this kind, which are of the form

$$\frac{dy}{dx} = f(x, y),$$

where $f(x, y)$ is a quotient of two polynomials in x and y. When in 1890 Picard reported on their decision, he praised the winner, the young Paul Painlevé, for being thoroughly familiar with the theory of functions, and this was very apparent from the memoir, which sprawled in five parts over the *Annales de l'ÉNS* in 1891–92.

Painlevé considered a larger class of differential equations, those that are of the form $F(x, y, y') = 0$, where $y' = \frac{dy}{dx}$. The difficulty with equations of this kind is that they may have both fixed and mobile singular points. A singular point is a point where the solution ceases to be holomorphic: it is fixed if its position is determined entirely by the function F; it is mobile if its position varies with the arbitrary constant of integration. Painlevé noted that Fuchs in his (1884) had given necessary and sufficient conditions for all the singular points to be fixed, and had then gone on to investigate when all the solutions are single-valued functions of x; this depends on the nature of the fixed critical points, a topic studied by Fuchs in 1866. But, said Painlevé, Poincaré had then taken up Fuchs's work and showed, unexpectedly, that when all the singular points are fixed then either the solutions are algebraic functions of x, or the differential equation can be solved by integration, or it is a case of Riccati's differential equation. Moreover, these cases can be distinguished by considering the equation $F(x_0, y, y') = 0$ for a fixed but arbitrary value of x_0. The equation represents an algebraic curve in the variables y and y' and has a genus p that is independent of the choice of x_0. The three cases described by Poincaré occur, respectively, when the genus p satisfies $p > 1$, $p = 1$, $p = 0$. Painlevé therefore proposed to consider the remaining case, where at least some of the singular points are mobile.

The judges noted that Painlevé's investigations, although they went far into the problem of when the solutions to a differential equation are algebraic, left the problem incomplete. This was, perhaps, the occasion for one of their number, Poincaré, to take it up in his (1891e). He remarked, with weak scholarship, that the problem had been first posed by Darboux twenty years ago, and then not much had been done with it. He observed that when the solution is algebraic, it can be written in the form $f + C\varphi = 0$, where, following Darboux, f and φ are homogeneous functions in x, y, z of the same degree. He looked for those values of the parameter C for which the solution is reducible, and found there was a finite number of such values. On the further assumption that the fixed singular points of the differential equation are all distinct he found that he could improve on some of Painlevé's results. He noted that in any case it was enough to find an upper bound on the degree of any algebraic integral, for then the problem reduced to one, as he put it, of pure algebra, but that for this it was necessary that the integral also be irreducible—and this was the hitherto unresolved difficulty that he set out to solve. Painlevé, he remarked, had made considerable progress with the problem of determining when a differential equation admits an algebraic solution of a given genus; he claimed to offer a complete solution of this problem in all but the simplest cases, where, he said, he believed his methods could probably be made to work with a little more effort. This intervention by Poincaré was, however, less significant than the achievements of Painlevé over the years on what became his chosen topic, the study of nonlinear differential equations all of whose singular points are fixed, and those for which the movable singular points are at worst poles. In particular, he showed that second-order differential equations with no movable singularities belong to one of fifty types, of which just six require new functions for their solution: these functions are known today as the Painlevé transcendents.

FUNCTION THEORY OF SEVERAL VARIABLES

If the development of single-variable complex function theory was one of the great success stories of the 19th century, the attempts to develop an analogous theory for several variables must largely be regarded as failures, or at best partial. In fact, even to define an analytic function of several variables was not easy. Weierstrass defined a function to be analytic if it can be given a uniformly convergent Taylor series development in some neighborhood of any ordinary point, and this was the definition Poincaré used in his doctoral thesis (1879, li). For other French mathematicians, a function of several complex variables was analytic at a point if and only if it was analytic in each variable separately at that point (Jordan 1893), and while a function on Weierstrass's definition was automatically analytic on the French definition the converse was not fully established until Hartogs (1905).[6] Whereas the theory of elliptic functions had always been a spur to the single variable theory, the theory of Abelian functions was much more difficult and much less suggestive. Riemann's brilliant exploitation of the connections between harmonic functions and complex functions was complicated in higher dimensions, as we shall see, by the fact that the generalization of the Cauchy–Riemann equations yields a family of conditions that not every harmonic function will satisfy. Cauchy's theory of residues in one variable had no straightforward generalization to more variables. Partly as a result, the theory of a single variable is dominated by the elucidation of the zeros, poles, and essential singularities of a function; in several variables the concept of a singular locus is much harder to analyze. In one variable the Riemann mapping theorem establishes that any simply connected domain whose boundary is a circle is the domain of an analytic function; even by 1900 no such theorem was known for functions of more variables.

It seemed to Weierstrass, who dominated the subject of complex function theory in the second half of the 19th century, that there could be only one way forward: the theory of power series in one and several variables, with its attendant branches of convergence theory and analytic continuation. But even in his hands it made little progress, a fact not helped by his great reluctance to publish outside the lecture theater. His four-semester lecture series always climaxed with his current version of the theory of Abelian functions, a forbidding account that might impress but hardly inspire, and he won few mathematicians to the cause of advancing and deepening it. None of Frobenius, Fuchs, Kovalevskaya, Mittag-Leffler, or Schwarz did so.

One can easily imagine that the topic would catch Poincaré's attention. It is the subject of his doctoral thesis, in which Weierstrass's name is never mentioned, and of quite a few challenges to Weierstrassian orthodoxy thereafter.

Weierstrass had been able to prove what is nowadays called the Weierstrass preparation theorem or the Weierstrass factorization theorem[7] that an analytic function $F(u; z_1, z_2, \ldots, z_n)$ that vanishes at the point

---

[6] And even this result seems not to have been widely known; see Bieberbach (1921, 518).

[7] In lectures from 1860, lithographed notes from 1879, and printed as Weierstrass (1886).

$Z = (0; 0, 0, \ldots, 0)$ of its domain, but for which $F(u; 0, 0, \ldots, 0)$ is not identically zero, can be written in a suitable neighborhood of the point Z as a product

$$F(u; z_1, z_2, \ldots, z_n) = \left(u^m + A_1 u^{m-1} + \cdots + A_m\right) G(u; z_1, z_2, \ldots, z_n),$$

$$(7.1)$$

where the functions A_r are analytic in $z_1, z_2, \ldots, z_n$ throughout the neighborhood and vanish at $(0, 0, \ldots, 0)$, and the function G, which is uniquely determined by F, is analytic and never vanishes in the neighborhood. This result was the first significant insight into the nature of the zero set of a complex function of several variables. It generalizes the result that an analytic function $f(z)$ of one variable can always be written uniquely in the form $f(z) = (z - z_0)^n g(z)$, $g(z_0) \neq 0$, which shows that the zeros of the function f form a discrete set. For a function of several variables it implies that locally the zero set is given by a polynomial in u with coefficients that are analytic functions in the z's. Weierstrass also showed that any function that does not vanish identically at the origin can be reduced to one of the above form by a linear change of the variables.

Poincaré seems not to have known of the preparation theorem when writing his thesis, but he did define a function u of the variables $z_1, z_2, \ldots, z_n$ to be what he called an algebroid function (1879, lii) if it was given by an equation of the form

$$u^m + A_1 u^{m-1} + \cdots + A_m = 0,$$

where the functions A_r are analytic in $z_1, z_2, \ldots, z_n$ throughout the neighborhood and vanish at $(0, 0, \ldots, 0)$, and he established several of their properties.

Weierstrass had defined a function of several variables to be meromorphic on a domain if it was locally rational (i.e., of the form of a quotient of two holomorphic functions), and he claimed without proof in his (1879) that a function of several variables that is meromorphic everywhere is also rational. This is not obvious, for there is no claim that the functions forming the quotient are unique, and there is no uniqueness claim in the preparation theorem. It implies that if a function is meromorphic on a domain but not a rational function everywhere then its domain of definition cannot be the whole domain of the variables, thus opening the way to a study of lacunary functions of several variables. The claim was speedily proved by Hurwitz in his (1883), who used induction on the number of variables, and Poincaré, independently.

Poincaré in his (1883d) exploited the analogy with the theory of harmonic functions. He had already written down the consequences of the Cauchy–Riemann equations for a function of two complex variables in his (1883e). These imply that if $z_r = x_r + iy_r$ and $F = u + iv$, then these equations hold for the real part u (and similar ones for the imaginary part v):

$$\frac{\partial^2 u}{\partial x_r^2} + \frac{\partial^2 u}{\partial y_r^2} = 0, \qquad\qquad r = 1, 2, \tag{7.2}$$

$$\frac{\partial^2 u}{\partial x_r \partial x_s} + \frac{\partial^2 u}{\partial y_r \partial y_s} = 0, \qquad\qquad r = 1, 2, \; s = 1, 2, \tag{7.3}$$

$$\frac{\partial^2 u}{\partial x_r \partial y_s} = \frac{\partial^2 u}{\partial y_r \partial x_s}, \qquad\qquad r = 1, 2, \; s = 1, 2. \tag{7.4}$$

The first of these equations says that the function u is harmonic in each pair of variables (x_r, y_r), but the other equations are a restriction on the harmonic functions that can be the real (or complex) part of an analytic function, for there will be a unique harmonic function of two or more variables satisfying some given boundary conditions that is not the real or imaginary part. It follows that there can be no simple relationship between harmonic functions and holomorphic functions of several complex variables: Poincaré (1883f, 102) said that this provided the true explanation of the differences between complex functions of one complex variable and complex functions of two.

To prove the theorem, Poincaré now covered $\mathbb{C}^2$ by "hyperspheres" R_j (open balls of 4 real dimensions). Inside each of these the given meromorphic function F was representable as a quotient of two analytic functions $F = \frac{N_j}{D_j}$, where N_j and D_j never vanish simultaneously except at isolated points; the task is to show that these functions can be patched together to form a single function of the required kind on the whole domain. For each hypersphere R_j he used a Poisson integral argument to define a harmonic function u_j which takes the value $|D_j|$ on the hypersphere.[8] The same integral also defines a potential function outside the hypersphere, but these two potential functions do not take the same limiting values on the hypersphere itself, so they do not form parts of a continuous function on the whole space. However, Poincaré found that it was possible to define a harmonic function that took account of the failure in the continuity of u_j and the limiting values of $\frac{d|D_j|}{dr}$ as $r \to 1$, and (invoking a theorem of Kovalevskaya's (1875) on the uniqueness of solutions to certain partial differential equations) so to define a continuous function J_j^0 that was harmonic inside and outside the hypersphere.

This function J_j^0 had these properties: outside the ball, it tends to zero at infinity, and at an arbitrary point inside the ball where $F = \frac{N_j}{D_j}$, the difference $J_j^0 - \log|D_j|$ is analytic. When two or more hyperspheres R_j and R_k overlapped, a separate argument showed that functions of this kind were defined for the overlapping regions.

Next, to define the function Φ on the whole of $\mathbb{C}^2$ with the property that the function $\Phi - \log|D_j|$ was holomorphic, he said it was enough to apply Weierstrass's proof of Mittag-Leffler's theorem. Although the function Φ was not an analytic function, he claimed that one could always find an entire harmonic function G satisfying equations (7.2)–(7.4) and such that the difference $\Phi - G$ was the real part of an analytic function Ψ of two variables. Accordingly, the functions G_1 and G_2 defined by the equations

$$e^\Psi = G_1 \text{ and } Fe^\Psi = G_2$$

were entire, and the function F was their quotient everywhere, $F = \frac{G_2}{G_1}$.

Although Weierstrass did not mention Poincaré by name he nonetheless remarked in his (1886, 137) that the question was unresolved and some considerable difficulties seemed to lie in the path of a solution.

The goal of most mathematicians working in this difficult area was to produce a theory that would generalize the powerful theory of elliptic functions to functions of several variables. These new functions, known for historical reasons as Abelian functions, were extremely difficult to

---

[8] He remarked that Kronecker in his (1869) had suggested that such a generalization to four dimensions might be useful, but that he (Poincaré) would try to avoid the language of hypergeometry.

study. Riemann had created the right geometrical framework for them in a remarkable paper of 1857, which caused Weierstrass to withdraw his own attempt from publication; he was not to publish his response until 1879. Then, in a paper the next year (1880) he stated the results he hoped could be obtained, but he could not indicate how they might be proved. His results implied that a further, and complicated, argument will show that every $2p$-periodic function of p variables can be expressed as a quotient of theta functions of p arguments.

In a short paper (1883), Poincaré and Picard outlined a proof that one of Weierstrass's claims implies the others. They also remarked that the most basic of these had been claimed by Weierstrass although he had never published a proof.

There matters rested until in 1890 Appell (Appell 1890; 1891) used Weierstrass's preparation theorem and Poincaré's result that a meromorphic function on $\mathbb{C}^2$ is a quotient of two entire functions to give a proof of Weierstrass's claim about the relation between $2p$-periodic functions and algebraic curves of genus p, in the special case when $p = 2$. He avoided Weierstrass's unproved claims, but the restriction to $p = 2$ is testimony to how difficult this work was.[9] His theorem became valid for all p when Cousin generalized Poincaré's result to any dimension in Cousin (1895).

Weierstrass died in February 1897, and later that year Poincaré, in his (1897c), announced detailed proofs of the most basic of Weierstrass's claims, and went on to claim that every Abelian function can be written as a quotient of theta functions. He promptly published his proof of the second of these claims in his (1898f), but he published his proof of the more fundamental result only in his (1902j), in answer to a request from Mittag-Leffler. There he went back to his *Acta Mathematica* article (1883d) and modified it, not altogether convincingly by modern standards, to establish that the entire functions forming the quotient can be taken to be theta functions. In this way Poincaré proved that every $2n$-periodic function whose periods satisfy the Riemann conditions can be expressed by means of theta functions. This had been claimed long ago by Weierstrass, but no proof had ever been forthcoming, and evidently

---

[9] As late as 1921 Bieberbach could remark in his survey of the theory (p. 519) that he would restrict himself to examples of functions of two variables: "This must suffice for a theory that is so much in flux as the one presented here."

Poincaré was not pleased with this. He commented sharply at the start of part III of his paper that although he believed that Weierstrass had given the principles of his proof in lectures, "be that as it may, the proof has never been made public and his pupils, if they knew it, have communicated it to no-one." He then recalled that he and Picard had given a proof in 1883, entirely without knowing what Weierstrass had done.[10]

Poincaré also made sure not to invoke Cousin's theorem. This can be thought of as the generalization of Mittag-Leffler's theorem on the existence of a function with prescribed principal parts. So it is a question of constructing a function on a whole domain given information about it on a covering by overlapping domains. He claimed that it implies the validity of two theorems claimed by Weierstrass, one asserting the existence of a function with prescribed zeros and the other the representation of a function with prescribed zeros and prescribed poles as a quotient of two entire functions. Cousin's generalizations were not entirely straightforward, and he found he had to pose the second theorem in a different way to take account of the fact that the zeros of a function of several variables were no longer isolated points.[11] His arguments were entirely function-theoretic and avoided Poincaré's reliance on ideas from potential theory. However, as Gronwall (1917) showed in his important criticism of Cousin's paper, Cousin had indeed established a generalization of Mittag-Leffler's theorem to several variables, but the purported generalization of Weierstrass's theorem was flawed. The difficulty he confronted is captured by a remark of the German mathematician Friedrich Hartogs in his (1905), who pointed out that the question was not to be resolved by appealing to the Weierstrass preparation theorem, because the decomposition afforded by that theorem was far from being unique, so one could not easily patch the locally defined expressions together to get a globally defined function. What Cousin failed to deal with adequately, and Gronwall put right, was that there also needs to be a restriction on the nature of the domains. Nonetheless, what Cousin seems to have suspected and Gronwall later showed to be the case is that

[10] It transpired when Weierstrass's proof of this result appeared posthumously in 1903 in his *Mathematische Werke* (3), that the proofs were essentially identical. Appell, and later Picard, had by then given other proofs.

[11] For an analysis of Cousin's work and the subsequent developments in sheaf theory, see Chorlay (2010).

there is a genuine difference between complex function theory in one and several variables: the Mittag-Leffler theorem for one variable does, and the Weierstrass theorem for one variable does not, generalize to several variables without a restriction on the domains.

In 1887 Poincaré had published a paper (1887c) on the residues of double integrals. He set himself the task of generalizing Cauchy's theory to the new situation, and after observing that a few people had tried this before him with varying degrees of success, began to outline his own approach. He admitted with regret that, because the problem involves two complex variables and therefore four real ones, if one uses geometrical language about a four-dimensional space one will drive away most readers, but the alternative is to renounce the use of geometry entirely, and geometry had been a great help in the single integral case. Therefore he resorted to what he called a trick, and regarded curves and surfaces in real four-dimensional space as the images of curves and surfaces in a three-dimensional space. In particular, if a surface in the three-dimensional space was closed, he called its image closed, and he proposed to concentrate on closed surfaces.

This enabled him to come up with a generalization of Green's theorem, which he gave in the form

$$\int_C P\,dx + Q\,dy = \iint_S \left(\frac{\partial Q}{\partial x} - \frac{\partial P}{\partial y} \right) dx\,dy,$$

where the surface S is bounded by the curve C, to double integrals. He showed that it was possible to find a class of integrands with the property that the integral over any surface spanned by the curve is the same. Indeed, he showed how to generalize this result to integrals over a k-dimensional surface with $(k-1)$-dimensional boundaries.[12] He then showed that this class of integrands included the integrals of holomorphic complex functions of two complex variables and deduced results about Abelian integrals by considering integrals of rational functions of the form $\iint \frac{P(x,y)}{Q(x,y)R(x,y)}dx\,dy$. He also investigated what could be said about integrands that are not of this type and have singularities within the region considered, and found that the integral now depends on the

---

[12] The condition, in language later introduced by Élie Cartan, is that the integrand is closed; that is, it is the exterior derivative of a form. For a brief history of differential forms that discusses Poincaré's contribution in more detail than I can do here, see Katz (1999).

singular curves enclosed within the surface. He concluded the paper by showing that Liouville's theorem generalized to functions of two complex variables: as he put it, an entire function of two variables x and y that tends to a finite limit as x and y tend to infinity in whatever way is a constant.

Poincaré's reflections on his discoveries concerning double integrals of a complex function are surprisingly modest (*Oeuvres* 5, 329). He considered it too early to say if the new theory would be as fertile as Cauchy's, and admitted that it would have been easier to obtain some of the results in other ways. But he hoped for truly new results in Abelian function theory, a topic he was currently working on, and his remarks can be seen as another rueful admission by a mathematician that several complex variables were proving unexpectedly hard to study.[13] But it had an influence he did not comment upon: Volterra, in his (1889), wrote one of the first papers on linear functionals ("functions of lines," as he called them). In it he drew explicit attention to what he hailed as Poincaré's very beautiful memoir, and observed that his functions of lines corresponded in every case he had studied to double integrals studied in the manner Poincaré had described (Volterra 1889, 236).[14]

In the late 1890s Poincaré was brought back to the study of functions of two complex variables by the problem of small divisors in the perturbation problem in celestial mechanics. This concerns the effect one planet has on another, and when their periods are nearly commensurable, some terms in the series expansion of the perturbation function may become very large, and it becomes necessary to investigate their size without knowing the value of the preceding terms (this rules out using the method of successive recurrence to calculate the terms, which is typical in the use of power series methods). When Poincaré began this work in 1891 he observed that the best work done on it so far was by Flamme in his thesis, where he had applied an old technique of Darboux for calculating the values of functions for very large numbers. Darboux's method, however, was devised only for functions of a single variable, whereas the astronomical problem is a two-variable one, so Flamme had devised a way of adapting Darboux's approach that took each variable

---

[13] He returned to the study of multiple integrals in his work on topology; see chap. 8, sec. "Poincaré's Work, 1895 to 1905" below.
[14] On Volterra's work, see Vesentini (1992).

in turn. Poincaré proposed to extend Darboux's method directly to functions of two variables, and this he did.

To obtain useful results, it was also necessary to take note of the specifics of the problem: the inclinations of the orbits, and their eccentricities. If these are very small, ideally zero, the calculations are greatly simplified, and it becomes possible to determine whether these contributions are, after all, negligible. In 1897 a succession of papers by Poincaré played variants on this theme, the most important of which was a long exploration of how relations between the coefficients of the perturbation function are related to the periods of a double integral. For, in terms of the eccentric anomalies, u and v, of two bodies a distance D apart, if $x = e^{iu}$ and $y = e^{iv}$ then $x^2 y^2 D^2$ is a polynomial in x, y, say $F(x, y)$, and $\frac{1}{D} = \frac{xy}{\sqrt{F}}$, and the task is to develop his expression in positive and negative powers of x and y as a series $\sum A_{jk} x^j y^k$. The coefficients are given by

$$A_{jk} = \frac{-1}{4\pi^2} \iint \frac{x^{-j-1} y^{-k-1} \, dx \, dy}{\sqrt{F(x, y)}},$$

where the integral is taken along the circles $|x| = 1$, $|y| = 1$. The A_{jk} are transcendental functions of the coefficients of F, but Poincaré showed that they all depend on only finitely many of them.

Numerous special cases had to be dealt with, but the crucial theoretical fact Poincaré discovered was that the integrals of total differentials depending on $\sqrt{F}$ cannot be transcendental (*Oeuvres* 8, 82) "When Picard announced, for the first time, that the most general surface of its degree had no linear cycles, this fact was met with great astonishment. One will now be less astonished that the surface $z^2 = F$ also has none when the polynomial F is indecomposable."

When Poincaré returned to the subject of complex functions of several variables, in his paper (1907c), the subject had again been moved forward by Hartogs, in important work on what domains in $\mathbb{C}^n$ can be the domains of holomorphic functions (Hartogs 1907). Poincaré decided to study the nature of maps between domains in two complex variables, and although he considered his work to be incomplete, he was able to show conclusively that the boundaries of some domains are such that there can be no regular map between the interiors of these domains. This immediately implied that the Riemann mapping theorem cannot be extended to two complex dimensions, but Poincaré gave no explicit examples, which were only exhibited for the first time in Reinhardt (1921).

Poincaré began by distinguishing two distinct ways of asking for the existence of a map in the complex function theory of a single complex variable. The one he called the local problem, starts with two copies of $\mathbb{C}$, the first of which contains a curve ℓ upon which there is a point m and the second of which contains a curve $\mathcal{L}$ upon which there is a point M, and asks for an analytic function, regular in a neighborhood of m that maps m to M and ℓ to $\mathcal{L}$. The second, or extended, problem, starts with two copies of $\mathbb{C}$, the first of which contains a closed curve ℓ bounding a domain d and the second of which contains a closed curve $\mathcal{L}$ bounding a domain D, and asks for an analytic function that maps ℓ to $\mathcal{L}$ and d to D. As he remarked, the local problem is always solvable and indeed in infinitely many ways, whereas the extended problem has a unique solution via the Dirichlet principle.

However, as Poincaré now proceeded to show, the analogous problems for analytic functions of two complex variables behave very differently. The local problem starts with two copies of $\mathbb{C}^2$, the first of which contains a three-dimensional "surface" (a hypersurface) s upon which there is a point m and the second of which contains a three-dimensional hypersurface S upon which there is a point M, and asks for an analytic function regular in a neighborhood of m that maps m to M and s to S. The extended problem starts with two copies of $\mathbb{C}^2$, the first of which contains a closed hypersurface s bounding a domain d and the second of which contains a closed hypersurface S bounding a domain D, and asks if there is a regular function that maps s to S and d to D.

As Poincaré showed, in this setting the local problem will not always have a solution because it asks for three functions that satisfy four differential equations. So the local question is one about types of surfaces, classified according to their groups of analytic automorphisms, for, if two surfaces correspond under an analytic automorphism, their groups are necessarily conjugate, and so the surfaces belong to the same class. In particular, if a surface s admits only the identity analytic automorphism, then the local problem has at most one solution, else the automorphism can be used to generate a second solution. Next Poincaré invoked Lie's theory of transformation groups to obtain all the relevant groups. He cited Lie's *Theorie der Transformationsgruppen* (vol. 3),

and Campbell's *Introductory Treatise* (1903) to establish that there are
27 possible groups, and showed explicitly that for most groups there is
a hypersurface having that group as its analytic automorphism group,
but some groups correspond to two-dimensional surfaces. It follows that
there are hypersurfaces that are not analytically equivalent, and so the
main result of the paper is established, but Poincaré's account was very
unspecific. The group of the hypersurface (hypersphere) with equation
$z\bar{z} + z'\bar{z}' = 1$ was described explicitly (in para. 7), but all Poincaré did by
way of exhibiting a hypersurface with a different group was to indicate
how its equation could be found by means of Lie's theory.

On the other hand, the extended problem in two complex dimensions
does have a solution. In fact, as Poincaré noted, an affirmative answer
follows directly from a theorem of Hartogs, and Poincaré sketched his
own proof of that result.

POINCARÉ'S APPROACH TO POTENTIAL THEORY

Poincaré developed an interest in potential theory in the late 1880s when
he took up the broad theme of mathematical physics. His first major
paper on potential theory was his (1890d). He later explained what
motivated him in an analysis of his own scientific work in a memoir
written in 1901 (published as his 1921a): in theoretical physics it is
important to get the physics right, while in applied mathematics it
is important to get the mathematics right. More soberly, in applied
mathematics it is often the author's intention not to find new results but
to find a valid way of getting known results. In particular, he suggested
that several different problems in physics led to one or other of the three
particular partial differential equations involving the Laplacian that were
of great importance for mathematical physics:

$$\Delta u = ku, \quad \Delta u = k\frac{\partial u}{\partial t}, \quad \Delta u = k\frac{\partial^2 u}{\partial t^2}.$$

It was usual, he said, to rely on more or less intuitive proofs of the
existence of solutions to these differential equations, based on an appeal
to Dirichlet's principle, and he admitted he did this when lecturing
on the subject. These observations, he said, although without value for

the mathematician, are of the right sort to satisfy a physicist because they leave the mechanism of the phenomena apparent. However, more rigorous arguments for the existence of solutions depended on convergence arguments: those of Schwarz, Neumann, and his own "method of sweeping out," which we shall look at shortly. However, this convergence was usually too slow, and the approximations involved too complicated for such approaches to yield effective numerical procedures.

In his (1890d) he had spelled out these views at greater length. The paper was published in the *American Journal of Mathematics* at the request of one of its editors, the American mathematician Thomas Craig, a professor at Johns Hopkins University, who kept up an extensive correspondence with Poincaré and was assiduous in soliciting papers from him. In it, Poincaré noted, among other examples, that the steady state distribution of heat in a body and the problem of the distribution of electric charge in the surface of a conductor lead to the Dirichlet problem, that problems in optics and in fluid flow lead to a common system of partial differential equations, and that problems in elasticity theory lead to more complicated equations that are yet of the same kind. Unfortunately, he then observed, the first common property of all these problems is their extreme difficulty. Not only could they seldom be solved completely, but it was only with great effort that one could show that they could be solved rigorously at all. In which case, he asked rhetorically, was it necessary to solve them at all, when the physicists were dealing with the issue quite well, guided as they are by experiment? He answered that one could not be content with the lack of a rigorous proof; analysis itself should be able to solve such problems. Any rigorous solution is, of course, a solution, and even if crude nonetheless teaches us something. But was it not needlessly pedantic to seek the rigorous solution of equations that had only been established by approximate methods and which rested on imprecise experimental foundations? His answer was no: How could one be sure that something less than a rigorous proof was not actually flawed? Had one the right to say that something inadequate for mathematics was yet good enough for physics?—the line was impossible to draw. One could not, as a mathematician, settle for less, and in any case many of these equations had applications not only in physics but also in pure mathematics (for example, he observed, Riemann himself had based his magnificent theory of Abelian functions on his use of Dirichlet's principle).

Sweeping Out

Poincaré then turned to the Dirichlet problem. He noted that Schwarz had given a good account of the two-dimensional problem with his so-called alternating method, which shows how to solve the problem for two overlapping regions if it can be solved for the two regions separately, and therefore shows how to solve the problem for a variety of regions. He added that various ad hoc methods had been given that generalized this method to three dimensions, and that Neumann and Gustave Robin (see Robin 1886) had given solutions that apply to any convex region.

He then put forward his own solution, which became known as the method of *balayage* or "sweeping out." It relied on the use of Green's functions. A Green's function is the mathematical version of the potential associated with a point mass or a point charge. Given a closed surface S and a point P with coordinates (a, b, c) inside it, the Green's function relative to S and concentrated at P is a function U that is finite and harmonic everywhere except at the point P where it becomes infinite in such a way that the function

$$U(x, y, z) - \frac{1}{\sqrt{((x - a)^2 + (y - b)^2 + (z - b)^2)}}$$

is finite, and which vanishes everywhere on the surface S. The Dirichlet problem that asks for a function V that is harmonic inside S and takes values on S given by the function f is solved by the function

$$V(P) = \frac{1}{4\pi} \int_S f \frac{dU}{dn} \, d\sigma.$$

So the solution to a Dirichlet problem is obtained by finding a suitable Green's function.

To do this, Poincaré took up the idea of inversion in a sphere. Given a sphere S with center O and radius R, one says that a point P has as its inverse point Q the point on the half line from O through P such that $OP \cdot OQ = OR^2$. The map that switches P and Q, and so switches the inside and the outside of the sphere is called inversion, and it is an anticonformal map. Poincaré considered a Green's function relative to the sphere and concentrated at P. It is the potential function associated to a unit mass placed at P and a mass of $-\sqrt{\frac{OQ}{OP}}$ placed at Q. The contribution to $\frac{dU}{dn}$ of a point M on the sphere is given by

$$\frac{R^2 - OP^2}{R} \frac{1}{MP^3}.$$

If the point P is inside the sphere, the potential of a unit charge at P is equal to $\frac{1}{MP}$ at P.

Poincaré now supposed that a unit charge was distributed over the sphere in such a way that the charge density at each point M on the sphere varied inversely with MP^3, and he argued that the potential W of this distribution at points M' outside the sphere was equal to $\frac{1}{M'P}$ while at points M'' inside the sphere the potential was less that $\frac{1}{M''P}$. It follows, he said, that a function equal to the Green's function U inside the sphere and zero outside the sphere is harmonic everywhere except at P and on the surface of the sphere, where it is continuous but its normal derivative jumps by $\frac{R^2 - OP^2}{R \cdot MP^3}$. Crucially, this function can be regarded as the potential function associated to a unit charge at the point P or as the potential function associated to a charge density on the sphere given by $-\frac{R^2 - OP^2}{4\pi R \cdot MP^3}$, which differs from the formula above only by a change of sign. Therefore the function V that was harmonic inside the sphere and took prescribed values on the sphere given by a function V^0 was defined by the equation

$$V(M'') = \int_S \frac{1}{4\pi R} \frac{V^0(R^2 - OP^2)}{M''P^3} \, d\omega.$$

Moreover, its maximum and minimum values lie between those of the given V^0, which can be assumed to be positive.

Poincaré now proposed to show how to solve the Dirichlet problem for the region outside an isolated charged conductor. This conductor has an arbitrary shape, provided that it has a tangent plane at every point and distinct principal curvatures—restrictions that Poincaré found it necessary to impose to secure various convergence arguments, as we shall see. He sketched an argument to show that the region R outside the conductor can be filled with an infinite number of spheres S_j of various sizes in such a way that each point of the region is inside at least one of these spheres. He labeled these spheres arbitrarily $S_1, S_2, \ldots, S_j, \ldots$. The task is to define a harmonic function on the region R which tends to the value 1 at points (x, y, z) at points nearer and nearer the surface of the conductor and which tends to the value 0 as (x, y, z) moves further and further away.

His method employed the crucial equivalence described above. He considered a sphere S_j and the electric charge it contained, and he supposed the charge distribution was switched, point by point, for the equivalent one entirely on the surface of the sphere. There is no effect on the potential outside the sphere, which is unchanged, but the potential inside the sphere is reduced. This operation he called "sweeping out" the sphere. He started with a large external sphere Σ that surrounds the conductor and has a charge distribution on it that gives rise to a potential V_0. At least one of the spheres S_j meets Σ and contains some of the charge distributed over Σ. Poincaré let S_1 be such a sphere and swept it out. The potential function becomes V_1, and there is no charge inside S_1. He now swept out S_2, an operation that can put some charge back inside S_1. Now he swept out S_1, S_2, and S_3. In order to sweep out every sphere infinitely often, Poincaré now swept them out in this order

$$S_1, S_2; S_1, S_2, S_3; S_1, S_2, S_3, S_4; S_1, \ldots .$$

If the nth sweeping out operation empties the sphere S_k the potential resulting function V_n agrees with the preceding one, V_{n-1}, outside S_k and inside S_k it is less: $V_n \leq V_{n-1}$. So everywhere one has $V_n \leq V_{n-1}$. Poincaré drew attention to the fact that in this process negative charge never occurs, so the decreasing sequence of V_n's is bounded below at every point and so tends to a limit, a function Poincaré called V.

Consider now the jth sphere S_j. It is swept out infinitely often, say at times $\alpha_k, k = 1, 2, \ldots$, and each time there is no charge in its interior and the corresponding potential function V_{α_k} is harmonic. Because the sequence of values of the V_{α_k} tends to a limit, Poincaré was able to use Harnack's theorem from 1887 to deduce that the limit function V is also harmonic. But every point of the region R lies in at least one sphere so, said Poincaré, there is a harmonic function defined everywhere on R. Because everywhere one has $V_0 > V > 0$, and V_0 tends to 0 at points arbitrarily far from the conductor, it follows that V tends to 0 at points arbitrarily far from the conductor.

It was harder to show that the potential function $V(x, y, z)$ tends to 1 as the point (x, y, z) tends to a point M, say, on the conductor. Poincaré argued that the assumptions about the shape of the conductor allowed him to define a sphere that touches the conductor at M and otherwise lies entirely inside the conductor. A sweeping out argument (omitted here; see *Oeuvres* 9, 44) then allowed him to reach the desired conclusion. This part of the paper concluded with an indication of how the result could be obtained under weaker assumptions on the boundary.

He next said that he had already described an approach to such problems in the *Comptes rendus* that was not entirely satisfactory but was nonetheless worth recalling. He first considered a solid, homogeneous, isotropic body isolated in an infinite region and cooling by radiating heat.[15] The surface temperature everywhere outside the body is 0, the temperature at time t at an interior point with coordinates (x, y, z) is a function $V(x, y, z, t)$. Inside the body the temperature is governed by the heat equation

$$\frac{dV}{dt} = a^2 \Delta V, \tag{7.5}$$

where a is a constant determined by the conductivity of the body, and on the boundary

$$\frac{dV}{dn} + hV = 0, \tag{7.6}$$

where h is a positive constant determined by the ability of the body to radiate heat.

To solve this problem, Poincaré looked for an infinite sequence of functions $U_1, U_2, \ldots, U_n, \ldots$, and an increasing sequence of constants $k_1, k_2, \ldots, k_n, \ldots$, so that inside the body

$$\Delta U_n + k_n U_n = 0$$

and on the surface

$$\frac{dU_n}{dn} + hV = 0.$$

---

[15] Mawhin (2006/2010, 262) notes that Heinrich Weber had given a very similar argument in the two-dimensional case in his (1869), of which Poincaré was unaware.

Poincaré imposed the further requirement that the integral $\int U_n^2 d\tau$ taken over the whole body be finite (it can be taken to be 1).

Under these conditions he felt enabled to solve the problem after the manner in which Riemann had solved the Dirichlet problem. He observed that there is a class of functions F for which $A(F) = \int_D F^2 = 1$ and

$$B(F) = h \int_{\partial D} F^2 d\sigma + \int_D \|\nabla F\|^2$$

is positive and never zero. He considered a function U_1 for which $B(F)$ took a minimum value and showed by a Green's theorem argument that there was then a k_1 such that the original problem was solved. In fact, k_1 is the minimum value of $\frac{B(F)}{A(F)}$. A relatively simple, similar argument led Poincaré to construct a series of required functions $U_1, U_2, U_3, \ldots$, and associated constants $k_1, k_2, k_3, \ldots$, in which each function U_j is used to construct the next one, but, of course, a convergence argument which requires the calculation of each term recursively is not one that can be used in practice. He admitted that his arguments were open to the same objections that had been raised to the use of the Dirichlet principle, but said that they were good enough for now and he would return to this difficulty later in the paper.

Poincaré also observed that the functions he obtained agreed with those Gabriel Lamé, a French mathematician of the previous generation, had obtained in the special cases when the solid is either a sphere or a parallelepiped (these are regions that can be easily subdivided into many similar regions). An argument I omit allowed Poincaré to show that the k_j increase to infinity with j, but it relied on approximating the region D by polyhedra whose faces are all parallel to the coordinate planes, and so he looked for a more general argument. This led him to his first version of what became called the Poincaré inequality, which we shall pick up below. He also gave an upper bound for k_j in terms of j.

Poincaré now returned to the original problem, and drew attention to some awkward features. For example, it may be that at the initial instant, equation (7.6) does not hold, although it does hold ever after. More troublingly, the solution function V cannot be written as a power series in the time variable t, for, as he showed, if it could be written that way the temperature at a point inside the body would be independent of the shape of the body, which is absurd. This culminates (p. 92) with an estimate of the mean error and remarks about the implausibility of rapidly oscillating functions. Indeed, by proposing that the idea of a solution be weakened Poincaré was onto a better idea than he or his contemporaries realized. Poincaré had noted that it was difficult to prove the convergence of the Fourier series that represented the solution of a problem. He wrote the difference between the solution function u and the Fourier series $\sum_1^n a_i v_i$, where the functions v_i are a basis of solutions, as

$$R := u - \sum_1^n a_i v_i.$$

He took from Chebychev the idea of considering the integral $S := \int R^2$ as a measure of the error, and commented,

> We have therefore shown that the average error tends to zero, but not that R tends to zero (with increasing n). This can perhaps suffice for the moment. In fact, how can S tend to zero but R not? It must be the case that R oscillates more and more rapidly as n increases, in such a way that for n very large R takes very different values at points that are very close together. No physicist believes that if such a state exists at an initial moment it can persist. This is what leads me to be content for the moment with the preceding considerations.[16] (1890d, 273, in *Oeuvres* 9, 92)

The idea that the mean error may tend to zero without the series converging to a solution is also rich in significance for theoretical physics.

In part 4 Poincaré established that the functions U_n and all their derivatives are bounded inside the body by a constant that can be determined in advance. Finally, in part 5 Poincaré returned to his less than rigorous Riemann-style argument, and observed that its validity—and that of all the subsequent deductions—depended on the continuity of the function U_1, which he was unable to establish. It would therefore be necessary, he said, to proceed as he had done in solving the Dirichlet problem, but the matter would be even more complicated and, he said, he had not done this. Instead, he offered some thoughts about what could be done by invoking a molecular model of heat diffusion. This would

---

[16] In fact, in 1899 Zaremba showed that Poincaré's "solution" to the problem was indeed a solution.

produce a very large system of linear equations, but they had a symmetry that suggested there might be a quadratic form present that would allow one to write the equations as sums of squares (this argument recalls his work on the stability of rotating fluid masses; see chap. 5, sec. "Rotating Fluid Masses" above). This might lead to a rigorous solution, but not one, he said, of any practical value. And in any case, the methods he had described in sections 2, 3, and 4 were rigorous from a physicist's point of view, although one could hope that some sort of passage to the limit would render them analytically satisfactory as well.

The Palermo Paper, 1894

Poincaré began his second major paper on the differential equations of mathematical physics, his (1894e), with a further review of the range of situations where the Laplacian is central (specifically citing the problem of the vibrating drum), and of the use of Green's functions.[17] He concentrated on the problem

$$\Delta v + \xi v + f = 0,$$

where v is required to be a function with continuous second derivatives in the interior of a domain D and which satisfies the equation

$$\frac{dv}{dn} + hv = 0$$

on the boundary. In these equations ξ and h are given constants and f is a given function.

He took his lead from papers by Schwarz (1885) and Picard (1893). Schwarz had used the calculus of variations to solve a mathematical problem that in physical terms amounted to finding the fundamental frequency of a vibrating drum. This automatically raised the question of what the other frequencies of the drum were, so that any vibration of the drum could be expressed as a weighted sum of these basic frequencies, but the problem was very difficult and the only progress made with it for some time was Picard's calculation of the second fundamental frequency.

---

[17] This account follows Mawhin (2006/2010), to which the reader is referred for many details.

Schwarz had proposed looking for solutions to these equations in the form of series in the parameter ξ,

$$v = v_0 + v_1\xi + v_2\xi^2 + \cdots,$$

where the v_j are unknown functions to be determined. This leads to a sequence of differential equations for the v_j,

$$\Delta v_j + v_{j-1} = 0.$$

Schwarz had found a series of integrals W_k,

$$W_{2n} = \int_D v_n^2, \quad W_{2n-1} = \int_D \|\nabla v_n\|^2,$$

with the property that the quotient $\frac{W_{n+1}}{W_n}$ is increasing as n increases. The growth of W_n with n controls the growth of v_n, and Schwarz's methods, suitably generalized from the simpler case that Schwarz had considered, enabled Poincaré to show that the series for v converges absolutely and uniformly for ξ within a certain range determined by the size and shape of D, and therefore that the series does indeed represent a solution of the given problem. To go further, Poincaré recalled his inequality from his (1890d), which he rederived in a new and, as he put it, "more manageable" form. It now gave him numbers L_p that he could use as upper bounds for the quotient $\frac{W_{n+1}}{W_n}$, specifically, $\frac{W_{n+1}}{W_n} < \frac{1}{L_p}$. When D is a three-dimensional domain his proof showed that $L_p \approx p^{2/3}$ as $p \to \infty$.

In paragraph IV Poincaré considered a sequence of p equations of the form

$$\Delta v + \xi v + f_j = 0,$$

with the same boundary conditions as before, and their corresponding solutions w_j. He let $\alpha_1, \alpha_2, \ldots, \alpha_p$ be arbitrary coefficients and regarded $[\alpha_1, \alpha_2, \ldots, \alpha_p]$ as the homogeneous coordinates of a point in $(p-1)$-dimensional projective space. He set

$$f = \alpha_1 f_1 + \alpha_2 f_2 + \cdots + \alpha_p f_p,$$
$$v = \alpha_1 w_1 + \alpha_2 w_2 + \cdots + \alpha_p w_p,$$

so

$$\Delta v + \xi v + f = 0.$$

This allowed Poincaré to deduce that as n increases there is a domain δ_n in $(p-1)$-dimensional projective space where the inequalities $\frac{W_{2n}}{W_{2n-1}} \leq \frac{1}{L_p}$ hold. Because δ_{n+1} is contained in δ_n these domains all have a common part, δ, which, Poincaré noted, could reduce to a single point. Poincaré deduced that when the α's specify a point in δ the series for v is a solution whenever $|\xi| < L_p$.

Poincaré now defined the functions u_j to satisfy the equation

$$\Delta u_j + \xi u_j + v_{j-2} = 0, \quad j = 2, 3, \ldots, p,$$

and deduced from the equation $\Delta v_j + v_{j-1} = 0$ that

$$u_j = v_{j-1} + \xi v_j + \xi^2 v_{j+1} + \cdots.$$

He set $w = \alpha_1 v + \alpha_2 u_2 + \alpha_3 u_3 + \cdots + \alpha_p u_p$, developed it in powers of ξ, say

$$w = w_0 + w_1 \xi + w_2 \xi^2 + \cdots,$$

and deduced p linear equations between v and the u's on the one hand and w and the v's on the other.

Next, he showed directly from those equations that v can be written as a quotient $v = \frac{P}{Q}$, where Q is a polynomial in ξ of degree $p-1$:

$$\alpha_p - \alpha_{p-1} \xi + \cdots + \alpha_1 (-\xi)^{p-1},$$

and P is another polynomial in ξ but with coefficients that involve w and $v_0, v_1, \ldots, v_{p-2}$. But as w is holomorphic in ξ for $|\xi| < L_p$ and the v's do not depend on ξ, this function is therefore meromorphic for $|\xi| < L_p$ and "as I can take p as large as I wish, the function will be meromorphic in the whole plane" (1894, 144). Poincaré checked that the function P satisfies the original differential equation, has continuous first and second derivatives in D, vanishes on the boundary of D, and satisfies the equation

$$\Delta P + \xi P + fQ = 0.$$

In paragraph 5 Poincaré showed that the poles of the meromorphic function $v = v(\xi)$ with modulus less than L_p are given by the roots of the equation $Q = 0$, which means that there are only finitely many within any given distance of the origin. If k is such a root then

$$\Delta P_k + k P_k = 0.$$

Since the function P_k vanishes on the boundary of D he called it a harmonic function relative to the boundary and k the characteristic number of the function. Poincaré now deduced that all the roots of the equation $Q = 0$ are simple, and so the P_k are the residues "which is the reason I called these functions harmonic," he said, invoking the analogy with a vibrating drum with which he had opened the paper. However, to avoid confusion the modern terminology of eigenfunction and eigenvalue will be preferred here. He then showed that the function v cannot be holomorphic in the entire ξ plane, for if it were, the series $W_0 + W_1\xi + W_s\xi^2 + \cdots$ would be, which it is not. It follows that eigenfunctions exist, and with some more work Poincaré showed that all the eigenvalues are real and positive and they increase indefinitely with p.

In the next part of the paper Poincaré studied the maximum principles involved in such problems. He was very aware that these further explorations were not rigorous except in a very special case, but

> I believed, however, that these results, incomplete as they are, are not absolutely deprived of interest, and I decided to publish them. I will be happy if this publication could provoke some new research in the subject. (1894e, 178)

Within two years, Poincaré published another long paper (1896b) on the subject, in which he compared the problems of Neumann and Dirichlet. Like the Dirichlet problem, the Neumann problem gives information about the distribution of charge of mass on a surface, and asks for the corresponding potential function. In the Neumann problem, it is supposed that there is one distribution on the inside of the surface that gives rise to a potential V nearby and another one on the outside that gives rise to a potential V'. The problem is to extend these functions V and V' to a potential W which is harmonic everywhere except on the surface, where it will be discontinuous, but is to have a continuous normal derivative on the surface. The problem supposes there is a function Φ on the surface and asks for V and V' such that

$$V - V' = \lambda(V + V') + 2\Phi$$

for some values of the constant λ. As Poincaré noted, if $\lambda = 1$ the equation is $V' = -\Phi$, which is the Dirichlet problem for the exterior, and if $\lambda = -1$ the equation is $V = \Phi$, which is the Dirichlet problem for the interior, so the Dirichlet problem is a special case of the Neumann problem. The conundrum facing him was that his method of sweeping out is a good convergence proof, and applicable to a general surface, but it is much less use for calculating with than Neumann's method, which, as Neumann had given it, only applied to convex surfaces.

He ended by sketching Neumann's solution to the problem, which gave the answer in the form of an infinite series of functions, and observed that Neumann's proof that the series converged made essential use of the convexity of the surface, which Neumann had been forced to assume. Poincaré proposed to remove this apparent dependence on the convexity of the surface, and he showed that in fact both Neumann's solution and Robin's applied to any simply connected surface which has a tangent plane everywhere and at each point has two distinct radii of curvature, and for which the given function Φ is infinitely differentiable.

In 1896 Poincaré extended these ideas to apply to the motion of the oceans in his (1896e). The problem is a very complicated one, in which the gravitational attraction of the sun and the moon, the rotation of the earth, the inertia of the seawater, and the presence of the continents must all be taken into account. The motion can be analyzed, he remarked, as the sum of several motions with different periods, and the motion with longest period (between two weeks and six months) had been successfully treated by Thomson and Tait in their (1879), when the problem was one of statics. If there were no continents, he went on, this motion could be analyzed by using spherical harmonics, and Thomson and Tait had shown how to adapt that technique in the presence of the continents but ignoring the rise and fall of the tides. Poincaré proposed to take account of the continents and the tides, which he did by introducing some modified spherical harmonic functions that took account of the continents. He also looked at motions with much shorter periods but only by ignoring up- and down-welling effects, and here he found analogies with his Palermo paper and the analysis of the motion of a membrane under a constant tension.

Poincaré formulated the first problem as one in potential theory, and followed the methods of his (1894e) quite closely. He treated it as an eigenvalue problem and used the methods of complex function theory to find the eigenvalues. He remarked (p. 79) that Thomson and Tait's conclusions were only accurate if the earth was taken not only to be a solid but one more rigid than steel, and he wondered, inconclusively, if they had taken sufficient account of the self-attraction of the water. As for the second problem, the much more rapid oscillations require the techniques of dynamics, and Poincaré again relied on the methods of his (1896e) to obtain what we would call the eigenfunction expansions of the functions describing the motion of the waves. He reduced the problem to the motion of a membrane under variable tension to which he believed the methods of his (1894e) applied mutatis mutandis.

Poincaré was to return to the subject a decade later in the final volume of his *Leçons de mécanique céleste* (1905b), where he noted that great progress had been made in part by the introduction of Fredholm's method in the theory of integral equations and in part because of the detailed observational work of several American writers that had changed what he called the "physiognomy of the results." In fact, both Thomson and Tait, and Poincaré obtained results of lasting significance. To quote a modern authority, what is called the Kelvin wave "is the most characteristic wave form of tides in the deep ocean" and the so-called Poincaré waves "support rotating currents, nodal points, and other known characteristics of oceanic tides."[18]

Conclusions

In the first three of these papers Poincaré treated the Dirichlet principle heuristically as a way to explore problems before sketching novel approaches that should lead to his newly discovered results. In so doing he came close to meeting the demands he had set himself. He certainly came close to delivering rigor. He not only claimed, correctly, that any rigorous solution is, of course, a solution, and even if crude nonetheless teaches us something, he provided a theory that explained how the eigenvalues

[18] Both quotations in Cartwright (1999, 84).

depend on the geometry of the body. Finally, his new theory went some way to showing that the earlier methods, which indeed offered something less than a rigorous proof, were not actually flawed.

In the event his approach soon inspired the new methods of Fredholm and Hilbert that speedily eclipsed it.[19] Poincaré was very impressed with Fredholm's innovative approach, and adopted it in his own work, but it was also taken up by Hilbert and the people around him, and this large school, which recovered quickly from the First World War, was the one that developed this family of ideas furthest, and in due course dominated the writing of the history of this part of mathematics. The lasting contribution of Poincaré, however much it is sometimes passed over, was to show how many of the important problems in mathematical physics are problems about eigenfunctions and eigenvalues in some space of functions. Indeed, what particularly impressed Hadamard (1921) was that Poincaré had introduced—and solved—the corresponding eigenvalue problem for the Neumann problem, when nothing suggested that it be introduced (*Oeuvres* 11, 235), and he commented that these fundamental functions had sprung "fully armed from the head of the analyst."

THE *SIX LECTURES* IN GÖTTINGEN, 1909

In October 1908 Hilbert set about inviting Poincaré to visit Göttingen to give a series of six lectures. They were to be the first lectures given with the interest earned from the money attached to the Wolfskehl Prize, with which the Göttingen Royal Society of Science had recently been endowed. Paul Wolfskehl had practiced as a doctor until multiple sclerosis made it impossible for him to continue. He then switched to mathematics, which he studied in Bonn in 1880 and then in Berlin from 1881 to 1883, where he took lectures from Kummer, the renowned number theorist, and he seems to have lectured at the Technische Hochshule in Darmstadt in the late 1880s until even that became too difficult for him. In 1905 he wrote a will leaving 100,000 marks for "whomsoever first succeeds in proving the Great Theorem of Fermat," and he asked the Göttingen Society to hold this money in trust and serve as judge for the awarding of the prize; he died on 13 September 1906. On 27 June 1908 the conditions for the

[19] See Dieudonné (1981).

Wolfskehl Prize endowment were laid down by the society and the terms under which it would be administered were published in several journals. The prize was to be valid until 13 September 2007, and almost at once a flood of attempts began, none of which were successful until it was solved by Andrew Wiles in 1995.[20]

Hilbert and Poincaré eventually settled on the period between 22 and 28 April 1909, and Poincaré offered to talk on the applications of Fredholm's methods and the reduction of Abelian integrals. Hilbert replied in late February suggesting that "Poincaré week" would be widely advertised outside Göttingen and something with a theme from mathematical physics or astronomy, and something of a logical or philosophical character would also be welcome. There would also be a meeting of the Mathematical Society that week, and on the 30th a celebration of the anniversary of Gauss's birthday, when a trip to one of the mountains Gauss had used in his measurement of the angle sums of triangles was planned and a Gauss tower would be inaugurated. But also they were still very much preoccupied with the sudden death of Minkowski (on 12 January 1909) who, Hilbert wrote, had been "his dearest and truest friend, and a thousand times more than a brother."

Poincaré replied that his lectures on Fredholm's work would cover applications to the motion of the tides and Hertzian waves, and that he would lecture in French unless asked well in advance to speak in German. In his next letter he gave the titles of five of his lectures, and added:

> I am still recovering from an accident which befell me last year in Rome, and I am obliged to take some precautions. I can drink neither wine nor beer, but only water. I can attend neither a banquet, nor a prolonged meal.
>
> This has made me hesitate to accept your invitation, but I think you will be able to arrange things accordingly. I think that I will have the opportunity to see your colleagues in circumstances other than banquets, and I hope that in these conditions I will have the pleasure of getting to know them. I will be very pleased to have the opportunity of seeing you.

With these arrangements in place, Poincaré traveled to Germany.

---

[20] For more on the Wolfskehl Prize, see Barner (1997).

Fredholm Equations

As we have seen, Poincaré wrote several papers in the 1890s on the partial differential equations of mathematical physics. This had led him to consider such problems as integral equations and to approach them via a theory of infinitely many linear equations. An integral equation—the name is due to du Bois-Reymond in 1888—is an equation of the form

$$\varphi(s) = f(s) + \lambda \int_a^b K(s,t) f(t) \, dt, \qquad (7.7)$$

where $f(t)$ is the unknown function that is to be found, $\varphi(s)$ is a given continuous function on $[a,b]$, $K(s,t)$ is another given function called the kernel, and the solution is expected to exist for a spectrum of possibly complex values of λ. Poincaré had also shown that they could be considered fruitfully as eigenvalue problems, and he had indeed shown that they had a spectrum of infinitely many eigenvalues. Important as these papers were, they did nothing to suggest that there was a fundamentally simple theory of integral equations quite unlike the complexities of the potential theory that Poincaré's work addressed. That breakthrough was due to the insight of a young Swedish mathematician, Ivar Fredholm.

Fredholm was a student of Mittag-Leffler's. He became a lecturer in mathematical physics at the University of Stockholm in 1898, the year he wrote his doctoral thesis, when he was 32. The next year he traveled to Paris, where he met Poincaré and other French mathematicians, and on his return he published a preliminary account of his work in 1899, another for domestic consumption (1900) and a detailed account in *Acta Mathematica* in 1903.

The paper (1900) begins,

> In his deep research (*Acta Mathematica*, vol. 20) on the convergence of the well-known Neumann method in potential theory, M. Poincaré considered the Dirichlet problem as a particular case of another problem, which he called the Neumann problem.... Neumann solved this problem by expanding the unknown function in powers of the parameter λ. But it results from the research of M. Poincaré that V is a meromorphic function. Thus it is clear that the Neumann series cannot converge for all values of λ. But because

we know that a meromorphic function can always be expressed as the quotient of two entire functions, it seemed natural to me to look for these entire functions directly. (Mawhin 2006/2010, 274)

The other publications give few ideas of how Fredholm came to formulate his ideas, but in a lecture of 1909 he indicated some sources, on the basis of which Dieudonné (1981) gave a very plausible reconstruction of Fredholm's first line of thought.

Fredholm acknowledged two sources: Volterra's idea of passing from an infinite system of linear equations to an integral equation, and von Koch's work on infinite determinants. Dieudonné suggested (1981, 99–102) that Fredholm first replaced the integral by its approximation as a Riemann sum of n terms, thus getting a system of n linear equations:

$$\varphi(y_j) = f(y_j) + \frac{\lambda(b-a)}{n} \sum_{k=1}^{n} K(y_k, y_j) f(y_k), \quad 1 \le j \le n. \qquad (7.8)$$

The guiding metaphor is to think of a finite system of such equations in vector form as

$$\varphi(y_k) = \sum_{j} (I + cA) f(y_j),$$

where A is the matrix $(K(y_k, y_j))$, and solve them by finding $\det(I + cA) \neq 0$ and, when it is nonzero writing

$$f(y_k) = (I + cA)^{-1} \varphi(y_j)$$

and then letting $n \to \infty$, to get $f(t)$ as an integral expression involving $\varphi(t)$.

Fredholm analyzed the way the determinant of that system behaved as n tended to infinity using von Koch's theory, which writes the expansion of the determinant as an infinite series $\Delta(\lambda)$. Then he proved the uniform convergence of the series on any compact set in the plane first using a simple estimate of his own devising, and by using a better inequality concerning determinants that Hadamard had established in 1893.

Fredholm next expanded the finite system of equations by Cramer's rule, and again let $n \to \infty$. He introduced an infinite series $\Delta(s, t; \lambda)$ for the minors that come from a further use of von Koch's series. This allowed him to write

$$\Delta(s,t;\lambda) = K(s,t)\Delta(\lambda) - \lambda \int_a^b K(s,\xi)\Delta(\xi,t;\lambda)\,d\xi.$$

He then defined the function

$$\Phi(s) = \varphi(s)\Delta(\lambda) - \lambda \int_a^b \Delta(s,\xi;\lambda)\varphi(\xi)\,d\xi,$$

from which it follows by endlessly eliminating $\Delta(s,t;\lambda)$ that

$$\Phi(s) = \lambda \int_a^b K(s,t)\Phi(t)\,dt = \varphi(s)\Delta(\lambda).$$

Therefore, if $\Delta(\lambda) \neq 0$ then

$$f(s) = \frac{\Phi(s)}{\Delta(\lambda)}$$

is a solution of the original integral equation (7.7). His account concluded with an investigation of the conditions on the nonvanishing of $\Delta(\lambda)$.

Fredholm's first publication on the subject was a note in the *Comptes rendus* in 1899. He then wrote to Poincaré to thank him for publishing it, and added some remarks on his work. In it, he explained how a generalization of Dirichlet's problem could be approached by methods used by Neumann to obtain a functional equation for a function $u(x)$ of the form

$$u(x) + \lambda \int_0^1 u(y)f(x,y)\,dy = v(x),$$

where $f(x,y)$ is a function that is bounded except at $(0,0)$, $v(x)$ is given function, and λ is a parameter. He had, he said, been able to obtain a result that it seemed would be very useful: the solution $u(x)$ could be written as a quotient of two entire functions of λ, and both the numerator and the denominator could be given explicitly. It was necessary to ensure that the values of λ were such that the denominator did not vanish, and he showed how this problem could be analyzed.

We do not have any other letters in either direction between Fredholm and Poincaré, but it is certain that Poincaré was very impressed by

Fredholm's approach, because he proceeded to adopt it in his own work on numerous occasions, notably, in Göttingen.

Göttingen

Famously, the other mathematician who took up Fredholm's approach was Hilbert. He concentrated on the case where the kernel is symmetric, $K(s,t) = K(t,s)$, and in a series of powerful papers starting in 1904 he built Fredholm's insight into a rich theory. When the kernel is symmetric the corresponding λ's are all real and the corresponding eigenfunctions are mutually orthogonal in a sense Hilbert made precise; all this closely resembled the theory of quadratic forms but for infinitely many variables, as Hilbert explicitly remarked. He also devoted the Göttingen seminar to this work, and drew in a number of his best students—this is the furnace in which the concept of Hilbert space was forged—and was led to formulate a clean, abstract theory of operators of various kinds on suitable infinite-dimensional vector spaces that included the theory of integral equations as a central but special case.

This vast and important new domain of linear operators on a Hilbert space was very much a Göttingen specialty, and so it was possibly something of a shock to them, and even an affront, when Poincaré accepted an invitation to give six lectures there in 1909 (published the next year as his 1910h) and chose to concentrate on Fredholm's theory of integral equations. Indeed, the first of the lectures was little more than a run through the basic theory, making a connection at the end to the theory of infinitely many linear equations in the work of Hill, von Koch, Hilbert and others—a list that cannot have gone down too well with the Göttingen mathematicians in his audience.[21]

The second lecture (1910b) applied Fredholm's theory to the motion of the tides. Stereographic projection allows the surface of the earth to be represented on a plane, so the motion of the tides can be represented by a function of x and y. If the internal forces caused by the displacement of the water are neglected, the function satisfies an inhomogeneous

---

[21] Veblen wrote to Birkhoff on 25 December 1913 that Hilbert "struck me as being both urbane and magnanimous, although the stories one hears do not bear this out—for example, the stories told from the German point of view about Poincaré's visit to Göttingen put Hilbert and the others in a rather bad light." Quoted in Barrow-Green (2011, 43).

Figure 7.1. David Hilbert (1862–1943). Source: Hermann Minkowski, *Briefe an David Hilbert* (Springer, 1973).

linear second-order partial differential equation; the land is modeled as if it has a vertical wall. Poincaré then set about converting this equation for the motion into an integral equation. First he recalled "the methods of Hilbert and Picard" for solving the differential equations using Green's functions. This gave the solution in the form of an integral, but its kernel becomes infinite and seemingly defies Fredholm's methods; Poincaré described how this difficulty could be overcome. He observed that Kellogg had already shown how to do this using the methods of potential theory, and then he sketched another method that relied on Cauchy's residue calculus. The third lecture (1910a) applied Fredholm's methods to the study of Hertzian waves and wireless telegraphy, with particular attention to the problem of how it was possible to send radio signals across the Atlantic; this has already been discussed in chapter 2, section "Contemporary Technology" above, but Poincaré noted that his present results required improvement.

The fourth lecture (1910i) was on a different subject: the reduction of Abelian integrals and the theory of Fuchsian functions. The first, more traditional formulation, formulates it as a system of Abelian functions in p variables with $2p$ periods that is expressible as a system of Abelian functions in q variables with $2q$ periods, with $q < p$. Poincaré described this difficult theory in the case where the two systems of Abelian functions both arose from algebraic curves, and then analyzed the question again by considering the Fuchsian polygons and the system of Fuchsian functions that correspond to the two algebraic curves. When the curve C_p reduces to the curve C_q and the map from C_p to C_q is $n - 1$ the corresponding group G_p is a subgroup of G_q of index n. It follows that the polygon for C_p is composed of n copies of the polygon for C_q. As Poincaré put it,

> The above general considerations now allow us to deduce a series of beautiful and important theorems on the geometry of circular-arc polygons, as well as on the geometry of algebraic curves.

Poincaré illustrated this point by examples too complicated to explain here.

The fifth lecture (1910j) was on transfinite numbers. Poincaré opened with a paradox he attributed to Jules Richard: there can only be a countable set of objects that are definable, but there are uncountable sets, such as the continuum. Poincaré then explained why these results were not mutually contradictory. This led him to the idea of predicative and impredicative definitions, and he said that Russell had given a good example of an impredicative definition: the least integer not definable in less than one hundred German words. Zermelo, said Poincaré, had argued that impredicative definitions should not be rejected, because so much mathematics depends on them, but Poincaré offered an example to suggest that impredicative definitions could be replaced by predicative ones. He brought his lecture to an end with some remarks about Bernstein's theorem (stated above, see chap. 1, sec. "Poincaré among the Logicians") and Zermelo's introduction of the well-ordering principle. Poincaré objected to the latter on the grounds that, as even Zermelo admitted, a well ordering of the real numbers could not be stated in finitely many words, and was therefore, in his opinion, empty, meaningless words. Here, he said, was the reason for the battle over Zermelo's "almost brilliant" theorem, and he ended his lecture by saying that he could speak about the problem for hours without solving it.

Up to now Poincaré had given his lectures in German, but he gave his final lecture (1910c) in French. He explained that he had been able to give his previous lectures because, although as it were "lame," he could walk with crutches—the mathematical formulas he used. But in this lecture he would use no formulas, and must therefore walk in his own language. The subject was the new mechanics, which was replacing the old Newtonian mechanics that was being demolished by such people as Max Abraham (who was in the audience) and Lorentz in Holland. In the new mechanics nothing could travel faster than light. So, if one imagines a steady impulse being given to a body, the effect of that impulse on the speed of the body diminishes the faster the body is traveling. The conclusion has to be that the mass of the body is increasing with its speed. The principle of relativity, which asserts that all motion is relative, presents itself in a way that cannot be disputed—here he noted with chagrin the recent incident in which "all the reactionary French journals had tried to prove that the sun went round the earth and that in the struggle between the Inquisition and Galileo only Galileo had been wrong."

Now, said Poincaré, in the new mechanics the principle of relativity implies that it was impossible to say that two observers in different places could have clocks that told exactly the same time. They could only know that they kept their local times equally. If the two observers were in a state of constant relative motion away from each other, each would see the others clocks run slow and lengths appear contracted. Curiously, Poincaré abstained from any formulas, even at this point; but what is still more curious is that he retained the ether as the medium through which radio waves travel. The relativistic phenomena he described are there as consequences of the principle of relativity.

Next Poincaré turned to the idea that the new mechanics was based on the view that all matter is fundamentally electrical. "An arbitrary body," he wrote, "being nothing but an assemblage of electrons, it will be enough for us to consider just the latter." Radio waves are created by a certain kind of motion of electrons that generates an electromagnetic field that disperses energy through the ether. The apparent mass of the electron increases with its speed, and the inertia of the ether tends to infinity as this speed approaches that of light. All of this, Poincaré now observed, causes problems for planetary astronomy: Newton's laws had assumed the mass of a body was constant, but this was now only true for bodies at rest; gravitational attraction could not be instantaneous. The issue had been tackled by Lorentz, he said, who had showed that the energy lost by the electrons that form the earth as it travels round the sun and through its electromagnetic field was so slight that while it would indeed ultimately result in the earth falling into the sun this catastrophe would not happen for millions of milliards of centuries. The fastest moving planet is Mercury, and its loss of energy in this way was the greatest. Could this explain the precession of the orbit of Mercury, which exceeded classically based estimates by 38" a year? The new mechanics reduced this discrepancy to 32" which gave no reason to accept the new theory but was at least no reason to reject it.

Poincaré ended his lecture with a reminder that Newtonian mechanics was still the right mechanics for our daily life. Yet some pedagogues in France, he said, seemed to have nothing better to do than to tell their students that the old mechanics was on its way out, and he concluded with the opinion that the new mechanics could not be understood without a good grounding in the old.

Poincaré's rivalry with Hilbert surfaced again in 1910. When the Hungarian Academy of Sciences had awarded its first Bolyai Prize to Poincaré, in 1905, it took the unusual step of making an honorable mention of Hilbert as well. From Szenassy's account, based on the records of the Hungarian Academy of Sciences[22] we learn that the Hungarians on the committee were Gyula König and Gusztáv Rados, and the external members were Darboux and Klein. The two candidates were Poincaré and Hilbert, supported equally. "Finally Poincaré was chosen, but the decision requested the same appraisement for the 2 scientists in the minutes made for the presidency. The other point of interest is that in 1905 it was Felix Klein who should have assessed the 2 candidate's merits 'for all the sessions' but because of his indisposition this difficult task fell upon Gusztáv Rados." So it is no surprise that when it met again in 1910, with Poincaré as the rapporteur, it decided that this was Hilbert's turn. Poincaré's report (1911c) is an impressive summary of Hilbert's many impressive achievements. He began with Hilbert's work on invariant theory, noting how many pages of Gordan's came down to so few lines of Hilbert's, and going on to observe that in his account of the transcendence of e Hilbert had again replaced arduous papers with a

---

[22] Available at http://vmek.oszk.hu/03200/03286/html/tallozo1/contrib.html.

proof of "astonishing simplicity." Then came an account of Hilbert's work in number theory, the theory of Galois fields and the origins of class field theory, and his solution of Waring's problem. This last caught Poincaré's attention because it demonstrated "a new way of introducing continuous variables into the theory of numbers," and so spoke to concerns Poincaré had learned from Hermite.

Then came Hilbert's work on the foundations of geometry, which Poincaré surveyed much as he had done when reviewing the book in 1902. He gave a careful and lengthy summary of the various types of axioms Hilbert presented, the methods used, and their consequences: the types of coordinates that can be used in various systems, and some of the unexpected geometries that arise when the Archimedean axiom is abandoned. Poincaré also noted that in his ICM address of 1900 Hilbert had called for an investigation into the status of continuity in Lie's theory of groups, and had shown that there is no complete embedding of the non-Euclidean plane into three-dimensional space.

Hilbert had then turned to study integral equations, which Poincaré said was a field invented by Fredholm and this prize now honored the author of important improvements. One wonders indeed if Poincaré had not held out for Fredholm, because he now explained that the prize recognized all of Hilbert's achievements and gave the clear impression that on this particular topic the laurels went to the inventor. So Poincaré explained quickly what Fredholm had done, and then singled out as Hilbert's contribution the attention to symmetric kernels, and the work (with Schmidt) on the completeness of the corresponding function spaces. All this led to an account of Hilbert's work on the Dirichlet principle, and the laudation ended with noting a few of Hilbert's smaller contributions to number theory and then praising his clarity, rigor, and influence.

8

Topology

TOPOLOGY BEFORE POINCARÉ

Topology became one of the central, and most fundamental, branches of mathematics in the 20th century, very much as a result of Poincaré's pioneering achievements—which means, of course, that it was not regarded as a truly significant way of approaching mathematical problems before he began. Even the name of the subject shows this evolution, Poincaré, like many before him, called it "analysis situs," after an analogy with some vague ideas of Leibniz about a subject more general than Euclidean geometry. The name "topology," meaning the study or *logos* of places (*topoi*) was introduced by Listing in 1847, but does not seem to have caught on until German mathematicians started using it in the early 20th century as they became caught up in understanding the implications of Poincaré's work.

There are two branches to the subject. One, often called point-set topology today, grew out of the study of subtle problems in mathematical analysis and eventually put down foundations underneath classical real and complex analysis. It became the natural setting for questions about the meaning of continuity of functions and mappings between mathematical objects, as considerations of continuous functions moved from the theory of a single real variable to other settings: several variables, complex variables, functions defined on spaces other than $\mathbb{R}$ or $\mathbb{R}^n$. In the 1870s Cantor worked on questions to do with integration and the behavior of trigonometric series. This led him to discover the concepts of nowhere dense subsets of an interval in the real line, limit sets and perfect sets—a decade later he was to see in this work the clue to his work on transfinite sets. In 1887 Harnack's paper on potential theory deepened people's ideas about the possible complexity of the interior of a region with a boundary by exploring the question of whether any two points in it can be joined by a sequence of overlapping disks. But this late 19th-century point-set topology was only one of a number of disciplines within mathematics that had to do with properties of sets of points: there were other questions that had to do with the concept of integrability and eventually became formulated in Lebesgue's measure theory around 1900; there were questions about Cantor's newly discovered infinite sets of different sizes; and there were questions about sets as the very foundations of mathematics that Dedekind was most prominently associated with at the time. All these were unclear, posed interesting, even challenging and highly counterintuitive problems, and overlapped in ways that were by no means well understood. This branch of topology never had to push for its acceptance, but it did have to fight for its independence as a separate subject and not a mere branch of analysis. That is largely a 20th century story, and will not be pursued here. Poincaré was to have a range of views about these issues: sympathetic to some, indifferent to others, and hostile to a few.

Whatever French mathematicians may have thought of Cantor's early work is not clear, but when he took up his new theory of transfinite sets in the 1880s the effect was to polarize the mathematical community. Mittag-Leffler was one of those who was attracted to the subject, but he soon discovered that his French contacts were not. When on 19 January 1883 he wrote to Hermite to propose that someone translate Cantor's major papers into French because, he believed, the new generation of French mathematicians would find them useful, Hermite replied on 5 March that "Mr. Poincaré judges that almost all French readers will have absolutely no inclination for the simultaneously philosophical and mathematical research of Mr. Cantor, where the arbitrary plays too great a part, and I do not think he is mistaken." Mittag-Leffler held out the hope that perhaps Poincaré himself would find them useful, but Hermite became if anything more negative in his opinion. A French translation was made of part of Cantor's *Grundlagen einer allgemeinen Mannigfaltigkeitslehre*,[1] but Hermite told Mittag-Leffler that while reading it "Picard never stopped cursing the author" (20 April 1883). However, a year later Picard began to change his mind after he met Cantor in person. By then Poincaré had been finding Cantor's ideas relevant to his own work on automorphic functions, and eventually Mittag-Leffler's optimism was vindicated when

---

[1] In *Acta Mathematica* (2, 381–408).

Appell and Poincaré proposed Cantor for membership of the SMF—he was elected unanimously on 1 April 1885.[2]

The second branch is called algebraic topology. It grew out of Riemann's study of many-valued function of a complex variable; one of its early successes was a complete classification of surfaces, and it became the natural way to frame, and sometimes solve, questions about higher-dimensional manifolds that mathematicians were beginning to ask, again under the influence of Riemann. It can appear superficial, even frivolous (hills and dales?) but was saved by its importance in the theory of surfaces, and it is difficult in an unusual way: nothing has prepared a mathematician for it in their previous work. It is this branch that Poincaré did most to develop; his work in topology was mostly concentrated in algebraic topology, and he was crucial to its development.[3]

In his (1851) and again in his (1857) Riemann had been interested in surfaces, because, at least informally, they form the possible domains of a complex function. This was a remarkable enlargement of the reach of that subject, which had hitherto been confined to complex functions defined on some or all of the complex plane. He took this step because he was interested in the problem of understanding Abelian functions, which are obtained as the integrals of expressions of the form $f(x, y) dx$, where $f(x, y)$ is a polynomial in the complex variables x and y that are connected by a polynomial equation $g(x, y) = 0$. Riemann regarded this equation as defining a surface spread out over the complex x plane, which generally has as many points over each value of x as the degree of the polynomial g in y would indicate, but at some points, where these values cease to be all distinct, has fewer. This gave him the idea of studying branched coverings of the (Riemann) sphere, and he showed that each such covering corresponds to a class of polynomial curves $g(x, y) = 0$.

This led him to two important ideas. One was that the surfaces defined by equations of the form $g(x, y) = 0$ are characterized by a single number, which he called their order of connectivity (and was soon to be lastingly replaced by the *genus* of the curve). This is the smallest number of curves that can be drawn on the surface so that it is always divided into separate pieces. It is 1 on the sphere, 3 on the torus, and generally of the form

---

[2] See *Bull. SMF* (1885, 13, 87).
[3] For a rich account of the context, and of the impact of Poincaré's work on later mathematicians see Epple (1999).

$2p + 1$ where p is the genus of the curve. Riemann also argued that two surfaces have the same order of connectivity if and only if there is a continuous one-to-one map from each onto the other. The second idea Riemann came to was that while all surfaces with order of connectivity $2p + 1$ (with $p > 1$) are topologically equivalent, they form a family of different surfaces from the standpoint of complex analysis of dimension $3p - 3$. This pointed the way to a higher-dimensional "manifold" of dimension $3p - 3$ that somehow parameterized the space of all surfaces of the appropriate order of connectivity. Riemann, and his close Italian colleague Enrico Betti, with whom he spent a lot of time in the last years of life (he died near Lake Maggiore) also had some ideas about how to characterize the three-dimensional generalization of surfaces.

After Riemann's work, A. F. Möbius in Germany and Jordan in France looked again at the classification of surfaces. The ones Riemann had considered were inevitably orientable, and Möbius showed in his (1865), as Listing had done before him, that there were also nonorientable surfaces—the eponymous Möbius band is the simplest example. Gradually it was realized that some familiar objects shared this strange property: the real projective plane, for example. He considered two surfaces to be the same if there were one-to-one maps in each direction between them that mapped nearby points to nearby points, and he showed that if two surfaces are equivalent then they have the same number of boundary components. He then showed that every surface without boundary arises by gluing together two surfaces with the same number of boundary components along their boundaries, but he did not remark that if each surface has p boundary components then the surface obtained in this way has an order of connectivity of $2p + 1$ in Riemann's sense.

Jordan came to most of the same results independently in his (1866) by studying closed curves drawn on a surface. He considered curves that start and finish at the same fixed point, and investigated when one such curve can be deformed into another without moving the fixed start and endpoint. He found that in general each surface has a certain number, $2p$, of curves that cannot be deformed into each other in this way, but that any other curve can be deformed into a sequence of these basic curves.

All of this work was in some measure imprecise and had to be reworked as the point-set branch of topology steadily revealed greater and greater need for care. Thus Jordan himself raised the question of whether every simple closed curve in the plane divides it into precisely two pieces

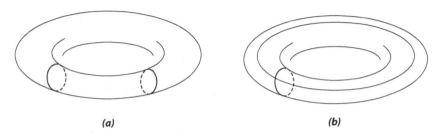

(a)   *(b)*

Figure 8.1. Equivalent curves, *left*, and inequivalent curves, *right*, on the torus.
Source: drawn by Michele Mayor Angel.

(an inside and an outside) and noted that he was unable to prove that the answer was in the affirmative. But the broad outlines of the topological theory of surfaces proved to be correct, although it was not much of a guide to what could happen in higher dimensions. Indeed, apart from an inspired but very difficult analysis by Picard in the 1880s on the nature of curves on an algebraic surface (a manifold of real dimension 4) nothing was done about curves on higher-dimensional manifolds until Poincaré took up the subject in 1895.

One crucial distinction was obscured by the comparative simplicity of the situation concerning surfaces. Riemann had regarded two closed curves on a surface as equivalent if together they formed the boundary of a piece of the surface. Two curves of latitude on the sphere are equivalent in this sense, as are the curves on the torus in figure 8.1 (*left*), but not those in figure 8.1 (*right*). In the terminology that Poincaré was to introduce later, equivalent curves are said to be homologous. Jordan's idea of equivalence was later to be called homotopy, and the distinction between homology and homotopy is exactly what Poincaré's finest work on the topology of three-dimensional manifolds was designed to explore; this is what forms the misnamed Poincaré conjecture, as we shall see below (in sec. "The Fifth Supplement").

Poincaré had met the classification of surfaces by their genus in two separate areas of his early mathematical work: flows on surfaces (where his whole approach was more and more topological; see chap. 4, sec. "Flows on Surfaces" above), and Fuchsian groups. Three-dimensional "manifolds" were largely implicit in his work on Kleinian groups, but he began to encounter other kinds of examples in his work on the three body problem in the late 1880s. They arose as subsets of real four-dimensional space defined by an invariant integral $F(x_1, x_2, y_1, y_2) = 0$, and it was clear that very little was known about them.

POINCARE'S WORK, 1895 TO 1905

Poincaré first turned his attention fully to topology—analysis situs, as he called it—in 1892, but his major paper "Analysis situs" was published in 1895 (in the centennial volume, as Poincaré was later pleased to observe, of the *Journal of the École polytechnique*). He recognized he had to meet three demands: to establish workable foundations for the subject; to show that it was intimately connected to significant problems in mathematics; and to make some progress with at least some of those problems. In the short note (1892e) he observed that the fragmentary remarks that Riemann had made on the subject had been extended to n-dimensional manifolds by Betti, who had provided numbers $p_1, p_2, \ldots, p_{n-1}$ that measure what he called the order of connectivity of the manifold, and these numbers had been shown to be far from "useless and [part of] a futile game... by our colleague M. Picard in pure analysis."[4] But if Betti had made a good start many questions remained to be asked, such as: Did these "orders of connectivity" characterize a manifold in the way that Riemann had shown that p_1 characterized two-dimensional (orientable, boundaryless) manifolds?

Poincaré used the note to show that this attempt at characterization fails even for three-dimensional manifolds. His way forward was to generalize Riemann's idea that a surface could be regarded as spread out or covering a region of the plane (or, perhaps better, a sphere). To start with that case, the surface defined by an algebraic function $f(z, w) = 0$ generally has a fixed number, k say, of points corresponding to each value z_0 of z: these are the k roots of the equation $f(z_0, w) = 0$ considered as a polynomial equation in w of degree k. For certain values of z_0 this equation has repeated roots, and these correspond to branch points of the surface. Poincaré's insight, which he had exploited obscurely in his attempt on the uniformization theorem in his (1883f), was to start with a circle in the plane and to consider what corresponds to it upstairs, on the surface. Informally, in a multistory building one can imagine

---

[4] The best account of the history of the manifold concept is still Scholz (1980).

Jack walking in circles on the ground floor while his wife, Jean, walks around in such a way as to stay vertically above her husband. It might be that when Jack returns to his starting point Jean returns to hers, but if Jack has walked around underneath staircases between the upper floors then Jean may wind up on another floor. By making Jack walk on every possible circle on the ground floor and recording every path taken by Jean, a picture of the upper floors can be established. More formally, to each circle in the z plane starting and finishing at z_0 say, and to each choice of point (z_0, w_0) such that $f(z_0, w_0) = 0$ there is a path on the surface composed of points (z, w) where z is on the circle in the plane and (z, w) is on the surface $f(z, w) = 0$. In later terminology, the paths obtained in this way are said to be lifts of the circle down below.

It is also clear that one lifted path on a surface may be followed by another, and may be undone by walking it in the opposite direction, so the collection of lifted paths forms a group, always a guarantee for Poincaré that something important was happening. Moreover, as Poincaré said,[5]

> It is clear that if two surfaces can each be transformed to the other by a continuous transformation, then their groups are isomorphic. The converse, though less evident, is again true for closed surfaces, so that *what defines a closed surface, from the viewpoint of "Analysis situs," is its group*.(1892e, 633)

So the question became: "do two surfaces with the same Betti numbers always have the same group?" For surfaces the answer was yes, but for three-manifolds in four-dimensional space the answer, as Poincaré concluded the note by showing, was in general, no. His examples will be described below.

It should now be clear that the task of creating an analysis situs for n-dimensional manifolds was threatened from the start by obscurity. What should an n-dimensional manifold be taken to be? How can it be handled with anything like rigor? How will the implicit appeal to geometrical intuition in two and three dimensions fare under such a generalization? Poincaré gave his answers to these questions in the major paper "Analysis situs" (1895a).

---

[5] The translations in this chapter are taken from Stillwell's translations of Poincaré's papers on topology, Poincaré (2011).

Mathematicians were already familiar with some manifolds. The sphere in three-dimensional space, given by the equation $x^2 + y^2 + z^2 = 1$, is a two-dimensional manifold. If one brings in a second surface, say the cone given by the equation $x^2 + y^2 - z^2 = 0$, and asks for the points common to both surfaces, one obtains two circles. In n dimensions a hypersurface is obtained as the points that satisfy an equation of the form $f(x_1, x_2, \ldots, x_n) = 0$, and this should define an $(n-1)$-dimensional manifold (Poincaré turned a blind eye to the possibility of singularities at this stage). More generally, the set of points in n-dimensional space common to p hypersurfaces given by equations $f_j(x_1, x_2, \ldots, x_n) = 0, 1 \leq j \leq p$, should form a "manifold" of dimension $n - p$. Poincaré noted that one also wants to include manifolds defined by inequalities. For example, the solid ball in three-dimensional space is a three-dimensional manifold, which can be thought of as the points satisfying either the equation $x^2 + y^2 + z^2 = 1$ or the inequality $x^2 + y^2 + z^2 < 1$. This allows one to study k-dimensional manifolds with $(k-1)$-dimensional boundaries. Poincaré also stipulated that the functions that defined the manifold be continuously differentiable and that no $p \times p$ subdeterminant of the Jacobian, the $p \times n$ matrix $(\frac{\partial f_j}{\partial x_k})$, vanish. This condition geometrically guarantees the existence of a tangent space at each point of the manifold and shows that locally a point on the manifold has $n - p$ degrees of freedom.

But, said Poincaré, a more basic intuition derives from the intuitive appreciation of a piece of surface as a region described by two coordinates, and a region of space as being described by three coordinates. For example, the (two-dimensional) sphere is described with the coordinates $(\cos\theta \sin\phi, \sin\theta \sin\phi, \cos\phi)$. By the same token, pieces of an n-dimensional manifold in $\mathbb{R}^p$ should be described by n coordinates, and this is done by specifying $p > n$ functions $\theta_j(x_1, x_2, \ldots, x_n), 1 \leq j \leq p$, of the n coordinate variables. As before, the manifold may also be limited by some inequalities. This captures the sense that locally an n-dimensional manifold is something described by n degrees of freedom. What gives this approach its great range is that one may imagine these pieces to overlap, so that an n-dimensional manifold was a sprawling network of overlapping pieces. This was familiar to mathematicians from Weierstrass's use of analytic continuation to describe the domain of an analytic function, and Poincaré promised to show that there were manifolds on this definition that were not manifolds on the earlier definition. And since even he felt the need for some precision, he now restricted his definition by requiring that the coordinate functions θ_j be analytic functions, although he somewhat weakened this by going on to show that every manifold on the first definition is a manifold on the second definition, which involves blurring the distinction between analytic and continuously differentiable. The proof is an exercise in the implicit function theorem.

A manifold defined by one set of coordinates is locally orientable, because an order has been specified on the directions obtained on varying each coordinate variable in turn. If this arbitrary order is changed it can happen that the sign of the Jacobian changes in which case the manifold is said to be oppositely oriented. If a manifold is made up of pieces, each oriented in some way, it can be that the orientations either agree, or can be altered so as to agree, in which case the entire manifold is said to be orientable, or that this cannot be done and the manifold is said to be nonorientable. The first case is typified by the cylinder and the sphere, the second by the Möbius band and the real projective plane. Poincaré showed that any manifold defined in the first manner is necessarily orientable, and since the Möbius band is not orientable he deduced that some objects that are manifolds on the second definition are not manifolds on the first definition.

In paragraph 5 of (1895a) Poincaré turned to the definition of the orders of connectivity, which he called the Betti numbers. He considered a manifold, W, of dimension p with a submanifold V of dimension q which has a boundary of dimension $q - 1$ composed of a certain number λ of manifolds: $v_1, v_2, \ldots, v_\lambda$. This, he said, he would write as

$$v_1 + v_2 + \cdots + v_\lambda \sim 0.$$

He went on,

More generally, the notation

$$k_1 v_1 + k_2 v_2 \sim k_3 v_3 + k_4 v_4$$

where the k are integers and the v are manifolds of $q - 1$ dimensions will denote that there exists a manifold W of q dimensions forming part of V, the boundary of which is composed of k_1 manifolds similar to v_1, k_2 manifolds similar to v_2, k_3 manifolds similar to v_3 but oppositely oriented, and k_4 manifolds similar to v_4 but oppositely oriented.

Relations of this form will be called homologies.

The word homology was taken from projective geometry, where it signifies that two figures are equivalent under a projective transformation. He noted that homology relations can be combined like ordinary equations.

To define the Betti numbers in paragraph 6, Poincaré said that a family of manifolds, all of the same dimensions and forming part of V are linearly independent if no sum of them with integral coefficients forms a homology. Then, if there are $P_m - 1$ closed manifolds[6] of m dimensions which are linearly independent and form part of V, but not more than $P_m - 1$, Poincaré said that the order of connectivity of V with respect to manifolds of m dimensions—in short, the mth Betti number—is equal to P_m. This generalizes Riemann's idea that a surface was $(2p + 1)$-fold connected if any system of $2p$ cuts rendered it simply connected, so a $(2p + 1)$-fold-connected surface has Betti numbers $P_0 = 2$, $P_2 = 2$, and more significantly $P_1 = 2p + 1$.

Poincaré then explained how to study m-dimensional integrals on a manifold, and then looked at how two submanifolds can meet. In the simplest case, two orientable manifolds V_p and V_{n-p} of dimensions p and $n - p$ in Euclidean n-dimensional space meet in finitely many points. At each point bases for the tangent vector spaces V_p and V_{n-p} can be chosen that respect the orientations on the two manifolds separately, and together they provide a basis for the tangent vectors to the ambient $\mathbb{R}^n$ at the point p. This basis either respects the orientation on $\mathbb{R}^n$ or it does not; if it does, Poincaré assigned the number $S = +1$ to the way the manifolds met at that point, otherwise he assigned the number $S = -1$. Poincaré proved that the number is independent of all the choices made, provided the manifolds are all orientable. When, for example, $p = 1$ these numbers provide a way of saying that the curve V_1 crosses the hypersurface V_{n-1} going from one side to the other that distinguishes whether the curve is entering or leaving any n-dimensional region that V_{n-1} bounds. It also provides a way of counting the number of times the curve crosses the

---

[6] A manifold is closed if it does not have a boundary.

hypersurface that keeps track of the entering and leaving, as Poincaré went on to show.

Poincaré considered what happens when the orientable manifolds V_p and V_{n-p} are merely subsets of some higher-dimensional manifold, W. He concentrated on the case where W is an n-dimensional region inside another n-dimensional region U bounded by some $(n-1)$-dimensional manifolds $V_1, V_2, \ldots, V_k$, and showed that if V_1 is a closed curve then the sum of the numbers S is zero, confirming the intuition that the curve enters W as often as it leaves it. Poincaré then established the converse, which is to show that if the $(n-1)$-dimensional manifolds $V_1, V_2, \ldots, V_k$ do not bound W it is possible to find a curve that starts and finishes in the part of U that is exterior to W and where the sum of the S numbers is nonzero.

Finally, Poincaré dropped the condition that $p = 1$ and looked at the general case of submanifolds of dimensions p and $n-p$ of an n-dimensional manifold, and showed that it made sense to talk about their intersection number. Since the bounding or nonbounding question is expressible in terms of homologies and this involves Betti numbers, Poincaré was able to deduce that $P_p = P_{n-p}$. Therefore, he wrote,

> *Consequently, for a closed manifold the Betti numbers equally distant from the ends of the sequence are equal.*

> This theorem has not, I believe, been announced previously; nevertheless it was known to various people who made applications of it. (para. 9)

This remarkable result, suitably given the security of a rigorous proof, is known as the Poincaré duality theorem.

In the next section Poincaré explained how to construct examples of three-dimensional manifolds. The construction, as he knew very well, was analogous to the construction of surfaces from Fuchsian polygons, or, even more simply, of a torus from a parallelogram. In each case one starts with a piece of surface homeomorphic to a disk, and with a boundary divided into $4k$ edges that are congruent in pairs. One identifies the corresponding pairs of edges, and obtains a surface (of genus k). In the Fuchsian case attention has to be paid to the way pieces of the polygon are arranged at the vertices. This information is provided by the way the corresponding Fuchsian group arranges copies of the polygon so as to tile the non-Euclidean disk.

In the simplest case, we shall think of a parallelogram in the Euclidean plane as having two long sides of length 1 parallel to the x-axis, and two short sides, one of which goes from the origin to the point represented by the complex number τ. There is a group consisting of pairs (m, n), m, n $\in \mathbb{Z}$. This group acts on the plane in this way: the element (m, n) sends the point $x + iy$ to the point $x + iy + m + n\tau$, which shifts the point m units parallel to the long side and n units parallel to the short side of the parallelogram. In this way the parallelogram tiles the Euclidean plane. We shall say that two points are equivalent if one is sent to the other by an element of the group. A surface is now formed by regarding equivalent points as the same, so it can be thought of as consisting of one example of each family of equivalent points. We get a good look at this surface by starting with all the points inside the original parallelogram. Plainly, none of these points are equivalent to any other. Equally, each point inside a copy of this parallelogram is equivalent to exactly one point inside the original one. That leaves points on the boundary of the parallelogram. Here we notice that points on any edge parallel to the long side are equivalent to two points, one on the upper edge of the original parallelogram and one on the lower edge. These two points are themselves equivalent, so we get a better picture of our surface by imagining them as having been glued together; this makes for a cylinder-shaped surface. The same is true for points on the shorter edges, which are now circles, and when they are identified the result is the familiar torus. The four vertices of the original parallelogram have all been identified to a single point, and this point, like every other point on the torus, is surrounded by a small disk.

Another way of thinking of such surfaces, which goes back to Riemann, is to think of them as a polygon with the curious property that when a moving point leaves it by crossing an edge it reenters the polygon at the corresponding point on the corresponding edge. This leaves the final surface harder to see, but perhaps easier to study, and it avoids talk of stretching and gluing. Either way, the message is that a polygon with edges identified in pairs is a surface. Any edges which are not paired off remain as boundaries of the surface.

To obtain examples of three-dimensional manifolds, Poincaré now proposed to start with a cube and identify its faces in pairs. The second way of thinking of surfaces is the one that generalizes most intuitively. The aim is to wind up with a point set in which every point can be surrounded by a small ball lying entirely in the set, or, perhaps by a solid hemisphere, as is the case with the cube itself (points inside it can be surrounded by solid balls, points on a face by solid hemispheres). So to see if a cube with faces identified in pairs is a three-dimensional manifold one looks to see if every point in it is surrounded in this fashion.

There is no problem with points inside the cube. Points on a face or on an edge also generally cause no trouble. A difficulty can arise when a face is identified with itself, and it can happen that the resulting space is not a manifold. But if only distinct faces are identified with another, two corresponding points were surrounded by solid hemispheres that now glue together to form a solid ball, and if the face is not identified with another, the point remains on the boundary of the resulting object. Any rule for gluing faces together will glue edges together, and attention must be paid to the orientation on any edge that is glued to itself, which can happen, or again the result may not be a manifold. The situation with the original eight vertices of the cube is, however, more likely to cause trouble. It might be, for example, that the eight vertices of the cube are all glued together in a single point surrounded nicely by eight octants. Or it might be that the eight vertices glue together in two families of four, each family fitting round a common point the way the faces of a tetrahedron fit around its center. In such cases, the resulting object is indeed a three-dimensional manifold. But it might be that the identified vertices are not completely surrounded, and in such a case the resulting object is not a three-dimensional manifold.

Poincaré investigated this requirement on the vertices, and showed that it yielded a simple arithmetical test that must be satisfied by the number of faces, edges and vertices after the identifications have been made if the identification space is to be a three-dimensional manifold. It followed that of the five simple examples he gave of a cube with identified faces, four yielded a three-dimensional manifold and one did not. He then gave a rather more complicated construction, in which the cube was made to fill out space under the action of a group, and which gave him infinitely many examples of three-dimensional manifolds.

Now that he had created an abundance of three-dimensional manifolds Poincaré turned to the question of how they might be distinguished or shown to be the same. In paragraph 12, he introduced what he called the fundamental group, as follows. He imagined that there was a family of many-valued differentiable functions on the manifold with the property that when a point is taken round a loop in the manifold and returns to its original position these functions all return to their original values if the loop can be shrunk to a point without leaving the manifold, but otherwise they may not. These functions allowed Poincaré to distinguish between kinds of loop, and each loop gives rise to a permutation of values of the functions at the starting point. Poincaré now argued that one could add two loops by doing first one and then the other (noting that the order in which this is done matters) and so the collection of all loops at a point gave rise to a group of permutations of these many-valued functions. This group is what he called the permutation group of the manifold.

Then, in a significant mistake, he claimed that if a loop is part of the boundary of a two-dimensional manifold in a three-dimensional manifold, it can be shrunk to a point (Poincaré was to correct this mistake in the fifth supplement). It follows, he said, that the study of loops in a three-dimensional manifold reduces to the study of homologies, with the single difference that in the equation

$$k_1 C_1 + k_2 C_2 \equiv k_3 C_3 + k_4 C_4,$$

where the curves C_j are closed curves starting and finishing at the same point, one could not change the order of the terms, for example, to

$$k_1 C_1 + k_2 C_2 \equiv k_4 C_4 + k_3 C_3,$$

and so from the equation $2C \equiv 0$ one cannot deduce $C \equiv 0$. The group of all loops based at a given point Poincaré called the fundamental group of the manifold; the permutation group is a representation of it.

Next, Poincaré applied these ideas to the manifolds he had constructed from cubes, where in each case it was easy enough to find a collection of loops with the property that every loop was a sum of these. His intention was to see whether the fundamental groups of these manifolds were different, in which case the corresponding manifolds could not be homeomorphic. He observed that in the case of surfaces the Fuchsian group used to construct the surface was also the fundamental group of the surface, and added that these groups are isomorphic if and only if the surfaces have the same genus, so the fundamental group, like the Betti number, serves to distinguish nonhomeomorphic surfaces.

He then asked about the situation in three dimensions and promised to exhibit examples of nonhomeomorphic three-dimensional manifolds with the same Betti numbers. This he did by taking the infinite class of group actions that filled the space with copies of a cube and showing that while their groups were all distinct the corresponding three-dimensional manifolds had their Betti numbers between 2 and 4. So there were many nonhomeomorphic three-dimensional manifolds with the same Betti numbers. This being the case, he then asked if the fundamental group sufficed to distinguish between manifolds of the same dimension. (He also asked if every group can arise as the fundamental group of a manifold.) He called a manifold simply connected if its fundamental group vanished, so spheres are simply connected, as are disks, but his use of the term was more than usually imprecise.[7]

In paragraph 15 he discussed other ways of constructing manifolds, by finding them as the images under a map of other manifolds. As one of his illustrations he supposed that one had a manifold W and a finite group G that acts on it (this means that for every point P of W and every element g of the group G there is a point gP of W). Then for each point P of W Poincaré considered the set of points $\{gP : g \in G\}$ and looked to see if these sets could be taken as the individual points of another manifold. To take an example in his spirit but simpler, let the manifold W be the n-dimensional sphere, the set of points $\mathbf{x} = (x_0, x_1, \ldots, x_n)$ such that $x_0^2 + x_1^2 + \cdots + x_n^2 = 1$, the group G the group with two elements, and let the nonidentity element map $(x_0, x_1, \ldots, x_n)$ to the diametrically opposite point $-\mathbf{x} = (-x_0, -x_1, \ldots, -x_n)$. Then each pair of points $\{\mathbf{x}, -\mathbf{x}\}$ forms a point in n-dimensional projective space, and one can ask whether it is orientable or not. Poincaré's argument about a more complicated situation shows that in this case n-dimensional projective space is orientable if and only if n is odd.

Paragraph 16 concludes the paper with a generalization of the Euler characteristic to the case of n-dimensional manifolds.[8] The Euler formula for a convex polyhedron (a sphere) asserts that $V - E + F = 2$, where V, E, and F stand for the number of vertices, edges, and faces of the polyhedron. For surfaces of higher genus, this was modified by Jordan (1866) to say $V - E + F = 3 - P_1$, where P_1 is the Betti number of the

---

[7] See Stillwell's commentary in Poincaré (2010).
[8] See Dieudonné (1989, 26–28).

polyhedron. Dieudonné claimed that for convex higher-dimensional polyhedra the prevailing idea was that there should be an alternating sum over the number of j-dimensional faces, but that no one before Poincaré had tackled the general case. Poincaré saw the problem in strictly topological terms: the various "polyhedra" had to be simply connected (in this case, disk shaped) but could be curvilinear. However, his understanding of what this entails was to be unduly naive, and we shall not attempt to follow all the details but rather to notice the issues that were thrown up in this fashion.

The analogy is with a surface that has been divided into a certain number of faces that are bounded by a certain number of edges that in turn are bounded by a certain number of vertices. Poincaré supposed that V is a p-dimensional manifold, divided into α_p p-dimensional manifolds v_p which are not closed; the α_{p-1} boundaries v_{p-1} of these manifolds are $(p-1)$-dimensional manifolds which are also not closed; the α_{p-2} boundaries v_{p-2} of these manifolds are $(p-2)$-dimensional manifolds, and so on, until α_0 points are reached that are the boundaries of the α_1 edges. Furthermore, he assumed that the manifolds v_j are all simply connected. He now proposed to compute the number

$$N = \alpha_p - \alpha_{p-1} + \alpha_{p-2} + \cdots + (-1)^{p-1}\alpha_1 + (-1)^p\alpha_0.$$

In paragraph 16 he supposed a subdivision of the kind described was given, and proposed to consider looking at two refinements of this subdivision and the subdivision that was obtained as further refinement common to each of these two. He proposed to show that the alternating sum was the same for each of these subdivisions and was therefore independent of any subdivision and a property of the manifold. To keep count of what can happen, Poincaré observed that in the two-dimensional case an edge separates two faces, but a vertex may belong to many faces. In higher dimensions similar phenomena can occur, and they complicate the process of reconstructing a subdivision from any one of its refinements, so Poincaré sketched an argument to show how the awkward cases can be avoided by a process of suppressing suitable submanifolds and annexing adjacent regions. As he observed, the process has to be carefully managed to ensure that all the submanifolds remain simply connected at every stage, but, he said, if that can be done then the alternating sum can be computed for a simply connected polyhedron.

Indeed, it is enough to take an n-dimensional tetrahedron, for which the sum is 2 if n is even and zero if n is odd.

"This being so," he boldly said, "I shall now establish our theorem in a complete and rigorous fashion by supposing it true for all manifolds of less than p dimensions." He proposed that any manifold together with a subdivision, can be embedded in a suitably high-dimensional Euclidean space, and as a result be given a suitably fine subdivision, one essentially obtained by dividing the ambient Euclidean space into very many, very small parallelepipeds and considering their intersections with the original, given subdivision. He then argued that the original subdivision could be reconstructed from the refined one, and that in the process the number N did not change; it was therefore a property of the manifold and not of the subdivision.

A further calculation, using Poincaré duality, showed that the number N was zero if the manifold had odd dimension and depended on the Betti numbers if the dimension was even. He went on to show that indeed for a two-dimensional manifold one has $\alpha_0 - \alpha_1 + \alpha_2 = 3 - P_1$, where α_j is the number of j-dimensional submanifolds in the subdivision, P_1 is the Betti number of the polyhedron, and for a three-dimensional manifold the analogous result is $\alpha_0 - \alpha_1 + \alpha_2 - \alpha_3 = P_2 - P_1$, where P_2 is the second Betti number of the manifold. By his duality theorem, $P_2 = P_1$, and so the alternating sum vanishes. He ended the paper with the claim that analogously in all higher dimensions, for a p-dimensional manifold,

$$N = P_{p-1} - P_{p-2} + \cdots + P_2 - P_1, \quad \text{if } p \text{ is odd,}$$

and

$$N = 3 - P_1 + P_2 - \cdots + P_{p-1}, \quad \text{if } p \text{ is even.}$$

It may be the right of the first person to enter, or create, a field of mathematics to do only the simplest or most general cases, or to use the most intuitive or naive methods that later mathematicians would find in need of repair, if not replacement, but seldom can that privilege have been exercised so boldly as it was on this occasion. Poincaré's arguments about refining subdivisions and reconstructing subdivisions from their refinements cannot be made to work—which does not mean that his conclusions were incorrect. Being counting arguments, they depend on every number being finite, but there is no reason at all to prevent infinite intersections appearing: consider, for example, the intersections of the graph of $x \sin(1/x)$ with the x-axis. At one point Poincaré suggested that problems like this could be avoided by a slight deformation of the manifolds in question, and this is not true either. And yet, as Dieudonné noted (1989, 27), Poincaré's approach has to be seen as a forerunner of the later, successful proofs by Brouwer and Alexander. Poincaré's calculation of the higher-dimensional Euler number is obscure even in dimension 2; in dimension 3 his arguments "are in fact unsupported by any proof" (Dieudonné 1989, 28), and for all higher dimensions are nothing more than a claim—and yet they too have a modern ring and a lasting influence, as we shall now see.

In the case of a two-dimensional manifold with a subdivision, Poincaré assigned integers at random to the vertices, and to each (directed) edge he assigned the difference of these integers. There are $\alpha_0 - 1$ of these differences, and α_1 edges, so there are $\alpha_1 - \alpha_0 + 1$ linear relations between these differences. Now consider a closed curve on the manifold that consists entirely of edges. The sum of the differences on such a contour will be zero. The important closed contours on a non-simply connected, two-dimensional manifold are the $P_1 - 1$ contours that divide it into simply connected pieces, and as one traces one of these contours, say C, one crosses successively from one face of the subdivision into another until one returns to the starting point. Poincaré supposed that the crossing points were always finite in number and are, say, $a_1, a_2, \ldots, a_n$. Each arc, say $a_j a_{j+1}$, lies entirely in one face and is homologous to the sum of the edges of the face that join a_j and a_{j+1}. But then, as one deals with the arc $a_{j+1} a_{j+2}$ one cancels part of the edges by tracing it in the opposite direction. In the end, one is left with a closed contour, homologous to C and consisting of complete edges. Since there are $P_1 - 1$ of the closed contours which we began with, we end up with $P_1 - 1$ of these closed contours composed of edges, and so $P_1 - 1$ more relations between the differences. "We thus obtain $\alpha_2 + P_1 - 1$ closed contours consisting of edges." said Poincaré, and "I claim that all possible closed contours are combinations of these." Let us grant his claim. It follows that there are $(\alpha_2 + P_1 - 1) - 1$ distinct relations between the differences, and therefore $\alpha_1 - (\alpha_2 + P_1 - 1) + 1$ arbitrary differences, and so

$$\alpha_0 - 1 = \alpha_1 - (\alpha_2 + P_1 - 1) + 1,$$

from which it follows that

$$N = \alpha_0 - \alpha_1 + \alpha_2 = 3 - P_1,$$

as was to be shown. This argument strongly suggests to a present-day algebraic topologist the method of 0-cochains and 1-coboundaries which became fundamental to arguments about the existence of vector fields on manifolds, and of course Poincaré had been the first to show that the only two-dimensional manifold to admit a continuous, nowhere-vanishing vector field was the torus, for which the Euler number is zero.

This "fascinating and exasperating paper," to quote Dieudonné again (1989, 28) was eventually to do several things. It certainly suggested to mathematicians that the subject of higher-dimensional topology could be done. In order to avoid the problems raised by Poincaré's naivety it also suggested to them that a combinatorial approach in terms of cell complexes could be the way forward. With this approach one specifies a set of vertices, a subset of the set of pairs of vertices that define the edges, a subset of the set of triples of vertices that define the faces, and so on, taking care that faces only meet in edges, for example. It then becomes an open question as to whether every n-dimensional manifold is homeomorphic to a suitable cell complex, but this restriction allows for a much greater degree of rigor.

The rectangle in figure 8.2 defines a torus, with a subdivision given by 7 hexagons (incidentally establishing that the torus is 7-colorable). In the subdivision, we have 9 vertices, to which we arbitrarily assign the numbers 1 to 9, so there are 8 differences. We have 16 edges, with their difference numbers shown in bold, and there are $16 - 9 + 1 = 8$ linear relations between these numbers. There are two elements of the homology basis, so $P_1 - 1 = 2$, and so there are $7 + 3 - 1 = 9$ closed contours consisting of edges (they are the 7 for the hexagonal faces and one containing the edge 12 and another the edge 23). This gives a further $9 - 1 = 8$ relations between the differences on the edges, for a total of $16 - 8$ arbitrary differences, and so

$$9 - 1 = 16 - 9 + 1 \quad \text{or} \quad 9 - 16 + 7 = 2 - (3 - 1) = 0,$$

confirming that the Euler number is zero.

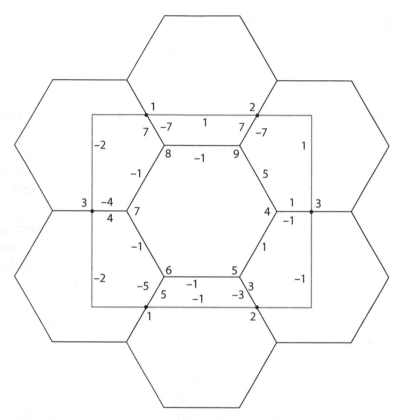

Figure 8.2. Computing the Euler number of the torus. Source: drawn by Michele Mayor Angel.

The First Supplement

The first critical reader of Poincaré's long paper was the Danish mathematician, Poul Heegaard. In his book (1898) he showed by an example that Poincaré's duality theorem was wrong.

Poincaré replied the next year in his (1899b). He now realized, he said, that there were two ways to define the Betti numbers and they did not always agree. One could suppose given an n-dimensional manifold M^n and a number of p-dimensional submanifolds $V_1, V_2, \ldots, V_k$ that do not form the boundary of a submanifold of dimension $p + 1$ in M^n, but are such that there is a further p-dimensional submanifold V_{k+1} such that the submanifolds $V_1, V_2, \ldots, V_k, V_{k+1}$ form the boundary of a submanifold of M^n. In this case, Betti had defined the pth Betti number of M^n as $k + 1$.

Or, one could work with homologies, and write

$$V_1 + V_2 + \cdots + V_k \sim 0$$

to mean that the $V_1, V_2, \ldots, V_k$ form the boundary of some $(p+1)$-dimensional submanifold of M^n. In this setting the same submanifold V_j may occur more than once, so one may have, for example, $2V_1 + 3V_2 \sim 0$. Submanifolds are said to be independent if they are not connected by any homology, and if there are k independent p-dimensional submanifolds but not $k+1$, the Betti number on this second approach is defined to be $k+1$.

However, Poincaré now said,

> This second definition, which is the one I adopted in "Analysis situs," does not agree with the first.

More precisely, on the first approach, as Heegaard had shown, there is a three-dimensional manifold in which the middle Betti numbers are not equal, indeed $P_1 = 2$, $P_2 = 1$, whereas on the second definition $P_1 = 1 = P_2$. Nor can this be dismissed as the simple consequence of different definitions, because, as Poincaré went on to say, after giving examples to illustrate the point,

> The proof that I have given in "Analysis situs" seems to apply equally well to the two definitions of Betti number; therefore it must have a weak point, since the preceding examples show adequately that the theorem is not true for the first definition.

He agreed that Heegaard had identified the weak point of the proof: "If conversely the homology $\sum_j V_j \sim 0$ does not hold, then we can trace a closed curve V' on it such that $\sum_j N(V_j, V') \neq 0$, but it is not certain that this curve is an intersection of the manifolds V_j."

To rectify matters, and to defend his definition of the Betti numbers, Poincaré reworked his earlier analysis, concentrating on the case of three-dimensional manifolds in Euclidean four-dimensional space and, more importantly for the later growth of the subject, working in markedly more combinatorial terms. Manifolds are now built up out of, or presumed to be divisible into, cells, much in the way a building is divided into rooms (corridors included). Rooms (three-dimensional cells) meet in walls (two-dimensional cells) that themselves meet in edges (one-dimensional cells) that end in vertices (zero-dimensional cells). Conversely, a three-dimensional manifold can be specified by listing its vertices (0-cells), specifying which of these are joined by edges (1-cells), which edges span walls (the 2-cells), and which walls span rooms (the 3-cells). There is no requirement that these cells be straight or flat.

To connect this approach with his earlier, more analytic or perhaps naively geometrical one, Poincaré had to show that such a schema (as he called it) will correspond to a manifold provided certain conditions are met that guarantee that every point of the schema has a neighborhood in the schema that is homeomorphic to a three-dimensional ball. It would also be helpful if, as Poincaré asked, two manifolds with the same schema are homeomorphic.

Putting these questions on one side, Poincaré proceeded to show that one can always refine a schema so that all the cells are simply connected (i.e., disk shaped) and then that the data defining a schema can be used to find homologies. In this way he was able to give a new, more rigorous proof of the duality theorem, by means of a theory of dual subdivisions, but, as he observed at the start of paragraph IX, "The fundamental theorem is now established, by a proof which differs essentially from that on p. 46 of 'Analysis situs.'" This did not satisfy him, because he also wanted to recover some of the preliminary theorems that had led up to this result, specifically, "the necessary and sufficient condition for the existence of a manifold V such that $\sum_j N(V, V_j) \sim 0$ is that there is no homology $\sum_j V_j \sim 0$." And this he did, again by his combinatorial or arithmetic methods. Only now did Poincaré return to the question of whether a manifold admits a subdivision into a schema. He ended this supplement with a proof that satisfied him, but has not stood the test of time.[9]

Poincaré sent his second supplement to the London Mathematical Society, who published it in 1900. The chief novelty here was what Poincaré called torsion. In the first supplement he had noted that a homology might take the form $2V_1 + 3V_2 \sim 0$, but he had explicitly commented that when one had a homology of the form $3V_1 + 3V_2 \sim 0$, where all the coefficients are the same, one can divide through and write $V_1 + V_2 \sim 0$. Now he noticed that this may not always be true. It

---

[9] Stillwell, in Poincaré (2010, xiii), says that the first rigorous proof that every differentiable manifold has a polyhedral subdivision is Cairns (1935).

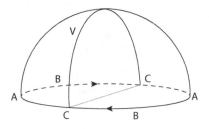 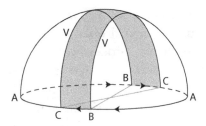

Figure 8.3. Torsion in projective space. Source: drawn by Michele Mayor Angel.

can happen that, for example, $2V \sim 0$ although it is not the case that $V \sim 0$. The simplest example is that of a closed curve in the projective plane (see fig. 8.3). It is possible to draw such a curve V that does not separate the projective plane and is not a boundary, but whose double, $2V$, separates the projective plane into a disk and a Möbius band, both of which it bounds. When this occurs Poincaré said that the manifold has torsion, and manifolds for which $nV \sim 0$ implies $V \sim 0$ he called manifolds without torsion.

Poincaré showed that torsion can be detected by the combinatorial methods of his first supplement, and that the two definitions of the Betti numbers agree if and only if the manifold has no torsion. He then gave illustrative examples of three-dimensional manifolds with and without torsion, including Heegaard's example. Sarkaria (1999, 150–151) points out that this involves the presence of a copy of $\mathbb{RP}^3$ in the complex hypersurface $z^2 = xy$ and that Poincaré did not notice that this is the same as example 5 in the "Analysis situs" paper and therefore diffeomorphic to $\mathbb{RP}^3$ "though it is all but apparent from the cell subdivision which he uses."

Poincaré's combinatorial methods involved writing down matrices T_q (which he called tables) that expressed the relationship between the cells of dimension $q - 1$ and those of dimension q, and then reducing these matrices to a canonical form by rules that on the one hand had a natural geometrical interpretation and on the other a simple algebraic character.[10] The geometry allowed for the creation or replacement of some cells by others that did not break any of the defining condition for a schema (such as removing an edge between two faces if the result

---

[10] The algebraic problem had already been solved by the English mathematician Henry Smith. Unaware of this, Poincaré rederived the same result.

is the creation of a new, simply connected face). The algebra took the form of standard operations on the rows in the table. He noticed that the reduction process would end with a matrix of a certain form if the manifold was orientable: it would be diagonal, the diagonal entries would be $(d_1, d_2, \ldots, d_r, 0, 0, \ldots, 0)$, and $d_1 | d_2 \ldots | d_r$. Now he noticed that when the manifold was orientable the quotients d_k / d_{k-1} would be 0 or ± 1—and there would be no torsion; but if torsion was present the manifold would be nonorientable. This led Poincaré to remark,

> In order to avoid making this work too prolonged, I confine myself to stating the following theorem, the proof of which will require further developments:
>
>   Each polyhedron which has all its Betti numbers equal to 1 and all its tables T_q orientable is simply connected, i.e. homeomorphic to a hypersphere. (para. 6)

This is the first appearance of what became known as the Poincaré conjecture, and in the fifth supplement he showed that in this form it is false. In this form it amounts to claiming that homology alone can tell if an orientable manifold is homeomorphic to a sphere of the same dimension, and as we shall see that Poincaré was to realize that this is not the case and formulate a much more challenging opinion.

I will not describe the third or fourth supplements in any detail, because they are difficult technical applications to contemporary work on complex algebraic geometry. The third (1902g) is devoted to the study of the complex surfaces defined by equations of the form $z^2 = F(x, y)$, where F is a polynomial in x and y. Poincaré had been led to this by the study of perturbations in celestial mechanics (see chap. 7, sec. "Function Theory of Several Variables" above). For each fixed value of $y = y_0$ the equation $z^2 = F(x, y_0)$ defines a hyperelliptic Riemann surface, the genus of which is the same for almost all values of y_0, and which is understood as a branched covering of the Riemann sphere branched over a fixed number of points. Poincaré thought of it as described by a certain Fuchsian polygon in the unit disk, and this made it possible for him to study the whole complex manifold obtained as y varies, for, if y is taken on a loop and returned to its starting point the polygon will change but finish in its original form, except that the vertices of the polygon may have been permuted. This allowed Poincaré to show that the singular points on the complex surface are conical singularities, and to deduce, using a result of Picard's—who had shown in Picard and Simart (1887, I, 85–93) that a general complex surface has $P_1 = 1$—that the fundamental group of the complex surface must be finite. Indeed, if the polynomial $F(x, y)$ is irreducible, Poincaré showed that the fundamental group of the surface is the identity.

In the fourth supplement (1902i) Poincaré applied his topological ideas to Picard's work on a smooth complex two-dimensional (real four-dimensional) manifold, and succeeded after much work in computing P_2 of such a manifold.[11] Here Poincaré also gave another proof of his duality theorem. Both of these papers connect to Poincaré's later work (1910) on the theorem of Castelnuovo and Enriques, in algebraic geometry (see chap. 9, sec. "Algebraic Geometry" below).

The Fifth Supplement

Fame, or perhaps notoriety, attaches, however, to the fifth and final supplement (1904a). The paper opens with the claim of a great achievement:

> I construct an example of a manifold, all Betti numbers and torsion coefficients of which equal 1, but which is not simply connected.

And it closes with this remark about a three-dimensional manifold V:

> One question remains to be dealt with: Is it possible for the fundamental group of V to reduce to the identity without V being simply connected?

This question, which is not a conjecture, grew into the Poincaré conjecture: Is a three-dimensional manifold with vanishing fundamental group homeomorphic to a three-dimensional sphere? The answer, after a number of profound innovations in mathematics, was only shown to be in the affirmative by Grigori Perelman in 2003, and then by methods than not even Poincaré could have imagined.

Although the supplement is principally about three-dimensional manifolds, Poincaré began by explaining a way of understanding a general k-dimensional manifold. He imagined that it was sitting inside an

[11] For an account, see Scholz (1980, 365–371).

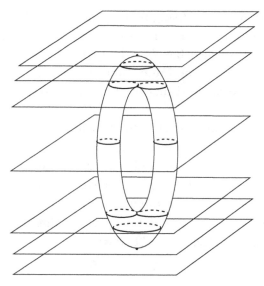

Figure 8.4. Slices through a torus. Source: drawn by Michele Mayor Angel.

n-dimensional manifold and then supposed that this manifold was filled out by an infinite family of nonintersecting $(n-1)$-dimensional manifolds, which cut the given k-dimensional manifold into $(k-1)$-dimensional slices. As with the surface considered in the third supplement, the idea was that most of these slices would not change significantly as one varied the height of the slice, but there would be a finite number of points (called singular points) where the topology of the slice would change. This approach has become standard under the name of Morse theory, after the mathematician Marston Morse, who was to use it to great effect in from 1925 onwards, and it has acquired a standard introductory example, which is well worth looking at.

Slicing a Torus

Consider a hollow tire or inner tube centered on a fictitious horizontal axis, and sliced by planes parallel to the ground, as in figure 8.4. Planes that are too low or too high will not meet the tire, but there is a lowest one that does, and this plane meets the torus in a point (the lowest singular point). Slices that are slightly higher meet the torus in loops that together fill out the bowl-shaped lower part of the tire. As the height of the slices rises the loops acquire two dimples that eventually join up and the slice through the tire takes a figure-eight shape, which it does at the second singular point. Above that height the slice becomes two circles, and it remains that way until the process is reversed near the top: the two circles turn into a figure eight at the height of the third singular point, then into a loop, then into a point at the top where the fourth and final singular point is, and no higher slice meets the tire at all.

This means that the torus can be regarded as made up of a bowl (at the base) upon which rests a saddle-shaped region where it branches into two tubes. These tubes come together further up, at an inverted saddle, and the whole torus is finished off with an inverted bowl.

If one now looks at the slices near to the lower saddle point, the ones below the saddle point and above it each cut the torus in nonsingular curves, whereas the slice through the saddle cuts the torus in a curve that crosses itself at the saddle point. However, the pair of curves above and below the central slice are different: what one sees as the slices rise from the lowest point of the torus is one curve forming a figure eight and then splitting into two. This is important information about how the pieces fit together.

More formally, at each of the singular points the plane is a tangent to the surface, and at such points the surface can be expressed as the graph of a function $z = f(x, y)$ where x and y are Cartesian coordinates in the plane and the z-axis points vertically upwards. When this is done, the surface near to a slice has an equation of the form

$$z = ax^2 + bxy + cy^2 + \text{higher-order terms},$$

because the x–y plane is a horizontal tangent plane. Furthermore, at the lowest singular point the surface is locally of the form $z = ax^2 + by^2 + \text{higher-order terms}$, and at the first saddle point the surface is locally of the form $z = ax^2 - by^2 + \text{higher-order terms}$, where in each case the coefficients a and b are positive.[12] If the higher-order terms are neglected, as they can be when very close to each of the four points in question, the result is that near the points where the torus is like a bowl it has an equation of the form $z = ax^2 + by^2$ (at the lowest singular point) or $z = -ax^2 - by^2$ (at the highest), and at the saddle points the corresponding equation looks like $z = ax^2 - by^2$ or $z = -ax^2 + by^2$. So Poincaré decided to concentrate on the behavior of the quadratic term.

---

[12] From now on in this chapter unless otherwise stated a and b are positive.

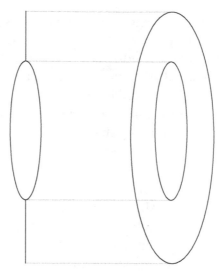

Figure 8.5. The skeleton, *left*, of a torus, *right*. Source: drawn by Michele Mayor Angel.

How these pieces fit together is difficult to see. Poincaré had the idea of hanging them on what he called a skeleton. This is a record of the number of distinct pieces each slice cuts the three-dimensional manifold into. Each piece is recorded as a point, the records are stacked up like the slices, and the result is a family of lines that separate or come together. If we return to the two-dimensional case briefly, the skeleton of a torus (see fig. 8.5) is a line segment that splits into two curves that join up and terminate in a further line segment.

If we look at a torus to which a lumpy sphere or bulb has been attached, as in figure 8.6, we see that the slices initially meet the torus, then they meet the sphere as well, then the slice through the torus splits into two, and so on. The feature to note is that as a point moves steadily upwards on the skeleton it either meets another point, or it has a choice to make, or it dead ends.[13]

Poincaré found it more useful to replace the plane slices with slices made by a family of concentric spheres centered on a singular point. This enabled him to identify the singular points that corresponded to dead ends very easily. Consider the quadratic term corresponding to the lowest singular point of the torus, $ax^2 + by^2 = 0$, which is simply the point $(0, 0)$,

---

[13] The skeletons have an amusing coincidental resemblance to Feynman diagrams.

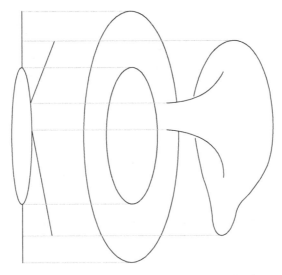

Figure 8.6. The skeleton of a torus with a bulb attached. Source: drawn by Michele Mayor Angel.

and its intersection with the circle $x^2 + y^2 = 1$. The intersection is empty. Now consider the intersection of the quadratic terms corresponding to a saddle corresponding to the equation $ax^2 - by^2 = 0$ and the circle $x^2 + y^2 = 1$. This intersection has four points in it, because $ax^2 - by^2 = (\sqrt{a}x + \sqrt{b}y)(\sqrt{a}x - \sqrt{b}y) = 0$ and the corresponding locus is a pair of straight lines through the origin. This says that the surface leaves the sphere in two pairs of directions. Now, said Poincaré, all the intersections are bounded, so the curves that leave must also return. He labeled the directions $1, 2, 3, 4$ and recalled that at the saddle the branches 1–3 and 2–4 cross (see fig. 8.7), and now he observed that below the saddle point the slice is a single loop, whereas at slices above the saddle point the resulting curve has two pieces. This explains how the curves must join up.

Given a saddle point, P, on an arbitrary surface, it can happen that the curve that exits in direction number 1 rejoins at direction 2, 3, or 4. To consider what can happen, Poincaré considered that P has four curves coming out of it cut out by the slice through P, which has height $t = 0$. He picked a point on each one and close to P; these he labeled $1, 2, 3, 4$. Now, by varying the height of the slice slightly on either side, cutting both higher and lower, the point 1 is moved up and down and sweeps out a strip. The points 2, 3, and 4 do the same, and define four strips coming

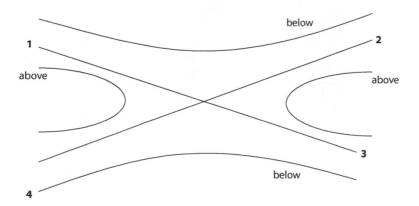

Figure 8.7. Curves near a saddle. Source: drawn by Michele Mayor Angel.

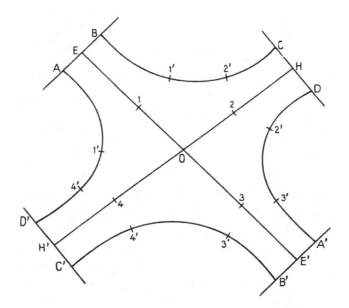

Figure 8.8. A familiar polygon. Source: Poincaré, *Oeuvres* (6, 443).

out of a neighborhood of P. The neighborhood of P defined by these four strips and the slices of height $+t_0$ and $-t_0$ for a small value of t_0 can be cut out and flattened, whereupon it looks like figure 8.8.

Poincaré now had what for him was a very familiar figure. As he put it,

We thus obtain a polygon, analogous to a Fuchsian polygon, whose edges are conjugate in pairs, two conjugate edges corresponding to two sides of the same cut. (para. 3)

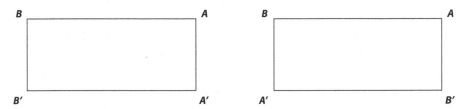

Figure 8.9. Identifications for a cylinder, *left*, and a Möbius band, *right*. Source: drawn by Michele Mayor Angel.

He continued,

> Let AB and $A'B'$ be two conjugate edges, so that the vertex A is conjugate to A', and the vertex B to B'. If in moving from A to B the interior of the polygon is (say) on the left, while in moving from A' to B' it is on the right, we say that the conjugation is *direct*. If, on the contrary, the interior is on the left when going from A to B and also in going from A' to B', we say that the conjugation is *inverse*. (This convention may seem surprising at first, but it will be justified on reflection.) That being given, if all pairs of edges are directly conjugate (as is the case for Fuchsian polygons) the corresponding surface is orientable. If the conjugation is inverse for one pair of edges then the surface is non-orientable.

The required reflection can be obtained by considering the familiar polygons that correspond to the torus, which is orientable, and the projective plane, which is not, as Poincaré immediately explained (see fig. 8.9).

In the case at hand,

> We have a singular point at O; the lines CHD, AEB, $C'H'D'$, $A'R'B'$ represent part of the boundary of the polygon. The two lines $E\,103\,E'$ and $H\,204\,H'$ which cross at O are the maps of the curves $w(0)$. The lines $B\,1'2'C$, $B'\,3'\,4'\,C'$ are the map of the curve $w(t)$ for $t < 0$. The lines $D\,2'3'A'$ and $D\,4'\,1'A$ are the map of the curve $w(t)$ for $t > 0$. By hypothesis, this curve $w(t)$ is closed onto itself in such a way that the branch $1'$ joins the branch $3'$ and the branch $2'$ joins the branch $4'$. Then A must be pasted to A', i.e. A is conjugate to A', and similarly B to B', C to C', D to D'. Thus in our polygon AB is conjugate to $A'B'$ and CD to $C'D'$. The conjugation is inverse, so our surface is non-orientable.

The algebraic argument is the one that most easily generalizes to three-dimensional manifolds. A three-dimensional manifold embedded in a space of four or more dimensions may be sliced by a family of parallel hyperplanes. The slices will be two-dimensional surfaces, and at points where the topology of the slices changes the hyperplanes will be tangent to the surface. If one chooses suitable (x, y, z)-coordinates in the hyperplane and a t-axis perpendicular to the hyperplane, the surface locally will be of the form $t = \pm(x^2 \pm y^2 \pm z^2)$. So the three-dimensional manifold will be made up of the three-dimensional analogues of bowls and saddles, joined together in some way. The bowls show up as dead ends in the skeleton, as before, but what happens at the analogues of the saddles is more complicated.

Poincaré had analyzed the two-dimensional case by looking at the intersection of a circle and a quadratic cone in two variables. In the three-dimensional case this led him to consider the intersection of a sphere with a quadratic cone in three variables. Such a cone contains a family of two-sheeted hyperbolas inside it and a family of one-sheeted hyperbolas outside it, and it is necessary to see how these surfaces join up. What replaces the saddle-shaped regions with four strips leaving it in the two-dimensional case is now a ball-shaped region with tubes leaving it.

Poincaré imagined an ellipse, E, at the neck of the hyperboloid of one sheet, noted that it vanishes as the slice passes through the singular point and the hyperboloid of one sheet turns into a hyperboloid of two sheets. He then asked if E separated the surface; either it does or it does not. If it does, the skeleton bifurcates at the critical point, otherwise it does not. If bifurcation does not occur, it is necessary to consider both the orientable and nonorientable cases. Poincaré concentrated on the orientable case. To track the change in the topology of the two-dimensional slices Poincaré observed that the topology of a surface was measured by its Betti numbers, the number of linearly independent closed curves that can be drawn on the surface under the relation of homology. So what one has to look for is closed curves that disappear, or become homologous to zero as the slice passes through the singular point. Such curves must pass through the ellipse E, and by counting the number of signed intersections one can tell if the curve is homologous to one that never meets E. The homology of such curves is not affected as E vanishes, but curves that are only homologous to curves that meet E are affected. These closed curves will vanish.

As for the closed curves that survive the transition from the one type of hyperboloid to the other, they may become homologous to zero although they were not before. In this case the original curve either bounded a region on the surface, or it did so together with E. Poincaré deduced that

> when t passes from $-\varepsilon$ to $+\varepsilon$ certain cycles can disappear, but new cycles cannot appear; certain cycles can become homologous to zero, but no cycle can lose this property, so that the Betti number can decrease, but not increase. (para. 2)

A consideration of the cases when E is or is not homologous to zero showed that when it is homologous to zero there are no new homologies and the number of cycles remains the same and when it is not homologous to zero the number of distinct cycles diminishes by 2, and the new surface is $(2p - 1)$-fold connected.

To study the surfaces more precisely, Poincaré once again considered a surface as given by a Fuchsian polygon with conjugate pairs of edges. The corresponding Fuchsian group is the fundamental group of the surface. The boundary curves of the Fuchsian polygon are copies of the homology basis $\{C_1, C_2, \ldots, C_{2p}\}$ arranged in the order

$$C_1, C_2, C_1^{-1}, C_2^{-1}, \ldots, C_{2j+1}, C_{2j+2}, C_{2j+1}^{-1}, C_{2j+2}^{-1}, \ldots.$$

The only relation on homotopy is therefore that

$$C_1 + C_2 - C_1 - C_2 + \cdots + C_{2j+1} + C_{2j+2} - C_{2j+1} - C_{2j+2} + \cdots = 0,$$

and this gave Poincaré a way to illustrate cycles that were homologous to zero but not homotopic to zero when $p > 1$.

Now, the cycles C_j and C_k meet if and only if $j = k \pm 1$, and every cycle is homologous to one of the form $\sum_j x_j C_j$, where the coefficients x_j are integers, so the cycles $\sum_j x_j C_j$ and $\sum_k y_k C_k$ meet at

$$x_1 y_2 - x_2 y_1 + x_3 y_4 - x_4 y_3 + \cdots$$

points, which is a quadratic form with integer coefficients. Poincaré investigated what happens when a different homology basis is chosen, $\{C_j'\}$, and showed after a lengthy argument that the surface can be mapped to itself by a homeomorphism taking the first homology basis to the second (more precisely, C_j to C'_j) if and only if the corresponding quadratic forms are the same. From this he deduced that a cycle of the form $\sum_j a_j C_j$ is simple (does not intersect itself) if and only if the integers a_j are relatively prime. This is true because the condition on the coefficients is exactly what is needed for there to be a change of basis that makes the given cycle a basis element. As Poincaré noted, the homology question is easier than the homotopy question, which he put this way: When is a given cycle equivalent to zero? He explained (para. 3) that equivalence of one loop with another had been defined to require that the two loops start and end at the same point. This definition, he explained in paragraph 4, can be replaced with a weaker one, which he called improper equivalence, in which the endpoints of the loops are allowed to move. If the endpoints are joined by a curve α then two loops C and C' are improperly equivalent if and only if the loops C and $\alpha + C - \alpha$ are equivalent.

Poincaré now considered several ways of depicting a cycle on a surface in terms of the corresponding Fuchsian polygon. In one, the polygon is moved around in the non-Euclidean disk by the action of the Fuchsian group, and a closed curve on the surface corresponds to an arc joining a point in a polygon to any one of its conjugates. In another, each piece of the arc just described that lies in a conjugate polygon is replaced by the conjugate piece in the original polygon. Each of these pieces can then be seen to be improperly equivalent to some combination of the fundamental cycles on the boundary of the polygon.

The simplest closed curve on the surface that does not bound a disk is one drawn by starting at an arbitrary point M in the interior of the polygon, going to a point N on a boundary arc of the polygon, and then from the corresponding point N' on the conjugate arc back to the point M. This curve is equivalent to the cycle separating the arcs to which N and N' belong. More complicated closed curves can be written, up to improper equivalence, as sums of these, although not uniquely.

Poincaré now turned to the principal task of the paper, which was to exhibit a manifold with the Betti numbers and torsion of a three-dimensional sphere but that is not homeomorphic to a sphere. He supposed there was a three-dimensional manifold embedded in some high-dimensional Euclidean space and considered slices of it parameterized by a parameter t. For small values of t the slice is a solid ball, but at a critical value t_1 the slice acquires a cone point and thereafter two new cycles appear. Poincaré considered the one that resembles the ellipse round the neck of a hyperboloid of one sheet. As t increases from t_1 this cycle spreads out from a point and steadily fills out a disk. He next considers what happens at the next critical point t_2. The new cycle that appears around the neck here is disjoint from the one already considered, and it remains so as t increases, so the corresponding disks are also disjoint. Poincaré now supposed that the process passes through p critical points at which new cycles are created and stops without any cycles being destroyed. The net result is that the three-dimensional manifold can be cut open along the p disks (he imagined removing an infinitesimal neighborhood of them) and the result is that the three-dimensional manifold becomes a solid ball with p pairs of disks marked on its surface. The original three-dimensional manifold is recovered by gluing these disks together in the appropriate way.

The modern name for such a three-dimensional manifold is a handlebody. It is presented here as the interior of a surface of genus p together with the surface itself, and in paragraph 5 Poincaré claimed (incorrectly) that he had shown that given two homeomorphic closed surfaces (which necessarily have the same genus) the corresponding solids are also homeomorphic.

To obtain a three-dimensional manifold V with the required properties, Poincaré now supposed that in the preceding construction $p = 2$. He imagined that two such three-dimensional manifolds V' and V'' had been constructed, each bounded by a surface of genus 2, and he now glued them together along this surface. In terms of the slices, they are two-dimensional disks in V' until some $t_1 < \frac{1}{2}$, then they are tori in V' between t_1 and some $t_2 < \frac{1}{2}$, and finally they are double tori in V' between t_2 and $t = \frac{1}{2}$, at which time the second three-dimensional manifold is glued on. He called this slice W. Then at some t_3 in $\frac{1}{2} < t_3 < 1$ they become tori in V'' again before becoming disks in V'' at $t_4 < 1$. It is not supposed that the halves above and below $t = \frac{1}{2}$ are identical, and in particular the important cycles and the disks described above meet the slice at $t = \frac{1}{2}$ in four different ways.

The surface W is visible on the handlebody as that part of the sphere that lies outside all the disks, together with the boundaries of the disks. As a surface in its own right it can be supposed to be obtained from an octagon by identifying sides in pairs, and this means that there are

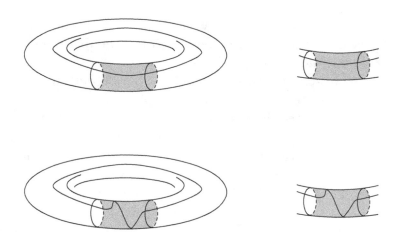

Figure 8.10. Torus with a twist. Source: drawn by Michele Mayor Angel.

four cycles on it that between them generate the homotopy group of the surface and form a homology basis. But it is also presented as a sphere with four circles on it that are to be identified in pairs. The natural question is to ask if the two sets of four curves are equivalent, and after a lengthy but not difficult argument Poincaré showed that if the homology basis on W is chosen, as it can be, to consist of simple disjoint curves then the answer is yes. It follows that the three-dimensional manifold V can be analyzed by looking at the homology bases on W obtained from V' and V''. Here it is important to remember that a surface may be homeomorphic to itself in ways that affect the homology basis nontrivially. For example, one may cut out, twist, and replace a region, as figure 8.10 shows. So the cycles obtained from W may look very different on V' and V''.

Poincaré also showed that every cycle in the three-dimensional manifold V is homotopic to a cycle in W, so all questions about the homology and homotopy of V can be transferred to W. Homology without division, as Poincaré called it, is fairly easy to study. Each cycle from V' is a combination of cycles forming a homology basis in W, as is each cycle in V''. The precise combination will depend on the nature of the homeomorphism between the boundary of V' and the boundary of V''. Poincaré required that it be such that the homology of V be that of the three-dimensional sphere, and then that homology with division also be trivial (so V has no torsion).

He let the cycles from V' be part of a homology basis for W and called them C_1 and C_3. The other boundary arcs he called C_2 and C_4. The cycles

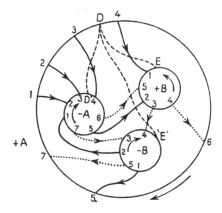

Figure 8.11. Two cycles on an octagon. Source: Poincaré, *Oeuvres* (6, 494).

K_1'' and K_2'' from V'' must be simple and disjoint, and they will appear in the Fuchsian octagon as curves running between the edges.

Poincaré described the curves he drew (see fig. 8.11) to represent the cycles in this way:

> Here is the explanation of the figure: the perimeter of R_0 is represented by the four circles $+A, -A, +B, -B$; the circles $+A$ and $-A$ are conjugate and correspond to $K_1' = C_1$; the circles $+B$ and $-B$ are conjugate and correspond to $K_2' = C_3$; the cycles K_1'' and K_2'' are represented by the arcs of curves running between points on the perimeter of R_0.

> The arcs which represent K_1'' are shown as unbroken lines; those which represent K_2'' are dotted. The arrows indicate the sense in which they are traversed.

> The points where the arcs meet the circles $\pm A$ and $\pm B$ are designated by numbers; these numbers also tell us which points are conjugate; thus the point $+B5$ is conjugate to $-B5$, $+A5$ to $-A5$. (para. 6)

He then expressed the cycles K_1'' and K_2'' in terms of C_1, C_2, C_3 and C_4, and found that

$$K_1'' \equiv 3C_2 + C_1 + C_2 - C_3 + C_4 - C_2 - C_4 - C_3 + 2C_4,$$
$$K_2'' \equiv -2C_4 + C_3 - C_2 - C_4 - C_3 + 2C_4 - C_2.$$

Then, using the homologies obtained from V'', which say that $K_1'' \equiv 0$ and $K_2'' \equiv 0$, and the equation for the edges of the octagon, which says

$$C_1 + C_2 - C_1 - C_2 + C_3 + C_4 - C_3 - C_4 \equiv 0,$$

he deduced that

$$-C_2 + C_4 - C_2 + C_4 \equiv 0, \quad 5C_2 \equiv 0, \quad 3C_4 \equiv 0.$$

But these relations, said Poincaré at the end of paragraph 6,

> are the relations of the structure in which the substitutions C_2 and C_4 generate the icosahedral group. We then know that $C_2 \equiv 0, C_4 \equiv 0$ cannot be deduced, so a fortiori these equivalences cannot be deduced from [the above equations] alone. We therefore have two cycles on V which are not equivalent to zero, so V is not simply connected.

This establishes that homology with torsion is not enough to discriminate between three-dimensional manifolds. He immediately asked:

> Is it possible for the fundamental group of V to reduce to the identity without V being simply connected?

and, after a few clarificatory remarks about what the question meant, concluded,

> However, this question would carry us too far away.

Thus was born the Poincaré conjecture, which, it can be seen, was not a conjecture at all but a question.[14] We have no evidence about Poincaré's opinion of the likely answer. Moreover, it only concerned three-dimensional manifolds—ironically, as a conjecture about manifolds of any dimension $n > 2$ it was first done for $n \geq 5$, then for $n = 4$, and for $n = 3$ last—and to be really pedantic, three-dimensional manifolds that split along a surface of genus 2.

There is no doubt that this final supplement is a remarkable piece of work, and it has invited speculation as to how even Poincaré was able

---

[14] For a highly accessible account of what happened in the study of three-manifolds generally, see O'Shea (2007).

to find, in the new and uncharted field of three-dimensional topology a manifold with just the properties he wanted. If anything, the mystery is deeper for specialists in what became a flourishing field at the end of the 20th century, who were very familiar with Heegaard diagrams and the construction of three-manifolds by gluing together the surfaces of handlebodies. They discovered that many such constructions lead to unexpected groups, some infinite, some trivial (when the resulting manifold is S^3), and wondered just how Poincaré had been so lucky. Thus Gordon, in his account of three-dimensional topology to 1960, writes,

> It is clear that in order to arrive at his example of a nonsimply-connected homology sphere, and also in investigating his question, Poincaré must have done a good deal of experimentation with Heegaard diagrams, of genus 2 and presumably higher genus also. In particular, he must have come across many nonstandard diagrams of S^3, and realized that they did indeed represent S^3. Thus he must have been aware that such diagrams can be quite complicated. (1999, 462)

But it is possible to wonder if there was not a more direct path open to Poincaré.[15] He would surely have said to himself that the fundamental group, G say, has to have the property that its Abelianization is the trivial group (such groups are called perfect). Furthermore, the construction that glues one handlebody to another by a homeomorphism of the surfaces requires that the group G has the same even number of generators as relations. This would have given him two tasks: to find such groups, and for each one to find a suitable pair of handlebodies and an identification of their boundaries. The group theory search might well have begun with some well-known groups (I am not aware that the specific task had been addressed already by 1905), and as it happens there is an immediate candidate: A_5, the alternating group on five objects, which is also the group of direct symmetries of a dodecahedron.

For, label the top face of a regular dodecahedron A, B, C, D, E, going clockwise as seen from above. Let a be the rotation that in permutation notation is (A, B, C, D, E). Let b be the rotation of order three about the vertex A that maps E to B. Then by drawing the top six faces as a net in the plane and chasing points, we see that $a^4ba^{-1}b = 1$ and $ab^{-1}ab^2 = 1$,

[15] I again thank Saul Schleimer for many helpful discussions.

reading the maps from right to left. Now define $c = a^{-1}b$ and notice that $c^2 = 1$, because c switches A and E.

The group of direct symmetries of the dodecahedron has the presentation

$$\{a, b, c : a^4ba^{-1}b = 1, \ a^5 = 1, \ b^3 = 1, \ c = a^{-1}b\},$$

and in this group $ab^{-1}ab^2 = 1$. The group is, of course, not trivial, but it is when Abelianized by adding the relations $ab = ba, bc = cb, ca = ac$, for then the relation $a^4ba^{-1}b = 1$ becomes $a^3 = b^{-2}$ and the relation $b^{-2}a^{-1}ba^{-1} = 1$ becomes $b = a^2$, which, together with $a^5 = 1$ imply that $a = 1$ and so $b = 1 = c$.

However, this group does not have the same number of generators and relations. But, if the generator c is dropped, attention shifts to the group G presented by

$$\{\alpha, \beta : \alpha^4\beta\alpha^{-1}\beta = 1, \ \alpha\beta^{-1}\alpha\beta^2 = 1\}.$$

This group has two generators and two relations, and what is more it is perfect, because adjoining $\alpha\beta = \beta\alpha$ allows us to write the first relation as $\alpha^3 = \beta^{-2}$ and the second relation as $\beta^{-1} = \alpha^2$, from which we deduce that $\alpha^3 = \alpha^4$ so $\alpha = 1$ and therefore $\beta = 1$ as required. It cannot be trivial, because it maps onto the dodecahedral group by adding the relation $(\alpha^{-1}\beta)^2 = 1$. It is even a well-known group, $SL(2, \mathbb{Z}/5\mathbb{Z})$, the 2×2 matrices with entries integers modulo 5 and determinant 1, which was important in the theory of modular functions.[16]

Now, it is just these relations and this argument for nontriviality that occur in Poincaré's fifth supplement, so it is entirely plausible that he came to the groups this way. He then had to check that the generators can be realized as paths without self-intersections on an octagon with identified edges, and there we can imagine Poincaré needing a certain amount of trial and error before he succeeded. But by starting with a perfect group he had the relations he needed, and he had the idea of using a handlebody whose boundary is a twofold torus. Only now did he need to be lucky.

---

[16] This group, known today as the binary icosahedral group, is not isomorphic to the extended icosahedral group, which also has order 120, that Klein studied in his (1884).

9

Interventions in Pure Mathematics

NUMBER THEORY

One of the most important subjects in 19th-century mathematics was number theory. This had been rewritten by Gauss in his *Disquisitiones Arithmeticae* (1801) on the back of considerable work by Euler, Lagrange, and Legendre, and had become something of a German specialty. Almost every major German mathematician of the period sought to contribute to it—Weierstrass is the most notable exception—and they collectively gave the subject a strong algebraic aspect, although Jacobi, Dirichlet, and Riemann also developed profound analytical methods.

The simplest topic in this rapidly growing field was that of binary quadratic forms. These are expressions of the form $ax^2 + bxy + cy^2$, where a, b, and c are integers, and the two central questions are: For a given a, b, c what integers m satisfy the equation $ax^2 + bxy + cy^2 = m$ when x and y are integers?—m is then said to be represented by the form—and conversely, given m what quadratic forms with a given value of $(b^2 - 4ac)$—the discriminant of the form—can represent m? Gauss had shown how to answer such questions more or less completely, and his solution, building on work by Lagrange and Legendre, involved the idea of the equivalence of two binary quadratic forms.

Two forms are equivalent if each can be obtained from the other by a transformation of the form

$$x' = \alpha x + \beta y, \quad y' = \gamma x + \delta y,$$

where $\alpha, \beta, \gamma, \delta$ are integers and $\alpha\delta - \beta\gamma = 1$. Such transformations leave invariant the quantity $b^2 - 4ac$, which explains why it was mentioned above, and why it is called the discriminant. More importantly, these transformations are critical to the study of how many essentially different (i.e., inequivalent) binary quadratic forms there are of a given discriminant. Dirichlet had been able to find a formula for the number of inequivalent binary quadratic forms of a given discriminant. This number is called the class number, and Dirichlet's class number formula was a profound and difficult result that drew on his work in establishing another major result in number theory. This was the theorem that in every arithmetic progression $a, a+b, a+2b, \ldots$, where a and b have no common factors, there are infinitely many primes. However, the class number formula cannot be inverted to give the number of forms having a given class number. The simplest cases in the theory are those where all quadratic forms of a given discriminant are equivalent, and Gauss had done extensive calculations which allowed him to conjecture when this was (in the case when the discriminant is negative), but no one at the time was able to show if his list was complete.[1]

The success of the theory of binary quadratic forms rested not only on their relative simplicity, but on Gauss's discovery that equivalence classes of quadratic forms of a given discriminant form a group—a discovery that was all the more remarkable because it was made some decades before there was an explicit concept of a group to guide his research. This structure enabled Gauss to deal with the general theory of these forms, but the whole subject of forms of higher degrees and forms involving more than two variables was left largely untouched, and subject to only piecemeal developments (some of them by Gauss and Legendre).

Instead, other branches of number theory blossomed. One of these, also algebraic and with a source in Gauss's work, is the theory of algebraic integers. These are numbers that satisfy a monic polynomial with integer coefficients, $x^n + a_1 x^{n-1} + \cdots + a_n = 0$. Another, that again Gauss had predicted but in this case done less to make public, was the connection between elliptic functions and the theory of numbers. Both are largely German stories, in which the most prominent names are those of Jacobi, Dirichlet, Kummer, Kronecker, and Dedekind. Less appreciated by historians until recently, however, were the developments in France

[1] It was, as Heegner was the first to show in 1952, but obscurities surrounding his proof led to Baker in 1966 and Stark in 1967 giving independent proofs. A historical account is forthcoming from Patterson and Schappacher.

that Hermite was responsible for, and in view of his dominant position in French mathematical life in the second half of the 19th century and his influence on the young Poincaré it is to them that we now turn.

Hermite always stressed the importance that Gauss's *Disquisitiones Arithmeticae* had for him.[2] Goldstein (2007, 381) more precisely observes that Hermite was most influenced by the long and difficult fifth section of that book, where Gauss described his new theory of quadratic forms, and she then quotes Poincaré's remarks about Hermite on the occasion of the jubilee celebrations for Hermite that marked his 70th birthday:

> You have never stopped cultivating the highest parts of the mathematical sciences, those where pure number reigns: analysis, algebra, arithmetic....You made your first discoveries in the nascent theory of algebraic forms and, while successively attacking all the interesting questions in arithmetic, you enlarged and illuminated with a new light the admirable structure raised by Gauss. (1893a)

She then shows that Hermite's originality took him from some ideas of Gauss into the uncharted domain of quadratic forms in any number of variables, and with coefficients that were no longer required to be integers, a bold gambit in number theory. Hermite had defined a determinant for an arbitrary quadratic form in $n + 1$ variables $x_0, x_1, \ldots, x_n$ and real coefficients and investigated its least nonzero positive value as a function of n and this determinant. He then redefined the nature of the search for the best representatives of inequivalent forms (the so-called canonical forms) and developed techniques for finding them. There is no question that he was the leader in this field. Goldstein (2007, 398) also picks out an important way in which his methods differed from those of Gauss: where Gauss pursued periodic phenomena into the underlying cyclic structures, "Hermite, on the other hand, focused on the finitely many transformations which bring out the periodicity."

When Poincaré began to publish in 1880 he did so in three fields at once: on the curves defined by real differential equations; on linear differential equations in the complex domain; and on the number theory of forms in several variables. He also published some shorter papers

---

[2] As Goldstein (2007) has pointed out, this is already interesting; Hermite could simply have said "number theory," but deliberately singled out the importance of Gauss. For an account of Hermite's views on mathematics generally, see Goldstein (2011).

which are worth discussing, and a lengthy one (1880b) which it will be worth describing first, because it helps fill out the picture of what he had learned.

In the introduction to this paper Poincaré observed that he would show how it was possible to extend the use of parallelogram lattices from the case of definite quadratic forms, where it was well known, to indefinite forms. For this purpose he introduced a new geometrical representation for the points of a lattice, and for the quadratic form associated with a lattice. In part I of the paper Poincaré summarized the familiar theory of lattices, some of which went back to the work of Eisenstein, as he observed. He described the multiplication of lattices by the matrix multiplication of their matrix representatives, and outlined how their arithmetic properties could be defined. He discussed how to show when two lattices are equivalent, and how they can be reduced to canonical forms. He discussed the norm of a lattice, established a theory of greatest common divisors and least common multiples of two lattices, and even defined prime lattices.[3]

In part II, Poincaré looked at the geometrical representation of complex numbers by points, and in part III he looked at the representation of forms by lattices. Matters became interesting when he looked at the reduction of forms to canonical forms. In the classical case due to Lagrange, there is as simple reduction theory for positive definite forms, but, as Gauss described, the reduction theory of negative definite forms involves looking at what Gauss had called the periods. In this case Poincaré gave a geometrical account of the periods in terms of what he called ambiguous triangles (*triangles ambigus*).

In part IV of the paper, Poincaré turned to the deepest and most confusing part of Gauss's theory of quadratic forms, the composition theory. The basic idea is that if a form $ax^2 + 2bxy + cy^2$ represents a number m and the form $a'x^2 + 2b'xy + c'y^2$ with the same discriminant represents a number m' then there is a form $Ax^2 + 2Bxy + Cy^2$ also with the same discriminant that represents mm'. Gauss had been the first person to define this for forms of arbitrary discriminant, but his method for producing the numbers A, B, C from a, b, c and a', b', c' was obscure.

---

[3] Much of this is close to Dedekind (1877) which, however, is not mentioned. Dedekind did not discuss lattices, and dealt with these questions from a more abstract and algebraic standpoint.

He had, however, been able to show that equivalence classes of forms of the same discriminant make up what, in later language, would be called a finite Abelian group, and to use the character theory of this group to provide a deep insight into what forms can represent what numbers. Poincaré proposed to give a simpler geometrical theory of composition. To do this he defined a new way to multiply lattices, which he called their second product, and which, necessarily, is commutative. If one lattice consists of all the numbers $Am + Bn$, where A and B are complex numbers, and a second lattice consists of all the numbers $A'm + B'n$, where A' and B' are complex numbers, then a natural candidate for the product of the two lattices is the set of all numbers

$$AA'\mu_1 + AB'\mu_2 + A'B\mu_3 + BB'\mu_4.$$

This is in fact a lattice, and in a few pages Poincaré showed that this second product of two lattices corresponding to two quadratic forms is a lattice that corresponds to the Gaussian composite of the two quadratic forms. The second product can therefore be written in the form $A''m + B''n$, but Poincaré did nothing to make the calculation of A'' and B'' in terms of A, B, A', and B' any easier to carry out.

In the final part of the paper Poincaré looked at the representation of ideal complex numbers. He showed how to decompose an ideal number into prime factors, and concluded the paper with a proof that this factorization was essentially unique.

In 1881 he gave two papers to the meeting of the Association française pour l'avancement des sciences in Algiers. The first of these, on arithmetic invariants, will be noted below, because it makes a good introduction to his later and longer paper on the subject, his (1905f). His second paper, (1881b), takes as its starting point the observation that while Hermite had drawn attention to the group that mapped an indefinite ternary quadratic form to itself, the properties of these transformations were not fully known. He then showed how these transformations can be regarded as isometries of the non-Euclidean disk, and indicated how this helps to illuminate the reduction of these forms to canonical form.

In his first substantial number theory paper, (1881g), Poincaré began by observing that Hermite had completely and elegantly solved the problem of finding canonical representatives for quadratic forms, and offered an extension of Hermite's methods to forms of higher degree.[4] Poincaré must have written to Jordan early on about this work, because there is a letter from Jordan to Poincaré dated 27 January 1880 in which Jordan says that the results described in a letter from Poincaré are

> certainly interesting, especially if your solutions permit you not only to find the group of substitutions that reproduces a given form up to a factor but to find, for a given number of variables, all the various groups and the corresponding forms. This question is very much the order of the day at this moment, M. Klein has written numerous memoirs about it in the *Mathematische Annalen*. But he has restricted himself to finite substitution groups in two variables.

In the published paper, "Not wishing to sacrifice clarity to generality," as he put it (p. 200), Poincaré emphasized the examples of cubic forms in 3 or 4 variables. These are homogeneous expressions of degree 3, such as $x^3 + y^3 + 3xyz$, and he noted the earlier work on them by a number of German mathematicians: Hesse, Aronhold, Clebsch, and Steiner.

Poincaré considered the effect of a linear change of variables on a given form. If the form is denoted by F and the change of variables by T then the transformed form was denoted FT by Poincaré. One can follow one change of variables by another, T' say, and Poincaré reminded his readers that in general FTT' and $FT'T$ are different. His interest was in the different conditions it was useful to impose on the transformations T. They could be unitary (they have determinant 1), real (they have real coefficients) or integral (integer coefficients). Two forms F' and F'' were equivalent, he said, if there is a third form F and transformations T' and T'' such that $F' = FT'$ and $F'' = FT''$. This equivalence is algebraic, real, or arithmetic if the corresponding transformations are, respectively, unitary, real, or integral.

In the first part of the paper, Poincaré confined himself to the algebraic theory. Forms in three homogeneous variables define a curve in the projective plane; forms in four homogeneous variables define a surface in projective space, and the transformations define what Poincaré called a homological relation (a term in use in projective geometry

---

[4] Hermite had looked for all transformations of a ternary quadratic form to itself in 1853 in papers in *JfM* (47); see his *Oeuvres* (I).

that Chasles had introduced). Now it is always possible to consider one of these transformations as changing the triangle or tetrahedron of reference, and Poincaré observed that the effect on a transformation T of a change of the reference frame given by S was to produce the transformation $\Sigma = S^{-1}TS$. He therefore set himself the task of classifying the possible transformations where T and $\Sigma = S^{-1}TS$ are regarded as equivalent.

This took him into the territory opened up by Jordan in his *Traité* (1870), which Poincaré did not mention by name, perhaps because these ideas had become standard by now in Paris, along with the method for finding them. The simplest canonical form is the diagonal one:

$$\begin{pmatrix} \lambda_1 & 0 & 0 & 0 \\ 0 & \lambda_2 & 0 & 0 \\ 0 & 0 & \lambda_3 & 0 \\ 0 & 0 & 0 & \lambda_4 \end{pmatrix}.$$

Poincaré put a transformation into the first category if the numbers $\lambda_1, \ldots, \lambda_4$ and all their powers were distinct; into a second category if the numbers were distinct but some of their powers were not (for example, $1, -1, 2, 3$); into a third category if the numbers were not all distinct but it could be regarded as a power of a transformation in the second category; and finally into a fourth category otherwise. He then showed, still following Jordan, that some transformations cannot be reduced to diagonal form but have off-diagonal elements, for example,

$$\begin{pmatrix} \lambda_1 & 0 & 0 & 0 \\ 0 & \lambda_2 & 0 & 0 \\ 0 & 0 & \lambda_3 & 0 \\ 0 & 0 & \mu & \lambda_3 \end{pmatrix}.$$

He then gave a classification of cubic ternary forms into seven families. The first four families corresponded to forms with a nonzero discriminant, and represented irreducible nonsingular curves. This fitted precisely into Hesse's classification of these curves according to the way their equations could be reduced by choosing a triangle of reference with particular geometrical properties, derived from the fact that they have 9 inflection points, of which 3 are real. These two families were distinguished by whether their equations could or could not be written as a sum of three cubes.

The third family had discriminant zero, but not all its invariants vanished, and corresponded to curves with a double point and only three inflection points. Their equations could be put in the form

$$6\alpha x y z + \beta(x^3 + y^3).$$

The fourth family consisted of cubic curves with a cusp. These four families correspond to the families in Plücker's analysis of cubic curves.

The remaining families correspond to the decomposable curves. They are represented by $\alpha z(z^2 + \beta xy)$, by $\alpha y(z^2 + \beta xy)$, and by the completely decomposable forms studied by Hermite.

Poincaré then reminded his readers of the theory of the invariants and covariants of these curves, before going on to consider the transformations that map a curve in one of the seven families to itself. These correspond to self-transformations of the coordinate frame that respect symmetries of the canonical curve. He noted that in any given case these transformations form a group. The paper ended with a table of the cubics in four variables that are indecomposable, do not reduce to forms in three variables, and have nontrivial self-transformations.[5] When Hermite wrote to Poincaré about this paper on 4 June 1880 (he was presenting it to the Académie des sciences at Poincaré's request) he said that the reduction to seven families was "excellent, if not, perhaps, new," and he urged Poincaré to look through the literature of the last decade to see if it had not been discovered already. But other sections of the memoir were unquestionably original, such as the simultaneous reduction of forms. Then, Hermite offered advice that was only to show how different he was from Poincaré. He urged him to make an explicit investigation of the reduced forms, because calculation can reveal what no one could otherwise see or predict. This was never to be Poincaré's way.[6]

In the second part of the paper, (1882e), Poincaré turned to the real and integral theories, which give a finer classification of the forms.

---

[5] The paper closely resembles the original *pli cacheté*, without an excursion of forms in n variables.

[6] Poincaré's thesis (1879) shows no sign of any influence of Hermite, so their first significant contact was with this work on number theory.

It was necessary, he said, to discuss how one could know if two given forms are equivalent, to distribute the forms into classes, orders, and genres (as Gauss had done for binary quadratic forms), and to find the integral transformations that are self-transformations of a given form. He proposed to do this by following the methods of Hermite, which he now sketched, noting where the papers by Korkine and Zolotaroff entered the account.[7] He also picked up a result due to Jordan that established that there is only a finite number of classes of forms with integer coefficients that are algebraically equivalent to a given form with nonzero discriminant, and enriched it by showing that there can be infinitely many classes of forms only if all the invariants of the form vanish.

The corresponding part of the *pli cacheté* makes a more generous acknowledgment of Poincaré's sources. It begins:

> Being given two forms in a given number of variables, discover if they are arithmetically equivalent.
>
> This problem has been solved for a long time for the case of binary quadratic forms and [illegible] for definite quadratic forms in an arbitrary number of variables. M. Hermite was the first to introduce continuous variables into this question; he thus came to an elegant method that he applied to arbitrary quadratic forms and to forms that decompose into linear factors, but which apply without any change to an entirely arbitrary form; the applications of this method to non-quadratic and indecomposable forms has been studied by M. Jordan (see *Comptes rendus*, session of 5 May 1879). Even though the question has been treated almost completely, I think that it will not be useless to look at it from a new point of view.

Hadamard observed of this work (1921, 168) that the problem of indecomposable forms "disappeared, in this sense, [and] an idea of rare simplicity gave the rule applicable to all problems of this sort at a stroke." What was left, he explained was a purely algebraic problem of reducing the form to a canonical type, and then a problem about the arithmetic group. Poincaré's methods were a mixture of the algebraic and the geometric, and considered the finer notions of equivalence in each

---

[7] Jordan's letter had told him where to find them.

of the seven families he had described in the first part. He described his results for the first six families in a lengthy table that concluded the paper, the seventh family (of decomposable forms) having been dealt with conclusively by Hermite. He then remarked that the same principles dealt with all forms, but even for cubic forms in four variables the extreme variety of cases made it not worthwhile to go into detail. Instead, he offered the thought that the principal cause of the difference in the properties of the various families was due to the fact that some were, and others were not, reproduced by certain transformations. This showed, he said, the important role that was played by purely algebraic considerations about linear transformations in the arithmetic study of forms, and this was why he had opened the paper with a complete study of the groups of transformations that map a given cubic form in four variables to itself. Group theory, it seemed, was the key—in its way a modest victory for Gauss's view over Hermite's, or perhaps for the union of the two.

Poincaré's paper (1885a) is notable for bringing together the ideas of Dedekind and Hermite. Poincaré addressed the question: When does the binary form

$$F = F(x, y) = B_n x^n + B_{n-1} x^{n-1} y + \cdots + B_0 y^n$$

represent the number N?

To tackle it, he reminded his readers how Dedekind had introduced the idea of an algebraic integer and its norm in Dedekind (1877). An algebraic integer satisfies a monic polynomial with integer coefficients, say

$$x^n + a_1 x^{n-1} + \cdots + a_n = 0. \qquad (9.1)$$

Sums and products of algebraic integers are again algebraic integers, so if α_1 satisfies (9.1), then any expression of the form

$$m(\alpha_1) = m_1 \alpha_1^{n-1} + \cdots + m_n,$$

where $m_1, \ldots, m_n$ are ordinary integers, is an algebraic integer. This generalizes the idea of a complex number $a + bi$, because $a \pm bi$ are the two roots of $x^2 - 2az + a^2 + b^2 = 0$. Indeed, if a and b are integers then $a + bi$ is called a complex or Gaussian integer. An equation such as (9.1) has n roots which are said to be mutually conjugate: $\alpha_1, \alpha_2, \ldots, \alpha_n$, and the norm N of an algebraic integer $m(\alpha_1)$ is the product of an algebraic integer and its conjugates, thus:

$$N(m(\alpha_1)) = (m_1\alpha_1^{n-1} + \cdots + m_n)(m_1\alpha_2^{n-1} + \cdots + m_n) \cdots (m_1\alpha_n^{n-1} + \cdots + m_n).$$

This generalizes the idea of the norm of a complex number $a + bi$, which is $(a + bi)(a - bi) = a^2 + b^2$.

In later terminology, the collection of all algebraic integers of the form $m_1\alpha^{n-1} + \cdots + m_n$ for a given algebraic integer α form a *ring*. The key concepts in Dedekind's presentation of the theory of algebraic integers were those of modules and ideals. A module is a set of algebraic integers in a ring that are of the form $m_0 + m_1x_1 + m_2x_2 + \cdots + m_nx_n$, where $m_1, \ldots, m_n$ are ordinary integers and $x_1, \ldots, x_n$ are algebraic integers (called the generators of the module). A module is an ideal if it is closed under multiplication by arbitrary algebraic integers in the ring.

Since the expression $B_nx^n + B_{n-1}x^{n-1}y + \cdots + B_0y^n$ factors completely into linear factors when complex numbers are allowed, simple algebraic substitutions allowed Poincaré to reduce his question to this: When is a given number represented by a form

$$(x - \alpha_1y)(x - \alpha_2y) \cdots (x - \alpha_ny) = N(x - \alpha_1y)?$$

He therefore asked, more generally, when is a given number a norm of an algebraic integer? If this algebraic integer is $m_1(\alpha_1)$ then its norm is a form, Ψ, in the variables $m_1, m_2, \ldots, m_n$, and the question is: When are there ordinary integers $x_1, x_2, \ldots, x_n$ such that $\Psi(x_1, x_2, \ldots, x_n) = N$? Poincaré now showed that these questions reduce to determining all the ideals of a given norm, and then deciding if two forms that decomposed as a product of linear factors were themselves equivalent. Since, he said, this second question had been completely solved by Hermite, he addressed himself to the first. He observed that it is enough to look for principal ideals, because one is interested in products of the form

$$(x_0 + \alpha_1x_1 + \cdots + \alpha_1^{m-1}x_{m-1})(m_0 + \alpha_1m_1 + \cdots + \alpha_1^{m-1}m_{m-1}),$$

where $m_0, m_1, \ldots, m_{m-1}$ are undetermined integers, and they form a principal ideal.

Dedekind had shown that a module may have infinitely many sets of generators, and that the algebra involved in finding a particularly suitable set of generators is very like that of changing the basis in a vector space (the only difference being the restriction to integer coefficients). Poincaré therefore recognized that he was once again in the situation of his earliest work on the arithmetic of forms, and he applied himself to the task of producing canonical forms for modules in a ring and for recognizing when a module was also an ideal. He then worked his way up through a sequence of simple cases until he could show how to find all ideals whose norm was prime, then those with norm a prime power, and finally those whose norms are products of prime powers.

Fuchsian Functions and Arithmetic

Poincaré's next investigation into arithmetic, his (1887a), connected his previous work on ternary forms with his study of Fuchsian groups. He began by introducing his cast of characters. First, he looked at the ternary quadratic form $\Phi = y^2 - xz$. The linear transformations that map this form to itself either have an eigenvalue $+1$ or an eigenvalue -1, and those with the positive one can all be written in the form

$$\begin{pmatrix} \delta^2 & -\delta\gamma & \gamma^2 \\ -2\delta\beta & \alpha\delta + \beta\gamma & -2\alpha\gamma \\ \beta^2 & -\alpha\beta & \alpha^2 \end{pmatrix},$$

where $\alpha, \beta, \gamma, \delta$ are four arbitrary quantities such that $(\alpha\delta - \beta\gamma)^2 = 1$. Poincaré restricted his attention to the case when they are all real and moreover $\alpha\delta - \beta\gamma = 1$.

Next, he recalled the Fuchsian transformations

$$z \mapsto \frac{\alpha z + \beta}{\gamma z + \delta},$$

and he observed that the groups of 3×3 matrices defined above and of Fuchsian transformations are isomorphic.

He then observed that if T is a linear transformation of the variables x, y, z that does not map the form Φ to itself, then the new form $F = \Phi T$ is mapped to itself by all the transformations of the form $\Sigma = T^{-1}ST$, and so the group of self-transformations of F and of Φ are isomorphic (indeed, conjugate).

Finally he supposed that the coefficients of the form F are integers and that the coefficients of the self-transformations of F are likewise integers. This led, he remarked, to discontinuous groups that had already attracted the attention of arithmeticians, notably Hermite. The corresponding Fuchsian functions he had earlier called arithmetic Fuchsian functions, and he now proposed to study them. In particular, he aimed to show that these functions satisfied a theorem that could be thought of as generalizing the addition theorem for elliptic functions, which more general Fuchsian functions do not.

To establish this result, Poincaré examined the Fuchsian groups that are associated to different (ternary quadratic) forms, and divided them into four families according as they have no elliptic or parabolic elements, elliptic but no parabolic elements, parabolic but no elliptic elements, or both elliptic and parabolic elements. He showed that it was always possible to determine which of these four cases any given ternary form belonged to. He then looked at the geometry of the corresponding Fuchsian polygon—the number of its sides, the size of its angles, whether the vertices lie inside or on the boundary of the non-Euclidean disk— and showed that in each case this information was determined by the arithmetic properties of the form F.

Poincaré next supposed that he had a form F with its corresponding Fuchsian group G and a transformation S that did not belong to G. In such a case, he said, there should be no algebraic relation between a Fuchsian function f for G and the function f^S, where $f^S(z) = f(zS)$. He then proved that for an algebraic relation between $f(z)$ and $f(zS)$ to exist it was necessary that the groups G and $S^{-1}GS$ be what he called commensurable, which means that there is a group H that is a subgroup of finite index of both G and $S^{-1}GS$ (it is enough to consider their intersection $G \cap S^{-1}GS$). Informally, this means that the two groups have a lot of elements in common.

He observed that the classical example of this was provided by the modular function $J(z)$, which is invariant under the group Γ of all transformations of the form

$$z \mapsto \frac{\alpha z + \beta}{\gamma z + \delta},$$

where $\alpha, \beta, \gamma, \delta$ are integers and $\alpha\delta - \beta\gamma = 1$. The transformation S given by $z \mapsto z/n$ is not in this group, but the relationship between $J(z)$ and $J(z/n)$ is governed by the celebrated modular equation. He then verified that the groups Γ and $S^{-1}\Gamma S$ are commensurable. The elements of $S^{-1}\Gamma S$ are of the form

$$z \mapsto \frac{\alpha z + \beta/n}{\gamma n z + \delta},$$

where, as before, $\alpha, \beta, \gamma, \delta$ are integers and $\alpha\delta - \beta\gamma = 1$. So the required subgroup of both Γ and $S^{-1}\Gamma S$ consisted of elements of the form

$$z \mapsto \frac{\alpha z + \beta}{\gamma z + \delta}$$

where $\gamma \equiv 0 \bmod n$. A repeat of the same argument showed that $J(z)$ and $J(pz/n)$ are algebraically related, and more generally that $J(z)$ and $J\left(\frac{\alpha z + \beta}{\gamma z + \delta}\right)$ are algebraically related.

Poincaré now introduced a whole class of arithmetic Fuchsian functions that obey a similar algebraic relationship, thus generalizing the modular function J. He observed that there are three groups of interest that map a given ternary form to itself: those where the corresponding matrices have real, rational, or integer coefficients. Each group gives rise to a corresponding Fuchsian group; in the third case Poincaré called the corresponding Fuchsian group the principal group. He now looked at the group of matrices with rational coefficients that preserved a given form, remarked that the Fuchsian group in this case is not discontinuous, and chose an element S in it with rational, nonintegral coefficients. This element gave rise to a Fuchsian transformation s that was not in the principal Fuchsian group, and by his earlier results this means that a Fuchsian function for the principal group is algebraically related to its transform by s. In a nod to Hermite, he concluded the paper by remarking that one could hope in this way to find arithmetic transcendents that rendered the same service in the theory of algebraic equations that the modular function does for equations of the fifth degree. Much later, in 1900, in the speech he gave at the celebrations of Hermite's seventieth birthday, Poincaré blurred the distinction between his early work and Hermite's and asked rhetorically, "Your groups of similar transformations, are they not in effect discontinuous groups, must they not give rise to uniform transcendents that are used in the theory of linear differential equations?"

It seems best to defer Poincaré's most famous paper on arithmetic to the next section and to present next the paper he wrote in 1905 and which he himself connected to his earliest work in 1879. This is his (1905f), which is extended and corrected in his (1912b). As a footnote observes, Poincaré posted this paper on 7 July 1912, the day he went into hospital for his fatal operation.

In this paper he connected these ideas to the papers of Dirichlet on the determination of the class number, and claimed applications and connections to the theories of Fuchsian and Abelian functions as well as to "a certain number of new transcendental functions more or less apparent in the theory of elliptic functions" and indeed to Fredholm's function. The link, he said, was certain arithmetic properties they had in common with each other and Dirichlet's investigations.

He began with a simple example. There was, he said, no algebraic invariant associated with the expression $ax + by$. That is to say, there is no single-valued function of a and b that is unaltered when a linear transformation in x and y is applied to $ax + by$. But there is an arithmetic invariant when the transformations are restricted to those with integer coefficients, namely

$$\Phi_k = \sum_{m,n} \frac{1}{(am + bn)^k},$$

where the summation omits the case $(m, n) = (0, 0)$ and k is an integer greater than 2 (this condition ensured the convergence of the series).

This function is closely connected to the Weierstrassian elliptic function, as Poincaré now reminded his readers, together with the series for g_2 and g_3 recently redescribed by Hurwitz (and essentially by Eisenstein before him), names Poincaré did not mention. Then he introduced a general automorphic function

$$\phi(a, b) = \phi(\alpha a + \beta b, \gamma a + \delta b),$$

where, as ever, $\alpha, \beta, \gamma, \delta$ are integers and $\alpha\delta - \beta\gamma = 1$. For such functions one has

$$\phi(a, b) = b^{-k} \phi\left(\frac{a}{b}, 1\right),$$

and more generally

$$\phi(\alpha a + \beta b, \gamma a + \delta b) = (\gamma a + \delta b)^{-k} \phi\left(\frac{\alpha a + \beta b}{\gamma a + \delta b}, 1\right).$$

On setting $b = 1$ this becomes

$$\phi\left(\frac{\alpha a + \beta}{\gamma a + \delta}, 1\right) = (\gamma a + \delta)^k \phi(a, 1),$$

which shows that when k is an even integer the function ϕ is a thetafuchsian function corresponding to the modular group Γ.

This raises the question of whether this function can be written as a thetafuchsian series, and Poincaré showed in detail that this can be done. This gave him a general form for the representation of the arithmetic invariants, in which the functions $\Phi_k(a, 1)$ appear as limiting cases although they cannot be represented in this manner themselves. Poincaré was led in this way to introduce the series

$$\sum \frac{1}{(a\alpha + b\beta)^{2k}} e^{2i\pi \frac{\gamma a + \delta b}{\alpha a + \beta b}},$$

which are a new and important type of arithmetic invariant. Unfortunately, as Poincaré noted, they have the disadvantage that they can vanish identically.

One of Poincaré's lifelong questions arose from his discovery of Fuchsian functions. Prior to their discovery, a certain kind of algebraic integral, the elliptic integral, had led to a theory of elliptic functions, which belonged to single-variable complex function theory, but all more complicated algebraic integrals led to the much less well-developed theory of Abelian integrals and Abelian functions, which belonged to the complex function theory of several variables. Fuchsian functions arose every time Abelian functions did, but they were functions of a single variable, so it was very natural to ask if they could not be used just as well if not better. The problem was that the series defining them were opaque: they could nontrivially sum to zero, and even their convergence properties were hard to establish.

In the second of his long memoirs of 1882 on Fuchsian functions, Poincaré had spent a lot of time developing their theory, showing, in particular, how they were connected to some more tractable expressions which he had called thetafuchsians because they played an analogous role to the theta functions in the theory of elliptic functions. He had also produced a related class of expressions, which he denoted Λ, and the thetafuchsians and these lambdafuchsian functions played a central role in his paper (1905f).

The functions $\Lambda(z)$ are defined this way. For a given Fuchsian group, two Fuchsian functions $X(z)$ and $x(z)$ are chosen, and $\Lambda(z)$ satisfies

$$\Lambda\left(\frac{az+b}{cz+d}\right) = \left(\frac{dx}{dz}\right)^{-h} X(z)$$

and $\Lambda\left(\frac{az+b}{cz+d}\right) = (cz+d)^{2h} \Lambda(z)$. Poincaré now supposed that the function $\Lambda(z)$ had some poles (none of them at the vertices of the fundamental polygon for the group), and stated the theorem that the number of linear relations between the residues of $\Lambda(z)$ was equal to the number of linear relations that exist between thetafuchsian functions of the form $\theta(z) = \left(\frac{dx}{dz}\right)^{h+1} X$. He pointed out that he had shown in (1882b) that this latter number was equal to the dimension of the space of thetafuchsian series, which was how he had been able to show that every thetafuchsian function is representable by a thetafuchsian series.

His argument was that $\frac{d^{2h+1}}{dx^{2h+1}} \Lambda(z)$ is a thetafuchsian function, which could be set equal to $Y(y) \left(\frac{dx}{dz}\right)^{h+1}$, where x and y are connected by an algebraic equation. The equation $Y = 0$ is a differential equation in X, which has these functions as a basis of solutions:

$$z^j \left(\frac{dx}{dz}\right)^h, \quad j = 0, 1, \ldots, 2h.$$

Next, Poincaré introduced a function $M(z)$ with the property that $\frac{d^{2h+1}}{dx^{2h+1}} M(z) = \theta(z)$, so

$$M(z) = X \left(\frac{dx}{dz}\right)^{-h}, \quad \theta(z) = Z \left(\frac{dx}{dz}\right)^{h+1}.$$

This implies that Z is a Fuchsian function and therefore a rational function of x and y that is treated as given. It follows that the solution of the differential equation $Y = Z$ is known, because the solutions of $Y = 0$

are known. From this Poincaré deduced that the fundamental property of the function $M(z)$ is

$$M\left(\frac{az+b}{cz+d}\right) = M(z)(cz+d)^{-2h} + (cz+d)^{-2h} P(z),$$

where $P(z)$ is a polynomial of degree $2h$.

The fundamental properties of the functions θ, Λ, and M enabled Poincaré to proclaim this analogy:

> The functions θ, or rather the rational functions Z on which they depend, play the role of algebraic expressions that are to be integrated, the functions M play the role of the Abelian integrals themselves, finally, the polynomials P play the role of the periods. (1905f, 104)

The analogy was trivial when $h = 0$ because $M(z)$ is then straightforwardly an Abelian integral. When $h > 0$ three more pages of work allowed Poincaré to deduce that the number of Abelian integrals of the first and second kinds were both $p - 1$ and the number of arbitrary coefficients in the periods was twice the number of integrals of the first kind, thus establishing the analogy.[8] Poincaré remarked that his results had analogies with theorems about the exchange of argument and parameter in the theory of Abelian integrals, and that the relations between the residues of $\Lambda(z)$ correspond to the theorem of Riemann–Roch.

When Poincaré turned to the implications of his approach for number theory, he noted that Dirichlet had made great use of the invariant

$$J(s) = \sum_{m,n} \left(am^2 + 2mn + cn^2\right)^{-s}, \quad (m,n)^{-1} \neq (0,0),$$

and observed that a similar analysis ought to yield results for other invariant functions, such as $F(q) = \sum_{m,n} q^{am^2+2mn+cn^2}$. He observed that the expression

$$\theta = \sum e^{i(mx+ny)} q^{am^2+2mn+cn^2}$$

---

[8] Integrals of the first kind are holomorphic, those of the second kind have simple poles.

defines the familiar theta function, and that the expression equals $F(q)$ when $x = 0 = y$, so $F(q)$ will be connected to Abelian functions.

Now, the function θ satisfies the heat equation

$$\frac{d\theta}{dt} = a \frac{\partial^2 \theta}{\partial x^2} + 2b \frac{\partial^2 \theta}{\partial x \partial y} + c \frac{\partial^2 \theta}{\partial y^2},$$

and this suggested to Poincaré that the analogy with the theory of heat diffusion could guide his research here, and in that way he was led to show that when q was very nearly 1,

$$F(q) \approx \frac{2\pi}{Et},$$

where $q = e^{-t}$ and $E^2 = 4(ac - b^2)$. This shows that for such values of q the invariant $F(q)$ depends solely on the discriminant, $(b^2 - ac)$.[9]

Poincaré then defined $\Phi(u) = F(e^{-2\pi i u})$ and deduced that

$$F\left(\frac{-1}{E^2 u}\right) = i E u \Phi(u) \text{ and } \Phi(u + 1) = \Phi(u).$$

The first of these resembles a familiar theta function identity discovered by Jacobi and used by, among others, Riemann in his study of the zeta function. The two together establish that $\Phi^2(u)$ is a thetafuchsian function for the group generated by

$$u \mapsto u + 1 \text{ and } u \mapsto \frac{-1}{E^2 u}.$$

This group, he remarked, is made of two kinds of substitutions:

1. $u \mapsto \dfrac{\alpha u + \beta}{E^2 \gamma u + \delta}$, $\alpha, \beta, \gamma, \delta \in \mathbb{Z}$, $\alpha\delta - \beta\gamma E^2 = 1$;

2. $u \mapsto \dfrac{E^2 \alpha u + \beta}{E^2 \gamma u + E^2 \delta}$, $\alpha, \beta, \gamma, \delta \in \mathbb{Z}$, $E^2 \alpha\delta - \beta\gamma = 1$;

and the first of these form a congruence subgroup of the Fuchsian group of index 2.

[9] Dirichlet had established a similar property for $J(s)$.

Poincaré concluded this part of the paper by showing how his approach simplified one of the cases of Dirichlet's class number formula, and offered the promise that pursuing this analogy would lead to new theorems in analysis. In the final section of the paper he then sketched the implications of his approach in the difficult case of indefinite quadratic forms. Meyer gave this paper a lengthy review in the *Jahrbuch*, and concluded that this work appears, in the proper sense, as an extension of Dirichlet's fundamental ideas and methods in the theory of quadratic forms. The paper (1912b) opens with a useful summary of the previous paper, gives some corrections, and goes into more detail on the construction of the thetafuchsian functions.

In his commentary on Poincaré's work in arithmetic, André Weil (1954) remarked that Poincaré distinguished two cases, one where the fundamental polygon has vertices on the boundary of the disk and the other where it does not. The first case is special, but it means that the corresponding functions are periodic (because the group contains parabolic elements) and so the methods of Fourier analysis can be used to effect, as Hecke had demonstrated. The second case is simpler function-theoretically, but the number theory had proved very much harder to understand. Poincaré's (1912b) ends with some remarks about the automorphic forms for congruence subgroups, with explicit calculations for the modular group (where Fourier methods can apply). The series expansions Poincaré obtained have coefficients that involve Bessel functions, and once again he expressed the confident opinion that his more general approach would yield new results. He was to be proved right when these series were worked up by Kloosterman in 1929 into results about quadratic forms in four variables.[10]

Arithmetic Properties of Algebraic Curves

Poincaré's best known paper in number theory is his (1901f). In line with what he had begun to think about in 1880 he began by observing that a number of properties of binary quadratic forms could be ascribed directly to the way these forms transform under the action of a group of

---

[10] For this and hints of the later history of Poincaré series and Kloosterman sums, see Kowalski (2006/2010).

linear transformations in the variables. In this spirit, he said, he offered more of a program than a genuine theory, and one that was based on the idea that problems in indeterminate analysis might be found to be connected by using the classification of algebraic curves in terms of their birational transformations with rational coefficients. Oddly enough, Poincaré did not cite the paper Hilbert and Hurwitz (1891) where several important points were established for the first time. In particular, they had observed that the study of rational points on a curve is a problem that is invariant under birational transformations with rational coefficients, and so Diophantine problems about curves should be classified by the genus of the curve and not its degree. They had also shown how to solve all such problems for rational curves of any degree (these are curves of genus 0).

First, Poincaré quickly reviewed the birational theory of curves. A line in this approach is given by an equation of the form $ax + by + cz = 0$ with rational coefficients, and any two lines are obviously birationally equivalent. As for conics, a conic with a rational point is birationally equivalent to a straight line (by stereographic projection) and therefore if a conic has a rational point it has infinitely many. So all conics with a rational point belong to one birational equivalence class, and Gauss had given a test in his *Disquisitiones Arithmeticae* to show if a conic has a rational point. The same section of that work also shows, said Poincaré, how to assign conics with no rational points to whichever class they belong.

Poincaré next discussed cubic curves, starting with the rational curves (those that have a double point). These curves are also birationally equivalent to a line, by projection from the double point. A slightly more complicated argument shows that the rational curves of any degree are equivalent to a line. He then turned to elliptic curves, nonsingular cubic curves, which have genus 1, and studied the distribution of the rational points on such curves.

A nonsingular cubic curve is in fact a group. One picks an arbitrary point O to be the identity, and the group operation consists of joining two points P and Q on the cubic by a line that meets the cubic again at a point R; one then joins R to O by a line that meets the cubic again at the point S. The sum of P and Q is said to be the point S.

Poincaré did not observe this, but reminded his readers that it is possible to attach an *argument* to each point of an elliptic curve in such a way that the sum of the arguments of three collinear points is a constant that is defined up to a period, and he proposed to define the argument in such a fashion that this sum is zero. This made it possible for Poincaré, given two rational points on a cubic curve and the line joining them, to compute the argument of the point where the line meets the curve again, and also for him to find the argument of the point where the tangent to the rational curve at a rational point meets the curve again. In particular, if M_0 is a rational point with argument α and one draws the tangent to it, meeting the cubic again at M_{-1}, then the argument of M_{-1} is -2α. The tangent to the curve at M_{-1} meets the curve again at M_1 with argument 4α, and if one carries on in this fashion the points will have arguments of the form $(3n + 1)\alpha$, which is never zero unless α is a rational multiple of a period of the curve. Similarly, all the points constructible in this fashion from two points with arguments α and β have arguments of the form $\alpha + 3n\alpha + p(\beta - \alpha)$.

In general, all the points obtainable from $q + 1$ points with arguments $\alpha, \alpha_1, \alpha_2, \ldots, \alpha_q$ have arguments of the form

$$\alpha + 3n\alpha + p_1(\alpha_1 - \alpha) + p_2(\alpha_2 - \alpha) + \cdots + p_q(\alpha_q - \alpha).$$

Poincaré now asked if it was possible to find a number q and points with arguments $\alpha, \alpha_1, \alpha_2, \ldots, \alpha_q$ such that every rational point on the cubic was obtainable from these points in this fashion. This number might depend on the choice of points, but there would be a least number q for which there were suitable points, and he proposed to call the number $q + 1$ the *rank* of the cubic. He was not able to show that this is always the case, and this result was established for the first time in Mordell (1922). Because a cubic curve is itself a group, a fact that Poincaré did not comment upon explicitly, this amounts to showing that the rational points on an elliptic curve form a finitely generated group that consists of $q + 1$ copies of the integers and a finite part that was only much later classified by Mazur (1977).[11]

Poincaré ended his paper with a few tentative remarks about how his approach might be extended to the arithmetic of algebraic curves of higher genus.

---

[11] Indeed, Poincaré did not actually show that the rank is a birational invariant, which it is; see Schappacher (1991).

LIE THEORY

Most of Sophus Lie's research was into transformations depending on parameters. At various times in his life he entertained high hopes for the enterprise; it was to be a Galois theory of differential equations.[12] He was a very productive mathematician—his *Gesammelte Abhandlungen* run to six volumes, and there are also several two- or three-volume books on particular topics. But being a Norwegian who came to mathematics comparatively late, he never found writing German easy, and when he was persuaded that his intuitive, geometric style was not easy to follow he switched to an uncomfortable mode of writing that brought out the analytical problems inherent in his work without making them clear. For these reasons he was never highly regarded in Berlin either by Weierstrass or his successor, Frobenius, even though one of their school, Wilhelm Killing, did a considerable amount of work trying to sort out what Lie had done. Lie returned the favor, writing on one occasion about Killing's papers, "the correct theorems in them are due to Lie, the false ones due to Killing," and they "contain not so many results that are correct and new. Proved, correct and new are even fewer."[13] This was doubly unfair: not only were Lie's papers obscure and at times wrong, but Killing was making an important contribution to the theory opened up by Lie. Lie was capable of complaining that no one appreciated his work and also that people took it up.

But if he was dismissed in Berlin, he had friends elsewhere. One was his friend from his student days in Berlin and Paris, Felix Klein, and the other was a group of mathematicians in Paris, Poincaré among them. Klein and Lie had passed through Berlin in 1869 where they found themselves drawn together with a common liking for geometry and a shared sense of not belonging in the rigorous analytic and algebraic atmosphere of the place. They then traveled to Paris together to meet Camille Jordan and other young French mathematicians, but Klein's stay was cut short by the outbreak of the Franco-Prussian War (in which he briefly served in the Prussian army and fell seriously ill from typhus). They kept in close touch, and when Klein moved from Leipzig to Göttingen in 1886 he was able to arrange for Lie to be his successor, over predictable objections from

---

[12] On Lie's idée fixe, see Hawkins (1991).
[13] Quoted in Schmid (1982, 179).

Figure 9.1. Sophus Lie (1842–1899). Source: Hermann Minkowski, *Briefe an David Hilbert* (Springer, 1973).

Weierstrass who found Lie neither good enough nor, indeed, German (moreover, Schwarz was the rival candidate). Lie stayed there until 1898 when he returned to his native Norway, by which time a pervasive distrust of mathematicians, linked perhaps to pernicious anemia, seems to have settled over him, and he disputed with many people, publicly attacking Klein in one of his books.

The Paris group was diffuse. Lie had traveled there for a second time in 1882, when he met Poincaré, and the two men got on well if distantly. It was Mittag-Leffler who introduced them by letter, at Lie's request, and Lie wrote back to him that he was keen to meet Poincaré and to read his memoirs on differential equations, saying "I have ordered Poincaré's Memoirs but unfortunately they are sold out" (Stubhaug 2002, 286). Lie hoped that his theory of what he called transformation groups would fit well into Poincaré's of linear differential equations. On his visit to Paris he got on well with Hermite and Darboux, and gave a successful talk at the SMF which improved his reputation (Stubhaug 2002, 296). After his return to Norway on 1 December he wrote to Mittag-Leffler in positive terms about his trip, remarking in particular that "Poincaré is difficult to understand as he often speaks indistinctly. Although I understood him to the extent that on several occasions I got a lively impression of his unusual Mental Powers" (quoted in Stubhaug 2002, 398).

A brief exchange of letters between Lie and Poincaré followed Lie's return, but nothing more came of it, and in a letter Lie wrote to Klein in September 1883 he remarked that "Poincaré impressed me on several accounts; however, I did not succeed in entering into real scientific contact with him." The benefits of Lie's trip to Paris did not begin to accrue until he took up his professorship in Leipzig, when Darboux, Picard, and Poincaré began to encourage some of the best students at the École normale supérieure to go to Leipzig to study with him. This not only recognized the importance and fecundity of his ideas—and Lie was very sensitive to slights—it brought him people willing to polish and adapt his ideas with more precision and skill than he had himself. Lie had already acquired Friedrich Engel as an assistant, and he now acquired Georg Scheffers as well, and they helped him to write several of his most important books.

More French support was on display in 1892 when he was elected to the Académie des sciences, where he replaced the recently deceased Kronecker. He returned to Paris in 1893, writing with more than mere politeness to Poincaré that "once again the centre of gravity for mathematics is Paris and France," and on one evening he dined with Poincaré at Jordan's home. He came again in 1895 when the École normale supérieure celebrated its centenary and Lie was invited to speak about Galois and his significance for the development of mathematics (Stubhaug 2002, 329).

Lie's death on 18 February 1899 was the occasion for much grief and international commemorations. In Paris, at a meeting organized by the Académie des sciences and the Institut de France, the poet Gunnar Heiberg, aware that Lie had stood in the world alongside such eminent Norwegians as Ibsen, Grieg, and Munch, compared him to Poincaré, who, he said, was becoming a "Universal Genius in the great Riches of Mathematics," and although he did not look like Lie, yet

there was something similar about him: There was a similar Anxiety of Struggle about them both. For over Sophus Lie, for all his Huge Presence, with his big Head and prominent Frontal Lobe, there was an eternal Anxiety....[and for Poincaré]: His Back is curved, and almost humped. His Head is dense—filled in a way. Calm, brown melancholy Eyes, which suddenly, either when he speaks to someone, or is sitting alone, begin to search, flit back and forth, to play. He is searching for something. Something dark that the Eyes will search out and illuminate. Then a while later, they are again calm and melancholy.[14] (Stubhaug 2002, 451)

Hawkins' fine book, *Emergence of the Theory of Lie Groups*, describes at length the long journey from Lie's work to the establishment of the theory of real and complex Lie groups and Lie algebras in the work of Killing, Élie Cartan, and Hermann Weyl.

Despite his admiration for Lie, Poincaré took up Lie's ideas only after Lie's death. In 1900 he tackled one of the more important problems in the theory as it then stood, and when he came to publish at length in a paper that formed part of a volume celebrating the work of Sir George Stokes, he found that he had been beaten to some of the ideas by Campbell and Schur. Nonetheless, he said, he hoped that his ideas had enough in them still to merit publication.

Lie had typically considered a transformation from a space to itself given by an expression of the form $y = f(x, a_1, \ldots, a_n)$, where x and y belong to the space and the a's are parameters. If one thinks of the transformation as, say, the rotation of a sphere about a fixed axis, then the single parameter a would stand for the angle of the rotation. Lie imagined that one could follow one transformation by another, to obtain

$$f(f(x, a_1, \ldots, a_n), b_1, \ldots, b_n) = f(x, c_1, \ldots, c_n), \qquad (9.2)$$

where the value of the parameters c are determined by the values of the a's and b's. In the simple case of rotations of a sphere about an axis any pair of parameters determine a third, but in the general setting this may fail; when it held Lie said the transformations formed a group. This is not the modern use of the term, and I shall call it a group germ instead.[15]

---

[14] Figure 9.2 is from a drawing by Käte Popoff whose husband Kyrille attended Poincaré's lectures on celestial mechanics in 1907–08.

[15] Nor did it occur to Lie to insist on the existence of a set of parameters $a_1, \ldots, a_n$ that correspond to the identity transformation (the transformation that moves no point of the space), nor did he demand that every transformation have an inverse. Caveat lector.

Figure 9.2. Poincaré lecturing, 1908. Source: Académie des sciences, dossier Poincaré, from a drawing by Käte Popoff, reproduced in Stubhaug, *The Mathematician Sophus Lie* (Springer, 2002, p. 411) and on the cover of the journal *Philosophia Scientiae*.

Lie had also considered the infinitesimal version of these transformations as the parameters vary by an arbitrarily small amount. Such an infinitesimal transformation can be written in the form $y = x + t\xi(x)$, where t is arbitrarily small and ξ depends on the parameters a. Lie often thought of these infinitesimal transformations as directional derivatives, because any smooth function φ of the x's, which will be regarded now as k-tuples $x_1, \ldots, x_k$, can be given a Taylor series expansion

$$\varphi(y_1, \ldots, y_k) = \varphi(x_1, \ldots, x_k) + t\left(\xi_1\frac{\partial\varphi}{\partial x_1} + \cdots + \xi_k\frac{\partial\varphi}{\partial x_k}\right).$$

He therefore denoted $\xi_1 \frac{\partial}{\partial x_1} + \cdots + \xi_k \frac{\partial}{\partial x_k}$ by X, which he called an infinitesimal operator, and wrote

$$\varphi(y_1, \ldots, y_k) = 1 + t X \varphi(x_1, \ldots, x_k).$$

One of his standard examples (Lie 1893, 6) was the transformations of the form

$$y = \frac{a_1 x + a_2}{a_3 x + a_4},$$

where x and y lie on the projective line and the four parameters a_1, a_2, a_3, a_4—or rather their three ratios $a_1 : a_2 : a_3 : a_4$—satisfy $a_1 a_4 - a_2 a_3 \neq 0$. These transformations do form a group in the modern sense of the term. He found the corresponding infinitesimal transformations by setting $a_1 = a_4(1 + e_2)$, $a_2 = e_1 a_4$ and $a_3 = -e_3 a_4$, so the transformation becomes

$$y = \frac{(1 + e_2)x + e_1}{1 - e_3 x} = x + e_1 + e_2 x + e_3 x^2$$

to first order in the e's. He sometimes wrote this, by analogy with the Taylor expansion, as

$$y = x + \frac{\partial}{\partial x} + x \frac{\partial}{\partial x} + x^2 \frac{\partial}{\partial x}.$$

Lie focused on the passage from the group germ to the infinitesimal transformations. Composition of the infinitesimal operators proceeds according to rules studied by Poisson and Jacobi. One defines the product (later called the Lie bracket) of two operators A and B by $[A, B] = AB - BA$ and for any two infinitesimal operators X_j and X_k corresponding to the same group germ, Lie proved that he could write

$$[X_j, X_k] = \sum_l c_{jkl} X_l, \tag{9.3}$$

where the c_{jkl} are constants and the X_l run through a basis for the infinitesimal operators corresponding to the same group germ. He also claimed the converse, that whenever there are partial differential operators $X_1, \ldots, X_n$ such that an expression like the one above holds for all pairs of these operators, there is a group for which the given operators are infinitesimal operators. In this setting, the constants c_{jkl} must satisfy some trivial identities. This result amounts to claiming that from (9.3) one can reconstruct the transformations (9.2) and give the rule for determining the $c_1, \ldots, c_n$ as functions of the $a_1, \ldots, a_n$, and $b_1, \ldots, b_n$.

It is this converse, known as Lie's third fundamental theorem, that attracted the most interest. Lie himself had given two proofs, one for what was later seen as a special case when the group has no center, and one in the language of differential equations. In the 1890s Friedrich Schur had given a different proof, one that made systematic use of a notation that captured the linear aspects of the problem, and showed that the way the c's depended on the a's and b's was determined by an infinite series of a universal form (independent of the parameters) in which the coefficients were polynomial functions of the a's and b's. Schmid (1982, 178) notes that Schur, a former student of Weierstrass, had a careful, analytic style quite unlike Lie's, and quotes Engel as observing that Lie and Schur had very different ideas of what was easy and what was not. Schur, for example, never spoke of infinitesimal transformations, and he was conscientious about checking when the infinite series converged.

In 1901 the Oxford mathematician John Edward Campbell also gave a proof of the theorem, one that took its cue from the idea that the exponential of an infinitesimal transformation is a group element, and so the problem is connected with the problem of finding Z when

$$e^X e^Y = e^Z. \tag{9.4}$$

This is straightforward when X and Y commute, but they do not in general. But unlike Schur he did not deal with the convergence of the power series he exhibited.

Poincaré also claimed to have reduced the problem to the solution of simple differential equations, a process that could, he said, be carried out in finite terms. His argument is long and technical, and cannot properly be described here.[16] In essence, he first elaborated what could be said about the polynomials and power series in variables $X_1, \ldots, X_n$

---

[16] See Schmid (1982) and Grivel (2010).

that satisfy the formal identities

$$[X_j, X_k] = X_j X_k - X_k X_j = \text{a linear combination of } X_1, \ldots, X_n.$$

In this framework he then obtained the solution of Campbell's problem (9.4) of writing Z in terms of X and Y by showing that it could be written as an infinite series. To obtain the convergence of the power series he made an ingenious use of the residue calculus of complex function theory. When he later commented on this work in his (1901c) he made the interesting observation that the only transcendental functions that enter the construction of the finite transformations are exponential functions. He therefore took up a purely formal point of view, abstracting, as he put it, completely from the "matter" of the group,[17] and found formulas applicable to all isomorphic groups. But these formulas gave the relations that exist between the parameters in the transformations, and from these relations he could construct a group isomorphic to the one he was studying. This not only established Lie's third theorem, it did so without the need for the differential equations that Lie had introduced. And, he concluded, the relationship between his formulas, some of Lie's, and some of Killing's, was likely to be most instructive.

In his (1906) Felix Hausdorff took up the question, noting the work of Campbell and a more recent paper (Baker 1905), and offering an account that he regarded as a considerable simplification of Poincaré's. This it is on the algebraic side; the convergence condition is established much as Poincaré had proceeded. Hausdorff's argument is the one that has met with the approval of later writers, and the result is today known as the Campbell–Baker–Hausdorff formula. Scharlau (2001, 464) notes that Hausdorff only heard of Baker's paper in a letter from Engel when he was finishing his own, and he wrote back to Engel that "sadly I notice that its content overlaps my work in large part. Evidently the problem has been hanging in the air... but it was Baker's–the only one that remained quite unknown to me–that has the most contact with mine."[18]

Poincaré again displayed his liking for form over matter in what has been known, since Cartan and Eilenberg (1956) and more precisely since

---

[17] A phrase he had used earlier in discussing how we use our innate concept of a group to construct space in his (1898c); see chap. 1, sec. "Non-Euclidean Geometry Comes to France" above.

[18] See Scharlau's commentary in Hausdorff's *Gesammelte Werke* (4, 461–465).

Bourbaki in the 1960s, as the Poincaré–Birkhoff–Witt theorem, which is a major result about the structure of a Lie algebra.[19] Ton-That and Tran write: "But by a careful analysis of Poincaré (1899f) one must conclude without a shade of doubt that Poincaré had discovered the concept of the universal algebra of a Lie algebra and gave a complete and rigorous proof of the so-called Birkhoff–Witt theorem."

It is only possible to be brief here about what Poincaré accomplished. The setting is that of a finite-dimensional family of vector fields $X_1, X_2, \ldots, X_n$—Poincaré called them infinitesimal transformations—with a multiplication, so

$$[X_j, X_k] = \sum_l c_{jkl} X_l,$$

where the so-called structure constants c_{jkl} satisfy certain identities that follow from the properties of the multiplication. This defines a Lie algebra $\mathcal{L}$. In this setting, Poincaré drew attention to the largest algebra one can usefully construct from these symbols. One starts by forming the formal, noncommutative, associative, polynomial algebra $\mathcal{A}$ with the variables $X_1, X_2, \ldots, X_n$, where the coefficients are either $\mathbb{R}$ or $\mathbb{C}$ (later writers showed how to work with an arbitrary field). In this algebra $\mathcal{A}$, consider the subset of elements that are of the form

$$P(XY - YX - [X, Y])Q, \tag{9.5}$$

where $[X, Y]$ denotes the multiplication in $\mathcal{L}$ and P and Q are arbitrary elements of $\mathcal{A}$. An element of $\mathcal{A}$ is said to be equivalent to 0 if it is a linear combination of elements of the form (9.5), and two elements A and A' in $\mathcal{A}$ are said to be equivalent if their difference $A - A'$ is equivalent to 0. The set of equivalence classes is the universal enveloping algebra of $\mathcal{L}$. Poincaré gave a characterization of the best form for an element in a given equivalence class—he called them regular elements—and showed they have a striking degree of symmetry among their monomial entries; a detailed statement of this result is the Poincaré–Birkhoff–Witt theorem. It follows that there is a canonical injective map from $\mathcal{L}$ to

---

[19] See Ton-That and Tran (1999) for a thorough account of Poincaré's work and the convoluted history of the names for this theorem.

its universal enveloping algebra, and therefore any Lie algebra over a field is isomorphic to a Lie subalgebra of an associative algebra and some problems about Lie algebras can be transferred to problems about associative algebras.

ALGEBRAIC GEOMETRY

In 1910 and 1911 Poincaré published two papers on the subject of curves and integrals on an algebraic surface that united two decades of parallel studies by French analysts and Italian geometers, and was regarded by one of those Italians almost thirty years later as displaying "the indelible mark of his universal genius."[20] Indeed, his proof survived the collapse of the earlier Italian version, and remained the only valid proof until Mumford (1966).

One of Riemann's great achievements in the 1850s had been to produce a theory of complex algebraic curves. This was the first time the variables in polynomial equations, of the form $f(x, y) = 0$ had properly been allowed to be complex variables, and Riemann showed how to use this theory not only for geometrical purposes but also analytical ones in the theory of Abelian functions (these are the functions obtained by integrating the everywhere finite, or holomorphic, integrands on the curve). It is this theory that Poincaré illuminated and expanded with his theory of Fuchsian groups and functions. But there was another direction in which Riemann's theory was to be taken, and which Poincaré concerned himself much less with until 1909, and that was the theory of complex algebraic surfaces. A typical algebraic surface has a polynomial equation of the form $f(x, y, z) = 0$, where x, y, and z are complex variables. Such an equation describes a surface in $\mathbb{C}^3$, which may well intersect itself along curves and have some isolated singular points. A surface expressed in this form invited Riemann's successors to generalize the theory of algebraic curves, and this was the approach taken in the 1860s by Clebsch and his followers Alexander Brill and Max Noether.

Brill and Noether had begun their mathematical careers in their (1874) by retreating from Riemann's theory of curves to the theory

---

[20] Castelnuovo (1938, Bologna ICM, I, 196) quoted in Poincaré, *Oeuvres* (6, 178, no. 1).

of curves in the projective plane, and in their theory a plane curve $f(x, y) = 0$ of degree n having d double points, k cusps, and no other singular points was said to have genus $(n-1)(n-2)/2 - d - k$. The number $(n-1)(n-2)/2$ is the estimate of the number of everywhere-finite integrands on a nonsingular curve of degree n, which is given by the number of linearly independent expressions of the form $\frac{g(x,y)}{f_y(x,y)}$ where $g(x, y)$ is a polynomial of degree at most $n - 3$. Singular points more complicated than double points and cusps were a challenge, and the hope was that they would be combinations of the simple ones made up in ways one could disentangle.

Brill and Noether had deliberately started with the geometry rather than the function theory, in which the genus of a curve is defined as the number of linearly independent holomorphic integrands in the curve, because they felt that Riemann's ideas were confusing and his reliance on the Dirichlet principle misplaced. It was better, in their view, to follow the approach of Clebsch and Gordan (1866) and establish solid foundations in geometry and then develop the analytical implications by using a theorem such as Abel's theorem on the integrals on an algebraic curve. Abel's theorem can mean many things (see Kleiman 2004), but the central idea is obtained by generalizing the addition theorem for elliptic integrals. The theorem concerns integrals of the form $\int_a^u f(z)\,dz$, where $f(z)$ is an expression of the form $\left((1 + e^2 z^2)(1 - c^2 z^2)\right)^{-1/2}$, and this integral, up to multiplication by a constant, is the only integral that is holomorphic everywhere on the curve $w^2 = f(z)$. The theorem says that

$$\int_a^u f(z)\,dz + \int_a^v f(z)\,dz + \int_a^w f(z)\,dz = 0,$$

when the upper endpoints u, v, w lie on a straight line. So the theorem says that a sum of holomorphic integrals on the curve vanishes if the upper endpoints have a certain simple geometric property, and the important point is that the above equation is true whatever straight line is chosen. The remarkable generalization that Abel pioneered concerns holomorphic integrands on an arbitrary algebraic curve, but it still singles out families of upper endpoints that are cut out on the curve by families of other curves, not necessarily straight lines. So it says that holomorphic integrands on a curve can be studied by looking at families of points on that curve which depend, in simple cases linearly, on the parameters determining a family of curves; in short, holomorphic integrands on a curve are studied by families of points on the curve that are determined geometrically in some way. The hope is that all the analytical problems are transferred to problems in the geometry of plane curves. Riemann's fundamental insight that the number of linearly independent holomorphic integrands on a nonsingular curve is equal to its topological genus is lost in their approach, which could only see birational invariants, not topological ones, and the concept of genus was retained only in reference to integrands.

What might be called the algebraic approach to the theory of surfaces likewise sought to establish foundations in projective geometry. In the hands of Noether and Cayley a surface defined by an equation of degree n is characterized by a number D that, to quote Cayley (1871, 527), is $(n-1)(n-2)(n-3)/6$ minus various terms, each term having to do with a singularity of some kind, the last of which "= the number of certain singular points on the cuspidal curve the nature of which I do not completely understand and which is here taken to be = 0." The honesty of that remark is commendable, but also indicative of the problems facing this approach. Here, the number $(n-1)(n-2)(n-3)/6$ is a formal estimate of the number of independent everywhere finite integrands (i.e., holomorphic 1-forms) on the surface. The simple singularities on a surface are double curves with distinct tangent planes where the surface self-intersects and isolated triple points. They are easy to understand, and all but the simplest algebraic surfaces have them; they were first called "normal" singularities by Salmon, and the name stuck.[21]

It had been one of Cayley's first contributions to this theory in his (1871) to drive a wedge between the algebraic or geometric and the analytical parts of the theory of surfaces. He did this by exhibiting a class of surfaces, which includes the ruled surfaces (surfaces with a one-parameter family of straight lines), for which the above number is negative, and therefore cannot be the dimension of a space of any kind. This had forced mathematicians to define two sorts of genus for an algebraic surface: the geometric genus p_g and the arithmetic genus p_a. The geometric genus is the dimension of the space of holomorphic

[21] On the study of the singularities of algebraic surfaces in the 1890s see Gario (1989).

2-forms or double integrals on the surface; these are integrals of the form

$$\iint \frac{Q(x,y,z)\,dx\,dy}{f_z},$$

where Q is a polynomial of degree $n-4$ when the surface is defined by an equation $f(x,y,z)=0$ of degree n and $f_z = \frac{\partial f}{\partial z}$. The arithmetic genus is the number defined by Cayley:

$$\frac{(n-1)(n-2)(n-3)}{6} - (n-4)d + 2t + p - 1.$$

Here $(n-1)(n-2)(n-3)/6$ is the geometric genus of a nonsingular surface of degree n, and $(n-4)d + 2t + p - 1$ is the number derived from the number of double points, d, and triple points, t, on a double curve of genus p on the surface.[22] The geometric and the arithmetic genera of a surface are both birational invariants.

By the 1890s the *algebraic* geometers like Brill and Noether, who were mostly German, were met by algebraic *geometers*, who were mostly Italian, and who sought to build up a birational theory of surfaces in terms of families of curves on the surface. At first, their leader was Corrado Segre in Turin. His influential and lengthy memoir of 1894 redescribed the Brill–Noether theory of algebraic curves in the language of birational geometry, deliberately with a view to assisting those then engaged in forging the new theory of algebraic surfaces. In this theory the aim is to generalize the Brill–Noether theory of a curve in the projective plane that is cut by families of other curves into interesting families of points, to a theory about a surface in some higher-dimensional projective space that is cut by a family of other surfaces into interesting families of curves. Segre, however, ran into difficult problems with his own idea of how to proceed, and he was soon overtaken by Enriques and Enriques' close friend Guido Castelnuovo. Castelnuovo had studied mathematics under Veronese, and briefly with Segre, before going to Rome, where he was professor from 1891 to 1935. Enriques graduated with distinction from Pisa in 1891 when he was 20, and went to study under Castelnuovo. Their association was famously fruitful, and his grasp of geometry was likewise deepened by time spent with Segre. In 1902 they were joined by Francisco

[22] This is known as Noether's formula; see Griffiths and Harris (1978, 618–628).

Severi, another former student of Segre, when he became an assistant to Enriques, before becoming professor at Padua.

A topic of the greatest importance in the theory of algebraic surfaces was that of families of other surfaces that cut the given surface in curves of certain kinds. The best understood case was when the family depended in a linear fashion on some parameters; the family was then said to form a linear series, and it would be of interest to know if the family contained all and only curves of such and such a kind. It would then be said to be a complete family. But one could be presented with examples of a complete family consisting of all surfaces of a particular kind, and then it was a question as to whether this family was linear.

Between 1900 and 1914 Enriques and Castelnuovo wrote a series of papers that culminated in a birational classification of the possible types of algebraic surface. They had discovered in the 1890s that the arithmetic and geometric genera do not suffice to classify surfaces; in particular it is not the case that when they both vanish the surface is birationally equivalent to the simplest surface, the complex projective plane. They were able, however, to introduce a family of other genera (the so-called plurigenera) that were birationally invariant numbers and could be used to refine the classification. They also needed tools that made no mention of the degree of a surface, because that is not a birational invariant (the same surface can be represented by equations of different degrees). Their way forward had been to study families of surfaces that meet the surface they wish to study in particular ways—by passing through the double curves on the surface, for example. Gradually they picked off identifiable families of surfaces. In 1905 Enriques showed how to characterize surfaces with $p_g = 0$ and $p_a < 0$ into two kinds, according as $p_a = -1$ or $p_a < -1$. The first kind resembles a family of elliptic curves, the second kind are ruled surfaces. Finally, in 1914 Castelnuovo and Enriques published two articles in the *EMW*, the second of which culminated in a classification of algebraic surfaces that still bears their names.

This Italian work was matched on the analytical side by the contemporary work of Picard, who had devoted many years to generalizing Riemann's theory of Riemann surfaces to a theory of integrals on algebraic surfaces. Indeed, so close was his work to that of the Italians that the first of several surveys of the field that they wrote, the "Résultats nouveaux," appears as an appendix in the second volume of Picard and Simart (1907). The French approach, in which Marie-Georges Humbert was also prominent, can be seen as an attempt to generalize yet another approach to curves to the case of surfaces: in this instance, Poincaré's. They defined a class of surfaces (the "hyperelliptic surfaces") by means of automorphic functions in two variables. The hyperelliptic surfaces became the subject of long and detailed studies by Enriques and Severi, for which they were awarded the French Prix Bordin in 1907.

Throughout the period the surfaces for which the geometric and arithmetic genera differ, the so-called irregular surfaces, were of particular interest. In 1904 Severi showed that the existence of simple integrals of the first kind (holomorphic 1-forms) implied that the surface was irregular, and in the same year Enriques established the converse. Almost at once Severi then showed that $r - q = p_g - p_a$, where q and r are the numbers of linearly independent simple integrals of the first and second kinds, respectively (this was among the papers for which he was awarded the first Guccia medal in 1908). Finally, first Castelnuovo and then Severi gave different proofs that showed that $q = p_g - p_a$ and $r = 2(p_g - p_a)$, although their arguments hinged on a result of Enriques' that was not wholly convincing.[23] This result, and the steps leading up to it, are given considerable prominence in the "Résultats nouveaux," because it made a connection between the geometrical theory of the Italians and the analytical or transcendental theory in France, which gave it considerable importance. As Castelnuovo and Enriques remarked (Picard and Simart 1906, 495): "This result is therefore the fruit of a long series of researches, to which the transcendental methods of M. Picard and the geometrical methods used in Italy contributed equally." It is this theorem that Poincaré proved analytically in his (1910g), when he showed that the irregularity is the maximum number of everywhere-finite simple integrands (holomorphic 1-forms) on the surface.

It is evident from the history of this discovery that there is no obvious reason why it should be true. However, although it is difficult, it is possible to identify the different aspects of the problem that Poincaré identified. He laid particular emphasis on the complex analysis. A key observation of Picard (1897, 116) was that a holomorphic 1-form on a surface can be

---

[23] Enriques' mistake was in claiming that a certain series he constructed was complete.

written in the form

$$\int \frac{P\,dx + Q\,dy}{\frac{\partial f}{\partial z}},$$

where P is a polynomial of total degree $n-2$ in x, y, z but of degree $n-3$ in x and z, and Q is a polynomial of total degree $n-2$ in x, y, z but of degree $n-3$ in y and z. So integrals of the form $\int (P/\frac{\partial f}{\partial z})\,dx$ define holomorphic integrands on the curve in the surface defined by $y = $ constant. But the contribution from Q is hard to understand, and involves the geometry of the surface. So, naively, the p_g-dimensional space of holomorphic 1-forms on the surface is split into a contribution to the holomorphic 1-forms on each curve where $y = $ constant and a part of a dimension determined by the geometry, which is how the arithmetic genus enters the story.

Another way to look for holomorphic 1-forms on a surface is to observe that there should be many of these on curves in the surface. Poincaré indeed began by imagining the complex surface S sliced up by the planes $y = $ constant. Except for a finite number of values of y, each such slice is an algebraic curve C_y of some genus, p say, and will have a p-dimensional family of holomorphic 1-forms. Their integrals define Abelian functions on the curve, where the integrals are taken over paths on the curve. As functions of their upper endpoints, the values of these integrals are defined up to an arbitrary multiple of the integrals taken around closed paths (called cycles) on the curve (the periods of the integrals). Since $2p$ cycles form a basis for the homology of the curve C_y, each Abelian integral travels with a list of its $2p$ periods that together form a ($p \times 2p$)-period matrix. These Abelian integrals and their period matrix, regarded as functions of y, are central to his analysis.

As the choice of value of y is varied, the curve will vary but usually not its genus. However, it can be expected that for certain values of y the curve C_y will have singular points, one or more of the cycles will shrink to a point (Poincaré called these the vanishing cycles), the genus will drop, one of the Abelian functions will become identically zero, and the period matrix will generally become singular; Poincaré called these points effective critical values.[24] He also dealt with what happens if one

---

[24] This recalls his "Morse-theoretic" analysis of three-manifolds.

of the Abelian functions becomes infinite. The behavior of the periods as functions of y—the parameter value of the slice—as y is taken on a loop around one of its singular values had already been studied by Picard, who obtained a differential equation for them that Poincaré relied upon.[25]

To the study of how Abelian integrals vary from slice to slice, Poincaré specified the limits and either the periods or the path of integration. He fixed the lower endpoint by a horizontal slice ($x = $ constant) but hoped to specify how the upper endpoint varied by drawing an arbitrary algebraic curve C on the surface that meets each C_y in p points.

To investigate how the periods of the Abelian integrals vary as y passes through a critical value, Poincaré defined certain sums of Abelian integrals which he called normal functions and denoted $v_1, v_2, \ldots, v_p$. Each $v_h(y)$ is a sum $\sum_k u_j(y_k)$ as y_k runs through the upper endpoints of the integral; that is, over the points where C_y meets the curve C. These normal functions are holomorphic except for the singular or critical points, where they may become infinite or cease to be single valued, but then there is a period Ω_j such that $v_j + \frac{1}{p}\Omega_j$ remains holomorphic, and the ratio $\frac{v_j}{\Omega_j}$ remains holomorphic in a neighborhood of the critical value. The normal functions depend on the choice of C, and are determined up to periods of the integrals. The periods also change at the critical slice, because there at least one cycle vanishes, and when y is taken on a circuit around a critical value, by hypothesis $\frac{v_j}{\Omega_j}$ remains holomorphic, but Ω_j is converted to $\Omega_j - \Omega'_j$, where Ω'_j is another period. The important point is that these normal integrals have a chance of defining 1-forms that are holomorphic on the entire surface.

To obtain the curve C, Poincaré brought in Abel's theorem. He observed that on an algebraic curve C_y of genus p with a basis of holomorphic integrands $du_1, du_2, \ldots, du_p$, for any given constants $v_1, v_2, \ldots, v_p$ there are points $x_1, x_2, \ldots, x_p$ on the curve such that

$$\int_0^{x_1} du_j + \int_0^{x_2} du_j + \cdots + \int_0^{x_p} du_j = pv_j, \quad 1 \le j \le p,$$

where O is a fixed but arbitrary point on the curve obtained by taking a plane section through the surface. Moreover, the x's are determined

[25] This differential equation is known today as the Picard–Fuchs equation; see Clemens (1980) for an attractive introduction.

uniquely by the v's unless the x's satisfy a particular condition (here suppressed). Now, as y varies, the v's will vary, as will the x's, and the x's will trace out a curve C on the surface. However, as y goes round a singular point, the condition that the v's are normal functions implies that when y returns to its starting point, so do the x's. So the curve C cuts the plane $y = $ constant in precisely p mobile points. But this does not make C into an algebraic curve, and in particular not a curve of degree p, because, as Poincaré said, the curve they trace out could pass through the lower points of the integrals, which will raise its degree above p. But Poincaré was able to show, using the well-developed theory of theta functions and Abelian functions, that even when this happens the curve C is nonetheless algebraic.

Poincaré then turned to investigate the normal functions and the curves that they define on the surface. In paragraph V he obtained this rather forbidding expression for his normal functions

$$v_h(y) = \sum_j \frac{\lambda_j}{2\pi i} \int_a^{y_j} \frac{\Omega_{hj}(u)\,du}{u-y} + c_h, \quad 1 \le h \le p.$$

Here the λ's are integers that depend on the path the jth integral takes from a to y_j, $\Omega_{hj}(u)$ is the period of the hth Abelian integral along the vanishing cycle δ_j, and the c_h are nonvanishing constants. Of crucial importance is the fact, as Poincaré proceeded to show, the normal functions form a q-dimensional family parameterized by the c_h.[26] Because the λ's are integers, if the system of normal functions is to depend continuously on q parameters the λ values must be constants. The corresponding normal functions behave appropriately at the vanishing cycles; as Poincaré put it, they have no effective critical values.

Up to now in this long argument, the dimension q has no geometrical or analytical significance. Poincaré next showed that the normal integrals in this q-dimensional family correspond to simple integrals of the first kind (i.e., without poles) on the surface. This had already been shown by Enriques, Castelnuovo, and Severi, but Poincaré now gave his own

---

[26] Strictly speaking, Poincaré only obtained the c_h as polynomials in y; it was Lefschetz (1924) who showed that they could be reduced to constants by suitable choices of the Abelian functions and the homology basis.

proof. A simple integral on the surface yields a simple integral in each curve C_y whose periods do not depend on y, so the existence of a q-dimensional space of simple integrals led quickly to the existence of a q-dimensional space of normal functions by looking at the periods. The converse was harder to establish, and required the theory of reducible algebraic integrals, which could be expected to play a role because for such Abelian integrals the period matrix splits into two blocks, which allows one to hope that the periods independent of y fill out one block and behave independently of the other periods. Poincaré's conclusion was that there was indeed a q-dimensional space of simple integrals of the first kind on the surface. Poincaré's argument also showed that the family of curves he had constructed was complete.

It remained to explain why $q = p_g - p_a$. This requires that the arithmetical genus be brought into the story, and Poincaré concluded his paper by showing that $p_g - q$ is indeed equal to the arithmetic genus as computed by Picard and Simart (*Théorie* 1906, 2, 438). He showed that $p_g - q$ was the dimension of the space of plane curves that are cut on each of the curves C_y by adjoint surfaces of order $n - 3$. These are surfaces that pass through the double curve and the triple points of the given surface, but otherwise cut each C_y in the right number of points to correspond to the zeros of a holomorphic 1-form.

In 1921 Severi gave a different proof of the result, and in so doing cast doubt on Enriques' account while also admitting that his new proof was not entirely satisfactory. Poincaré's paper was then taken up by Solomon Lefschetz in his *L'analysis situs et la géométrie algébrique* (1924), and also in his (1929), where he praised Poincaré for giving a new and entirely analytic treatment of algebraic surfaces. The problems are discussed in chapter V of Zariski's book *Algebraic Surfaces* (1935), the book that both gave an epitome of the Italian school of algebraic geometers and opened the way to the newer methods of arithmetic and algebraic geometry. The transition was momentous: Zariski (*Collected Papers*, 1972, I, xi) famously later said that "the price was my own personal loss of the geometric paradise in which I was living so happily theretofore," as he became convinced that "the whole structure must be done over again by purely algebraic methods." That this process, as far as Enriques' contribution above is concerned, has only been completed in 2011 is a measure of how sure was Poincaré's grasp. His proof has stood undisputed since it was first put forward, whereas Enriques' proof depended on a crucial lemma that G. Zappa (1945) showed by a counterexample to be false. Only in 1966 did David Mumford put the whole matter straight for algebraic surfaces defined over algebraically closed fields of any characteristic. Mumford (2011) is a reinterpretation of Enriques' later attempts at the main theorem, which it also instructively compares with Poincaré's (1910).[27]

---

[27] See also Babbitt and Goodstein (2011) for more information and references to the Italian literature mentioned above.

10

Poincaré as a Professional Physicist

WHATEVER THE REASONS that led to Poincaré taking up a chair in physics in 1886, he took his duties as a professor of physics not only very seriously but much more imaginatively than his predecessors, so much so that in barely a decade he had published a dozen volumes on different topics. While these largely took the form of lectures edited by some of his students they are nonetheless clear and speak with Poincaré's voice. The lectures on the interrelated topics of thermodynamics and probability are the most interesting apart from the accounts of electricity, magnetism, and optics, but all are worth at least a mention.

They appeared as follows:[1]

- WS, SS 1885–86, Kinematics, potentials and mechanics of fluids (1886a) and (1899a)

- WS 1887–88, The mathematical theory of light (1889a)

- ?? 1888–90, Electricity and optics (1890b)

- WS 1888–89, Thermodynamics (1892g)

- SS 1888–89, Capillarity (1895c)

- WS 1890–91, Elasticity (1892a)

- WS 1891–92, Mathematical theory of light: Diffraction and dispersion (1892f)

- SS 1891–92, Turbulence (1893e)

- WS 1892–93, Electric oscillations (1894b)

- WS 1893–94, Heat propagation (1895e)

---

[1] WS stands for winter semester, SS for summer semester.

- SS 1893–94, Probability (1896a)

- WS 1894–95, Newtonian potential (1899g), (1912a)

There is also Poincaré's *Cosmogonie* (1911a).

It was Poincaré's custom to publish on topics that caught his attention as he was preparing the lectures. Sometimes his interest ran to a few papers, as with telegraphy. On other occasions it became a long-lasting interest, as it did with the intertwined subjects of electricity, magnetism, and optics. Some are more like standard teaching fare, for example the lectures on kinematics and mechanisms provide an elementary account of the motion of solids, including rolling, and of fluids, with a far-from-rigorous account of Dirichlet's principle. The course on astronomy that Poincaré gave in 1907–08 (1906b) covered the systems of coordinates in use, measurements of times and angles, the movement of planets and comets, and the motion of the moon, and in the second semester looked at probability and the theory of errors in the context of planetary astronomy. But their impact on a physics community starved of up-to-date, insightful accounts of different branches of physics was considerable. Langevin (1913, 678) said that Poincaré "lectured successively on every part of physics in these courses, most of which were published and immediately exercised, in France as abroad, a considerable effect on the movement of ideas and the orientation of experimental research." And on the occasion of the centenary of Poincaré's birth, Prince Louis de Broglie said,

> Every young man of my generation who was interested in mathematical physics was sustained on the books of Henri Poincaré. Teaching at the Sorbonne being then somewhat out-of-date, Paul Langevin not yet having published his beautiful course at the Collège de France, it was in Poincaré's books that we could find, written in a perfect form, the recent theories of physics, and reading them was such that, many years later, one could see that they were well done. (Poincaré, *Oeuvres* 11.2, 62)

Electromagnetism and optics have been discussed at length in chapter 6, and among the other topics some are notable for combining a French tradition with a few pages where Poincaré just could not resist saying something that was surely too hard for his audience. Elasticity theory, which concerns the bending and vibrating of solids, began with

Cauchy's work and ended up with a discussion of Abelian functions. The lectures on capillarity went back to Laplace, but not Young, in a way that recalls Poincaré's comparison of French and British readers in his *E&O*, and got round to one of the hardest problems in contemporary mathematics: the study of surfaces of least area with a given boundary. These minimal surfaces, as they are called, are displayed by thin liquid films bounded by wire shapes. They take the shape that minimizes their energy, and this shape is one for which the sum of the principal curvatures is zero, so it can be discussed using the theory of capillarity. However, despite work by Riemann and Schwarz, mathematical solutions of the problem of finding the shape for any but the simplest of wire shapes were not known. Darboux, in his major book on the geometry of surfaces (Darboux 1887) regarded the minimal surface problem as one of the most important and difficult. Poincaré gave some of the better known results: the helicoid is such a surface, as is the catenoid, which is the surface spanned by two circles in parallel planes and whose centers are joined by a line perpendicular to those planes. He showed that there are in fact two catenoids satisfying these conditions, a fat one which is stable and a thinner, unstable one. He finished the chapter with a mention of Riemann's description of the minimal surface spanned by two arbitrary circles in parallel different planes, noting that Darboux's discussion of minimal surfaces gave a good account of Riemann's ideas on the subject but had not treated this one.[2] Poincaré ended the course with a discussion of the role of thermodynamics, a topic recently raised by Lord Kelvin and Pierre Duhem among others, and to whom, in the end, he deferred.[3]

The last lecture course to be discussed briefly was his course on vortices in the summer semester 1891–92, which led Poincaré to propose an interesting analogy with electrodynamics and once again to confront issues in topology. In the introduction Poincaré pointed out that the greatest single advance in this subject had been made by Helmholtz in 1858, and that this had opened the way not only to advances in meteorology that Helmholtz himself had made, but to the theory of vortex atoms of Thomson and Tait, and to an interesting analogy between

---

[2] Darboux treated this case in the second edition (1914, 461–464).
[3] Gibbs had published his thermodynamic account (1876–78) so obscurely it is hardly surprising that Poincaré did not mention it.

the equations of hydrodynamics and electrodynamics that was again due to Helmholtz.[4]

To follow Helmholtz's argument, consider first the motion of a solid ball. At any instant it has a direction of motion and a generally different axis about which it rotates. Helmholtz added to this description the observation, made earlier by Stokes, that if the solid ball is replaced by an infinitesimal element of a liquid it must also be supposed to be dilating along three mutually perpendicular axes (none of which need line up with the previous two). So the motion of a fluid may be looked at as that of infinitely many infinitesimal spheres each traveling in its own direction, rotating about its own axis, and perhaps changing in size. Helmholtz knew that some motions of an incompressible, nonviscous fluid, such as those impressed upon it by the motion of an immersed body, are given by a potential, but motion involving friction is not of this kind. A vector potential $\mathbf{v}$ by definition satisfies $\nabla \times \mathbf{v} = 0$, and the infinitesimal rotation of a fluid is given by $\omega/2$ where $\omega = \nabla \times \mathbf{v} = 0$, so there will be a vector potential only if there is no instantaneous rotation of the fluid.

Helmholtz now studied the motion of the fluid using the continuity equation $\nabla \cdot \mathbf{v} = 0$, which says that the volume of the fluid does not change, and Euler's equation for the motion of an incompressible fluid, which says

$$\rho \left(\frac{\partial \mathbf{v}}{\partial t} + (\mathbf{v} \cdot \nabla)\mathbf{v} \right) + \nabla(P + V) = 0,$$

where $\mathbf{v}$ is the velocity, ρ the constant density, P the pressure and V the potential due to external forces such as gravity. By taking the curl $(\nabla \times)$ of this equation, Helmholtz obtained

$$\frac{\partial \omega}{\partial t} + (\mathbf{v} \cdot \nabla)\omega = (\omega \cdot \nabla)\mathbf{v}.$$

Helmholtz's brilliance was in how he interpreted the resulting equation geometrically. By analogy with the way a flow line in a fluid is defined as a curve whose tangent at each point agrees with the direction of the velocity potential of the fluid at that point, Helmholtz defined a vortex line to be a curve in a fluid whose tangent at each point agrees with

---

[4] On Helmholtz's work on hydrodynamics, see Epple (1999) and Darrigol (2005).

the direction of the axis of rotation of the fluid at that point. He then deduced that if two points belong to the same vortex line at some moment then they always do, so the fluid can be regarded as filled out with these nonintersecting curves, and one can think of the flow as moving whole vortex lines around. Furthermore, it follows from his equation that the flow stretches the vortex lines in a manner proportional to the rotational velocity, so the flow must shrink any infinitesimal patch transverse to the vortex line. More work enabled Helmholtz to deduce that vortex lines must either be closed or end at the boundary of the liquid. Helmholtz's theory also made great use of the concept of a vortex sheet. These are surfaces made up of vortex lines that move en bloc with the motion of the fluid.

Poincaré expressed Helmholtz's theorem in this form. Let σ_0 be a closed curve in the liquid at time zero and σ the curve formed by the same molecules at some later time, then the integral $J = \int_\sigma \mathbf{v} ds$ is a constant independent of the time (so $\frac{dJ}{dt} = 0$). Helmholtz's vector form of this result was obtained via an application of Stokes' theorem. Poincaré's derivation has the virtue that it applies to any flow, whereas Helmholtz's derivation was restricted to flows in liquids.

Poincaré defined a vortex sheet to be a surface in the surface with the property that the integral J vanishes when taken on any closed curve in the surface, and he showed that a vortex sheet remains a vortex sheet as it evolves under the flow. He gave various indications of what could be said about vortex sheets and vortex tubes, noting that they might divide space up into several regions with complicated topologies, such as multiple tori, and when this happened integrals can be many valued and no longer depend solely on their endpoints. The opening chapter concluded with his versions of Helmholtz's original proof of his theorem and Kirchhoff's subsequent proof. He also noted an analogy with electrodynamics in the case when the liquid is at rest in an infinite space. In this case Helmholtz's equations have the same form as the equations for a magnetic field.

THERMODYNAMICS

The course on thermodynamics is interesting because of the controversies it aroused, although the content of the course (1892g) itself was not very advanced. Poincaré opened his account with a chapter on energy, the conservation of energy, and the impossibility of perpetual motion. This was followed by one with the surprising title of "The indestructibility of caloric," but Poincaré immediately explained that this old theory was entirely false, and came from a time when alongside the supposedly indestructible fluids there were other fluids, such as the electrical fluid. In its place he discussed what is meant by temperature, specific heat, and the processes of heating and cooling. Then came his account of the Carnot cycle, isothermal and adiabatic equilibrium, Clausius's law (which says that heat cannot pass from a colder body to a hotter one), and, on page 120, the second law of thermodynamics, which Poincaré stated this way: a heat machine cannot function with a single source of heat. Poincaré then applied these ideas to the theory of steam engines—it was still the age of the train—and to the construction of batteries, before concluding with Helmholtz's theory of reversible and irreversible phenomena and only a short paragraph about Boltzmann's ideas. When the lectures, as edited by Jules Blondin, were published in 1892 they were heavily criticized by Tait for conveying too much mathematics and not presenting the "true, (i.e. the statistical) basis" of the second law of thermodynamics.[5] Poincaré replied that he had deliberately chosen to avoid all mechanical explanations, which he found most unsatisfactory, a point he alluded to briefly in a later essay, (1893c).

Tait's criticism came from a substantial position that Poincaré was later to characterize as the English position (although Tait, like Maxwell, was a Scot), and which will lead us into a central concern of the whole topic. It is a familiar paradox that while the laws of dynamics are time-reversible, life never feels like that. The physical laws that rise to meet this challenge, and give what is often called a direction or an arrow to time, are those of thermodynamics, which has its own long and complicated history.[6] The first explicit statement in print that the laws of thermodynamics and those of mechanics conflict was made by Josef Loschmidt in 1876, but the issue had by then already been discussed in 1874 by Maxwell, Tait, and J. J. Thomson, who concentrated on the kinetic theory of gases and favored a probabilistic interpretation of thermodynamics. Loschmidt was a friend and supporter of Boltzmann, who had offered a significant mathematical interpretation of the second law of thermodynamics in a paper of 1872,

[5] Tait, *Nature* (45, 1892, 246).
[6] See, for example, Brush (1976).

in which he offered what he called his H-theorem. This said that a certain quantity, symbolized by H, could be defined for a mechanical system and would decrease monotonically with time and would be constant for the Maxwell distribution. Boltzmann commented, "It has thus been rigorously proved that whatever may have been the initial distribution of kinetic energy, in the course of time it must necessarily approach the form found by Maxwell."[7]

At least initially, Boltzmann seems to have regarded this theorem as a mathematical derivation of the second law of thermodynamics from the laws of mechanics. Loschmidt objected that if for some initial conditions the quantity H decreased then reversing all the velocities of the initial distribution would lead to H increasing, because the second setup is the time-reversed version of the first, and to deal with this Boltzmann reconsidered the assumptions he had made in order to derive his H-theorem. In 1872 these had included a degree of uniformity in the initial distribution of positions and velocities of the molecules of the gas; now he was prepared to allow that some initial distributions might lead to a short-term increase in the value of H, but these could be dismissed on the grounds that they were extremely unlikely. Although this reply does not get to the heart of the objection, no one had a better idea until Tait caught Boltzmann's attention in 1887. Now Boltzmann argued that any initial distribution would lead to an equilibrium distribution, but not because of his H-theorem, which no longer had acceptable assumptions.

In 1889, presumably as a consequence of the reading he had done in preparing to lecture on thermodynamics that semester, Poincaré had come across Helmholtz's theory of how the subject could be squared with mechanics. Helmholtz (1886) had offered a marriage of the two theories that covered reversible phenomena, and in (1889c) Poincaré professed himself in agreement with it. But then Helmholtz had turned to irreversible phenomena, and Poincaré was not persuaded that his account explained such processes as the flow of heat solely from hotter bodies to colder ones. Poincaré argued that when a system of particles is formulated in Hamiltonian dynamics, it will be time-reversible if and only if the coordinates $q_j = \frac{dp_j}{dt}$ enter not only in even powers but also in odd powers. Helmholtz had suggested ways in which this could happen, in terms of what are called hidden motions of the system, but Poincaré

---

[7] Quoted in Uffink (2004).

doubted that these ideas were compatible with a monotonically increasing function that could represent entropy. In particular, he showed that the idea of increasing entropy was incompatible with the principle of least action. On the other hand, the probabilistic assumptions Maxwell, Boltzmann, and others were making were murky and contentious, which is surely why Poincaré dismissed them, but in the second edition of the lecture notes on thermodynamics (1908f, para. 333) Poincaré added a final paragraph in which he said that all attempts of the kind Helmholtz and others had proposed must be abandoned and only those founded on statistical laws, such as the kinetic theory of gases, had any chance of success.[8]

His problem was not with entropy; in his paper (1898e) on the stability of the solar system he noted that tidal effects, such as the effect of the moon on the earth, of the sun on the planets, caused energy to be dissipated as heat, and eventually the whole solar system would reach a state of equilibrium in which everything would rotate with the same speed around the same axis, and even the celestial bodies would not escape Carnot's law. But it seems clear from his remarks to Tait that Poincaré could not have been happy with a probabilistic analysis of thermodynamics without a much better grounding for it. When he lectured on probability for the first time, in the summer semester of 1893–94, one of his main points was that the hardest part of probabilist's job was to establish the right measure of probability in any given problem (see sec. "Probability" below). Indeed, when he raised the subject of thermodynamics for a popular audience in his (1893c), he concluded that Maxwell had tried and failed to explain irreversible phenomena, and this, he suggested, was because the problem was so difficult that it could not be handled with complete rigor. It was therefore necessary to fall back on simplifying hypotheses, but he doubted that these were either legitimate or even consistent with one another. If the problem was never to be solved, "it would be a definitive condemnation of mechanism if the experimental laws could be other than approximate" (1893c, 537).

The most memorable consequence of Poincaré intervention in the study of thermodynamics came in 1896 when Zermelo (who was a student of Max Planck's at the time) noticed that Poincaré's recurrence theorem could be applied to thermodynamics, an application that, he suggested,

---

[8] See the further discussion in Brown, Myrvold, and Uffink (2009).

Poincaré had missed.[9] Zermelo first rederived the recurrence theorem for the sake, he said, of physicists who might find Poincaré's original paper too difficult. From it he deduced that if, as the theorem suggests, it is the fate of every system of mass points to return arbitrarily close to any initial configuration, then there can be no irreversible processes at work, and therefore no mechanical system, and in particular no gas, can be said to have a continually increasing entropy. As he put it:

> In such a system irreversible processes are impossible since (aside from singular initial states) no single-valued continuous function of the state variables, such as entropy, can continually increase; if there is a finite increase, then there must be a corresponding decrease when the initial state recurs. (Zermelo 1896)

It remained possible, Zermelo noted, that the real states of physical systems were always those that Poincaré had said occurred with only zero probability, but this seemed to him to defy our sense of causality.

Zermelo made no proposals about how Boltzmann's kinetic theory could be saved, so Boltzmann felt compelled to reply. He agreed that Poincaré's recurrence theorem was valid, but argued that it did not apply to the theory of heat, which was essentially statistical in nature. So, he said, entropy would increase almost all the time, but there would be times when it suddenly and briefly decreased. Such phases of the motion would, however, usually be a very long time apart. Poincaré's recurrence theorem would hold, but the recurrence time would usually be very long indeed, so long that there would be no disagreement with experience. "Zermelo's paper," he said, "shows that my writings have been misunderstood; nevertheless it pleases me for it seems to be the first indication that these writings have been paid any attention in Germany."

Zermelo replied very quickly, and before the year was out Boltzmann again replied. Boltzmann's views often lacked mathematical justification, and they were soundly criticized on those grounds by Zermelo, but the major part of his defense of his views rested on the hypothesis that the universe started from a very improbable state and was still in an improbable state. In such a case, he said, the direction of time should

---

[9] See Brush (1966, pp. 208–246) for translations of the papers by Zermelo and Boltzmann, lengthy extracts of which are also available at http://www.informationphilosopher.com/solutions/scientists/boltzmann/zermelo.html and Brush (1976, esp. pp. 630–640).

be defined as that which goes from the less probable state to the more probable one, and on this definition entropy will in general increase, Poincaré's theorem notwithstanding. With this, the exchange came to an end. Whatever the later developments of the theory of thermodynamics, one consequence of Poincaré's recurrence theorem seems paradoxically to have been the recognition of the importance of entropy for any physically meaningful definition of time.

PROBABILITY

The topic of thermodynamics led naturally to questions about the nature of probability, upon which Poincaré lectured in 1896; a second edition of the course was published in 1912.

In between these two dates he wrote a lightweight essay on chance, (1907b). Poincaré began by reviewing how we might define chance. One way is that it applies to deterministic systems too complicated to predict and in which small errors of observation can produce large errors in the final result, for example, when a pencil is balanced on its point and we want to know where it will fall. Deeper problems arise with the kinetic theory of gases. Sometimes it seems that the world becomes a tidier place, as when two adjacent bodies at different temperatures eventually arrive at the same temperature, but on other occasions increasing disorder is the norm. In all these cases, Poincaré imagined that it is overwhelming complexity that makes us say chance is at work.[10] How is it, then, that chance is sometimes subject to laws? Any sufficiently energetic mixing process is liable to produce a distribution of values that obeys Gauss's law of errors, the so-called normal distribution, and Poincaré explained this as a result of a simplifying hypothesis about the nature of the errors. If the errors are all small and equally likely, whatever their cause, then indeed their spread can be calculated. Poincaré never considered that an event may simply happen at random, which Peirce had speculated about, and which was to become the preferred model of radioactive decay. He also supposed that the deterministic process was continuous—and

---

[10] Indeed, automated gambling machines rely on deterministic processes too complicated to predict, and the house loses when astute, well-informed players armed with good programs know what to look for.

interestingly he was willing not only to give this hypothesis up but to argue against it when he wrote on the emerging quantum theory in late 1911.

The lecture courses naturally went into much more detail. Poincaré began his lecture course in 1896 by observing that it seemed almost impossible to define probability. One could suppose that the probability of an event occurring was the ratio of favorable outcomes to all outcomes, but one is quickly led to realize that this is no use if some outcomes are more likely than others. For example, suppose one is confronted with two identical urns, one containing n white balls and p black ones, the other containing n' white balls and p' black ones, and is asked to draw a ball from one of the urns: with what probability is it white? The answer, said Poincaré, might seem to be the total number of white balls divided by the total number of balls, or $\frac{n+n'}{n+p+n'+p'}$. Or one might argue that the probability of choosing the first urn is $\frac{1}{2}$ and so the probability of choosing a white ball from it is $\frac{1}{2}\frac{n}{n+p}$. Likewise the probability of choosing the second urn is $\frac{1}{2}$ and a white ball from it is $\frac{1}{2}\frac{n'}{n'+p'}$, so the probability of choosing a white ball is

$$\frac{1}{2}\frac{n}{n+p} + \frac{1}{2}\frac{n'}{n'+p'},$$

which is not equal to $\frac{n+n'}{n+p+n'+p'}$. It is the second argument that is correct. The first can be seen to be wrong by noticing that its value varies if both n and p are multiplied by the same amount, but that leaves the ratio $\frac{n}{n+p}$ unchanged.

Such paradoxes had been discussed before, by Joseph Bertrand in his *Calcul des probabilités* (1889) and Poincaré reproduced some of his examples. Matters became more interesting when, following Bertrand, he allowed the space of all possible outcomes to be infinite, and summation over all possible outcomes must be replaced by an integral. The probability that a function f takes a value between x_0 and x_1 is now given by the integral $\int_{x_0}^{x_1} f(x)\,dx$.

In this setting, one of Bertrand's paradoxes was this: given a circle of unit radius, estimate the probability that a randomly chosen chord is longer than the side of an inscribed equilateral triangle. With one approach he gave the answer $\frac{1}{3}$, with another $\frac{1}{2}$ (other plausible answers are also possible). As Poincaré remarked (1895, 97), the mathematician can choose between these hypotheses, but once a choice is made he must keep to it and not pass to another that contradicts it.

As an example, Poincaré considered the familiar problem of a stick of unit length broken in two places: When can the three pieces form a triangle? He argued that the problem can be represented geometrically by drawing an equilateral triangle of unit height and dropping the three perpendiculars from an interior point to the sides of the triangle. By considering areas, if these perpendiculars have lengths x, y, and z then $x + y + z = 1$. If we agree that the probability that x lies between x and $x + dx$ and y lies between y and $y + dy$ is $dx\,dy$, then the answer to the broken stick problem is given by comparing the area of the region of the triangle where

$$x < y + z, \quad y < z + x, \quad z < x + y,$$

with the total area of the triangle. Therefore the probability is $\frac{1}{4}$. This rather artificial problem had been discussed in German by Emanuel Czuber in his (1884), who saw, as Bertrand and Poincaré did later that the deeper question was what is meant by a random choice.

Poincaré was also struck by the fact that despite the seemingly great freedom to choose the law of probability in any problem, there were occasions when the answer was independent of the law. For example, the probability that any continuous function takes an irrational value is always 1, and that it takes a rational value is always infinitely small (as he put it). A more interesting example is given by a roulette wheel divided into a large number of equal parts, alternately colored red and black. The probability that a ball finishes up in a division of a given color depends on the wheel and on the initial conditions, all of which vary, but the probability nonetheless tends to $\frac{1}{2}$ as the number of divisions is increased to infinity.

Other topics followed, including Bayesian analysis and Gauss's law of errors and the method of least squares, with an application to the estimation of a function given some of its values, which was, of course, the problem that had provoked Gauss to invent the method in the first place, in order to locate the asteroid Ceres.

The second edition of this book, in 1912, did not elevate it to the status of one of Poincaré's major works, but it contained some interesting improvements. Particularly interesting is the theorem concerning how likely it is that two curves in the plane or on the sphere meet as one of them is moved around. This surprising result requires an interesting detour to explain.

It has its origin in Buffon's needle. Georges Louis Leclerc, count of Buffon, wrote a book in 1733, when he was 26, that was only published in 1777 in which, among other things, he considered what would happen if a needle of length ℓ was dropped onto a plane ruled by lines a length $a \geq \ell$ apart. In particular, how often will the needle lie across one of the lines? The answer, he said, was that the probability that the needle crossed a line was $\frac{2\ell}{a\pi}$. One locates the needle by the distance y of its center to the nearest line below it (so $0 \leq y < a$), and by the angle θ that it makes with the lines. The needle overlaps a line when either $y + \frac{\ell}{2}\sin\theta \geq a$ or $y - \frac{\ell}{2}\sin\theta \leq 0$, and the area of this region as a fraction of the area spanned by the range of possibilities ($0 \leq y \leq a$ and $0 \leq \theta \leq \pi$) works out as $\frac{2\ell}{a\pi}$. Computer simulations have confirmed that this estimate is actually quite good.

Buffon's needle problem was taken up by Laplace, but then it slept until Lamé took it up and generalized it to closed curves and polygons thrown on to the plane in lectures at the Université de Paris.[11] One of his students was Joseph-Émile Barbier, and he saw that these results could be extended to apply to arbitrary convex curves and indeed to unbroken curves. He published results about this in Liouville's *Journal des mathématiques* in 1860, the year he graduated from the École normale supérieure. The subject was then extended by Joseph Bertrand, who added a section entitled what he called "Crofton's theorem" to his *Traité* (1870), which he regarded as a generalization of Buffon's needle problem.

Crofton's theorem concerned a closed convex curve of length L enclosing an area of Ω and inside a larger region Ω'. If the region subtends an angle ω from a point P outside it, then

$$\int (\omega - \sin\omega)\,d\sigma = \frac{1}{2}L^2 - \pi\Omega.$$

Crofton proved this result by looking at "an infinite number of lines drawn at random in the plane," and it was the probabilistic formulation

---

[11] See Seneta, Parshall, and Jongmans (2001).

that caught the attention of Bertrand, and explains his allusion to Buffon's needle.

Morgan William Crofton, an Irishman who had graduated from Trinity College, Dublin, in 1848 and come to the attention of Sylvester, who was professor of mathematics at the Military College in Woolwich, London, and who in 1864 was able to get him a job as an instructor there. Sylvester had become interested in such questions as "Three points being taken at random inside a disc, what is the probability they form an acute-angled triangle?" Crofton took the subject up and developed it extensively. In 1868, with Sylvester's help, he published a short note in the *Comptes rendus* presented by Hermite, thus gaining himself a French audience, and a long account in the *Phil. Trans. Roy. Soc.*

Thereafter the subject developed independently in Britain and France. In 1885, the year after he retired from the professorship of mathematics at Woolwich (having succeeded Sylvester in 1870) Crofton published a remarkable essay on probability in the ninth edition of *Encyclopaedia Britannica* that more or less founded the new subject of geometric probability in English. He had proceeded in ignorance of Barbier's work, which only entered his purview when Sylvester mentioned it in a 20-page paper of 1890–91 in *Acta Mathematica*. By this time Barbier had died. He had obtained work with le Verrier as an assistant astronomer, but in 1865 he had succumbed to mental problems, and disappeared from scientific life until 1880, when he was found in a mental institution in Charenton–St. Maurice. Bertrand, who had become permanent secretary of the Académie des sciences in 1874, arranged for him to have some money and to resume publishing in the *Comptes rendus*, which he did from 1882 to 1887; he died in 1889.

We can now return to Poincaré's theorem.[12] He considered a mobile figure on a sphere, and asked for the possibility that it was in a given position. It is necessary first of all to say how one specifies its position. Poincaré supposed that one took an arbitrary point P on the figure, with initial position P_0 and final position P_1. Then there is a unique rotation of the sphere that brings P_0 to P_1, which is a rotation about an axis given by (α, β, γ) through an angle of 2θ. Poincaré set

$$\lambda = \cos\theta, \quad \mu = \alpha \sin\theta, \quad \nu = \beta \sin\theta, \quad \rho = \gamma \sin\theta,$$

---

[12] Poincaré, *Calcul des probabilités* (2nd ed., Paris 1912, chap. VIII).

and noted that it is enough to know three of them, because they satisfy

$$\lambda^2 + \mu^2 + \nu^2 + \rho^2 = 1.$$

He then proposed that the probability be estimated by means of the integral

$$\int \frac{d\mu\, d\nu\, d\rho}{\lambda},$$

which, as he explained, is the generalization to the three-dimensional sphere of the idea on the ordinary sphere that the probability of a point being in a given region is the ratio of the area of the region to the area of the sphere. Next he showed that this estimate of probability is not affected by the choice of initial point P, which is a calculation which shows that the estimate of probability is invariant under the group of rotations of the sphere, and comes down to an exercise in the formula for change of variables in an integral.

He then reformulated the problem in terms of a point P and an arc of a great circle through P, and showed that the probability that a point P lies in an area $d\sigma$ and that the arc of the great circle through P makes a given angle ω with a fixed great circle on the sphere is given by

$$\int \Phi\, d\sigma\, d\omega,$$

where Φ is a constant that is readily seen to be $\frac{1}{8\pi^2}$. So he wrote

$$\int \frac{d\sigma\, d\omega}{8\pi^2} = \int \frac{dx\, dy\, d\omega}{8\pi^2 z}.$$

Now he proposed the following game (para. 84). A fixed and a mobile curve are drawn on the sphere, and the contestant is promised as many francs as there are points of intersection of the two curves: What is the expectation? Poincaré answered that it is proportional to the lengths of the two curves. He argued as follows. First, the probability that a point P on the mobile curve lies in a given area $d\sigma$ is given by

$$\int \frac{d\sigma\, d\omega}{8\pi^2} = \frac{d\sigma}{4\pi}.$$

To take care of the orientation of the mobile figure once a point on it is fixed, he asked for the probability that a mobile circle on the sphere cuts a fixed one. He found that this was proportional to $\sin\theta\sin\theta'$, where the θ's are the angles the radii of each circle subtend at the center of the sphere. Next, he supposed that the fixed and the mobile curve are each made up of arcs of great circles, the one of n arcs of length ℓ and the other by n' arcs of length ℓ'. By the above argument the expectation of an intersection of the two curves is given by the product $nn'\ell\ell'$. If the fixed and mobile curves are now supposed to be approximated arbitrarily well by curves made up of arcs of great circles, then the probability becomes proportional to the lengths of the curves. The constant of proportionality can be found by considering the case of two great circles: they meet in two points, and so the constant is $\frac{1}{2\pi^2}$. So the expected number of points that two curves C and C' on the sphere of lengths $|C|$ and $|C'|$ have in common is given by $4|C|\cdot|C'|$.

Poincaré then gave similar arguments dealing with the question of when do two distributions of n stars on the celestial sphere have a number of stars less than a given angular distance apart, and then for the question of when n points moving according to the laws of Hamiltonian dynamics but with unknown initial positions will have given positions at a later time, a question that had implications for the origin and future state of the solar system.

11

Poincaré and the Philosophy of Science

To go back to our philosophy book; it is like rational principles or scientific laws, reality does conform to it more or less, but bear in mind that the great mathematician Poincaré is by no means certain that mathematics is rigorously exact. (Saint-Loup, discussing military strategy, in Proust, *Le côté de Guermantes*, I, 149, but see C. K. Scott Moncrief's translation in *The Guermantes Way*, I, 152.)

POINCARÉ: IDEALIST, SKEPTIC, OR STRUCTURAL REALIST?

The philosophical stance for which Poincaré is best remembered is his geometric conventionalism, but in his day he was also accused of being both too much and too little of a skeptic about science, and he was willing to charge others with undue skepticism. He was also said to be more of an idealist than a realist about the knowledge claims of science, and once called himself a pragmatic idealist (as opposed to a Cantorian realist).

There are four questions to ask about Poincaré's philosophy of science. The first concerns his distinction between geometric conventionalism and other forms of conventionalism within science; the second (theory change) considers how he defended geometric conventionalism when ideas about space–time challenged his preference for Galilean or Newtonian space and time; the third, more speculatively, considers the import of general relativity and the lack of support for his view that groups were more fundamental than spaces, just as form, to which mathematics applied, was more fundamental than matter; and the fourth looks at his lack of enthusiasm for talking about objects in science, which led to accusations of skepticism.

Geometric Conventionalism

Poincaré's geometric conventionalism is an epistemological position. It says (agnostically) that no experiment can decide between two geometrical descriptions of space because, firstly, no statement of the space we live in is about *space* directly but about objects in that space such as rigid bodies or light rays, and secondly any disparity between the purely geometrical statements (such as angle sums of triangles) can be put down to a mismatch between the theoretical entities and their presumed spatial representatives (light rays could be regarded as not straight). His conventionalism is the deduction that because there is no logical way to choose between these two descriptions a decision can only be taken on grounds of convenience, which he suggests will reflect the working of our minds. Poincaré wanted a class of knowledge claims that were not analytic, not Kantian synthetic a priori, and not subject to empirical revision in the manner of many scientific claims. The first two exclusions open the way to some role for experience; the third exclusion severely limits that role. He explicitly ruled out saying that a claim based on a convention was true; it was merely a reliable basis for action. People who took an opposite decision would not be wrong, merely inconvenienced to a greater or lesser degree.

Geometric conventionalism differed from other conventions that Poincaré found it necessary to adopt within physics. When he argued that the "law" of conservation of energy or the equality of action and reaction were not ideas that could be refuted he sometimes referred to them as being true. There was a conventional element in saying that the earth goes round on its axis, but the alternative was so extremely inconvenient, if not indeed inconsistent with the rest of physics, that Poincaré dismissed it out of hand. This leads to an important question of how to characterize Poincaré's distinction between geometric conventionalism and other conventions at work in science.

The fundamental distinction here is between our knowledge of the external world and our detailed knowledge of parts of physics. An infant growing up who could not sort out what is and what is not part of itself, who could not form a sense of rigid bodies and how they move, would be a helpless creature indeed. Every functioning child learns how to do this automatically—in the modern phrase, this ability is hardwired into us. So we acquire the ability to know where things are and find our way about in the world. This is very different from how we learn about Newton's laws and the conservation of energy, which usually involves shedding some deeply ingrained beliefs about the world. To use another modern term, folk physics is a poor guide to real physics. Geometric conventionalism is a theory about how we can have knowledge of the world at all, as well as a "Just So" story about how we come to be native Euclideans and not native non-Euclideans.[1]

The sharper question is how Poincaré distinguished between geometric conventionalism and the conventionalisms he found in physics proper. In his (1901e) he argued convincingly that none of the basic principles of mechanics (the principle of inertia, the principle of the equality of action and reaction, the principle of relative motion, and the principle of conservation of energy) could be established on the basis of experiments, but nonetheless they are approximately correct. Nor do we imagine more accurate experiments refuting them, indeed we believe that experience could suggest them but never reverse them, the reason being that we have been compelled, said Poincaré (p. 83), to use them as definitions. Force, for example, is defined as the product of mass and acceleration and "this is a principle which is henceforth beyond the reach of any experiment." These principles form a consistent and effective theory when taken together, they have been elevated, one might say in terms Poincaré used elsewhere, to apply to the form of things and not merely the matter. Each such principle or law is a convention, although far from arbitrary, "the truth of which we regard as absolute [because we have] freely conferred this certainty on it by looking upon it as a convention" (p. 86).

The concept of truth is one of the most vexed ones in philosophy. The allegedly absolute truth Poincaré was presenting here was not a statement of fact of the kind "There is an elephant in my garden" but a supposed law. There is excellent, but not logically compelling, experimental evidence for it, which is the best science can ever provide, but it is taken to be true because we have made it a convention, although far from an arbitrary one. This is a pragmatic theory of truth inasmuch as it says that to claim that a statement or law is true is merely to say that it will never let you down, but with the twist that theoretical mechanics

---

[1] Recall Poincaré's insistence on the psychological meaning of the classical geometries, and the very different nature of some of Hilbert's axiomatic geometries.

can be practiced only by those who adopt these conventions because they determine the definitions and the theorems in the subject.

From Poincaré's excursions into fictional universes (shrouded by clouds, or with planets all in circular orbits) it is clear that he supposed that had the world been remarkably different we would have come up with other conventional truths, but also, the world being as it is we have no realistic choice but to come up with the ones we have. This is another way in which the conventions Poincaré located within physics differ from geometric conventionalism. The supposed dichotomy between Euclidean and non-Euclidean geometry started life as a question in cosmology, but it was transferred by Poincaré to a question in epistemology. To forget this—as we shall deliberately do below—is to make it hard to see why Poincaré denied that truth came into one issue but was there in the other. For present purposes the more useful comparison is between conventions and hypotheses in physics, because that is a further source of the misunderstanding that conventions in physics are also true.

Poincaré discussed hypotheses in his (1900d), where he distinguished several kinds: natural and necessary (we may ignore the effects of distant galaxies), indifferent (continuous or discrete matter), and real generalizations, to be confirmed or refuted by experiment. The first two, in their different ways, are entirely pragmatic, but the third is important. Poincaré frequently became involved in issues where theory and experiment collided, most notably in the Crémieu–Pender controversy, but also with other experiments in the theory of electricity, magnetism, and optics, including radio waves. These branches of physics for him were mathematical theories, and a hypothesis of the third kind was both a conjectural theorem and a prediction in the laboratory. As a theorem it ought to be provable (and if obviously well beyond proof not worth discussing); as a prediction it ought to be testable (and if not, once again not worth discussing). It might be that the mathematicians or the experimenters would get there first, and the matter should not rest until there was enough agreement between the various camps. In the happy event that the hypothesis held up, it became true that, to take an example, light was an electromagnetic phenomenon.

A well-supported hypothesis usually differs from a convention by being a theoretical consequence of the fundamental laws of the relevant part of physics. Poincaré explicitly denied that hypotheses were lightly adopted (p. 116), and he had little sympathy for any particular theory of matter (electric fluids, vibrating atoms, and the like were little more than metaphors to him). He put his trust on two pillars: mathematical theorems and experimental findings. These are not so different. To interpret an experiment as vindicating Pender and refuting Crémieu, or as vindicating Maxwell and refuting Helmholtz, is to indulge in a considerable amount of theorizing. It can be that the theoretical work rests on assumptions the mathematicians cannot guarantee (such as Dirichlet's principle), but past a certain point no experiment has a theory-free interpretation. But there are occasions when a new branch of physics must make assumptions peculiar to it, and if they continue to hold up then these hypotheses are on their way to becoming conventions in the same sense as the conventions of mechanics. So the truth status of a hypothesis is that of a well-grounded generality. But because the conventions of mechanics were originally hypotheses subject to experimental verification, there lingers about them a feeling that they are true in the way well-grounded generalities are true, but Poincaré was explicit that they are true by convention because we have made them into definitions and axioms.

Space and Time or Space–Time

There were surely few people who believed that, confronted with a choice between Euclidean and non-Euclidean geometry, the correct alternative was the non-Euclidean one, but as Minkowski's ideas gradually circulated among the physics community more and more physicists came to believe that space–time was the reality and theories that separate space and time were incorrect (see Walter 1999). Poincaré however, did not. He preferred a theory in which space and time were separate and Lorentz contractions really occurred, and argued, quite correctly, that because there could never be conclusions on this view that were incompatible with conclusions derived from a theory of space–time it was simply a matter of convenience which theory was adopted. The issue for us is what this tells us about his conventionalism.

At the level of philosophy, this view had the support of Einstein, who deduced from the absence of truly rigid bodies in nature because of the effects of temperature, external forces, etc., that the original, immediate relation between geometry and physical reality appears destroyed, and we feel impelled toward the following more general view, which characterizes Poincaré's standpoint. Geometry (G) predicates nothing about the behavior of real things, but only geometry together with the totality (P) of physical laws can do so. Using symbols, we may say that only the sum of (G)+(P) is subject to experimental verification. Thus (G) may be chosen arbitrarily, and also parts of (P); all these laws are conventions. All that is necessary to avoid contradictions is to choose the remainder of (P) so that (G) and the whole of (P) are together in accord with experience. Envisaged in this way, axiomatic geometry and the part of natural law, which has been given a conventional status, appear as epistemologically equivalent.

Sub specie aeterni Poincaré, in my opinion, is right. (Einstein 1921)

Hilbert did not agree, although he does not seem to have gone out of his way to challenge geometric conventionalism until he lectured on what he called the fundamental questions of modern physics in 1921 and 1923, by which time he was thoroughly familiar with Einstein's general theory of relativity.[2] In these lectures he asked his listeners to consider what they would really do if they lived in a (two-dimensional) world in which the angle sums of triangles made of rigid rods were always less than π except at atomic distances, when it became arbitrarily close to π. He suggested that we would have two three-parameter abstract geometries, each with their metrics, "but it would be neither appropriate, nor simple, nor conventional, to choose the parameters so that the metric we would use would be the simplest, namely the Euclidean one" (p. 429). Were it to be adopted, Hilbert suggested, it would not correspond at all to our representation of the real world and would seem completely strange. Rather, in the real world one would choose the Euclidean metric only when the Euclidean transformations fitted the phenomena, which in the present situation would not be the case. Einstein's position on the issue should therefore be rejected.

These two contradictory positions each have something to recommend them, and each misunderstands Poincaré's position. Einstein is right

[2] See Hilbert (2009, 428–430).

that logically one can defend a geometry (G) by redefining (P). Hilbert is right that given different behavior of rigid bodies we might decide to stick with non-Euclidean geometry—we might even have evolved as non-Euclideans—but nor would Poincaré have denied this. Both have simplified the issue that Poincaré confronted by removing it from the epistemological setting where Poincaré located it and returning it as a question of physics. Poincaré's geometric conventionalism was not about how different mathematical theories might be interpreted in physical terms and brought to bear on any issue of that kind in the laboratory. Hilbert and Einstein wanted to know how people would choose between physical theories. We may find it odd that Poincaré preferred a Galilean theory of space and time, especially when we recall his insights into the nature and importance of the Lorentz group and its rotations about any axis in $\mathbb{R}^4$, but it is of a piece with his belief about how we, as infants, acquire our knowledge of the external world. As knowers of the world, we know about space, a propertyless arena filled with rigid bodies, and we must now learn that at speeds vastly in excess of anything we might have evolved to know about, these rigid bodies behave in an unexpected way. His geometric conventionalism can defy the fashionable shift to space–time at the price of some oddities in the calculations, because it is a preference for the Euclidean group of transformations of space that is hardwired into us.

The Import of the General Theory of Relativity

With the success of Einstein's general theory of relativity it has become clear that almost everyone who works on the topic finds Einstein's theory somehow better, which raises the question of why. Poincaré was convinced that talk about space simply reduced to talk about objects, but this reductionist position was not and is not widely shared and needs to be examined, and this in turn is connected to the distinction he made between those who start with space and a metric (Riemann, Helmholtz, Lie) and those who start with a group that defines the isometries (Poincaré himself).

It is clear that a century of mathematicians since Poincaré have had none of his enthusiasm for starting geometry by starting with a group, and little of his reluctance in talking about space. They define metric spaces (a topological space together with a "distance" function

d satisfying certain properties, crucially the triangle inequality: $d(AC) \leq d(AB) + d(BC)$) and Riemannian manifolds, and prove that the latter are examples of the former. They have no problem in talking about the distance between points in these spaces, and are not disconcerted to find no isometries of the space at all (other than the identity map). On this approach there is no conceptual or philosophical problem in assigning a distance to each pair of points provided the axioms for a distance function are satisfied, and it is a matter of luck whether there is a way of mapping AB onto $A'B'$ and thereby confirming that they have the same length. The reasons are clear: none of these mathematicians have interested themselves in the epistemological problem that detained Poincaré, and (I would guess) few of them have his qualms about saying that there is a distance between points independently of our ability to slide a measuring rod around. In this, if in nothing else, Russell may have been right after all.

But why do we talk about space in a way that Enriques supposed, and not in Poincaré's reductionist fashion? We imagine that our knowledge of the external world is taken from the familiar behavior of objects of about our size, and that this gives a good approximation to the real, Riemannian geometry of space. We construct good enough measuring bodies, ones that are rigid to a high degree of approximation, and this is precisely what Einstein went on to consider (in "Geometry and experience"). He supposed we would use the paths of light in empty space–time to construct clocks (and he noted that the uniform behavior of the spectral lines of atoms of each chemical type supports this analysis). His conclusion was:

> According to the view advocated here, the question whether this [space–time] continuum has a Euclidean, Riemannian, or any other structure is a question of physics proper, which must be answered by experience, and not a question of a convention to be chosen on grounds of mere expediency. Riemann's geometry will hold if the laws of disposition of practically-rigid bodies approach those of Euclidean geometry the more closely the smaller the dimensions of the region of space-time under consideration.

All the geometries under Einstein's consideration were locally Minkowskian. But his conclusion is that experience with these bodies can enable us to choose a geometry that measures how each finite region departs from being flat. A Poincaré-style response was ruled out because Einstein had put his geometry into the language of practically rigid bodies. It was not possible for him to articulate the alternative that we might have a different concept of geodesic and regard these bodies as behaving in unexpected, unrigid, ways. In Poincaré's simple dichotomy the Euclideans and the non-Euclideans each have their collective experience of rigid bodies which have taught them what straight lines are and what the angle sums of plane figures are. In Einstein's case the only alternative to using practically rigid bodies in order to say anything geometrical about space is an unmotivated and hard-to-defend liking for one classical geometry. (And why that one, why not a non-Euclidean geometry with your favorite negative value for the curvature?) On the other hand, there is, within the limits of experimental error, just one geometry that will be described by the metric one obtains by looking at the practically rigid bodies and taking as geodesics the paths of rays of light.

The later reception of Poincaré's geometric conventionalism was colored by the success of Hilbert's entirely axiomatic approach, the successes of Riemannian geometry within mathematics, and its apparent vindication as the right framework for fundamental cosmology. The epistemological, and especially the cognitive, framework Poincaré used was lost, and what remained, his conventionalism (minus the qualifying adjective "geometric"), was there to be adopted as a possible philosophy of science. Although enthusiasts for it in the Vienna circle found conventional choices to be made all over the place and vindicated by general theory of relativity, others (see Friedman 1996) have taken conventionalism to have met its end with the general theory of relativity. Poincaré had, of course, excluded general Riemannian geometries from considerations precisely because they do not admit rigid-body motions. In his opinion, they must be regarded as purely analytical, and "would not lend themselves to proofs analogous to those of Euclid." By this he meant they could not be presented as geometries of space intelligible through rigid body motions, and so lacked epistemological significance.

Entities

Poincaré, as has often been remarked, had little liking for objects in physics. He was taken to be an idealist, and unduly skeptical. Maxwell's bell ringers were proof for him that talk about electric fluids or electrons was simply talk. If it generated good physics, that was fine, but the ruins of failed theories consisted largely of discredited objects; only the relations described by mathematics survived the transition from one theory to another, as his remarks at the ICP in 1900 illustrate (see chap. 1, sec. "Paris Celebrates the New Century" above). He was not dogmatic on the point: electrons and atoms were admitted once talk about them became adequately rich and convincing, but he saw science as a collection of steadily improving theories about objects it was wise to be hesitant about. In his controversy with Le Roy, Poincaré set great store by the fact that science predicts, and does not merely describe, so we value theories that do this over theories that cannot, and are entitled to draw some sense of confirmation from a prediction that has been verified. He also argued that scientists are constructing a language, a way of talking to one another, and the ongoing sense of dialogue is what creates, or at least amplifies, a feeling of objectivity. And finally, when we get to the point where we do introduce a theory, in that case Newtonian dynamics, it does a very good job very easily. By comparison, any other theory of planetary motion, such as Copernicus's own, has nothing but trouble passing from the descriptive to the explanatory. This invites some final reflections about Poincaré's philosophy of science and mathematics, to which we now turn.

Structural Realism and Skepticism

If geometric conventionalism, in the form Poincaré proposed it, cannot be sustained as a view about space it still has some interest as a cognitive theory of how we are able to construct a concept of space and obtain knowledge of it. And it still makes sense to see if Poincaré's ideas about science (its hypotheses, its laws, its relation to abstract mathematics) do not still have something to say, and here the crucial test, as with any theory of science, is theory change. The two current philosophies of science best suited to grapple with theory change and the realist–idealist debate are structural realism and some form of Wittgensteinian skepticism about any kind of knowledge. The structural realist approach to Poincaré has been set out most recently in Zahar (2001); the second approach is one Heinzmann has developed,[3] and since guides to Wittgenstein are always needed I am happy to have been guided by Martin Kusch (2006; 2009).

---

[3] See his (2008), and also his (1995).

Because the skeptical challenge usually strikes anyone who has not heard it before as absurd, and most people who have bothered with it to be wrong and refuted, I shall take the structural realist case up first. This has the feature that the nature of mathematics is not questioned, and indeed Poincaré's views about the truth of abstract mathematics are much less articulated. But it is possible to be a skeptic about mathematics as well, and we can ask if Poincaré at least looked in that direction.

Poincaré's view has been called structural realism by Zahar. Structural realism, he explains, is the view that there is a structured and undivided reality of which the mind is part, that this structure is at least partly intelligible, and that successful theories are approximately true. This form of realism, he explained, implies that relations among universals explain, or capture, the ontological order of things while the *relata* themselves remain inaccessible to human knowledge. It was originally proposed as the best way of keeping alive the idea that there is something deeply right about science while admitting that our best scientific theories change, or have changed, quite radically. In fact, structural realism was put forward in Worrall (1989) very much with the example of Poincaré in mind, and one of the defenses of science that it wishes to preserve is what Putnam (1975, 73) once called the no-miracles argument: "The positive argument for realism is that it is the only philosophy that doesn't make the success of science a miracle." But structural realism also wishes to take note of the fact that theories do change and all sorts of objects and theories once clasped to the scientists's heart have later been discarded, and it does so by saying that the old and new theories have enough in common for it to be possible to say that the new one is an improvement of the other and not simply different (if, also, better).

But if the perception one hundred years ago that Poincaré was largely an idealist and antirealist is not either a change in terminology or simply wrong, it is worth looking more closely at the varieties of structural realism that now exist to see where, if at all, Poincaré can find a home among them. One version of structural realism is epistemological: it advises us to put our trust in what science says about the structure of the external world, but not its nature, or essence, or what it "really" is (this is Zahar's view, above). We shall never know about the objects in themselves, but we can say reliable things about the relations they enter into. This was explicitly Poincaré's view. Another view, due to French and Ladyman and called by Ladyman "ontic structural realism," dispenses with talk of references to objects and properties altogether and defends science in other terms (see Ladyman 2009). As Ladyman admits (2009, 14), there is something prima facie odd about talking about relations when talk about what is related is disallowed, and the oddness is increased when, after all, we are talking about science, our best form of knowledge about the world.[4] Nor does any accumulation of entangled particles, space–time foam, strings or branes, not even multiverses, seem likely to shift physicists from their belief that they are talking about stuff out there, and not some mathematical theory selected out of many because of some measurements taken with huge pieces of equipment. His way out is to observe that there can be *relata* that are not about primitive, irreducible objects, and all that structural realism denies is the existence of such primitive objects (or, if you prefer, that we can talk meaningfully about such things). Now, while modern debates about structural realism are rightly tied to modern scientific issues, it remains true that Poincaré not only did not believe that one could talk about the primitive constituents of reality, he doubted the very existence of things one could not talk about. So on this view, he was an ontic structural realist.

As put, there's nothing wrong with this, but I think it leaves out a lot. For better or worse, it does not discuss adequately Poincaré's conventionalism, which Zahar deals with elsewhere in his book (pp. 102–105). Here he argues that Poincaré was not a conventionalist but close to being a structural realist, because, he says Poincaré ascribed to the view that "convenience is the hallmark of high verisimilitude" (p. 105). This is only applicable to Poincaré's conventionalism in physics, not to his geometric conventionalism, but even so it rests on a pun. A high degree of verisimilitude is what promoted the hypotheses of Newtonian mechanics into the list of conventions, but verisimilitude had nothing to do with Poincaré's resistance to Minkowskian space–time: if that was not Poincaré's own personal convenience pure and simple it rested on his epistemology.

It might instead be instructive to look at Poincaré as a skeptic. I shall restrict skepticism here to the ideas associated with Kripke's reading of

---

[4] Available at James Ladyman (2009). Structural realism. *The Stanford Encyclopedia of Philosophy* (summer 2009 edition), ed. by Edward N. Zalta. http://plato.stanford.edu/archives/sum2009/entries/structural-realism/.

Wittgenstein, which have recently been defended by Kusch (2006; 2009). Kusch calls the view skeptics attack "meaning-determinism," and defines it as the assertion that talk about the meaning of terms is talk about the mental states of people (Smith knows what he means when he says X, he knows what the sign "+" means because he is in a certain mental state). It also asserts that the principle of classical realism holds: every declarative sentence has a meaning, which is the proposition it expresses, and true propositions correspond to facts. The skeptics claim that such talk is incoherent, and argue that nothing precludes one having failed to exclude some other meanings: Kripke's example is that "+" could obey rules for large numbers that someone used to adding only small ones could never have ruled out (such as $a + b = 5$ for all numbers $a, b > 100$). Faced with the challenge one either takes it on its own terms and refutes it, or accepts it but argues that it is harmless after all. In the case at hand acceptance means (if I can still use the word) that talk about mental states is replaced by talk about intersubjectivity.

Poincaré's position, I suggest, is close to that of the Kusch's skeptic. He agreed that we rely on the testimony of experts and on a shared communication with others; that we speak a shared family of languages, natural, scientific, mathematical, which work because of a shared set of conventions, and we have ideas about what we would do if our statements conflict or communication failed. As he put it in (1902h) (see *VS* pp. 292, 345): No discourse—no objectivity. His dispute with Le Roy and his imaginary discussions between Euclideans and non-Euclideans are far from the only occasions where Poincaré put his trust in the possibility of effective communication. None of this involves knowing about meanings or having particular mental states.

If talk about meaning proceeds from introspection ("I know what I mean by X") to a charitable interpretation of what everyone else is saying as being sufficiently like what one says oneself, then Wittgenstein's alternative says that "By their deeds ye shall know them." It is clear that Poincaré did not talk about meanings, and certainly not about mental states, which he disparaged as carriers of truth in his lecture to the psychologists, (1908d). He was reluctant to speak about eternally established facts, and placed great weight on actions and usage (for example, in measuring). Not only did he not think a list of facts was anything like as good as a theory (one cannot make predictions from a list of facts, science cannot progress by generalization and analogy from a mere list) he often openly doubted if today's facts would be accepted tomorrow. But was he a skeptic in the sense just described?

The usual charge leveled at skeptics is that they are relativists. This charge alleges, to quote from Kusch (2009, 19) again, that "people using different epistemic systems (consisting of epistemic standards) can 'faultlessly disagree' over the question whether a given belief is epistemically justified or not. Faultless disagreement in such a scenario is possible because (1) beliefs can be justified only within epistemic systems; (2) there are, and have been, many radically different epistemic systems; and (3) it is impossible to demonstrate by rational argument that one's own epistemic system is superior to all or most of the others." But it can surely be argued that conventionalism in physics, and even geometric conventionalism, is akin to a language game. Poincaré was evidently a relativist over the question of the geometry of space, because faultless disagreement between a Euclidean and a non-Euclidean is exactly what he said would happen, and he was a relativist again when it came to a choice between the Galilean and the Lorentz group in special relativity. The nub of these disagreements is the existence of two distinct epistemic systems.

Kripke's skeptic, of course, also finds that people place undue confidence in mathematics, and this would seem to be a decidedly modern position first held by the later Wittgenstein. Poincaré does not seem to have held a developed position on abstract mathematics before Hilbert's *Grundlagen der Geometrie* of 1899. Before then we can see only that he believed it was now rigorous as a result of the creation of modern analysis. To judge by his position in 1897 in Zürich, he seems to have had no problem with combinatorial or formal algebraic arguments in number theory or finite group theory, and to have accepted that the real number continuum was now properly handled and based on the natural numbers. His view therefore reduces to confidence in primitive mental capacities plus confidence in some formal machinery to do with the real numbers that drew on our experience of physics. He was clear that the continuum is a construction of ours, one that the calculus relies upon and that is essential in the mathematization of physics, because he noticed favorably the existence of du Bois-Reymond's alternative in (1893b). Had the classical continuum ceased to be essential, had a working alternative come along, we can be sure that Poincaré would not have defended the continuum on the grounds that it is "out there." But it seems clear that

Poincaré would simply have been unable to accept Kripke's view about "+" and would have said our understanding of addition was built into our understanding of the integers; it was a mental state par excellence.

When in 1909 Poincaré responded to Zermelo's first attempt at an axiomatization of set theory he observed that the essential thing to establish with an axiom system of any kind is that it is free of contradiction. This Hilbert did with his axioms for various geometries because he allowed appeals to arithmetic, but Zermelo's problem is that his axioms are intended to be truly fundamental, so there is nothing for him to appeal to. The only hope, then is that the axioms be self-evident, and this, as we saw, Poincaré does not find them to be. What, then, did Poincaré take the criteria for self-evidence to be? And what would a satisfactory skeptical position be? To take the second question first, as a useful mirror to hold up to Poincaré, Wittgenstein allowed that certainty was attainable (as in (56)) and on the other hand wanted to rule out endless doubting (115): if you tried to doubt everything you would not get as far as doubting anything.[5] The game of doubting itself presupposes certainty. But while he did speak of the certainty of mathematical propositions, he barely expressed a view about what that might be with anything like the clarity that he deplored attempts to bring in logic to do the job. Perhaps the clearest statements we have are those collected in the *Remarks on the Foundations of Mathematics*. There, in section III he queried the notion of self-evidence of a mathematical axiom system, and decided that it meant "that we have already chosen a definite kind of employment for the proposition without realising it. The proposition is not a mathematical axiom if we do not employ it precisely *for this purpose*. The fact, that is, that here we do not make experiments, but accept the self-evidence, is enough to fix the employment" (italics in original). And in V 14, he imagined what might happen given a system that was inconsistent but had never yet been made to generate a contradiction, and made the "good angel" defense: "Well, what more do you want? One might say, I believe: a good angel will always be necessary, whatever you do."

Poincaré had two objections to the direction he saw axiomatic theory would try to take. The first was the lack of self-evidence in its axioms, and this came up in the way Zermelo used the word "Menge" to identify

---

[5] These numbered remarks come from Wittgenstein's "On certainty" (1969).

a certain type of collection, the type about which we can reason. For Poincaré these would be sets with predicative definitions, so that each member has, as it were, its own entitlement to membership. For Zermelo, these were collections with a "definite" membership criterion, but by not requiring definiteness to mean predicativity Poincaré felt that Zermelo had not been careful enough: "But even though he has closed his sheepfold carefully, I am not sure that he has not set the wolf to mind the sheep." Predicative definitions permit clear checks on membership of a set and impose limits on the size of sets, or so Poincaré believed, which is why he rejected the well-ordering axiom; no set larger than the first uncountable set can be surveyed. This was Poincaré's second objection: set theorists spoke to him far too easily of very large sets. Since this was a consequence of their approach, he took it as evidence that their concept of a set was not self-evident, and accordingly rejected it.

Poincaré's deepest philosophical beliefs about mathematics and science as set out in his (1897d) and his (1909d) are these: There is no valid or clear distinction to be made between mathematics and physics because the two are so intimately entangled. If a distinction is to be made, the criteria for validity in science are pragmatic and social, and those for mathematics are its self-evidence. It is not hard to imagine him making a "good angel" defense, harder to imagine him demanding a certificate of logical consistency before investigating an interesting family of ideas. What mattered to him was the exercise of understanding: how theorems enabled the mathematician and the physicist to know what to look for and how to find it.

Taken literally, Poincaré's epistemological theory of the natural numbers and Kripke's skeptical position about "+" are incompatible. Even the positions about epistemology itself are incompatible: Poincaré believed some knowledge was certain, Kripke invites us to accept that we manage very well without certainty. So Poincaré cannot be a skeptic in Kripke's fashion. But if the foundations are different, the buildings erected upon them are strikingly similar, and bring out a way in which Poincaré's idealist position can be regarded as a refusal to give credence to objects we ought anyway to try to manage without.

Further insight into Poincaré's views may come from a renewed attention to the practice of mathematics, and in particular the nature of discovery in mathematics. No serious philosophy of mathematics can ignore or mistreat the role of discovery, for without it there would be no mathematics. As Poincaré said (1899d), even "the next generation of leading mathematicians will need intuition, for if it is by logic that one proves, it is by intuition that one invents" (in *Oeuvres* 11.1, 132). His views about even proof in mathematics had little to do with logical rigor, much more to do with action. As he put it in *L'avenir* (*S&M* 22, 373): "In mathematics rigour is not everything, but without it there is nothing." For Poincaré, a proof in mathematics was enabling; it rested on general ideas capable of wider application (by analogy); it showed not merely that some things are the case, it also explained *why* some things are the case; and in physics it dealt in relations that would survive changing beliefs or practices about objects. A good proof, for Poincaré, is a new and valid use of the terms it involves.

There is nothing "psychologistic" about this position. Understanding is measured, in Machian terms, by the economy of thought it yields. Doubtless fresh understanding comes with a feeling of excitement, doubtless mathematicians may understand something very well and still be stuck, and certainly there is an aesthetic element that leads mathematicians to prefer this or that topic to investigate. But a core component of it, in Poincaré's view, was the ability to generalize, to provide the context that explains why something is the case and so enables the mathematician to show that something new can also be proved. This aspect of the practice of mathematics may well fall outside the purview of mathematical logic and proof theory, but that does not make it a purely personal matter that even philosophers should not, or need not, touch.

Epple (1996), following Wittgenstein, distinguishes between discovery (an epistemological matter) and invention (a pragmatic one). The distinction is clear, and yet it may not matter. Almost all mathematicians, and here we must surely include Poincaré, bridle at the suggestion that they make mathematics up freely, like novelists. It is not, for them, a game with arbitrary rules to be varied at will. Rather, it is a tightly constrained activity, in which good discoveries are often hard won. A good new result (a theorem, an example, a counterexample, a method) must be presented rigorously, and this is a conceptual exercise conveyed through a mathematical language that obeys the rules (the axioms and accepted results) of the relevant domain. A heuristic, although informal, must ultimately yield new results, and perhaps good new proofs of old results. If mathematicians speak of their discoveries they must convey it in the language of invention, and if they have invented something others may yet refer to their discovery.[6] Epple offered an account of what he called mathematical pragmatics (1996, 563) that drew on Poincaré's essay on invention (1908d). The view of Poincaré presented here has emphasized the epistemological character of his work, but in the sense that knowing something involves reflecting on how you can know it and on what you can know about it. The test, for Poincaré, was then how to understand what he knew, and to make productive use of it. Not only was this how he came to decide what should be proved, his consideration of how certain mathematical facts have come about often suggested how others might be proved, or illuminated the point beyond which existing proofs could not be stretched. In these ways his work with concepts, when rigorous, can be studied with the tools of modern logic, but even when not it forms part of his creative life and so forms part of mathematical and scientific practice. His, and that of every productive mathematician and scientist.

---

[6] The distinction may be nothing more than a lazy feeling that one discovers objects but invents methods for finding them, as with patents.

12

Appendixes

ELLIPTIC AND ABELIAN FUNCTIONS

Elliptic functions arose in the 1820s as a class of complex functions connected with the study of elliptic integrals, that is, integrals of the form $\int_0^z \frac{dt}{\sqrt{(1-t^2)(1-k^2t^2)}}$, which had hitherto been regarded as impossible to evaluate. Although

$$w(z) = \int_0^z \frac{dt}{\sqrt{(1-t^2)(1-k^2t^2)}}$$

is an infinitely many-valued function of z that depends on the path of integration as it winds around the singular points $t = 1, -1, \frac{1}{k}, -\frac{1}{k}$, the inverse function $z = z(w)$ is a single-valued function with two periods, conventionally denoted $2\omega_1$ and $2\omega_2$ (which depend on k):

$$z(w) = z(w + 2\omega_1) = z(w + 2\omega_2).$$

So the values of an elliptic function are known once they are given on the parallelogram with vertices $0, 2\omega_1, 2\omega_2, 2\omega_1 + 2\omega_2$. This is closely analogous to the sine function: the integral $y(x) = \int_0^x \frac{dt}{\sqrt{(1-t^2)}}$ defines the multivalued function $y(x) = \arcsin x$ with the inverse function $x(y) = \sin y$, whose values are known once they are given on the interval $[0, 2\pi]$. In fact, the familiar trigonometric functions turned out to be special cases of the new elliptic functions, and although elliptic functions are complex functions of a complex variable they also proved to be useful in problems in applied mathematics—accurate theories of the simple pendulum, the spinning top and other rotating bodies such as planets fell under their scope. They called for, and received, a grounding in a general theory of complex functions of a complex variable which was a boost to that theory.

Elliptic functions are products of quotients of theta functions. A theta function,

$$\theta(z, \tau) = \sum_{n=-\infty}^{\infty} \exp\left(\pi i\, n^2 \tau + 2\pi i\, n\, z\right),$$

is periodic in $z \in \mathbb{C}$ and quasiperiodic in $\tau \in \operatorname{Im}(\tau) > 0$; that is, $\theta(z, \tau) = \theta(z + 1, \tau)$ and $\theta(z + \tau, \tau) = \theta(z, \tau)\exp(-\pi i\,\tau - 2\pi i\, z)$. So it is defined independently of the theory of elliptic functions, and could be used as the starting point of the theory, by defining the elliptic function

$$\prod \frac{\theta(z - a_1)\theta(z - a_2)}{\theta(z - b_1)\theta(z - b_2)}.$$

Periodicity in the z variable suggested that a theta function be given a Fourier series expansion, and this led Jacobi to discover the remarkable connections between theta functions and number theory. This arose because the number theory of quadratic forms is intimately tied to a theory of planar lattices, and elliptic functions, being almost by definition doubly periodic functions, come with a lattice built-in. Furthermore, elliptic functions and planar lattices each come in one-parameter families (a complex parameter that determines the aspect ratio of the lattice), and the way this parameter enters the story turned out to be controlled by a singularly important group of 2×2 matrices with integer entries and determinant 1 (the modular group). Unsurprisingly, elliptic function theory was the subject of regular lectures by Hermite at the École polytechnique in the 1870s, as it was in Berlin, where Weierstrass gave a very different account, and it also appeared prominently in the advanced textbooks of the period.

Beyond the theory of elliptic functions, which was well-understood, mainstream, advanced mathematics by the 1870s, lay the much less understood theory of Abelian functions that Riemann had tackled and was now under the care of Weierstrass in Berlin. The connection arises because an elliptic integral is of the form $\int \frac{dt}{u}$, where $u^2 = (1 - t^2)(1 - k^2 t^2)$ is what is called an elliptic curve. Replacing the curve by a general algebraic curve $g(u, t) = 0$ and then admitting a general integrand $f(u, t) = 0$ leads to an integral of the form $\int_0^z f(u, t)\,dt$. Now the trick of writing $\zeta = \varphi(z) = \int_0^z f(u, t)\,dt$ and studying $z(\zeta)$ does not lead to a function defined on the whole of $\mathbb{C}$, and the recognized approach, pioneered by Jacobi, was to see it as part of a theory of p periodic functions of p complex variables, where p is an integer determined by the curve $g(u, t) = 0$ (and known as the genus of the curve). These p functions define what is called an Abelian function, a function from $\mathbb{C}^p$ to $\mathbb{C}^p$ which has $2p$ periods. One of the many complications in this subject is that when $p \geq 4$ there are also functions from $\mathbb{C}^p$ to $\mathbb{C}^p$ with $2p$ periods that do not arise in this way.

MAXWELL'S EQUATIONS

The Maxwell–Heaviside Formulation

Let $\mathbf{E}$ denote electric field strength, and $\mathbf{B}$ denote magnetic field strength. Let ρ denote the electric charge density and $\mathbf{j}$ denote the electric current density (the rate at which charge flows through a unit area per second). Then

$$\nabla \cdot \mathbf{E} = \frac{\rho}{\epsilon_0}, \tag{12.1}$$

$$\nabla \times \mathbf{E} = -\frac{\partial \mathbf{B}}{\partial t}, \tag{12.2}$$

$$c^2 \nabla \times \mathbf{B} = \frac{\partial \mathbf{E}}{\partial t} + \frac{\mathbf{j}}{\epsilon_0}, \tag{12.3}$$

$$\nabla \cdot \mathbf{B} = 0. \tag{12.4}$$

Potential Functions

A function f is a (scalar) potential function for a vector field $\mathbf{V}$ if $\nabla f = \mathbf{V}$. So it is a necessary condition for a vector field $\mathbf{V}$ to be derived from a potential function that $\nabla \times \mathbf{V} = 0$. This condition is sufficient on any simply connected region.

A vector field $\mathbf{X}$ is a (vector) potential function for a vector field $\mathbf{V}$ if $\nabla \times \mathbf{X} = \mathbf{V}$. So it is a necessary condition for a vector field $\mathbf{V}$ to be derived from a vector potential function that $\nabla \cdot \mathbf{V} = 0$. This condition is sufficient on any region. A vector potential, if it exists, is not unique, and can be modified by adding any expression of the form $\nabla \cdot f$.

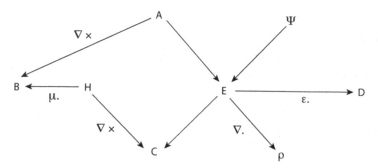

Figure 12.1. The ingredients of Maxwell's equations. Source: drawn by Michele Mayor Angel.

B admits a vector potential **A**; that is, $\mathbf{B} = \nabla \times \mathbf{A}$.

E does not admit a scalar potential (except, topology permitting, in the static case). Instead, $\mathbf{E} = -\nabla\phi - \frac{\partial \mathbf{A}}{\partial t}$.

The Maxwell Diagram

There are many ways to read this diagram. A pure mathematician might well see it as the consequence of knowing a vector field **A** and a potential function ψ, together with some constants[1] μ and ϵ, or alternatively as a puzzle if the given information is lower down the diagram and it is required to work upwards. A physicist might well also see it in one of two ways: either in terms of measurable quantities or in terms of more fundamental ones that drive the physics. A Maxwellian would start in the middle, with the magnetic induction **B** and the electromotive force **E**, these being the most obviously measurable quantities. A modern physicist would quickly try to locate what he or she calls Maxwell's equations:

- $\nabla \cdot \mathbf{E} = \dfrac{\rho}{\epsilon_0}$ is visible lower right;

- $\nabla \times \mathbf{E} = \dfrac{\partial \mathbf{B}}{\partial t}$ is not visible;

---

[1] "μ is the coefficient of magnetic permeability, which may be considered a scalar quantity, or a linear and vector function operating in **H**, according as the medium is isotropic or not" (Maxwell, *Treatise*, 1873, para. 614).

- $c^2 \nabla \times \mathbf{B} = \dfrac{\partial \mathbf{B}}{\partial t} + \dfrac{\mathbf{j}}{\epsilon_0}$ is the equation represented by the arrow from $\mathbf{E}$ to the dummy quantity $\mathbf{C}$;

- and $\nabla \mathbf{B} = 0$ is an automatic consequence of the arrow, top left.

The missing equation is not a simple mathematical consequence of what is in the diagram. It was first obtained by Heaviside after he reinterpreted Maxwell's equations for the electromotive force by restricting them to a stationary medium; then the missing "Maxwell equation" falls out readily.

Lorentz's Equations

- The magnetic force is represented by $\boldsymbol{\alpha} = (\alpha, \beta, \gamma)$. It satisfies $\nabla \cdot \boldsymbol{\alpha} = 0$, which says that there is no true magnetism.

- $\mathbf{F} = (F, G, H)$ is a vector potential satisfying $\nabla \cdot \mathbf{F} = 0$ and introduced so that one can write $\boldsymbol{\alpha} = \nabla \times \mathbf{F}$.

- The total current is represented by $\mathbf{u} = (u, v, w)$, which satisfies $\mathbf{u} = \frac{d}{dt}\mathbf{f} + \rho \boldsymbol{\xi}$, where ρ is some sort of density of electricity, $\mathbf{f} = (f, g, h)$ satisfies $\nabla \cdot \mathbf{f} = \rho$ and $\boldsymbol{\xi} = (\xi, \eta, \zeta)$ satisfies $\nabla \cdot \boldsymbol{\xi} = 0$.

- $4\pi \mathbf{u} = \nabla \times \boldsymbol{\alpha}$

- $\triangle \mathbf{F} = -4\pi \mathbf{u}$

- In Hertz's theory, the electromotive force of induction is represented by the line integral $P\,dx + Q\,dy + R\,dz$, and $\mathbf{P} = (P, Q, R)$ satisfies $\nabla \times \mathbf{P} = \frac{d}{dt}\boldsymbol{\alpha}$.

- The function ψ is a potential satisfying $\triangle \psi = -\frac{4\pi}{K_0}\rho$.

- $\dfrac{4\pi}{K_0} + \dfrac{d}{dt}\mathbf{F} + \nabla \cdot \psi = 0$

- $\dfrac{d\rho}{dt} = \nabla \cdot (\rho \boldsymbol{\xi})$

Poincaré stated that these equations are to be solved on the assumption that ρ and $\boldsymbol{\xi}$ are known.

GLOSSARY

This glossary is intended to define and explain some terms that some readers may know but others not. Terms in general use are not covered, nor are ones so obscure that they had to be explained where they occurred in the body of the book. For topics introduced in electromagnetic theory, see the above appendix.

The **Abelianization** (of a group G) is the largest Abelian group A for which there is a surjective (onto) map $G \to A$. It is obtained by imposing on G all the relations $ghg^{-1}h^{-1} = 1$ with $g, h \in G$.

A **birational** map between complex algebraic varieties is a rational map with an inverse that is also a rational map. A **rational** map is a holomorphic map defined on all but a subset of codimension 2 or more (where, informally, it has poles or is "infinite").

The **Cauchy–Riemann equations** for a function from one domain in $\mathbb{C}$ to another, $w = f(z)$, where $z = x + iy$ and $w = u + iv$, are

$$\frac{\partial u}{\partial x} = \frac{\partial v}{\partial y}, \quad \frac{\partial u}{\partial y} = -\frac{\partial v}{\partial x}.$$

The **center** of a group G is the set of elements that commute with every element of the group, $\{h \in G : gh = hg \; \forall g \in G\}$.

A complex function $w = u(x + iy) + iv(x + iy)$ from one domain in $\mathbb{C}$ to another is **holomorphic** if it satisfies the Cauchy–Riemann equations everywhere and the functions u and v are differentiable. It is **meromorphic** if it is holomorphic except at a set of isolated points at which the reciprocal of w tends to zero. The corresponding definitions for complex functions of two or more variables are more complicated and are omitted.

A sequence (a_n) **converges** to a if for all $\varepsilon > 0$ there is an N such that $n > N$ implies $|a_n - a| < \varepsilon$. Convergence can be required in several forms. It is **absolute** if the sequence $(|a_n|)$ converges. A series $(\sum_{n=1}^{\infty} a_n)$ **converges** if the sequence of partial sums $(\sum_{j=1}^{n} a_j)$ converges, and this convergence is said to **conditional** if it depends on the order of the terms. If the sequence (a_n) is a sequence of functions $a_n(x)$ converging to a function $a(x)$ on some domain, the convergence is **uniform** if the choice of N does not depend on x.

A **Green's function** is a function f on a domain (usually in $\mathbb{R}^2$ or $\mathbb{R}^3$) that is harmonic except at one point P in the domain where $f(Q) - \frac{1}{|Q-P|} \to 0$ as $Q \to P$.

A function $H(q_1, \ldots, q_n, p_1, \ldots, p_n, t)$, where each q_j and p_j is a function of the time, t, satisfies **Hamilton's equations** if

$$\frac{dq_j}{dt} = \frac{dH}{dp_j}, \quad \frac{dp_j}{dt} = -\frac{dH}{dq_j}, \quad j = 1, 2, \ldots, n.$$

A three-dimensional **handlebody** is a three-dimensional manifold whose boundary is a two-dimensional manifold and that can be cut along embedded disks to form a three-dimensional ball.

A **harmonic function** is a function $f(x_1, x_2, \ldots, x_n)$ that satisfies Laplace's equation:

$$\frac{\partial^2 f}{dx_1^2} + \frac{\partial^2 f}{dx_2^2} + \cdots + \frac{\partial^2 f}{dx_n^2} = 0.$$

An **indirect definition** is one that defines a mathematical object by its properties, typically by a system of axioms that specify what can be said about it.

An **integral equation** is an equation of the form $f(x) = \int_a^b K(x, t)\varphi(t)\, dt$ for an unknown function φ, where f and K (the **kernel**) are given functions. This equation is a Fredholm equation of the first kind. A Fredholm equation of the second kind is an integral equation of the form $\varphi(x) = f(x) + \lambda \int_a^b K(x, t)\varphi(t)\, dt$, where the solution will exist only for certain values of λ that form the **spectrum** of the kernel K.

An **invariant integral** for a flow carrying a region T at one time to a region T' at another is an integral that takes the same value when taken over either region. For example, if a flow is volume preserving, the volume of a region is given by an invariant integral.

Invariants of polynomial expressions are expressions in the coefficients of the expression that are unaltered by linear changes in the variables, or alter only by a power of the determinant of the transformation. For example, the polynomial $ax^2 + bxy + cy^2$ has the invariant $b^2 - 4ac$. If the invariant is allowed to involve the variables as well as the coefficients, it is called a covariant.

A **manifold**, informally, is a topological space that is locally Euclidean: it is composed of overlapping regions that are homeomorphic to n-dimensional balls in Euclidean n-space. It is therefore a space on which functions may be defined and the operations of the calculus and differential geometry may be carried out. Examples include surfaces, and the sets defined as $\{x \mid f(x) = 0\}$ where f is a differentiable function on $\mathbb{R}^n$. Modern mathematics offers a more precise definition that excludes certain counterintuitive candidates. The **boundary** of a manifold is points in it that have neighborhoods homeomorphic to n-dimensional hemispheres in Euclidean n-space. For example, the points on the boundary of the disk $\{(x, y) \mid x^2 + y^2 \leq 1\}$ are the points of the unit circle: $\{(x, y) \mid x^2 + y^2 = 1\}$. A manifold is said to be **closed** if it is compact and without boundary.

The **Peano axioms** for the natural numbers are (1) 1 is a number; (2) if a is a number then the successor of a, written $a + 1$, is a number; (3) for any two numbers a and b , $a = b$ if and only if $a + 1 = b + 1$; (4) no number has 1 as a successor; (5) if k is a class containing the number 1, and if k has the property that if x is a number belonging to the class k then $x + 1$ is in the class k, then every number is in the class k.

Point sets: an **accumulation point** or **limit point** P of a subset X of a topological space S is a point such that every open set containing P contains at least one point of X other than P itself. The **derived** set of X is the set of accumulation points of X. A set is **perfect** if it contains all its accumulation points and if every point of it is an accumulation point. The closure of a set is the union of a set and its derived set. A set X is a **nowhere-dense subset** of S if its closure has an empty interior.

A **ring** R is a set of elements that form an Abelian group under addition, that can be multiplied together, and where the distributive law, $a(b + c) = ab + ac$ for all a, b, c in the ring, is satisfied. The ring is commutative if the multiplication is commutative. A set of elements A in R is an **ideal** if it is closed under subtraction (a and b in A implies $a - b$ in A) and if it is closed under multiplication by elements of the ring (a in A and r in R implies ar in A). The ideal is **principal** if every element in it is a multiple of a single element, so it is of the form $A = \{ar \mid r \in R\}$ for some element a.

The **principle of least action** (Hamilton's principle) applies to a mechanical system with states of kinetic energy T and potential energy V. The Lagrangian of the system is $L = T - V$, and Hamilton's principle asserts that the actual evolution of the system from time t_0 to time t_1 minimizes the **action** integral $\int_{t_0}^{t_1} L \, dt$ among all possible evolutions—more precisely, the actual evolution is a stationary value of the action integral. It can be shown that this requirement is equivalent to saying that the system obeys Newton's laws of motion.

A **well-ordered set** is a set, S, with a total ordering, $<$, (for any two points $x, y \in S$ precisely one of the following statements is true: $x < y, x = y, y < x$), and such that every nonempty subset T of S has a least element (an element $t \in T$ such that $t < x$ for all $x \in T$).

References

For papers included in editions of collected works the latter's page numbering has been used.

Standard abbreviations of book and journal titles have been used, including:

Acta	*Acta Mathematica*
AHES	*Archive for History of Exact Sciences*
Amer. J. Math.	*American Journal of Mathematics*
AMS	American Mathematical Society
Bull. AMS	*Bulletin of the American Mathematical Society*
	Likewise *Trans. AMS, Notices AMS*
Cahiers	*Cahiers du séminaire d'histoire des mathématiques*
CR	*Comptes rendus hebdomadaires des séances de l'Académie des sciences*
EMW	*Encyclopädie der mathematischen Wissenschaften*
Göttingen Nachr.	*Nachrichten der K. Gesellschaft der Wissenschaften zu Göttingen. Mathematisch–Physikalische Klasse*
HMath	*American Mathematical Society and London Mathematical Society. Series in the history of mathematics.* Providence, RI.
Jahrbuch	*Jahrbuch über die Fortschritte der Mathematik.*
J. de math.	*Journal de mathématiques pures et appliquées*
JDMV	*Jahresbericht der Deutschen Mathematiker-Vereinigung*
J. Éc. poly.	*Journal de l'École polytechnique*
JfM	*Journal für die reine und angewandte Mathematik*
LMS	London Mathematical Society
Math. Ann.	*Mathematische Annalen*
Phil. Trans. Roy. Soc. London A	*Philosophical Transactions of the Royal Society of London A*
Rend. Palermo	*Rendiconti del Circolo matematico di Palermo*

ARTICLES AND BOOKS BY POINCARÉ

Armstrong, H. E., M. Foster, F. Klein, T. P. Köppen, H. Poincaré, A. W. Rücker, B. Schwalbe, and E. Weiss (1899). International catalogue of scientific literature: Report of the Provisional International Committee. *Science 10*, 482–487.

Nabonnand, P. (Ed.) (1999). *La correspondance entre Henri Poincaré et Gösta Mittag-Leffler*. Basel: Birkhäuser.

Poincaré, H. (1874). Démonstration nouvelle des propriétés de l'indicatrice d'une surface. *Annales de mathématiques 13*, 449–456.

Poincaré, H. (1878). Note sur les propriétés des fonctions définies par les équations différentielles. *J. Éc. poly. 45*, 13–26. In *Oeuvres 1*, xxxvi–xlviii.

Poincaré, H. (1879). *Sur les propriétés des fonctions définies par les équations aux différences partielles*. Ph. D. thesis, Université de Paris. *Oeuvres 1*, xlix–cxxxi.

Poincaré, H. (1880a). Sur les courbes définies par une équation différentielle. *CR 90*, 673–675. In *Oeuvres 1*, 1–2.

Poincaré, H. (1880b). Sur un mode nouveau de représentation géométrique des formes quadratiques définies ou indéfinies. *J. Éc. poly. 47*, 177–245. In *Oeuvres 5*, 117–180.

Poincaré, H. (1881a). Mémoire sur les courbes définies par une équation différentielle (1ère partie). *J. de math. 7*, 375–422. In *Oeuvres 1*, 3–44.

Poincaré, H. (1881b). Sur les applications de la géométrie non euclidienne à la théorie des formes quadratiques. *Association française pour l'avancement des sciences 10*, 132–138. In *Oeuvres 5*, 267–274.

Poincaré, H. (1881c). Sur les courbes définies par les équations différentielles. *CR 93*, 951–952. In *Oeuvres 1*, 85–86.

Poincaré, H. (1881d). Sur les fonctions abéliennes. *CR 92*, 958–959. In *Oeuvres 4*, 299–301.

Poincaré, H. (1881e). Sur les fonctions fuchsiennes. *CR 92*, 333–335. In *Oeuvres 2*, 1–4.

Poincaré, H. (1881f). Sur les fonctions fuchsiennes. *CR 93*, 301–303. In *Oeuvres 2*, 29–31.

Poincaré, H. (1881g). Sur les formes cubiques ternaires et quaternaires I. *J. Éc. poly. 50*, 190–253. In *Oeuvres 5*, 28–72.

Poincaré, H. (1882a). Mémoire sur les courbes définies par une équation différentielle (2nde partie). *J. de math. 8*, 251–296. In *Oeuvres 1*, 44–84.

Poincaré, H. (1882b). Mémoire sur les fonctions fuchsiennes. *Acta 1*, 193–294. In *Oeuvres 2*, 169–257.

Poincaré, H. (1882c). Sur la théorie des fonctions fuchsiennes. *Mémoires de l'Académie nationale des sciences, arts et belles lettres de Caen*, 3–29. In *Oeuvres 2*, 75–91.

Poincaré, H. (1882d). Sur les fonctions uniformes qui se reproduisent par des substitutions linéaires. *Math. Ann 19*, 553–564. In *Oeuvres 2*, 92–104.

Poincaré, H. (1882e). Sur les formes cubiques ternaires et quaternaires II. *J. Éc. poly. 51*, 45–91. In *Oeuvres 5*, 293–334.

Poincaré, H. (1882f). Sur les groupes discontinus. *CR 94*, 840–843. In *Oeuvres 2*, 38–40.

Poincaré, H. (1882g). Sur les séries trigonométriques. *CR 95*, 766–768. In *Oeuvres 4*, 585–587.

Poincaré, H. (1882h). Sur les transcendantes entières. *CR 95*, 23–26. In *Oeuvres 4*, 14–16.

Poincaré, H. (1882i). Théorie des groupes fuchsiens. *Acta 1*, 1–62. In *Oeuvres 2*, 108–168.

Poincaré, H. (1883a). Mémoire sur les groupes kleinéens. *Acta 3*, 49–92. In *Oeuvres 2*, 258–299.

Poincaré, H. (1883b). Sur certaines solutions particulières du problème des trois corps. *CR 97*, 251–252. In *Oeuvres 7*, 251–252.

Poincaré, H. (1883c). Sur les fonctions à espaces lacunaires. *Acta Societatis Scientiarum Fennicae 12*, 343–350. In *Oeuvres 4*, 28–35.

Poincaré, H. (1883d). Sur les fonctions de deux variables. *Acta 2*, 97–113. In *Oeuvres 4*, 147–161.

Poincaré, H. (1883e). Sur les séries trigonométriques. *CR 97*, 1471–1473. In *Oeuvres 4*, 588–590.

Poincaré, H. (1883f). Sur un théorème de la théorie générale des fonctions. *Bulletin de la Société mathématique de France 11*, 112–125. In *Oeuvres 4*, 57–69.

Poincaré, H. (1884a). Mémoire sur les fonctions zétafuchsiennes. *Acta 5*, 209–278. In *Oeuvres 5*, 209–278.

Poincaré, H. (1884b). Sur certaines solutions particulières du problème des trois corps. *Bulletin astronomique 1*, 65–74. In *Oeuvres 7*, 253–261.

Poincaré, H. (1884c). Sur les groupes des équations linéaires. *Acta 4*, 201–311. In *Oeuvres 2*, 300–401.

Poincaré, H. (1885a). Sur la représentation des nombres par les formes. *Bulletin de la Société mathématique de France 13*, 162–194. In *Oeuvres 5*, 400–432.

Poincaré, H. (1885b). Sur l'équilibre d'une masse fluide animée d'un mouvement de rotation. *CR 100*, 346–348. In *Oeuvres 7*, 14–16.

Poincaré, H. (1885c). Sur l'équilibre d'une masse fluide animée d'un mouvement de rotation. *Bulletin astronomique 2*, 109–118. In *Oeuvres 7*, 17–25.

Poincaré, H. (1885d). Sur l'équilibre d'une masse fluide animée d'un mouvement de rotation. *Bulletin astronomique 2*, 405–413. In *Oeuvres 7*, 26–33.

Poincaré, H. (1885e). Sur l'équilibre d'une masse fluide animée d'un mouvement de rotation. *Acta 7*, 259–380. In *Oeuvres 7*, 40–140.

Poincaré, H. (1885f). Sur les courbes définies par les équations différentielles (3ème partie). *J. de math. 4*, 167–244. In *Oeuvres 1*, 90–161.

Poincaré, H. (1885g). Sur les intégrales irrégulières des équations linéaires. *CR 101*, 939–941. In *Oeuvres 4*, 611–613.

Poincaré, H. (1885h). Sur un théorème de M. Fuchs. *Acta 7*, 1–32. In *Oeuvres 3*, 4–31.

Poincaré, H. (1886a). *La cinématique pure, mécanismes, potentiel, mécanique des fluides*, ed. by A. Guillet. Paris: Association amicale des élèves de la Faculté des sciences de Paris.

Poincaré, H. (1886b). Sur les courbes définies par les équations différentielles. *J. de math. 2*, 151–217. In *Oeuvres 1*, 167–222.

Poincaré, H. (1886c). Sur les intégrales irrégulières des équations linéaires. *Acta 8*, 295–344. In *Oeuvres 1*, 290–332.

Poincaré, H. (1887a). Les fonctions fuchsiennes et l'arithmétique. *J. de math. 3*, 405–464. In *Oeuvres 2*, 463–511.

Poincaré, H. (1887b). Remarques sur les intégrales irrégulières des équations linéaires (réponse à M. Thomé). *Acta 10*, 310–312. In *Oeuvres 1*, 333–335.

Poincaré, H. (1887c). Sur les résidus des intégrales doubles. *Acta 9*, 321–380. In *Oeuvres 3*, 440–489.

Poincaré, H. (1888). Sur une propriété des fonctions analytiques. *Rend. Palermo 2*, 197–200. In *Oeuvres 4*, 11–13.

Poincaré, H. (1889a). *Leçons sur la théorie mathématique de la lumière*, ed. by J. Blondin, M. Lamotte, and D. Hurmuzescu. Paris: Georges Carré.

Poincaré, H. (1889b). Sur le problème des trois corps et les équations de la dynamique avec des notes par l'auteur – mémoire couronné du prix de S. M. le Roi Oscar II. Printed in 1889 but not published.

Poincaré, H. (1889c). Sur les tentatives d'explication mécanique des principes de la thermodynamique. *CR 108*, 550–553. In *Oeuvres 10*, 231–233.

Poincaré, H. (1890a). Contribution à la théorie des expériences de M. Hertz. *CR 111*, 322–326. In *Oeuvres 10*, 1–5.

Poincaré, H. (1890b). *Électricité et optique*, ed. by J. Blondin and B. Brunhes. Paris: Georges Carré.

Poincaré, H. (1890c). Sur le problème des trois corps et les équations de la dynamique. *Acta 13*, 1–270. In *Oeuvres 7*, 262–479.

Poincaré, H. (1890d). Sur les équations aux dérivées partielles de la physique mathématique. *Amer. J. Math. 12*, 211–294. In *Oeuvres 9*, 28–113.

Poincaré, H. (1891a). *Elektricität und Optik*. Berlin: Springer.

Poincaré, H. (1891b). Les géométries non euclidiennes. *Revue générale des sciences pures et appliquées 2*, 769–774. Modified in *S&H*.

Poincaré, H. (1891c). Sur la résonance multiple des oscillations hertziennes. *Archives des sciences physiques et naturelles 25*, 609–627. In *Oeuvres 10*, 20–32.

Poincaré, H. (1891d). Sur le problème des trois corps. *Bulletin astronomique 8*, 12–24. In *Oeuvres 7*, 480–490.

Poincaré, H. (1891e). Sur l'intégration algébrique des équations différentielles du premier ordre et du premier degré. *Rend. Palermo 5*, 161–191. In *Oeuvres 3*, 35–58.

Poincaré, H. (1892a). *Leçons sur la théorie de l'élasticité*, ed. by É. Borel and J. Drach. Paris: Georges Carré.

Poincaré, H. (1892b). Les formes d'équilibre d'une masse fluide en rotation. *Revue générale des sciences pures et appliquées 3*, 809–815. In *Oeuvres 7*, 203–217.

Poincaré, H. (1892c). *Les méthodes nouvelles de la mécanique céleste*. Paris: Gauthier-Villars. 3 vols: vol. 2, 1893; vol. 3, 1899.

Poincaré, H. (1892d). Sur la polarisation par diffraction. *Acta 16*, 297–339. In *Oeuvres 9*, 293–330.

Poincaré, H. (1892e). Sur l'analysis situs. *CR 115*, 633–636. In *Oeuvres 6*, 189–192.

Poincaré, H. (1892f). *Théorie mathématique de la lumière II*, ed. by M. Lamotte and D. Hurmuzescu. Paris: Georges Carré.

Poincaré, H. (1892g). *Thermodynamique*, ed. by J. Blondin. Paris: Georges Carré.

Poincaré, H. (1893a). Discours au jubilé de M. Charles Hermite. In *Jubilé de M. Charles Hermite*, pp. 6–8. Paris: Gauthier-Villars.

Poincaré, H. (1893b). Le continu mathématique. *Revue de métaphysique et de morale 1*, 26–34. In *S&H*.

Poincaré, H. (1893c). Mécanisme et expérience. *Revue de métaphysique et de morale 1*, 534–537.

Poincaré, H. (1893d). Sur la propagation de l'électricité. *CR 117*, 1027–1032. In *Oeuvres 9*, 278–284.

Poincaré, H. (1893e). *Théorie des tourbillons*, ed. by M. Lamotte. Paris: Georges Carré.

Poincaré, H. (1894a). La lumière et l'électricité d'après Maxwell et Hertz. *Annuaire du Bureau des longitudes*, A1–A22. In *Oeuvres 10*, 557–569.

Poincaré, H. (1894b). *Les oscillations électriques*, ed. by C. Maurain. Paris: Carré et Naud.

Poincaré, H. (1894c). Réponse de M. H. Poincaré à M. Lechalas. *Revue de métaphysique et de morale 2*, 197–198.

Poincaré, H. (1894d). Sur la nature du raisonnement mathématique. *Revue de métaphysique et de morale 2*, 371–384. Modified in *S&H*.

Poincaré, H. (1894e). Sur les équations de la physique mathématique. *Rend. Palermo 8*, 57–156. In *Oeuvres 9*, 123–196.

Poincaré, H. (1895a). Analysis situs. *J. Éc. poly. 1*, 1–121. In *Oeuvres 6*, 193–288.

Poincaré, H. (1895b). À propos de la théorie de M. Larmor. *Éclairage électrique 3*, 5–13, 285–295. In *Oeuvres 9*, 369–394.

Poincaré, H. (1895c). *Capillarité*, ed. by J. Blondin. Paris: Georges Carré.

Poincaré, H. (1895d). L'espace et la géométrie. *Revue de métaphysique et de morale 3*, 631–646. Modified in *S&H*.

Poincaré, H. (1895e). *Théorie analytique de la propagation de la chaleur*, ed. by L. Rouyer and R. Baire. Paris: Carré et Naud.

Poincaré, H. (1896a). *Calcul des probabilités*, ed. by A. Quiquet. Paris: Gauthier-Villars.

Poincaré, H. (1896b). La méthode de Neumann et le problème de Dirichlet. *Acta 20*, 59–142. In *Oeuvres 9*, 202–272.

Poincaré, H. (1896c). Les rayons cathodiques et les rayons Röntgen. *Revue générale des sciences pures et appliquées 7*, 52–59. In *Oeuvres 10*, 570–583.

Poincaré, H. (1896d). Sur la vie et les travaux de F. Tisserand. *Revue générale des sciences pures et appliquées 7*, 1230–1233.

Poincaré, H. (1896e). Sur l'équilibre et les mouvements des mers. *J. de math. 105*, 57–102. In *Oeuvres 8*, 198–236.

Poincaré, H. (1897a). Les idées de Hertz sur la mécanique. *Revue générale des sciences pures et appliquées 8*, 734–743. In *Oeuvres 7*, 231–250.

Poincaré, H. (1897b). Sur la polarisation par diffraction. *Acta 20*, 313–355. In *Oeuvres 9*, 331–368.

Poincaré, H. (1897c). Sur les fonctions abéliennes. *CR 124*, 1407–1411. In *Oeuvres 4*, 469–472.

Poincaré, H. (1897d). Sur les rapports de l'analyse pure et de la physique mathématique. *Acta 21*, 331–341. In *Verhandlungen des ersten internationalen Mathematiker-Kongresses in Zürich*, ed. by F. Rudio, 81–90, and *VS*.

Poincaré, H. (1898a). La mesure du temps. *Revue de métaphysique et de morale 6*, 1–13.

Poincaré, H. (1898b). Les fonctions fuchsiennes et l'équation $\Delta u = e^u$. *J. de math. 4*, 137–230. In *Oeuvres 9*, 512–559.

Poincaré, H. (1898c). On the foundations of geometry. *Monist 9*, 1–43. Repr. in (Ewald 1996, 982–1012).

Poincaré, H. (1898d). Préface. In *Oeuvres de Laguerre*, ed. by C. Hermite, H. Poincaré, and E. Rouché, pp. 1:v–1:xv. Paris: Gauthier-Villars.

Poincaré, H. (1898e). Sur la stabilité du système solaire. *Annuaire du Bureau des longitudes*, B1–B16. In *Oeuvres 8*, 538–547.

Poincaré, H. (1898f). Sur les propriétés du potentiel et sur les fonctions abéliennes. *Acta 22*, 89–178. In *Oeuvres 4*, 162–243.

Poincaré, H. (1899a). *Cinématique et mécanismes, potentiel et mécanique des fluides*, ed. by A. Guillet (2nd ed.). Paris: Carré et Naud.

Poincaré, H. (1899b). Complément à l'analyse situs. *Rend. Palermo 13*, 285–343. In *Oeuvres 6*, 290–337.

Poincaré, H. (1899c). Des fondements de la géométrie; à propos d'un livre de M. Russell. *Revue de métaphysique et de morale 7*, 251–279.

Poincaré, H. (1899d). La logique et l'intuition dans la science mathématique et dans l'enseignement. *Enseignement mathématique 1*, 157–162. In *Oeuvres 11*, 129–133.

Poincaré, H. (1899e). *La théorie de Maxwell et les oscillations hertziennes*. Paris: Carré et Naud.

Poincaré, H. (1899f). Sur les groupes continus. *Transactions of the Cambridge Philosophical Society 18*, 220–255. In *Oeuvres 3*, 173–212.

Poincaré, H. (1899g). *Théorie du potentiel newtonien*, ed. by É. Le Roy and G. Vincent. Paris: Carré et Naud.

Poincaré, H. (1900a). Comptes rendus des séances du Congrès de philosophie, discussion. *Revue de métaphysique et de morale 8*, 556–561.

Poincaré, H. (1900b). La géodésie française. *Bulletin de la Société astronomique de France 14*, 513–521.

Poincaré, H. (1900c). La géodésie française (discours prononcé à la séance des cinq Académies le 25 octobre 1900). *Mémoires de l'Institut 20*, 13–25. In *S&M*.

Poincaré, H. (1900d). Les relations entre la physique expérimentale et la physique mathématique. *Revue générale des sciences pures et appliquées 11*, 1163–1175. Modified in *S&H*, chaps 9 and 10.

Poincaré, H. (1900e). Sur les principes de la géométrie; réponse à M. Russell. *Revue de métaphysique et de morale 8*, 73–86.

Poincaré, H. (1900f). Sur les rapports de la physique expérimentale et de la physique mathématique. In C.-E. Guillaume and L. Poincaré (Eds.), *Rapports présentés au Congrès international de physique, vol. 1*, pp. 1–29. Paris: Gauthier-Villars. Repr. as "Les hypothèses en physique" and "Les théories de la physique moderne" in *S&H*. English trans. *Monist 12*, (1902), 516–543.

Poincaré, H. (1901a). À propos des expériences de M. Crémieu. *Revue générale des sciences pures et appliquées 12*, 994–1007. In *S&H*, modified in 2nd ed.

Poincaré, H. (1901b). *Électricité et optique: La lumière et les théories électrodynamiques*. Paris: Carré et Naud.

Poincaré, H. (1901c). Quelques remarques sur les groupes continus. *Rend. Palermo 15*, 321–366. In *Oeuvres 3*, 213–260.

Poincaré, H. (1901d). Sur la stabilité de l'équilibre des figures piriformes affectées par une masse fluide en rotation. *Proceedings of the Royal Society of London 69*, 148–149. In *Oeuvres 7*, 159–160.

Poincaré, H. (1901e). Sur les principes de la mécanique. In *Bibliothèque du Congrès international de philosophie*, pp. 3: 457–494. Paris: Armand Colin. Modified in *S&H*.

Poincaré, H. (1901f). Sur les propriétés arithmétiques des courbes algébriques. *J. de math. 7*, 161–233. In *Oeuvres 5*, 483–548.

Poincaré, H. (1902a). Du rôle de l'intuition et de la logique en mathématiques. In E. Duporcq (Ed.), *Comptes rendus du IIe Congrès international des mathématiciens*, pp. 115–130. Paris: Gauthier-Villars. In *VS*.

Poincaré, H. (1902b). *La science et l'hypothèse*. Paris: Flammarion.

Poincaré, H. (1902c). Les fondements de la géométrie. *Journal des savants*, 252–271. English trans. *Bull. AMS 10*, 1903, 1–23.

Poincaré, H. (1902d). Notice sur la télégraphie sans fil. *Revue scientifique 17*, 65–73. In *Oeuvres 10*, 604–622.

Poincaré, H. (1902e). Notice sur la télégraphie sans fil. *Annuaire du Bureau des longitudes*, A1–A34.

Poincaré, H. (1902f). Rapport présenté au nom de la Commission chargée du contrôle scientifique des opérations géodésiques de l'equateur. *CR 134*(17), 965–972. In *Oeuvres 8*, 593–601.

Poincaré, H. (1902g). Sur certaines surfaces algébriques. Troisième complément à l'"analysis situs". *Bulletin de la Société mathématique de France 30*, 49–70. In *Oeuvres 6*, 373–392.

Poincaré, H. (1902h). Sur la valeur objective de la science. *Revue de métaphysique et de morale 10*, 263–293. Repr. with modifications, as "La science est-elle artificielle?" and "La science et la réalité" in *VS* 213–247 and 248–276.

Poincaré, H. (1902i). Sur les cycles des surfaces algébriques. Quatrième complément à l'analysis situs. *J. de math. 8*, 169–214. In *Oeuvres 6*, 397–434.

Poincaré, H. (1902j). Sur les fonctions abéliennes. *Acta 26*, 43–98. In *Oeuvres 4*, 473–526.

Poincaré, H. (1903a). L'espace et ses trois dimensions. *Revue de métaphysique et de morale 11*, 281–301, 407–429. Repr. in *VS*. English trans. "The notion of space" and "Space and its three dimensions".

Poincaré, H. (1903b). Rapport présenté au nom de la Commission chargée du contrôle scientifique des opérations géodésiques de l'équateur. *CR 136*(14), 861–871.

Poincaré, H. (1904a). Cinquième complément à l'analysis situs. *Rend. Palermo 18*, 45–110. In *Oeuvres 6*, 435–498.

Poincaré, H. (1904b). La terre tourne-t-elle? *Bulletin de la Société astronomique de France 18*, 216–217.

Poincaré, H. (1904c). Les définitions générales en mathématiques. *Enseignement mathématique 6*, 257–283. Modified in *S&M*.

Poincaré, H. (1904d). Les rayons N existent-ils ? Opinion de M. Poincaré. *Revue scientifique 2*, 682.

Poincaré, H. (1904e). L'état actuel et l'avenir de la physique mathématique. *Bulletin des sciences mathématiques 28*, 302–324. In *VS*.

Poincaré, H. (1904f). Rapport du prix Le Conte. *CR 139*, 1120–1122.

Poincaré, H. (1904g). Rapport présenté au nom de la Commission chargée du contrôle scientifique des opérations géodésiques de l'equateur. *CR 138*, 1013–1019.

Poincaré, H. (1905a). *La valeur de la science*. Paris: Flammarion.

Poincaré, H. (1905b). *Leçons de mécanique céleste*. Paris: Gauthier-Villars.

Poincaré, H. (1905c). Les mathématiques et la logique. *Revue de métaphysique et de morale 13*, 815–835. Modified in *S&M*.

Poincaré, H. (1905d). Rapport présenté au nom de la Commission chargée du contrôle scientifique des opérations géodésiques de l'equateur. *CR 140*, 998–1006.

Poincaré, H. (1905e). Sur la dynamique de l'électron. *CR 140*, 1504–1508. In *Oeuvres 9*, 489–493.

Poincaré, H. (1905f). Sur les invariants arithmétiques. *JfM 129*, 89–150. In *Oeuvres 5*, 203–265.

References 561

Poincaré, H. (1905g). Sur les lignes géodésiques des surfaces convexes. *Trans. AMS 6*, 237–274. In *Oeuvres 6*, 38–84.
Poincaré, H. (1905h). The principles of mathematical physics. In H. J. Rogers (Ed.), *Congress of Arts and Science, Universal Exposition St. Louis, vol. 1*, pp. 604–622. Boston: Houghton, Mifflin & Co. Repr. *Monist* 15, (1905), 1–24.
Poincaré, H. (1906a). À propos de la logistique. *Revue de métaphysique et de morale 14*, 866–868. Modified in *S&M*.
Poincaré, H. (1906b). *Cours d'astronomie générale*. Paris: École polytechnique.
Poincaré, H. (1906c). La fin de la matière. *Athenæum 4086*, 201–202.
Poincaré, H. (1906d). Le choix des faits. *Université de Paris 20*(13). In *S&M*. English trans. "The choice of facts," *Monist*, 19, (1909), 231–239.
Poincaré, H. (1906e). Les mathématiques et la logique. *Revue de métaphysique et de morale 14*, 294–317. In *S&M*.
Poincaré, H. (1906f). Les mathématiques et la logique (suite et fin). *Revue de métaphysique et de morale 14*, 17–34. In *S&M*.
Poincaré, H. (1906g). Sur la dynamique de l'électron. *Rend. Palermo 21*, 129–176. In *Oeuvres 9*, 494–550.
Poincaré, H. (1907a). La relativité de l'espace. *Année psychologique 13*, 1–17. In *S&M*.
Poincaré, H. (1907b). Le hasard. *Revue du mois 3*, 257–276. In *S&M*, 64–94. English trans. "Chance," *Monist*, 22, (1912), 31–52.
Poincaré, H. (1907c). Les fonctions analytiques de deux variables et la représentation conforme. *Rend. Palermo 23*, 185–220. In *Oeuvres 4*, 244–289.
Poincaré, H. (1907d). Rapport présenté au nom de la Commission chargée du contrôle scientifique des opérations géodésiques de l'equateur. *CR 145*, 366–370.
Poincaré, H. (1907e). Sur le récepteur téléphonique. *Éclairage électrique 50*, 221–224, 257–262, 329–338, 365–372, 401–404. In *Oeuvres 10*, 487–539.
Poincaré, H. (1908a). La dynamique de l'électron. *Revue générale des sciences pures et appliquées 19*, 386–402. In *S&M* as "La mécanique et le radium," "La mécanique et l'optique," and "La mécanique nouvelle et l'astronomie.".
Poincaré, H. (1908b). La dynamique de l'électron. In *S&M*, Bibliothèque de philosophie scientifique, pp. 215–272. Paris: Flammarion.
Poincaré, H. (1908c). L'avenir des mathématiques. *Revue générale des sciences pures et appliquées 19*, 930–939. Also in *Atti del IV Congresso internazionale dei matematici*, 1909, 167–182. Only partially in *S&M*, and in the English trans. "The Future of Mathematics," *Monist*, 20, (1910), 76–92 and *Science and Method*.
Poincaré, H. (1908d). L'invention mathématique. *Enseignement mathématique 10*, 357–371. In *S&M*. English trans. "Mathematical Creation," *Monist*, 20, (1910), 321–335.
Poincaré, H. (1908e). *Science et méthode*. Paris: Flammarion.

Poincaré, H. (1908f). *Thermodynamique,* ed. by J. Blondin (2nd ed.). Paris: Gauthier-Villars.

Poincaré, H. (1909a). *Conférences sur la télégraphie sans fil.* Paris: Éds. La lumière électrique.

Poincaré, H. (1909b). La logique de l'infini. *Revue de métaphysique et de morale 17,* 461–482. In *DP.*

Poincaré, H. (1909c). La mécanique nouvelle. *Bulletin mensuel de l'Association française pour l'avancement des sciences 38,* 39–48. Also in *Revue scientifique,* 12, 1909, 170–177.

Poincaré, H. (1909d). L'avenir des mathématiques. In *Atti del IV Congresso internazionale dei matematici, vol. 1,* pp. 167–182. Rome: Accademia dei Lincei. Also published in *Revue générale des sciences pures et appliquées,* 19, (1908), 930–939, and partially in *S&M.* English trans. "The Future of Mathematics," *Monist,* 20, (1910), 76–92 and *Science and Method,* 371–386.

Poincaré, H. (1909e). Réflexions sur les deux notes précédentes. *Acta 32,* 195–200.

Poincaré, H. (1910a). Anwendung der Integralgleichungen auf Hertzsche Wellen. In *Sechs Vorträge über ausgewählte Gegenstände aus der reinen Mathematik und mathematischen Physik,* Mathematische Vorlesungen an der Universität Göttingen, pp. 23–31. Leipzig/Berlin: Teubner. In *Oeuvres 10,* 78–88.

Poincaré, H. (1910b). Anwendung der Theorie der Integralgleichungen auf die Flutbewegung des Meeres. In *Sechs Vorträge über ausgewählte Gegenstände aus der reinen Mathematik und mathematischen Physik,* Mathematische Vorlesungen an der Universität Göttingen, pp. 12–19. Leipzig/Berlin: Teubner. In *Oeuvres 8,* 289–296.

Poincaré, H. (1910c). La mécanique nouvelle. In *Sechs Vorträge über ausgewählte Gegenstände aus der reinen Mathematik und mathematischen Physik,* Mathematische Vorlesungen an der Universität Göttingen, pp. 51–58. Leipzig/Berlin: Teubner.

Poincaré, H. (1910d). La morale et la science. *Foi et vie 13,* 323–329. In *DP 32–47.*

Poincaré, H. (1910e). Sur la diffraction des ondes hertziennes. *Rend. Palermo 29,* 169–259. In *Oeuvres 10,* 94–203.

Poincaré, H. (1910f). Sur la précession des corps déformables. *Bulletin astronomique 27,* 321–356. In *Oeuvres 8,* 481–514.

Poincaré, H. (1910g). Sur les courbes tracées sur les surfaces algébriques. *Annales scientifiques de l'École normale supérieure 27,* 55–108. In *Oeuvres 6,* 88–139.

Poincaré, H. (1910h). Über die Fredholmschen Gleichungen. In *Sechs Vorträge über ausgewählte Gegenstände aus der reinen Mathematik und mathematischen Physik,* Mathematische Vorlesungen an der Universität Göttingen, pp. 1–10. Leipzig/Berlin: Teubner. In *Oeuvres 3,* 547–554.

Poincaré, H. (1910i). Über die Reduktion der Abel'schen Integrale und die Theorie der Fuchs'schen Funktionen. In *Sechs Vorträge über ausgewählte Gegenstände aus der reinen Mathematik und mathematischen Physik,* Mathematische

Vorlesungen an der Universität Göttingen, pp. 33–41. Leipzig/Berlin: Teubner. In *Oeuvres 3*, 429–436.

Poincaré, H. (1910j). Über transfinite Zahlen. In *Sechs Vorträge über ausgewählte Gegenstände aus der reinen Mathematik und mathematischen Physik*, Mathematische Vorlesungen an der Universität Göttingen, pp. 33–41. Leipzig/Berlin: Teubner. In *Oeuvres 11*, 120–124.

Poincaré, H. (1911a). Leçons sur les hypothèses cosmogoniques, ed. by H. Vergne. Paris: Hermann.

Poincaré, H. (1911b). Préface. In *Jacques Lux, Histoire de deux revues françaises: La Revue bleue et la Revue scientifique*. Paris: Éditions des deux revues.

Poincaré, H. (1911c). Rapport sur le prix Bolyai, 18/10/1910. *Bulletin des sciences mathématiques 35*, 67–100.

Poincaré, H. (1911d). Sur la théorie des quanta. *CR 153*, 1103–1108. In *Oeuvres 9*, 620–625.

Poincaré, H. (1911e). Vue d'ensemble sur les hypothèses cosmogoniques. *Revue du mois 12*, 385–403.

Poincaré, H. (1912a). *Calcul des probabilités*, ed. by A. Quiquet (2nd ed.). Paris: Gauthier-Villars.

Poincaré, H. (1912b). Fonctions modulaires et fonctions fuchsiennes. *Annales de la Faculté des sciences de Toulouse 3*, 125–149. In *Oeuvres 2*, 592–618.

Poincaré, H. (1912c). La logique de l'infini. *Scientia (Rivista di scienza) 12*, 1–11. In *DP*. English trans. "Mathematics and logic".

Poincaré, H. (1912d). Les conceptions nouvelles de la matière. *Foi et vie 15*, 185–191. Repr. in *Le matérialisme actuel*, Paris, 1916.

Poincaré, H. (1912e). Les rapports de la matière et de l'éther. *Journal de physique et le radium 2*, 347–360. In *Journal de physique théorique et appliquée* (5) 2, 347–360, and *Les idées modernes sur la constitution de la matière*, 1913. Gauthier-Villars.

Poincaré, H. (1912f). L'espace et le temps. *Scientia (Rivista di scienza) 12*, 159–170. In *DP*.

Poincaré, H. (1912g). L'hypothèse des quanta. *Revue scientifique 50*, 225–232. In *DP*.

Poincaré, H. (1912h). Pourquoi l'espace a trois dimensions. *Revue de métaphysique et de morale 20*, 483–504. In *DP*.

Poincaré, H. (1912i). Sur la théorie des quanta. *Journal de physique et le radium 2*(1), 5–34. In *Oeuvres 9*, 626–653.

Poincaré, H. (1913). Preface. In *The Foundations of Science*, pp. 3–7. Lancaster PA: Science Press.

Poincaré, H. (1916). *Oeuvres de Henri Poincaré*, ed. by P. Appell and J. Drach, vol. 1. Paris: Gauthier-Villars.

Poincaré, H. (1921a). Analyse des travaux scientifiques de Henri Poincaré faite par lui-même. *Acta 38*, 1–135.

Poincaré, H. (1921b). Lettres à L. Fuchs (1880, 1881). *Acta 38*, 175–184. In *Oeuvres 11*, 13–25.

Poincaré, H. (1923). Correspondance d'Henri Poincaré et de Felix Klein (1881, 1882). *Acta 39*, 94–132. In *Oeuvres 11*, 26–65.

Poincaré, H. (1928). *Oeuvres de Henri Poincaré*, ed. by P. Appell and J. Drach, vol. 1. Paris: Gauthier-Villars.

Poincaré, H. (1950–1956). *Oeuvres de Henri Poincaré*, various eds. Paris: Gauthier-Villars.

Poincaré, H. (1953). Les limites de la loi de Newton. *Bulletin astronomique 17*(3), 121–269.

Poincaré, H. (1963). *Dernières pensées* (2nd ed.). Paris: Flammarion.

Poincaré, H. (1967). *New Methods of Celestial Mechanics*. Washington DC: NASA.

Poincaré, H. (1968). *La science et l'hypothèse*. Paris: Flammarion.

Poincaré, H. (1985). *Papers on Fuchsian Functions*, ed. and trans. by J. Stillwell. New York: Springer.

Poincaré, H. (1997). *Trois suppléments sur la découverte des fonctions fuchsiennes*, ed. by J. J. Gray and S. Walter. Berlin: Akademie.

Poincaré, H. (2001). *The Value of Science: Essential Writings of Henri Poincaré*. New York: Random House.

Poincaré, H. (2002). *L'opportunisme scientifique*, ed. by L. Rollet. Basel: Birkhäuser.

Poincaré, H. (2010). *Papers on Topology: Analysis Situs and its Five Supplements*, ed. and trans. by J. Stillwell. HMath.

Poincaré, H., G. Darboux, and P. Appell (1908). Rapport de MM. les experts Darboux, Appell et Poincaré. In *Affaire Dreyfus; La revision du procès de Rennes; Enquête de la chambre criminelle de la Cour de cassation, vol. 3*, pp. 500–600. Paris: Ligue française pour la défense des droits de l'homme et du citoyen.

Poincaré, H., M. Noether, and C. Segre (1908). Relazione del concorso internazionale per la "Medaglia Gucci". *Rend. Palermo 26*, 145–151.

Poincaré, H. and E. Picard (1883). Un théorème de Riemann relatif aux fonctions de *n* variables indépendantes admettant 2*n* systèmes de périodes. *CR 97*, 1284–1287. In *Oeuvres 4*, 307–310.

Poincaré, H. and A. Potier (1902). Sur les expériences de M. Crémieu et une objection de M. Wilson. *Éclairage électrique 31*, 83–93.

Poincaré, H. and F. K. Vreeland (1904). *Maxwell's Theory and Wireless Telegraphy*. New York: McGraw Publishing Co.

Walter, S., E. Bolmont, and A. Coret (Eds.) (2007). *La correspondance entre Henri Poincaré et les physiciens, chimistes et ingénieurs*. Basel: Birkhäuser.

OTHER AUTHORS

Ampère, A.-M. (1826). *Théorie des phénomènes électrodynamiques uniquement déduite de l'expérience*. Paris: Méquignon-Marvis.

Anantharaman, N. (2006/2010). On the existence of closed geodesics. In Charpentier, Ghys, and Lesne (2006/2010), 143–160.

Appell, P. (1890). Sur les fonctions de deux variables à plusieurs paires de périodes. *CR 110*, 181–183.

Appell, P. (1891). Sur les fonctions périodiques de deux variables. *J. de math. (4) 7*, 157–219.

Appell, P. (1925). *Henri Poincaré*. Paris: Plon-Nourrit et Cie.

Appell, P. (1956). Henri Poincaré, en mathématiques spéciales à Nancy. In Poincaré, *Oeuvres 11*, 139–145.

Arago, F. (1810). Mémoire sur la vitesse de la lumière, lu à la prémière classe de l'institut, le 10 décembre 1810. *CR 36*, 38–49.

Ashmore, M. (1993). The theatre of the blind: Starring a Promethean prankster, a phony phenomenon, a prism, a pocket, and a piece of wood. *Social Studies of Science 23*, 67–106.

Atten, M. (1998). La nomination de H. Poincaré à la chaire de physique mathématique et calcul des probabilités de la Sorbonne. *Cahiers 9*, 221–230.

Atti del IV Congresso internazionale dei matematici (1909). Accad. dei Lincei, Rome.

Babbitt, D. and J. Goodstein (2011). Federigo Enriques's quest to prove the "completeness theorem". *Notices AMS 58*, 240–249.

Baker, H. F. (1905). Alternants and continuous groups. *Proc. LMS (2) 3*, 24–47.

Barner, K. (1997). Paul Wolfskehl and the Wolfskehl Prize. *Notices AMS 44*, 1294–1303.

Barrow-Green, J. E. (1997). *Poincaré and the Three Body Problem*. HMath *11*.

Barrow-Green, J. E. (2011). An American goes to Europe: Three letters from Oswald Veblen to George Birkhoff in 1913/1914. *Mathematical Intelligencer 33*, 37–48.

Belhoste, B. (2003). *La formation d'une technocratie*. Paris: Belin.

Bellivier, A. (1956). *Henri Poincaré ou la vocation souveraine*. Paris: Gallimard.

Beltrami, E. (1868). Saggio di interpretazione della Geometria non-euclidea. *Giornale di Matematiche 6*, 284–312. French trans. by J. Hoüel, "Essai d'interprétation de la géométrie non-euclidienne," *Annales Ecole normale supérieure 6*, 251–288 (1869).

Bendixson, I. (1901). Sur les courbes définies par des équations différentielles. *Acta 24*, 1–88.

Berry, M. V. (1989). Uniform asymptotic smoothing of Stokes's discontinuities. *Proceedings of the Royal Society of London A, 422*, 7–21.

Bertrand, J. (1870). *Traité de calcul différentiel et de calcul intégral 2*. Gauthier-Villars.

Bertrand, J. (1889). *Calcul des probabilités*. Paris: Gauthier-Villars.

Bertrand, J. (1891). Review of Poincaré's *Électricité et optique*. *Journal des savants*, 742–748.

Bieberbach, L. (1921). Neuere Untersuchungen über Funktionen von komplexen Variabeln. *EMW II C 4*, 379–532.

Birkhoff, G. D. (1913). Proof of Poincaré's geometric theorem. *Trans. AMS 14*, 14–22. In *Collected Mathematical Papers 1*, 654–672, 3 vols. AMS 1950.

Boltzmann, L. (1891). *Vorlesungen über Maxwells Theorie der Elektricität und des Lichtes*. Leipzig.

Bolyai, Farkas (1832). *Tentamen juventutem studiosam in elementa matheosis purae, etc.* Maros-Vásérhely.

Bolyai, J. (1832). Appendix scientiam spatii absolute veram exhibens. In F. Bolyai (1832). French trans. by J. Houël, "La science absolue de l'espace," *Mémoires de la Société des sciences physiques et naturelles de Bordeaux 5*, 189–248 (1867). English trans. by G. B. Halsted, "Science Absolute of Space." Appendix in Bonola (1912).

Bonola, R. (1906). *La geometria non-euclidea*. English trans. by H. S. Carslaw, preface by F. Enriques, *History of Non-Euclidean Geometry*, Open Court, Chicago (1912). Dover repr. New York (1955).

Borel, E. (1901). *Leçons sur séries divergentes*. Gauthier-Villars.

Bottazzini, U. and J. J. Gray (2012). *Hidden Harmony—Geometric Fantasies: The Rise of Complex Function Theory*. Springer (forthcoming).

Boutroux, A. (2012). *Vingt ans de ma vie, simple vérité …; La jeunesse de Henri Poincaré racontée par sa soeur*, ed. by L. Rollet. Hermann.

Boutroux, É. (1874/1895). *De la contingence des lois de la nature. Thèse, etc.* Paris.

Boutroux, É. (1908). *Science et religion dans la philosophie contemporaine*. Paris: Flammarion.

Boutroux, P. (1908). Leçons sur les fonctions définies par les équations différentielles du premier ordre professées au Collège de France. Avec une note de *P. Painlevé*. Paris: Gauthier-Villars.

Boutroux, P. (1914/1921). Lettre de M. Pierre Boutroux à M. Mittag-Leffler. *Acta 38*, 197–201. Repr. in Poincaré, *Oeuvres 11*, 146–151.

Brewster, Sir D. (1855). *Memoirs of the Life, Writings, and Discoveries of Sir Isaac Newton*, 2 vols. Constable and Co.

Brigaglia, A. and G. Masotto (1982). *Il circolo mathematico di Palermo*. Edizioni Dedalo.

Brill, A. and M. Noether (1874). Ueber die algebraischen Functionen und ihre Anwendung in der Geometrie. *Math. Ann. 7*, 269–310.

Briot, Ch. and J.-C. Bouquet (1856a). Étude des fonctions d'une variable imaginaire. *J. Éc. poly. 21*, 85–132.

Briot, Ch. and J.-C. Bouquet (1856b). Recherches sur les propriétes des fonctions définies par des équations différentielles. *J. Éc. poly. 21*, 133–198.

Briot, Ch. and J.-C. Bouquet (1856c). Mémoire sur l'intégration des équations différentielles au moyen des fonctions elliptiques. *J. Éc. poly. 21*, 199–254.

Briot, Ch. and J.-C. Bouquet (1859). *Théorie des fonctions doublement périodiques et, en particulier, des fonctions elliptiques*. Paris: Mallet-Bachelier.

Brodén, T. (1905). *Bemerkungen über die Uniformisierung analytischer Funktionen*. Lund, Berling.

Brown, H. R., W. Myrvold, and J. Uffink (2009). Boltzmann's *H*-theorem, its discontents, and the birth of statistical mechanics. *Studies in History and*

Philosophy of Modern Physics 40, 174–191. Available at http://www.phys
.uu.nl/igg/jos/publications/harveywayneshpmp.pdf.

Bruns, H. (1887). Ueber die Integrale des Vielkörper-Problems. *Acta 11*, 25–96.

Brush, S. G. (1966). *Kinetic Theory*, vol. 2, *Irreversible Processes*. Pergamon Press.

Brush, S. G. (1976). *The Kind of Motion We Call Heat; A History of the Kinetic Theory of Gases in the 19th Century*, vol. 2. Studies in Statistical Mechanics, vol. 6. North Holland.

Buchwald, J. Z. (1985). *From Maxwell to Microphysics: Aspects of Electromagnetic Theory in the Last Quarter of the Nineteenth Century*. University of Chicago Press.

Buchwald, J. Z. (1994). *The Creation of Scientific Effects: Heinrich Hertz and Electric Waves*. University of Chicago Press.

Cairns, S. S. (1935). Triangulation of the manifold of class one. *Bull. AMS 41*, 549–552.

Calinon, A. (1886). Étude critique sur la mécanique. *Bulletin de la Société des sciences de Nancy (2) 7*, 87–180.

Campbell, J. E. (1901). Proof of the third fundamental theorem in Lie's theory of continuous groups. *Proc. LMS 33*, 285–294.

Campbell, J. E. (1903). *Introductory Treatise on Lie's Theory of Finite Continuous Transformation Groups*. Oxford: Clarendon Press.

Cartan, E. (1928). Sur la stabilité ordinaire des ellipsoïdes de Jacobi. *Proceedings of the International Congress of Mathematicians, Toronto 2*, 9–17.

Cartan, H. and S. Eilenberg (1956). *Homological Algebra*. Princeton University Press.

Cartwright, D. E. (1999). *Tides: A Scientific History*. Cambridge University Press.

Castelnuovo G. and F. Enriques (1907). Sur quelques résultats nouveaux dans la théorie des surfaces algébriques. In Picard and Simart (1907), 485–522.

Cauchy, A. L. (1843). Sur l'emploi légitime des séries divergentes. *CR 17*, 370–376. In *Oeuvres* (ser. 1) *8*, 1–24.

Cayley, A. (1859). On contour and slope lines. *Phil. Mag. 18*, 264–268. In *Collected Papers 4*, 108–111 (1891).

Cayley A. (1871). On the deficiency of certain surfaces. *Math. Ann. 3*, 526–529. In *Collected Papers 8*, 394–397 (1895).

Cerveau, D. (2006/2010). Singular points of differential equations: On a theorem of Poincaré. In Charpentier, Ghys, and Lesne (2006/2010), 99–112.

Chandrasekhar, S. (1989). The equilibrium and the stability of the Riemann ellipsoids, 2. In *Selected Papers 4, Plasma Physics, Hydrodynamic and Hydromagnetic Stability, and Applications of the Tensor-Virial Theorem*, 842–877. University of Chicago Press.

Charpentier, É., É. Ghys, and A. Lesne (eds) (2006). *L'héritage scientifique de Poincaré*. Belin. English trans. *The Scientific Legacy of Poincaré*, HMath 36 (2010).

Chasles, M. (1852) *Traité de géométrie supérieure*. Paris.

Chorlay, R. (2010). From problems to structures: The Cousin problems and the emergence of the sheaf concept. *AHES 64*, 1–73.

Clebsch, R.F.A. and P. Gordan (1866). *Theorie der Abelschen Functionen*. Leipzig: Teubner.

Clemens, C. H. (1980). *A Scrapbook of Complex Curve Theory*. Plenum PR.

Coffa, J. Alberto (1991). *The Semantic Tradition From Kant to Carnap: To the Vienna Station*, ed. by Linda Wessels. Cambridge University Press.

Comptes rendus du IIe Congrès international des mathématiciens, ed. by E. Duporcq. (1902). Paris: Gauthier-Villars.

Congress of Arts and Science, Universal Exposition St. Louis, 7 vols, ed. by H. J. Rogers. (1905). Boston: Houghton Mifflin.

Cooke, R. (1984). *The Mathematics of Sonya Kovalevskaya*. Springer.

Corry, L. (2004). *David Hilbert and the Axiomatization of Physics (1898–1918): From* Grundlagen der Geometrie *to* Grundlagen der Physik. Kluwer.

Courant, R. (1912). Über die Anwendung des Dirichlets Prinzipes auf die Probleme der konformen Abbildung. *Math. Ann. 71*, 145–183.

Cousin, P. (1895). Sur les fonctions de n variables complexes. *Acta 19*, 1–62.

Couturat, L. (1896). Études sur l'espace et le temps de MM. Lechalas, Poincaré, Delboeuf, Bergson, L. Weber et Évelin. *Revue de métaphysique et de morale 4*, 646–669.

Couturat, L. (1905). *Principes des mathématiques*. Paris: Alcan.

Couturat, L. (1912). For logistics; reply to M. *Poincaré*. *Monist 22*, 483–523.

Coxeter, H.S.M (1961). *Introduction to Geometry*. Wiley.

Crawford, E. (1985). *The Beginning of the Nobel Institution. The Science Prizes, 1901–1915*. Cambridge University Press.

Crowe, M. J. (1967). *A History of Vector Analysis: The Evolution of the Idea of a Vectorial System*. University of Notre Dame Press.

Czuber, E. (1884). *Geometrische Wahrscheinlichkeiten und Mittelwerte*. Teubner.

Damour, T. (2004). Poincaré, relativity, billiards and symmetry. In Gaspard, Henneaux, and Lambert (2004), 149–170.

Darboux, G. (1887). *Leçons sur la théorie des surfaces* (2nd ed. 1914). Paris: Gauthier-Villars.

Darboux, G. (1916). Éloge historique d'Henri Poincaré. In Poincaré, *Oeuvres 2*, vii–lxxi.

Darrigol, O. (2000). *Electrodynamics from Ampère to Einstein*. Oxford University Press.

Darrigol, O. (2004). The mystery of the Einstein–Poincaré connection. *Isis 95*, 614–626.

Darrigol, O. (2005). *Worlds of Flow. A History of Hydrodynamics from the Bernouillis to Prandtl*. Oxford University Press.

Darrigol, O. (2006). The genesis of the theory of relativity. *Einstein, 1905–2005*, ed. by T. Damour, O. Darrigol, B. Duplantier, and V. Rivasseau. Progress in Mathematical Physics *47*, 1–32. Birkhäuser.

Darwin, G. H. (1901). The stability of the pear-shaped figure of equilibrium of a rotating mass of liquid. *Vierteljahrsschr. Astr. Ges. 37*, 202–207.

Darwin, G. H. (1902). On the pear-shaped figure of equilibrium of a rotating mass of liquid. *Phil. Trans. Roy. Soc. A*, *198*, 301–331.

Davis, P. J. and D. Mumford (2008). Henri's crystal ball. *Notices AMS 55.4*, 458–466.

Debye, P. (1909). Näherungsformeln für die Zylinderfunktionen für grosse Werte des Arguments und unbeschränkt veränderliche Werte des Index. *Math. Ann. 67*, 535–568.

Dedekind, R. (1877). Sur la théorie des nombres entiers algébriques. *Bulletin des sciences mathématiques (2) 1*, 17–41, 69–92, 114–164, 207–248. Also separate ed. (1877). Paris: Gauthier-Villars.

Dedekind, R. (1888). *Was sind und was sollen die Zahlen?* pp. 335–392, Braunschweig, Vieweg & Sohn. In *Gesammelte mathematische Werke 3*, ed. by R. Fricke, E. Noether, O. Ore (1930). Braunschweig, Vieweg & Sohn.

Della Dora, V. (2010). Making mobile knowledges: The educational cruises of the *Revue générale des sciences pures et appliquées* 1897–1914. *Isis 101*, 467–500.

Diacu, F. and P. Holmes (1996). *Celestial Encounters: The Origins of Chaos and Stability*. Princeton University Press.

Dieudonné, J. (1981). *History of Functional Analysis*. North-Holland.

Dieudonné, J. (1989). *A History of Algebraic and Differential Topology, 1900–1960*. Boston: Birkhäuser.

DiSalle, R. (2002). Newton's philosophical analysis of space and time. *The Cambridge Companion to Newton*, ed. by I. B. Cohen and G. E. Smith. Cambridge University Press.

Domar, Y. (1982). On the founding of *Acta Mathematica*. *Acta 148*, 3–8.

Dugas. R. (1988). *A History of Mechanics*. Dover.

Duhamel, J.M.C. (1841). *Cours d'analyse de l'École polytechnique*. Paris.

Duhem, P. (1954). *The Aim and Structure of Physical Theory*. Princeton University Press.

Dulac, H. (1908). Détermination et intégration d'une certaine classe d'équations différentielles ayant pour point singulier un centre. *Bulletin des sciences mathématiques (2) 32*, 230–252.

Einstein, A. (1905). Zur Elektrodynamik bewegter Körper. *Annalen der Physik 17*. English trans. "On the electrodynamics of moving bodies." In *The Collected Papers of Albert Einstein, vol. 2, The Swiss Years: Writings, 1900–1909*, ed. by J. Stachel, D. C. Cassidy, J. Renn, and R. Schulmann, 275–310. Princeton University Press.

Einstein, A. (1921). Geometrie und Erfahrung. *Sitzungsberichte der Königlich Preussischen Akademie der Wissenschaften*, 123–130. Berlin. English trans. "Geometry and experience." In *The Collected Papers of Albert Einstein, vol. 7, The Berlin Years: Writings, 1918–1921*, 208–222. Princeton University Press.

Elkana, Y. (1974). *The Discovery of the Conservation of Energy*. Hutchinson.

Enriques, F. (1898). *Lezioni di geometria proiettiva*. Bologna: Zanichelli.

Enriques, F. (1905). Sulle superficie algebriche di genere geometrico zero. *Rend. Palermo 20* 1–36. In *Memorie scelte*, 2 vols, vol. 2, 169–204. Bologna: Zanichelli.

Enriques, F. (1906). *Problemi della scienza*. Bologna: Zanichelli. English trans. by K. Royce *Problems of Science*, Chicago, Open Court (1914).

Enriques, F. (1907). Prinzipien der Geometrie. *EMW III.1*, 1–129.

Enriques, F. and G. Castelnuovo (1914). Die algebraischen Flächen vom Gesichtspunkte der birationalen Transformationen aus. In *EMW III.2.1 C*, 674–768.

Epple, M. (1996). Mathematical inventions: Poincaré on a "Wittgensteinian" topic. In *Henri Poincaré: Science et philosphie*, 559–576, ed. by J. L. Greffe, G. Heinzmann, and K. Lorenz. Akademie, Blanchard.

Epple, M. (1999). *Die Entstehung der Knotentheorie*. Vieweg.

Ewald, W. B. (1996). *From Kant to Hilbert: A Source Book in the Foundations of Mathematics*. Oxford University Press.

Ferreirós, J. (2007). *Labyrinth of Thought*. Birkhäuser, Science Networks, Basel. (1st ed. 1999).

Feynman, R. P. (1966). *The Feynman Lectures on Physics*, 3 vols. Addison-Wesley.

Fitzgerald, G. F. (1891). M. Poincaré et Maxwell. *Nature 45*, 532–533.

Folina, J. (1992). *Poincaré and Philosophy of Mathematics*. Macmillan.

Fraenkel, A. (1923). *Einleitung in die Mengenlehre*, 2nd ed. Springer.

Fredholm, I. (1899). Sur une classe d'équations aux dérivées partielles. *CR 129*, 32–34.

Fredholm, I. (1900). Sur une nouvelle méthode pour la résolution du problème de Dirichlet. *Öfversigt Kongl. Vetenskaps-Akad. Förhandlingar 57*, 39–46. In *Oeuvres complètes* 61–69.

Fredholm, I. (1903). Sur une classe d'équations fonctionnelles. *Acta 27*, 365–390.

Freudenthal, H. (1954). Poincaré et les fonctions automorphes. In Poincaré, *Oeuvres 11*, 212–219.

Fricke, R. and C. F. Klein (1897, 1912). *Vorlesungen über die Theorie der automorphen Functionen*, 2 vols. Leipzig: Teubner.

Friedman, M. (1996). Poincaré's conventionalism and the logical positivists. In *Henri Poincaré: Science et philosophie*, 333–344, ed. by J. L. Greffe, G. Heinzmann, and K. Lorenz. Akademie, Blanchard.

Fuchs, L. I. (1866). Zur Theorie der linearen Differentialgleichungen mit veränderlichen Coefficienten. *JfM 66*, 121–160. In *Werke 1*, 159–204.

Fuchs, L. I. (1880). Über eine Klasse von Functionen mehrerer Variabeln [etc.] *JfM 89*, 151–169. In *Gesammelte mathematische Werke 2*, 191–212.

Fuchs, L. I. (1882). Über Functionen, welche durch lineare Substitutionen unverändert bleiben. *Göttingen Nachr.*, 81–84. In *Gesammelte mathematische Werke 2*, 285–288.

Fuchs, L. I. (1884). Ueber Differentialgleichungen, deren Integrale feste Verzweigungspunkte besitzen. *Sitzungsberichte der K. Preuss. Akademie* der Wissenschaften 699–710. In Gesammelte mathematische Werke 2, 355–368.

Galison, P. (2003). Einstein's Clocks, Poincaré's Maps: Empires of Time. London: Sceptre.

Gario, P. (1989). Resolution of singularities of surfaces by P. Del Pezzo. AHES 40.3, 247–274.

Gaspard, P., M. Henneaux, and F. Lambert (eds) (2004). Solvay Workshops and Symposia, vol. 2—Symposium Henri Poincaré.

Gauss, C. F. (1801). Disquisitiones Arithmeticae. G. Fleischer, Leipzig. In Werke 1, K. Gesellschaft der Wissenschaften. Göttingen.

Gibbs, J. W. (1876–78). On the equilibrium of heterogeneous substances. Transactions of the Connecticut Academy of Sciences 3, 1876, 108–248, and 1878, 343–524. In Collected Works 1, 55–353 (1948).

Gilain, Chr. (1991). La théorie qualitative de Poincaré et le problème de l'intégration des équations différentielles. In Gispert (1991), 215–242.

Ginoux, J.-M. (2011). Analyse mathématique des phenomènes oscillatoires non linéaires: Le carrefour français (1880–1940). Thèse de doctorat, Université Pierre et Marie Curie (Paris VI).

Ginoux, J.-M. and Chr. Gérini (2012). Henri Poincaré: Une biographie au(x) quotidien(s). Ellipses.

Ginoux, J.-M. and R. Lozi (to appear). Blondel et les oscillations auto-entretenues. AHES.

Gispert. H. (ed.) (1991). La France mathématique. Cahiers d'histoire et de philosophie des sciences 34. Société Mathématique de France.

Goldfarb, W. (1988). Poincaré against the logicists. In History and Philosophy of Modern Mathematics, ed. by W. Aspray and P. Kitcher, pp. 61–81. University of Minnesota Press.

Goldstein, C. (2007). The Hermitian form of reading the Disquisitiones. In Goldstein, Schappacher, and Schwermer (2007), 377–410.

Goldstein, C. (2011). Un arithméticien contre l'arithmétisation: Les principes de Charles Hermite. In Justifier en mathématiques, ed. by D. Flament and P. Nabonnand, 129–165. Éditions MSH.

Goldstein, C., N. Schappacher, and J. Schwermer (2007). The Shaping of Arithmetic After C. F. Gauss's Disquisitiones Arithmeticae. Springer.

Golé, C. and G. R. Hall (1992). Poincaré's proof of Poincaré's last geometric theorem. Twist mappings and their applications. IMA Math. Appl. 44, 135–151. Springer.

Gordon C. McA. (1999). 3-dimensional topology up to 1960. In History of Topology, 449–490. North-Holland.

Gray, J. J. (1991). Did Poincaré say "Set theory is a disease?" Mathematical Intelligencer, 13, 19–22.

Gray, J. J. (2000). Linear Differential Equations and Group Theory from Riemann to Poincaré, (2nd ed). Boston: Birkhäuser.

Gray, J. J. (2008). *Plato's Ghost: The Modernist Transformation of Mathematics*. Princeton University Press.

Gray, J. J. (2010). *Worlds Out of Nothing; A Course on the History of Geometry in the 19th Century*, (2nd rev. ed). Springer.

Gray, J. J. (2012, to appear). Poincaré replies to Hilbert: On the future of mathematics ca. 1908. *Mathematical Intelligencer*.

Griffin, N. (1991). *Russell's Idealist Apprenticeship*. Oxford: Clarendon.

Griffiths, P. and J. Harris (1978). *Principles of Algebraic Geometry*. Wiley.

Grivel, P.-P. (2006/2010). Poincaré, and Lie's third theorem. In Charpentier, Ghys, and Lesne (2006/2010).

Gronwall, T. H. (1917). On the expressibility of a uniform function of several complex variables as the quotient of two functions of entire character. *Trans. AMS 18*, 50–64.

Hadamard, J. (1893). Étude sur les propriétés des fonctions entières et en particulier d'une function considerée par Riemann (Mémoire couronné par l'Académie, Grand Prix des sciences mathématiques). *J. de math. (4) 9*, 171–215. In *Oeuvres 1*, 103–147.

Hadamard, J. (1897). Sur certaines propriétés des trajectoires en dynamique. *J. de math. (5) 4*, 331–387. In *Oeuvres 4*, 1749–1805.

Hadamard, J. (1898). Les surfaces à courbures opposées et leurs lignes géodésiques. *J. de math. (5) 4*, 27–73. In *Oeuvres 2*, 729–780.

Hadamard, J. (1906). La logistique et la notion de nombre entier. *Revue générale des sciences 17*, 906–909. In *Oeuvres 4*, 2145–2155.

Hadamard, J. (1921). L'oeuvre mathématique de H. Poincaré. *Acta 38*, 203–287. In *Oeuvres 4*, 1921–2005 and *Oeuvres 11*, 152–242.

Hadamard, J. (1968). *Oeuvres de Jacques Hadamard*, 4 vols. CNRS, Paris.

Halsted, G. B. (1897). Review of *The Foundations of Geometry*, by B.A.W. Russell. *Science 6*, no. 143, 487–491.

Halsted, G. B. (1904). *Rational Geometry: A Textbook for the Science of Space Based on Hilbert's Foundations*. John Wiley & Sons.

Halsted, G. B. (1905). The Bolyai Prize. *Science 22*, no. 557, 270–271.

Halsted, G. B. (1906). Report on the Bolyai Prize. *Science 23*, no. 594, 793–794.

Harman, P. M. (1992). Maxwell and Saturn's rings. In Harman and Shapiro (1992), 477–502.

Harman P. M. and A. E. Shapiro (eds) (1992). *The Investigation of Difficult Things: Essays on Newton and the History of the Exact Sciences in Honour of D. T. Whiteside*. Cambridge University Press.

Harnack, A. (1887). *Grundlagen der Theorie des logarithmischen Potentiales [etc.]* Leipzig: Teubner.

Hartogs, F. (1905). *Zur Theorie der analytischen Funktionen mehrerer unabhängiger Veränderlichen [etc]* Habilitationsschrift, München. Repr. *Math. Ann. 62* (1906) 1–88.

Hartogs, F. (1907). Über neuere Untersuchungen auf dem Gebiete der analytischen Funktionen mehrerer Variabeln. *JDMV 16*, 223–240.

Hausdorff, F. (1906). Die symbolische Exponentialformel in der Gruppentheorie. *Berichte der köngl. Sachsischen Gesellschaft der Wiss. zu Leipzig* (Math.-Phys. Klasse) *63*, 19–48. In *Gesammelte Werke 4*, 429–466 (2001).

Hawkins, T. (1991). Jacobi and the birth of Lie's theory of groups. *AHES 42*, 187–278.

Hawkins, T. (2000). *The Emergence of the Theory of Lie Groups: An Essay in the History of Mathematics, 1869–1926*. Springer.

Heegaard, P. (1898). *Forstudier til en topologisk teori for algebraiske Sammenhäng* Copenhagen, det. Nordiske Forlag Ernst Bojesen.

Heidelberger, M. (2009). Contingent laws of nature in Émile Boutroux. In Heidelberger and Schieman (2009), 99–144.

Heidelberger, M., and G. Schieman (eds) (2009). *The Significance of the Hypothetical in the Natural Sciences*. De Gruyter.

Heijenoort, J. van (1967). *From Frege to Gödel: A Source Book in Mathematical Logic 1879–1931*, Harvard University Press.

Heinzmann, G. (1995). *Zwischen Objektkonstruktion und Strukturanalyse: zur Philosophie der Mathematik bei Jules Henri Poincaré*. Vandenhoeck & Ruprecht.

Heinzmann, G. (2008). Poincaré wittgensteinien? In E. Rigal (ed.), *Wittgenstein. État des lieux*, 274–289. Paris: Vrin.

Heinzmann, G. and P. Nabonnand (2008). Poincaré: Intuitionism, intuition, and convention. In M. van Atten, P. Boldini, M. Bourdeau, and G. Heinzmann (eds), *One Hundred Years of Intuitionism (1907–2007)*. Archives Henri Poincaré, Nancy, 163–177.

Helmholtz, H. von (1868). Über die thatsächlichen Grundlagen der Geometrie. *Verhandlungen des naturhistorisch-medzinischen Vereins 4*, 197–202 and *5*, 31–32.

Helmholtz, H. von (1870). Über die Bewegungsgleichungen der Elektricität für ruhende Körper. *JfM 72*, 57–129.

Helmholtz, H. von (1886). Über die physikalische Bedeutung des Princips der kleinsten Wirkung. *JfM 100*, 137–166, 213–222.

Hermite, C. (1984, 1985, 1989). Lettres de Charles Hermite à Gösta Mittag-Leffler (1874–1883). *Cahiers du séminaire d'histoire des mathématiques 5*, 49–285; *6*, 79–217; *10*, 1–82.

Hertz, H. (1894). *Die Prinzipien der Mechanik in neuen Zusammenhange dargestellt.* Leipzig: Barth.

Hilbert, D. (1897). Die Theorie der algebraischen Zahlkörper (Zahlbericht). *JDMV 4*, 175–546. In *Gesammelte Abhandlungen 1*, 63–363. English trans. by I. Adamson *The Theory of Algebraic Number Fields*, with an introduction by F. Lemmermeyer and N. Schappacher (1998). Springer.

Hilbert, D. (1899). *Festschrift zur Feier der Enthüllung des Gauss–Weber–Denkmals in Göttingen [etc.] Grundlagen der Geometrie*. Leipzig.

Hilbert, D. (1901). Mathematische Probleme. *Archiv für Mathematik und Physik 1*, 44–63, 213–237. In *Gesammelte Abhandlungen 3*, 290–329.

Hilbert, D. (1905a). Über die Grundlagen der Logik und der Arithmetik. *Verhandlungen des dritten internationalen Mathematiker-Kongresses in Heidelberg*, 174–185. Leipzig: Teubner.

Hilbert, D. (1905b). Zur Variationsrechnung. *Göttingen Nachr.*, 159–180. In *Gesammelte Abhandlungen 3*, 38–55.

Hilbert, D. (1925). Über das Unendliche. *JDMV 36.1*, 201–215. English trans. "On the infinite." In van Heijenoort (1967), 367–392.

Hilbert, D. (1927). Die Grundlagen der Mathematik. In *Die Grundlagen der Mathematik*, 1–21. Leipzig: Teubner. English trans. "The foundations of mathematics." In van Heijenoort (1967), 464–479.

Hilbert, D. (2009). *David Hilbert's Lectures on the Foundations of Mathematics and Physics, 1891–1933*, ed. by T. Sauer and U. Majer. Springer.

Hilbert, D. and A. Hurwitz (1891). Ueber die diophantischen Gleichungen vom Geschlecht Null. *Acta Mathematica 14*, 217–224.

Hill, G. W. (1877). On the part of the motion of the lunar perigee which is a function of the mean motions of the sun and moon. Cambridge, Mass.: John Wilson & Son. Repr. in *Acta 8*, 1–36. In Hill (1905), 243–270.

Hill, G. W. (1878). Researches in the lunar theory. *Amer. J. Math. 1*, 5–27, 129–148, 245–251. In Hill (1905), 284–335.

Hill, G. W. (1905). *Collected Mathematical Works 1*. Washington: Carnegie Institute.

Hopf, E. (1930). Zwei Sätze über den wahrscheinlichen Verlauf der Bewegungen dynamischer System. *Math. Ann. 103*, 710–719.

Houzel, Chr. (2002). *La géométrie algébrique*. Paris: Blanchard.

Howard, D. (1998). Le Roy, Édouard Louis Emmanuel Julien. In E. Craig (ed.), *Routledge Encyclopedia of Philosophy*. London: Routledge.

Hunt, B. J. (1991). *The Maxwellians*. Cornell University Press.

Hurwitz, A. (1883). Beweis des Satzes, dass eine einwertige Funktion beliebig vieler Variabeln [etc.] *JfM 95*, 201–206. In *Math. Werke 1*, 147–152.

Indorato, L. and G. Masotto (1989). Poincaré's role in the Crémieu–Pender controversy over electric convection. *Annals of Science 46*, 117–163.

Israel, G. and L. Nurzia (1984). The Poincaré–Volterra theorem: A significant event in the history of the theory of analytic functions. *Historia Mathematica 11*, 161–192.

Jacobi, C.G.J. (1891). *Gesammelte Werke 7*. Königlich Preussischen Akademie der Wissenschaften, ed. by K. Weierstrass.

Jacobi, C.G.J. (1999). *Vorlesungen über analytischen Mechanik*, 2nd ed., ed. by H. Pulte. Vieweg.

Jeans, J. H. (1916–17). On the instability of the pear-shaped figure of equilibrium of a rotating mass of liquid. *Phil. Trans. Roy. Soc. A, 217*, 1–34.

Johansson, S. (1905). Über die Uniformisierung Riemannscher Flächen mit endlicher Anzahl Windungspunkte. *Acta Soc. Sci. Fennicae 33*, n. 7.

Johansson, S. (1906). Ein Satz über die konformen Abbildung einfach zusammenhängender Riemannscher Flächen auf den Einheitskreis. *Math. Ann. 62*, 177–183.

Jordan, C. (1866). Des contours tracées sur les surfaces. *Journal de mathématiques 2*, 11, 110-130. In *Oeuvres 4*, 91–110.

Jordan, C. (1870). *Traité des substitutions et des équations algébriques*. Paris: Gauthier-Villars.

Jordan, C. (1893). *Cours d'analyse de l'École polytechnique*. 2nd ed., vol. 1. Gauthier-Villars.

Jourdain, P.E.B. (1912). Couturat's "For logistics." Introductory note, M. Poincaré and M. Couturat. *Monist 22*, 481–523.

Kant, I. (1929). *Immanuel Kant's Critique of Pure Reason*, trans. by Norman Kemp Smith, 2nd ed., repr. 1970.

Katz, V. (1999). Differential forms. In *History of Topology*, ed. by I. M. James, 111-122. North-Holland.

Kleiman, S. (2004). What is Abel's theorem anyway? In O. A. Laudal and R. Piene (eds), *The Legacy of Niels Henrik Abel: The Abel Bicentennial, Oslo, 2002*, 395–440.

Klein, C. F. (1872). *Vergleichende Betrachtungen über neuere geometrische Forschungen (Erlanger Programm)*, Deichert, Erlangen. In *Gesammelte mathematische Abhandlungen 1*, 460–497 (1921). Springer.

Klein, C. F. (1876). Über lineare Differentialgleichungen, I. *Math. Ann. 11*, 115–118. In *Gesammelte mathematische Abhandlungen 2*, 302–306 (1922). Springer.

Klein, C. F. (1882a). *Über Riemanns Theorie der algebraischen Funktionen und ihrer Integrale*. Leipzig: Teubner. In *Gesammelte mathematische Abhandlungen 3*, 499–573 (1923). Springer. English trans. *On Riemann's Theory of Algebraic Functions and Their Integrals*. Cambridge: Macmillan and Bowes. Dover repr. New York (1963).

Klein, C. F. (1882b). Über eindeutige Funktionen mit linearen Transformationen in sich. *Math. Ann. 19*, 565–568. In *Gesammelte mathematische Abhandlungen 3*, 622–626.

Klein, C. F. (1884). *Vorlesungen über das Ikosaeder und die Auflösung der Gleichungen vom fünften Grade*. Leipzig.

Klein, C. F. (1921–1923). *Gesammelte mathematische Abhandlungen*, ed. by R. Fricke, A. M. Ostrowski, H. Vermeil, and E. Bessel-Hagen. 3 vols. Berlin: Springer.

Klein, C. F. (1923). Autobiography. *Göttingen Mitteilungen des Universitätsbundes Göttingen*, 5.1.

Klein, C. F. (1926–27). *Vorlesungen über die Entwicklung der Mathematik im 19. Jahrhundert*, 2 vols, ed. by R. Courant and O. Neugebaue. Springer, Chelsea reprint, 1948.

Kline, M. (1972). *Mathematical Thought from Ancient to Modern Times*. Oxford University Press.

Koebe, P. (1907). Über die Uniformisierung beliebiger analytischer Kurven (Zweite Mitteilung). *Göttingen Nachr.*, 633–669.

Koebe, P. (1908). Über die Uniformisierung beliebiger analytischer Kurven (Dritte Mitteilung). *Göttingen Nachr.*, 337–358.

Koebe, P. (1910). Über die *Hilbert*sche Uniformisierungsmethode. *Göttingen Nachr.*, 59–74.

Kovalevskaya, S. (1875). Zur Theorie der partiellen Differentialgleichungen. *JfM 80*, 1–32.

Kovalevskaya, S. (1885). Zusätze und Bemerkungen zu Laplaces Untersuchungen über die Gestalt der Saturnringe. *Astronomische Nachrichten 111*, 37–48.

Kowalski, E. (2006/2010). Poincaré and analytic number theory. In Charpentier, Ghys, and Lesne (2006/2010), 73–86.

Kronecker, L. (1869). Ueber Systeme von Funktionen mehrerer Variabeln. *Berlin Monatsberichte*, 159–193, 688–698. In *Werke 1*, 175–212, 213–226 (1895).

Kuhn, T. S. (1978). *Black-Body Theory and the Quantum Discontinuity, 1894–1912*. Oxford University Press.

Kusch, M. (2006). *A Sceptical Guide to Meaning and Rules: Defending Kripke's Wittgenstein*. Acumen.

Kusch, M. (2009). Kripke's Wittgenstein, on certainty, and epistemic relativism. In *The Later Wittgenstein on Language*, ed. by D. Whiting. Palgrave Macmillan.

Lagemann, R. T. (1977). New light on old rays: N Rays. *American Journal of Physics 45*, 281–284.

Langevin, P. (1904). The relations of physics of electrons to other branches of science. *Congress of Arts and Science 7*, 121–156.

Langevin, P. (1911). L'évolution de l'espace et du temps. *Revue de métaphysique et de morale 19*, 455–466.

Langevin, P. (1913). L'oeuvre d'Henri Poincaré, le physicien. *Revue de métaphysique et de morale 21*, 675–718.

Le Roy, E. (1907). *Dogme et critique*. Paris.

Lebon, E. (1909). *Savants du jour. Henri Poincaré, Biographie, bibliographie analytique des écrits*, 2ᵉ édition, entièrement refondue, 1912. Paris: Gauthier-Villars.

Lefschetz, S. (1924). *L'analysis situs et la géométrie algébrique*. Collections Borel, Paris.

Lefschetz, S. (1929). *Géométrie sur les surfaces et les variétés algébriques*. Mémorial des sciences mathématiques, XL. Paris: Gauthier-Villars.

Liapunov, A. M. (1884). The stability of ellipsoidal forms of equilibrium of a rotating liquid. *Mémoires de l'Université de Kharkow* (Russian) (French in 1904).

Liapunov, A. M. (1897). Sur l'instabilité de l'équilibre dans certains cas où la fonction des forces n'est pas un maximum. *Journal de mathématiques (5) 3*, 81–94.

Liapunov, M. A., (1907). Problème général de la stabilité du mouvement. Trans. by A. Davaux. *Annales de la Faculté des sciences de Toulouse (2) 9*, 203–474. Repr. (1947), Princeton University Press (Russian original, 1892).

Lie, S. (1893). *Theorie der Transformationsgruppen*, vol. 3. Leipzig: Teubner.

Listing, J. B. (1847). *Vorstudien zur Topologie*. Göttinger Studien.

Lobachevskii, N. I. (1840). *Geometrische Untersuchungen*. Berlin, repr. (1887) Mayer & Müller. French trans. by J. Houël, Etudes géométriques sur la théorie des parallèles. *Mémoires de la Société des sciences physiques et naturelles de Bordeaux 4*, 1867, 83–128. Paris: Gauthier-Villars. English trans. by G. B. Halsted, Geometric researches in the theory of parallels. Appendix in Bonola (1912).

Lorentz, H. A. (1892). La théorie électromagnétique de Maxwell et son application aux corps mouvantes. *Arch. Neerl. 25*, 363–551. In *Collected Papers 2*, 146–343.

Lorentz, H. A. (1895). *Versuch einer Theorie der elektrischen und optischen Erscheinungen in bewegten Körpern*. Leiden.

Lorentz, H. A. (1904a). *Maxwells elektromagnetische Theorie. EMW V 1.2*, 63–144, and Weiterbildung der *Maxwell*schen Theorie. Elektronentheorie. *EMW V 1.2*, 145–280.

Lorentz, H. A. (1904b). Electromagnetic phenomena in a system moving with any velocity less than that of light. *Proceedings of the Academy of Sciences, Amsterdam 6*. In *The Principle of Relativity*, ed. by A. Sommerfeld, 9–34. Dover repr. (1952).

Lorentz, H. A. (1915/1921). Deux mémoires de Henri Poincaré sur la physique mathématique. In Poincaré, *Oeuvres 11*, 247–261.

Love, A.E.H. (1915). The transmission of electric waves over the surface of the earth. *Phil. Trans. Roy. Soc. A, 215*, 105–131.

Lützen, J. (1984). Joseph Liouville's work on the figures of equilibrium of a rotating mass of fluid. *AHES 30*, 113–166.

Lützen, J. (1990). *Joseph Liouville, 1809–1882. Master of Pure and Applied Mathematics*. Springer.

Lützen, J. (2005). *Mechanistic Images in Geometric Form: Heinrich Hertz's Principles* of Mechanics. Oxford University Press.

Lux, J. (1911/12). *Histoire de deux revues françaises. La Revue bleue et la Revue scientifique, 1863–1911*. Préface de M. H. Poincaré.

MacDonald, H. (1902). *Electric Waves*. Cambridge University Press.

MacDonald, H. (1903). The bending of electric waves round a conducting obstacle. *Proceedings of the Royal Society of London 71*, 251–258.

Mach, E. (1883). *Die Mechanik in ihrer Entwickelung, historisch-kritisch dargestellt*. Leipzig: Brockhaus.

Mancosu, P. (1998). *From Brouwer to Hilbert*. Oxford University Press.

Martinez, A. A. (2009). *Kinematics: The Lost Origins of Einstein's Relativity*. Johns Hopkins University Press.

Martinez, A. A. (2011). *Science Secrets: The Truth about Darwin's Finches, Einstein's Wife, and Other Myths*. University of Pittsburgh Press.

Mawhin, J. (1994). The centennial legacy of Poincaré and Lyapunov in ordinary differential equations. *Rend. Palermo (2) 34*, 9–46.

Mawhin, J. (1996). The early reception in France of the work of Poincaré and Lyapunov in the qualitative theory of differential equations. *Philosophia Scientiae 1*, 119–133.

Mawhin, J. (2005). Alexandr Mikhailovich Lyapunov, thesis on the stability of motion. In *Landmark Writings in Western Mathematics, 1640–1940*, ed. by I. Grattan-Guinness, 664–676.

Mawhin, J. (2006/2010). Henri Poincaré and the partial differential equations of mathematical physics. In Charpentier, Ghys, and Lesne (2006/2010), 257–278.

Maxwell, J. C. (1856). On the stability of motion of Saturn's rings. In Maxwell (1890), *1*, 288–376.

Maxwell, J. C. (1865). A dynamical theory of the electromagnetic field. *Phil. Trans.* 155, 465–536. In Maxwell (1890), *1*, 526–597.

Maxwell, J. C. (1870). On hills and dales. In Maxwell (1890), *2*, 233–240.

Maxwell, J. C. (1873). *A Treatise on Electricity and Magnetism*, 2 vols. Oxford: Clarendon Press Series. Dover repr. (1954).

Maxwell, J. C. (1890). *The Scientific Papers of James Clerk Maxwell*, 2 vols, ed. by W. D. Niven. Cambridge University Press. Dover repr. (1965).

Mazur, B. (1977). Rational points on modular curves. *Modular Functions of One Variable, V*, 107–148. LNM 601. Springer.

Maz'ya, V. and T. Shaposhnikova (1998). *Jacques Hadamard, A Universal Mathematician*. HMath *14*.

McCormmach, R. (1967). Henri Poincaré and the quantum theory. *Isis 58*, 37–55.

Miller, A. I. (1973). A Study of Henri Poincaré's "Sur la dynamique de l'électron." *AHES 10*, 207–328.

Miller, A. I. (1981). *Albert Einstein's Special Theory of Relativity: Emergence (1905) and Early Interpretation (1905–1911)*. Addison-Wesley.

Minkowski, H. (1896). *Geometrie der Zahlen*. Leipzig.

Minkowski, H. (1908). Raum und Zeit. *JDMV* 75–88. In *Gesammelte Abhandlungen 2*, 431–444. English trans. in *The Principle of Relativity*, ed. by A. Sommerfeld. Dover repr. (1952), 73–91.

Möbius, A. F. (1865). Über die Bestimmung des Inhalts eines Polyeders. *Abhandlungen Sächsische Gesellschaft der Wissenschaften 17*. In *Gesammelte Werke 2*, 473–512.

Mordell, L. J. (1922). On the rational solutions of the indeterminate equations of the third and fourth degrees. *Proc. Cam. Phil. Soc. 21*, 179–192.

Mumford, D. (1966). *Lectures on Curves on an Algebraic Surface*, with a section by G. M. Bergman. Annals of Mathematics Studies, no. 5. Princeton University Press.

Mumford, D. (2011). Intuition and rigor in Enriques's quest. *Notices AMS 58*, 250–260.

Nabonnand, P. (2000). La polémique entre Poincaré et Russell au sujet du statut des axiomes de la géométrie. *Revue d'histoire des mathématiques 6*, 219–269.

Nauenberg, M. (2006/2010). Periodic orbits of the three-body problem. In Charpentier, Ghys, and Lesne (2006/2010), 113–142.

Newton, Sir I. (1687/1999). *The* Principia: *Mathematical Principles of Natural Philosophy*, ed. and trans. by I. B. Cohen and A. Whitman. University of California Press.

Noir, J. (2000). Écoulements d'un fluide dans une cavité en précession: Approches numérique et expérimentale. PhD thesis, l'Université Joseph Fourier – Grenoble I.

Nye, M. J. (1986). *Science in the Provinces*. University of California Press.

Osgood, W. F. (1900). On the existence of the Green's function for the most general simply connected plane region. *Trans. AMS 1*, 310–314.

O'Shea, D. (2007). *The Poincaré Conjecture: In Search of the Shape of the Universe*. New York: Walker & Co.

Parshall, K. H. (2006). *James Joseph Sylvester: Jewish Mathematician in a Victorian World*. Johns Hopkins University Press.

Pasch, M. (1882). *Vorlesungen über neuere Geometrie*. Leipzig: Teubner.

Peano, G. (1889). *Arithmetices Principia*. Turin: Bocca.

Picard, É. (1893). Sur l'équation aux dérivées partielles qui se présente dans la théorie de la vibration des membranes. *CR 107*, 502–507. In *Oeuvres 2*, 545–550.

Picard, É. (1896). *Traité d'analyse* 3 vols. Paris: Gauthier-Villars. Subsequent eds 1908, 1928.

Picard, É and G. Simart (1897, 1906). *Théorie des fonctions algébriques de deux variables indèpendants*, 2 vols. Gauthier-Villars, Chelsea, New York (1971).

Planck, M. (1915/1921). Henri Poincaré und die Quantentheorie. In Poincaré, *Oeuvres 11.1*, 347–356.

Pomeau, Y. (2006/2010). Henri Poincaré as an applied mathematician. In Charpentier, Ghys, and Lesne (2006/2010), 351–371.

Pulte, H. (2000). Beyond the edge of certainty: Reflections on the rise of physical conventionalism. *Philosophia Scientiae 4*, 47–68.

Putnam, H. (1975). *Mathematics, Matter and Method*. Cambridge University Press.

Rapports présentés au Congrès international de physique. (1900). Ed. by Ch.-É. Guillaume and L. Poincaré. Paris: Gauthier-Villars.

Répertoire bibliographique des sciences mathématiques (1894). Paris.

Reinhardt, K. (1921). Über Abbildungen durch analytische Funktionen zweier Veränderlichen. *Math. Ann. 83*, 211–255.

Ribot, Th. (1873). *L'Hérédité. Étude psychologique sur ses phénomènes, ses lois, ses causes, ses conséquences*. Paris.

Ribot, Th. (1881). *La Psychologie allemande contemporaine*, 2nd edition. Paris. English trans. by J. M. Baldwin (1886) *German Psychology of To-Day*, Charles Scribners Sons.

Riemann, G.F.B. (1851). *Grundlagen für eine allgemeine Theorie der Functionen einer veränderlichen complexen Grösse*. In *Gesammelte mathematische Werke*, 35–80.

Riemann, G.F.B. (1857). Theorie der Abel'schen Functionen. *JfM 54.* In *Gesammelte mathematische Werke,* 120–144.

Riemann, G.F.B. (1861). Ein Beitrag zu den Untersuchungen über die Bewegung eines flüssigen gleichartigen Ellipsoides. *Abhandlungen der Königlichen Gesellschaft der Wissenschaften zu Göttingen 9.* In *Gesammelte mathematische Werke,* 182–211.

Riemann, G.F.B. (1867). Ueber die Hypothesen welche der Geometrie zu Grunde liegen. *K. Ges. Wiss. Göttingen 13,* 1–20. In *Gesammelte mathematische Werke,* 304–319.

Riemann, G.F.B. (1990). *Gesammelte mathematische Werke, Wissenschaftliche Nachlass und Nachträge, Collected Papers.* Ed. by R. Narasimhan. Springer.

Robadey, A. (2004). Exploration d'un mode d'écriture de la généralité: l'article de Poincaré sur les lignes géodésiques des surfaces convexes (1905). *Revue d'histoire des mathématiques 10,* 257–318.

Robin, G. (1886). Sur la distribution de l'électricité à la surface des conducteurs fermés des conducteurs ouvert. *Annales scientifiques de l'École normale (3) 3.* Supplément, 3–58.

Rollet, L. (1997). Henri Poincaré et l'action politique – Autour de l'Affaire Dreyfus. Strasbourg: Séminaire de l'Institut de Recherche sur les Enjeux et les Fondements des Sciences et des Techniques.

Rollet, L. (2002a). Un mathématicien dans l'affaire Dreyfus: Henri Poincaré, Séminaire d'histoire des mathématiques—IHP 13 février 2002.

Rollet, L. (2002b). Un mathématicien au Panthéon, à propos de la mort d'Henri Poincaré. Preprint 127ème Congrès national des Sociétés historiques et scientifiques.

Rollet, L. (2010). De l'Algérie à Vesoul: Henri Poincaré ingénieur des mines. In *Construction: Festschrift for Gerhard Heinzmann,* 63–75, ed. by P. E. Bour, M. Rebuschi, L. Rollet. College Publications.

Rollet, L. and P. Nabonnand (2002). Une bibliographie mathématique idéale? Le Répertoire bibliographique des sciences mathématiques. *Gazette des mathématiques 92,* 11–25.

Roque, T. (2010). Stability of trajectories from Poincaré to Birkhoff: Approaching a qualitative definition. *AHES 65,* 295–342.

Rouché, E., and C. Comberousse (1864–1869). *Traité de géométrie élémentaire.* Paris, two subsequent eds.

Rüdenberg, L. and H. Zassenhaus (eds) (1973). *Hermann Minkowski – Briefe an David Hilbert.* Berlin: Springer.

Russell, B. (1897). *An Essay on the Foundations of Geometry.* Cambridge University Press.

Russell, B. (1899). On the axioms of geometry. In *Philosophical Papers 2,* 394–415.

Russell, B. (1903). *The Principles of Mathematics.* Cambridge University Press.

Russell, B. (1908). Mathematical logic as based on the theory of types. *Amer. J. Math. 30,* 222–262.

Russell, B. (1959). *My Philosophical Development.* Allen and Unwin.

Russell, B. (1998). *Autobiography*. Routledge.

Rynasiewicz, R. (2008). Newton's views on space, time, and motion. *The Stanford Encyclopedia of Philosophy* (fall 2008 edition), ed. by Edward N. Zalta. http://plato.stanford.edu/archives/fall2008/entries/newton-stm/.

Saint-Gervais, H. P. de, (2010). *Uniformisation des surfaces de Riemann*. Lyons: ENS Editions.

Sarkaria, K. S. (1999). The topological work of Henri Poincaré. In *History of Topology*, ed. by I. M. James, 123–167. North-Holland.

Schappacher, N. (1991). Développement de la loi de groupe sur une cubique. *Séminaire de théorie des nombres, Paris 1988/1989*, 159–184 (1991). Boston: Birkhäuser.

Scharlau, W. (2001). Kommentar. In Hausdorff, *Gesammelte Werke 4*, 461–465.

Schmid, W. (1982). Poincaré and Lie groups. *Bull. AMS 6.2*, 175–186.

Scholz, E. (1980). *Geschichte des Mannigfaltigkeitsbegriffs von Riemann bis Poincaré*. Boston: Birkhäuser.

Schwarz, H. A. (1871). Mittheilung über diejenigen Fälle, in welchen die Gaussische hypergeometrische Reihe $F(\alpha, \beta, \gamma, x)$ eine algebraische Function ihres vierten Elementes darstellt. *Verhandlungen der Schweizerischen Naturforschenden Gesellschaft* 74–77. In *Gesammelte mathematische Abhandlungen 2*, 172–174.

Schwarz, H. A. (1885). Ueber ein die Flächen kleinsten Inhalts betreffendes Problem der Variationsrechnung. *Acta Soc. Sci. Fennicae 15*, 315–362. In *Gesammelte mathematische Abhandlungen 1*, 223–269.

Schwarzschild, K. (1898). Die Poincaré'sche Theorie des Gleichgewichtes einer homogenen rotirenden Flüssigkeitsmasse. *N. Ann. Sternw. München 3*.

Seneta, E., K. H. Parshall, and F. Jongmans (2001). Nineteenth-century developments in geometric probability: J. J. Sylvester, M. W. Crofton, J.-É. Barbier, and J. Bertrand. *AHES 55*, 501–524.

Simonetta, M. and N. Arikha (2011). *Napoleon and the Rebel: A Story of Brotherhood, Passion, and Power*. Palgrave Macmillan.

Smirnov, V. I. and A. P. Youshkevitch (1987). Correspondance Liapunov–Poincaré. *Cahiers, 8*, 1–18.

Sommerfeld, A. (1896). Mathematische Theorie der Diffraction. *Math. Ann. 47*, 317–374.

Staudt, C.G.C. von (1847). *Geometrie der Lage*. Nürnberg.

Staudt, C.G.C. von (1856–1860). *Beiträge zur Geometrie der Lage*, 3 vols. Nürnberg.

Stubhaug, A. (2002). *The Mathematician Sophus Lie: It Was the Audacity of My Thinking*. Springer.

Stubhaug, A. (2010). *Gösta Mittag-Leffler: A Man of Conviction*. Springer.

Terrall, M. (2002). *The Man Who Flattened the Earth: Maupertuis and the Sciences in the Enlightenment*. University of Chicago Press.

Thomé, L. W. (1887). Bemerkung zur Theorie der linearen Differentialgleichungen. *JfM 101*, 203–208.

Thomson, Sir W. and P. G. Tait (1879). *Treatise on Natural Philosophy*, 2 vols. Cambridge University Press.

Tieszen, R. Poincaré in intuiton and arithmetic: une "saine psychologie." In *Construction: Festschrift for Gerhard Heinzmann* 97–106, ed. by P. E. Bour, M. Rebuschi, and L. Rollet. College Publications.

Ton-That, T. and T.-D. Tran (1999). Poincaré's proof of the so-called Birkhoff–Witt theorem. *Revue d'histoire des mathématiques 5*, 249–284.

Toulouse, Le Dr. (1909). *Henri Poincaré*. Michigan University reprint (2010).

Uffink, J. (2004). Boltzmann's work in statistical physics. *The Stanford Encyclopedia of Philosophy* (spring 2004 edition), ed. by Edward N. Zalta. http://plato.stanford.edu/entries/statphys-Boltzmann/#3.

Ullrich, P. (2000). *The Poincaré–Volterra Theorem: From Hyperelliptic Integrals to Manifolds with Countable Topology*. *AHES 54*, 375–402.

Verhulst, F. (2012). *Henri Poincaré: Impatient Genius*. Springer.

Vesentini, E. (1992). I funzionali isogeni di Volterra e le funzioni di variabili complesse. In *Convegno Internazionale in memoria di Vito Volterra, Atti dei Convegni Lincei 92*. Roma, Accademia Nazionale dei Lincei.

Vivanti, G. (1888a). Sulle funzioni ad infiniti valori. *Rend. Palermo 2*, 135–138.

Vivanti, G. (1888b). Ancora sulle funzioni ad infiniti valori. *Rend. Palermo 2*, 150–151.

Volterra, V. (1888). Sulle funzioni analitiche polidrome. *Rend. Lincei (4) 4*, 355–361. In *Op. Mat. 1*, 356–362.

Walter, S. (1999). Minkowski, mathematicians, and the mathematical theory of relativity, *The Expanding Worlds of General Relativity*, ed. by H. Goenner, J. Renn, T. Sauer, and J. Ritter, 45–86. Einstein Studies 7. Birkhäuser.

Walter, S. (2008). Hermann Minkowski's approach to physics. *Mathematische Semesterberichte 55*, 213–235.

Walter, S. (2009a). Hypothesis and convention in Poincaré's defense of Galilei spacetime. In Heidelberger and Schliemann (2009), 193–219.

Walter, S. (2009b). Henri Poincaré, theoretical physics, and relativity theory in Paris. In *Mathematics Meets Physics*, ed. by K.-H. Schlote, M. Schneider. Harri Deutsch.

Warwick, A. (2003). *Masters of Theory: Cambridge and the Rise of Mathematical Physics*. University of Chicago Press.

Watson, G. N. (1918). The diffraction of electric waves by the earth. *Proceedings of the Royal Society of London A, 95*, 83–99.

Weber, H. (1869). Ueber die Integration der partiellen Differentialgleichung: $\frac{\partial^2 u}{\partial x^2} + \frac{\partial^2 u}{\partial y^2} + k^2 u = 0$. *Math. Ann. 1*, 1–36.

Webster, A. G. (1966). *The Partial Differential Equations of Mathematical Physics*. Dover.

Weierstrass, K.T.W. (1879). Einige auf die Theorie der analytischen Functionen mehrerer Veränderlichen sich beziehende Sätze. (lith.) Berlin. First printed in Weierstrass (1886), 105–16. In *Math. Werke 2*, 135–188.

Weierstrass, K.T.W. (1880). Untersuchungen über 2r–fach periodische Funktionen. *JfM 89*, 1–8. In *Math. Werke 2*, 125–133.

Weierstrass, K.T.W. (1886). *Abhandlungen aus der Funktionenlehre*. Springer.

Weierstrass, K.T.W. (1903). Allgemeine Untersuchungen über 2n–fach periodische Funktionen von *n* Veränderlichen. MS in *Math. Werke 3*, 53–114.

Weierstrass K.T.W. (1894–1927). *Mathematische Werke*, 7 vols. Mayer and Müller, Berlin. Repr. Olms, Hildesheim.

Weil, A. (1954). Poincaré et l'arithmétique. *Oeuvres scientifiques, Collected Papers 2*, 189–196. In Poincaré, *Oeuvres 11*, 206–212.

Weyl, H. (1913). *Die Idee der Riemannschen Fläche*. Teubner, Leipzig. English trans. of 3rd ed, *The Concept of a Riemann Surface*. Addison-Wesley Pub. Co., Reading, Mass. (1955). Dover repr. New York (2009).

Whitehead, A. N. (1902). On cardinal numbers. *Amer. J. Math. 24*, 367–394.

Whittaker, E. T. (1899). Report on the progress of the solution of the problem of three bodies. *Report of the BAAS*, 121–159.

Whittaker, E. T. (1937). *A Treatise on the Analytical Dynamics of Particles and Rigid Bodies*, 4th ed. Cambridge University Press.

Wills, G. (ed.) (2011). *Augustine's Confessions: A Biography*. Princeton University Press.

Wilson, C. (2010). *The Hill–Brown Theory of the Moon's Motion*. Springer.

Wittgenstein, L. (1956). *Remarks on the Foundations of Mathematics*, ed. by G. H. von Wright, R. Rhees, G.E.M. Anscombe. English trans. by G.E.M. Anscombe. Basil Blackwell.

Wittgenstein, L. (1969). *On Certainty*, ed. by G.E.M. Anscombe and G. H. von Wright. Basil Blackwell.

Wood, R. (1904). The N-Rays. *Nature 70*, 530–531.

Worrall, J. (1989). Structural realism: The best of both worlds? *Dialectica 43*, 99–124. Repr. in *The Philosophy of Science*, 139–165, ed. by D. Papineau. Oxford University Press.

Yavetz, I. (1995). *From Obscurity to Enigma: The Work of Oliver Heaviside, 1872–1889*. Science networks historical studies, *16*. Birkhäuser.

Yeang, C.-P. (2003). The study of long-distance radio-wave propagation, 1900–1919. *Historical Studies in the Physical Sciences 33*, 369–404.

Zahar, E. (2001). *Poincaré's Philosophy: From Conventionalism to Phenomenolgy*. Open Court.

Zappa, G. (1945). Sull'esistenza sopra le superficie algebriche di sistemi continui completi infiniti. *Pont. accad. sci. acta.*

Zaremba, S. (1899). Sur l'équation aux dérivées partielles $\Delta u + \xi u + f = 0$ et sur les fonctions harmoniques. *Ann. de l'éc. norm. (3) 16*, 427–464.

Zariski, O. (1935). *Algebraic Surfaces*. Springer.

Zariski, O. (1972). *Collected Papers*, 4 vols, various eds. MIT Press.

Zermelo, E. (1896). Über einen Satz der Dynamik and die mechanische Warmetheorie. *Annalen der Physik 57*, 485–494; English trans. by Brush, 208 (1966).

Zermelo, E. (1904). Beweis, daß jede Menge wohlgeordnet werden kann. *Math. Ann. 59*, 514–516. In Zermelo (2010), 114–119.

Zermelo, E. (1908a). Neuer Beweis für die Möglichkeit einer Wohlordnung. *Math. Ann. 65*, 107–128. In Zermelo (2010), 120–159.

Zermelo, E. (1908b). Untersuchungen über die Grundlagen der Mengenlehre, I. *Math. Ann. 65*, 261–281. In Zermelo (2010), 188–229.

Zermelo, E. (1909a). Sur les ensembles finis et le principe de l'induction complète. *Acta 32*, 185–193. In Zermelo (2010), 236–251.

Zermelo, E. (1909b). Ueber die Grundlagen der Arithmetik. *Atti del IV Congresso internazionale dei matematici 2*, 8–11. In Zermelo (2010), 252–259.

Zermelo, E. (2010). *Collected Works, Gesammelte Werke*, ed. by H.-D. Ebbinghaus, C. G. Fraser, A. Kanamori. Springer.

Zhukovsky, N. E. (1876). *Math. Sbornik 8*, 163–238.

Zitarelli, D. (2011). The 1904 St. Louis Congress and the westward expansion of American mathematics. *Notices AMS 58*, 1100–1111.

Name Index

Subject Index

Abel's theorem, 499, 505

Académie des sciences, Paris, 101, 160–163, 168, 169, 175, 189, 193, 195, 201, 207, 279, 306, 389, 491, 522

Académie française, 194, 200, 201

Acta Mathematica, 27, 29, 74, 142, 161n.5, 162, 194, 196, 202, 209, 232, 240, 243, 245, 263n.2, 268, 270, 277–279, 282, 303, 310, 396, 418, 428n.1, 522

Aleph-one, 138, 143

Aleph-zero, 142, 143

Alsace-Lorraine, 18, 38, 154, 201

antinomy, 153

astronomy: planetary, 4, 6, 15, 30, 58, 73, 101n.34, 157, 162, 165, 169, 193, 207, 265, 277, 283, 289, 293, 424, 510, 387; stellar, 315

asymptotic solutions, single and double, 32, 33, 259, 275–278, 281–282, 290, 313n.8

atom, existence of, 9, 66, 70, 151, 359, 511, 534

atom of time, 152

automorphic functions, 3, 5, 14, 161, 246, 428, 481, 486, 503

Bertrand's paradox, 519

Betti numbers, 433, 435–437, 440, 441, 443, 446–447, 449, 450, 451, 458–459, 460

bibliography of scientific literature, 164, 165, 187n.31

birational geometry (also invariant, transformation), 236, 487, 488n.11, 500–502, 548

Brownian motion, 150, 357

Budapest, 194

Buffon's needle, 521–522

Bureau des longitudes, 4, 165, 181, 187, 188n.33, 194, 201

Caen, 159, 160, 209, 217, 221, 228, 232

Catholic philosophy of science, 24, 70, 148

circuits, open and closed, 172, 300, 322, 331

continuum: mathematical, 7, 8, 45, 68, 97, 99, 100, 299, 423, 528, 538; physical, 45, 68, 69, 96–98, 152

conventionalism, geometric, 8, 60, 83, 95, 206, 525–528, 530–531, 533, 534, 536, 538

convex domains, 116, 404, 414, 443–444

convex surfaces, 5, 191–192, 194–195, 197

cubic curve, 232, 247, 248, 474, 487–488

current, open or closed, 172–174

curve, algebraic, 115, 116, 118, 119, 207, 208, 227, 232, 247, 255, 258, 390, 423, 486–488, 498–501, 504–506, 544

definition, nature of, 12, 21, 35, 42, 43, 59, 63, 65, 66, 71, 79, 80, 90, 93, 97, 129, 130, 131–133, 136–143, 188, 196, 205n.45, 348, 394, 518, 529, 549, 550

differential equation: linear, 160, 207, 208, 211, 212, 220, 223, 225, 232, 233, 235, 243, 246, 268, 283, 388, 469, 481, 490; non-linear, 391; ordinary, 116, 159, 160, 209, 253, 280, 387, 388; partial, 5, 17, 23, 75, 116, 159, 162, 197, 203, 247, 277, 280, 380, 382, 395, 402, 403, 418, 421

dimension, definition of, 97–100

Dirichlet problem, 76, 116, 197, 249–251, 403–405, 408, 409, 413–414, 418

displacement currents, 173, 183, 199, 330, 349, 350

distance, 8, 9, 37, 44, 47, 49, 53, 56, 57, 78–82, 96, 98, 99, 108, 221, 358, 532

DP, xiii, 110–112, 140n.68, 141, 152, 200

Dreyfus affair, 19, 24, 165–169, 193

E&O, xiii, 171, 183, 197, 328, 333, 346, 347, 351, 372, 511

École des mines, 34, 158, 159, 165, 201, 217

École normale supérieure, 19, 34, 39, 154–157, 160, 491, 521

École polytechnique, 10, 19, 24, 34, 39, 154–158, 160, 162, 170, 193, 194, 199, 203, 207, 267, 333, 383, 432, 544

CPSIA information can be obtained
at www.ICGtesting.com
Printed in the USA
JSHW021909041122
32631JS00004B/6

9 780691 242033